10th International Conference on Chemical and Environmental Engineering (ICEE-10)

IOP Conference Series: Materials Science and Engineering
Volume 975

Cairo, Egypt
7 - 9 April 2020

ISBN: 978-1-7138-3204-1
ISSN: 1757-8981

Printed from e-media with permission by:

Curran Associates, Inc.
57 Morehouse Lane
Red Hook, NY 12571

CURRAN ASSOCIATES INC.
proceedings.com

Some format issues inherent in the e-media version may also appear in this print version.

Institute of Physics
Dirac House, Temple Back
Bristol BS1 6BE UK

Phone: 44 1 17 929 7481
Fax: 44 1 17 920 0979

techtracking@iop.org

Additional copies of this publication are available from:

Curran Associates, Inc.
57 Morehouse Lane
Red Hook, NY 12571 USA
Phone: 845-758-0400
Fax: 845-758-2633
Email: curran@proceedings.com
Web: www.proceedings.com

TABLE OF CONTENTS

IOP Conf. Series: Materials Science and Engineering **975** (2020) 011001 doi:10.1088/1757-899X/975/1/011001

Ministry of Defence
Military Technical College

I.C.E.E.2020

THE 10th INTERNATIONAL CONFERENCE OF THE MILITARY TECHNICAL COLLEGE

Abstracts of Scientific Papers,

The 10th International Conference
On Chemical and Environmental
Engineering
7 - 9 July, 2020

IOP Conf. Series: Materials Science and Engineering **975** (2020) 011001 doi:10.1088/1757-899X/975/1/011001

Military Technical College
Kobry El-Kobbah,
Cairo, Egypt

I.C.E.E.2020

10th International Conference on Chemical & Environmental Engineering
7 – 9 April, 2020.

CONTENTS

IOP Conf. Series: Materials Science and Engineering **975** (2020) 011001 doi:10.1088/1757-899X/975/1/011001

Military Technical College
Kobry El-Kobbah,
Cairo, Egypt

I.C.E.E.2020

10th International Conference
on
Chemical & Environmental
Engineering
7 – 9 April, 2020.

IOP Conf. Series: Materials Science and Engineering **975** (2020) 011001 doi:10.1088/1757-899X/975/1/011001

Military Technical College
Kobry El-Kobbah,
Cairo, Egypt

10th International Conference on Chemical & Environmental Engineering
7 – 9 April, 2020.

I.C.E.E.2020

PREFACE

The military Technical College is honored to organize the 10th International Conference on Chemical and Environmental Engineering (ICEE) in the period 7-9 July, 2020 sponsored by the Egyptian Ministry of Defense. The conference provides a valuable opportunity to exchange, update the knowledge and stimulate discussions on recent developments and research activities.

Accepted papers are arranged in the following sessions:

1. Nanotechnology and its chemical Applications (NCA)
2. Energetic Mat. and its Applications (EMA)
3. Chemical Engineering Applications (CEA)
4. Environmental Studies and Pollution Control (ESPC)
5. Water and Waste Water Treatment By Synthetic Polymers And Adsorbents (WWWT)
6. Nuclear and Radiometric Applications (NRA)
7. Extraction Techniques of Nuclear Materials (ETNM)

We are highly indebted to all conference contributors. Also; the effort of the scientific committee is highly appreciated.

IOP Conf. Series: Materials Science and Engineering **975** (2020) 011001 doi:10.1088/1757-899X/975/1/011001

MilitaryTechnical College
Kobry El-Kobbah,
Cairo, Egypt

10th International Conference
on
Chemical & Environmental
Engineering
7 - 9 April, 2020.

I.C.E.E.2020

Conference Committees

Higher Committee

- Maj. Gen. Prof. Dr. Ismail Kamal — *Chairman*
- Maj. Gen. Prof. Dr. Yehia Mehasseb — *Vice - Chairman*
- Maj. Gen. (R) Prof. Dr./Mohamed T. Sallam — *General Rapporteur*
- Brig. Assoc. Prof. Dr. Mohamed Misbah Elkhatib — *General Rapporteur Assistant*

Steering Committee:

- Maj. Gen. (R) Assoc. Prof. Dr. M.Kassem Abdel-Magid — *Rapporteur*
- Maj. Gen. (R) Assoc. Prof. Dr. Said M. A. Hassanein
- Maj. Gen. (R) Prof. Dr. Hosam E. Mostafa
- Brig. Gen. Prof. Dr. Ahmed A. Baraka
- Brig. Gen. Prof. Dr. Mohamed Misbah Elkhatib
- Brig. Gen. Prof. Dr. Mohamed Abo El fotoh Elsayed
- Brig. Gen. Assoc. Prof. Dr. Mohammed M. Gobbara
- Col. Dr. Hisham Gad El Kareem Tantawy
- Col. Assoc. Prof. Dr.Dr.Sherif Shaaban Elbasuney

Organizing Committee:

- Brig. Gen. Prof. Dr. Ahmed A. Baraka
- Brig. Gen. Prof. Dr. Mohamed Abo El fotoh Elsayed
- Brig. Gen. Assoc. Prof. Dr. Mohammed M. Gobbara
- Brig. Gen.Dr.Mohamed I. Awaad
- Col. Dr. Ahmed Farghal Tawfik
- Maj. Dr. Osama abo Zalat

Scintific Committee:

- Professor Dr./ Elsayed Mohamed abu El Ella
- Professor Dr./ Zainab Abdel Hamid
- Maj. Gen. (R) Assoc. Prof. Dr. Said M. A. Hassanein
- Maj. Gen. (R) Prof. Dr. Hosam E. Mostafa
- Brig. Gen. Prof. Dr. Ahmed A. Baraka
- Brig. Gen. Prof. Dr. Mohamed Abo El fotoh Elsayed
- Brig. Gen. Assoc. Prof. Dr. Mohammed M. Gobbara
- Brig. Gen.Dr.Mohamed I. Awaad
- Col. Assoc. Prof. Dr./ Mohamed Abdel-Rahman
- Col. Dr. Hisham Gad El Kareem Tantawy
- Col. Assoc. Prof. Dr.Dr.Sherif Shaaban Elbasuney
- Col. Dr. Ahmed Farghal Tawfik
- Col. Dr. Mokhtar Kotb
- Maj. Dr. Osama abo Zalat
- Lt. Col. Mahmoud Abd El Hafiez

IOP Conf. Series: Materials Science and Engineering **975** (2020) 011001 doi:10.1088/1757-899X/975/1/011001

MilitaryTechnical College
Kobry El-Kobbah,
Cairo, Egypt

10th International Conference on Chemical & Environmental Engineering
7 - 9 April, 2020.

I.C.E.E.2020

CONTRIBUTORS

I Egyptian Universities and Organizations

1. Assiut University, Faculty of Engineering, Mechanical Engineering Department
2. British University in Egypt, Elshorouk City, Cairo, Egypt.
3. Basic Engineering Science Department, Faculty of Engineering, Menoufia University, Shebin El-Kom, Egypt
4. Center of Metallurgy Research and Development Institute
5. Chemical Engineering Branch, Military Technical College, Cairo, Egypt
6. Chemical Engineering Department, Faculty of Engineering, Cairo University, Egypt
7. Chemical engineering department, faculty of engineering, Minia University, Minia, Egypt.
8. Chemical Engineering Department, Military Technical College,Cairo, Egypt
9. Department of Petroleum Applications, Egyptian Petroleum Research Institute, Nasr City, P.O. Box 11727, Cairo, Egypt.
10. Chemical engineering department,. British University in Egypt (BUE).
11. Chemistry Administration, Cairo, Egypt.
12. Chemistry Department, Faculty of Science, Alexandria University, Alexandria, Egypt.
13. Corrosion Control and Surface Protection Lab. Central Metallurgical R & D Institute, Cairo, Egypt forzeinab@yahoo.com
14. Cyclotron project, Nuclear Physics Department, Nuclear Research Center, Atomic Energy Authority, Egypt
15. Department of Botany, Faculty of Science, Zagazig University, Egypt.
16. Department of Chemistry, Faculty of Science, University of Mansoura, Egypt.
17. Department of Science and Mathematical Engineering, Faculty of Petroleum and Mining Engineering, Suez University, Suez 43521, Egypt
18. Department of Uranium Ores Processing, Production Sector, Nuclear Materials Authority
19. Egypt Nuclear and Radiological Regularity Authority (ENRRA)
20. Egyptian Armed Forces, Cairo, Egypt
21. Egyptian Atomic Energy Authority, Cairo, Egypt
22. Entomology Department, Faculty of Science, Cairo University, Egypt.
23. Environmental Geology, water and microbiology lab, Chemistry Administration
24. Faculty of petroleum and mining engineering, Suez University, Suez, Egypt
25. Geology Department, [3]ChemistryDepartment, Faculty of Science, Assiut University
26. Geology Department, Assiut University
27. Geology Department, Faculty of Science, Ain Shams University, Egypt.
28. Geology Departments, Faculty of Science, Sohag University, Sohag, Egypt.
29. Glass Research Department, National Research Centre, Dokki, Cairo, Egypt
30. Medical Laboratory Department, Faculty of Allied Medical Sciences, Pharos University in Alexandria, Egypt.
31. Metallurgical and Materials Engineering Department, Faculty of Petroleum and Mining Engineering, Suez University, Suez 43521, Egypt

IOP Conf. Series: Materials Science and Engineering **975** (2020) 011001 doi:10.1088/1757-899X/975/1/011001

MilitaryTechnical College
Kobry El-Kobbah,
Cairo, Egypt

I.C.E.E.2020

10th International Conference
on
Chemical & Environmental
Engineering
7 - 9 April, 2020.

32. Military Technical College, Kobry Elkobbah, Cairo, Egypt
33. Mining and petroleum Dep., faculty of Engineering, Al-Azhar University, Cairo.
34. Nuclear Materials Authority, P.O. Box 530 El-Maadi, Cairo, Egypt.
35. Oil and Green Chemistry Resea rch Centre / Enhanced Oil Recovery Laboratory, Suez University, Suez, Egypt
36. Ore Processing Department, Nuclear Materials Authority, P.O. box 530, Cairo, Egypt
37. PVT Laboratory, Production Department, Egyptian Petroleum Research Institute (EPRI), Cairo, Egypt.
38. Radiation measurements Branch, Main Defense Chemical Laboratory , Cairo-Suez Road - Almaza, Cairo Egypt
39. Research Department, Nuclear Materials Authority, Cairo, Egypt
40. School of Ships and submarines engineering, Military Technical College, Cairo, Egypt.
41. Science and Technology Center of Excellence (STCE), Egyptian Military Production Authority, Egypt.
42. Technical Research Center, Cairo, Egypt

II Arab and Foreign Universities and Organizations

1. Water Resources Research Center, DPRI, Kyoto University**, Japan**

IOP Conf. Series: Materials Science and Engineering **975** (2020) 011001 doi:10.1088/1757-899X/975/1/011001

MilitaryTechnical College
Kobry El-Kobbah,
Cairo, Egypt

I.C.E.E.2020

10th International Conference on Chemical & Environmental Engineering
7 - 9 April, 2020.

REVIEWERS

1	Amr M.I. Kany	Faculty of Science, Al-Azhar University
2	Sayed Ali El-Mongy	Nuclear Regulatory Authority (ENRRA)
3	R. M. Megahid	Nuclear Research Centre, Atomic Energy Authority
4	M. Mahmoud El-Oquer	Faculty of Science, Al-Azhar University
5	Ahmed A. Baraka	Chemical Engineering Department,MTC
6	Ahmed Farghal Tawfik	Chemical Engineering Department,MTC
7	Mohamed S. El-Deab	Faculty of Science, Cairo University
8	Hussein S. Hegazy	National Research Centre, Dokki, Giza, Egypt
9	Mohamed A. Gobbara	Chemical Engineering Department,MTC
10	M.Kassem Abdel-Magid	Chemical Engineering Department,MTC
11	Mohamed Abo-El fotoh Elsayed	Chemical Engineering Department,MTC
12	Mohamed I. Awaad	Chemical Engineering Department,MTC
13	Nabeel M. Abdel-Monem	Cairo University of Engineering
14	Sherif Fahmy Mahmoud	Faculty of Medicine, Al-Azhar University
15	Sherif Sh. El-Bassuny	Chemical Engineering Department,MTC
16	Zaynab Abdel-Hameed	Central Metallurgical R & D Institute, Cairo, Egypt
17	Hisham Tantawy	Chemical Engineering Department,MTC
18	Said A. Hassanein	Chemical Engineering Department,MTC
19	Mohamed Abdel-Rahman	Chemical Engineering Department,MTC
20	Mahmoud El-Ahdal	Nuclear Research Centre, Atomic Energy Authority
21	Manar Fathy Omran	Atomic Energy Authority, Cairo Egypt
22	Salah El Mofty	Cairo university Mining Engineering Department
23	M. Amin El-Sahir Sadek	British University in Egypt
24	Mahmoud Yossery	Chemical Engineering Department,MTC
25	Mona Mostafa Mahmoud	Faculty of physics Minia university
26	Samah Ibraheem Radwan	Atomic Energy Authority, Cairo Egypt.
27	Mahmoud Abd El Hafeez	Chemical Engineering Department,MTC
28	Mohamed Y. Hassaan	Faculty of science Azahar University
30	Salwa Abo El kassem Maqbol	Faculty of science Minia university
31	Mostafa Radwan	British University in Egypt
32	M. Mohamed El Dosokey	British University in Egypt
33	A. Salah El-Din El-Kholany	Pharos University in Alexandria, Egypt
34	Osama Ali Abuzalat	Chemical Engineering Department,MTC
35	Mohammad Husain Sammour	Chemical Engineering Department,MTC
36	Ahmed El sayed Azazy	Science and Technology Center of Excellence, Military production
37	Magdy M. Hassan Senna	Atomic Energy Authority, Cairo Egypt.

IOP Conf. Series: Materials Science and Engineering **975** (2020) 011001 doi:10.1088/1757-899X/975/1/011001

SCIENTIFIC SESSIONS

IOP Conf. Series: Materials Science and Engineering **975** (2020) 011001 doi:10.1088/1757-899X/975/1/011001

Military Technical College
Kobry El-Kobbah,
Cairo, Egypt

I.C.E.E.2020

10th International Conference on Chemical & Environmental Engineering
7-9 April 2020

Session No 1
Nanotechnology and its chemical Applications
(NCA)

Room: N
Date : Tuesday, 7/7/2020
Time : 11:00 – 12:45
Board: Prof. Dr. / Magdy Foad Abadeer — Cairo Univ.
Maj.Gen.(R)Prof.Dr./Hosam El Sayed Mostafa — MTC
Col. Dr. / Shreif Shaaban Elbasuney — MTC

001-ICEE-10	**Synthesis of High-Quality Multi-Walled Carbon Nano-Tubes (MWCNTs) by Injection Chemical Vapour Deposition; The effect of pyrolysis temperature on MWCNTs** Mohamed Abouelfotoh Elsayed; Omar Ibrahim Abdel aziz; Tamer Z.Wafy; Mahmoud Abdelhafiz
002-ICEE-10	**Synthesis and Characterization of $BaFe_{12}O_{19}$ Ferrite Nanoparticles from Iron Ore Waste and Their Possible Application in Water Treatment** Nehal A. Erfan*a, Salwa A. M. Abdel-Hameedb, Asma A. Mohammeda
003-ICEE-10	**Thermal Properties of Nano Tungsten - Ethylene Vinyl Acetate (EVA) Composites** M. Sobhy*, Alaa I Eid**, W. M. Khairaldien***
004-ICEE-10	**Catalytic Oxidation over Nanostructured Heterogeneous Process as an Effective Tool for Environmental Remediation** Rahma H. Thabet[1], Maha A. Tony[1*], Mai K. Fouad[2], Shakinaz A. El Sherbiny[2], I. A. Ali[3]

IOP Conf. Series: Materials Science and Engineering **975** (2020) 011001 doi:10.1088/1757-899X/975/1/011001

Military Technical College
Kobry El-Kobbah,
Cairo, Egypt

I.C.E.E.2020

10th International Conference on Chemical & Environmental Engineering
7-9 April 2020

001-ICEE-10

Synthesis of High-Quality Multi-Walled Carbon Nano-Tubes (MWCNTs) by Injection Chemical Vapour Deposition; The effect of pyrolysis temperature on MWCNTs

Mohamed Abouelfotoh Elsayed; Omar Ibrahim Abdel aziz; Tamer Z.Wafy; Mahmoud Abdelhafiz

Abstract

Production of high-quality Muti-Walled Carbon Nano-Tubes (MWCNTs) is essential for achieving desired fields, especially in medical and mechanical applications. This work illustrates the synthesis of high quality (MWCNTs) by Injection Chemical Vapour Deposition (Inj-CVD) method using the optimum concentration of Ferrocene dissolved in Toluene solution as a carbon precursor. The effect of pyrolysis temperature, which has a pivot impact on the synthesis process, was investigated. The structure defects, impurities, yield, and sample morphology, as well as the mean diameters of the MWCNTs, were analyzed using Raman spectroscopy, thermal gravimetric analysis (TGA) and scanning electron microscope (SEM), respectively. The results revealed that the change of quality, mean diameter and purity were observed when the pyrolysis temperature increase between 700°C to 850°C. High-quality carbon Nanomaterial was observed by Raman spectroscopy at a pyrolysis temperature of 700°C. While the Transmission electron microscope (TEM) was used to confirm that this product is MWCNTs with 24±5nm, 7.8±1.5nm outer and inner diameter, respectively. The Inj-CVD achieves excellent control of the carbon to catalyst ratio.

Keywords:
High-quality Carbon Nano-tubs, Effect of temperature on CNTs and Characterization of CNTs

[1]Military Technical College, Kobry Elkobbah, Cairo, Egypt

Military Technical College
Kobry El-Kobbah,
Cairo, Egypt

10th International Conferenc
on
Chemical & Environmental Engineering
7-9 April 2020

I.C.E.E.2020

002-ICEE-10

Synthesis and Characterization of $BaFe_{12}O_{19}$ Ferrite Nanoparticles from Iron Ore Waste and Their Possible Application in Water Treatment

Nehal A. Erfan*[a], Salwa A. M. Abdel-Hameed[b], Asma A. Mohammed[a]

Abstract

In this work hard magnetic glass ceramics has been fabricated using Egyptian Bahariya oasis ore; which contains not less than 51% of its weight as pure iron. High energy planetary ball mill was used to convert the prepared magnetic sample into nanoparticles. Different mechanical milling times were applied to study its effect on the properties of the produced magnetic glass ceramic. SEM, TEM, XRF, DTA, XRD, Raman spectroscopy and VSM analysis techniques were performed to characterize the prepared glass ceramic samples before and after exposing to mechanical milling process.
XRD revealed crystallization of barium hexa-ferrite as major phase with average crystallite size ~26 nm. Saturation magnetization Ms depicted 21.2 emu/g after 5 hours milling. The produces nanoparticles after 5 hours milling showed a good performance in removal of pollutants like methylene blue and Congo red from water by adsorption.

Keywords:

Magnetic glass ceramics; Ball milling; Barium hexa-ferrite; nanoparticles; Iron ore, Waste; Recycling

[a] Chemical engineering department, faculty of engineering, Minia university, Egypt.
[b] Glass Research Department, National Research Centre, Dokki, Cairo, Egypt

IOP Conf. Series: Materials Science and Engineering **975** (2020) 011001 doi:10.1088/1757-899X/975/1/011001

Military Technical College
Kobry El-Kobbah,
Cairo, Egypt

10th International Conference on Chemical & Environmental Engineering
7-9 April 2020

003-ICEE-10

Thermal Properties of Nano Tungsten - Ethylene Vinyl Acetate (EVA) Composites

M. Sobhy*, Alaa I Eid**, W. M. Khairaldien***

Abstract

Ethylene vinyl acetate (EVA) with different percentage of nano tungsten powder (0wt%, 30 wt%, 45 wt%, 60 wt%, and 70 wt%) composites(EVA/W nano-composites)were prepared by melt blending in a twin-screw brabender. The structural, morphological, thermal and mechanical properties of the prepared nano-composites were investigated using powder X-ray diffraction (PXRD), Scanning Electron Microscopy (SEM), and thermogravimetric analysis (TGA). It was found that the EVA composites show a higher thermal stability at high levels of tungsten loading.

Keywords:
EVA, Nano tungsten powder, higher thermal stability

*Science and Technology Center of Excellence (STCE), El- Salam City, Cairo, Egypt
**Center of Metallurgy Research and Development Institute
***Asiut University, Faculty of Engineering, Mechanical Engineering Department
Correspondence to: Maged Sobhy; e-mail: maged@stce-egypt.org

IOP Conf. Series: Materials Science and Engineering **975** (2020) 011001 doi:10.1088/1757-899X/975/1/011001

Military Technical College
Kobry El-Kobbah,
Cairo, Egypt

I.C.E.E.2020

10th International Conference
on
Chemical & Environmental Engineering
7-9 April 2020

004-ICEE-10

Catalytic Oxidation over Nanostructured Heterogeneous Process as an Effective Tool for Environmental Remediation

Rahma H. Thabet[1], Maha A. Tony[1*], Mai K. Fouad[2], Shakinaz A. El Sherbiny[2], I. A. Ali[3]

Abstract

Industrialization has led to a severe deterioration in water quality. Textile industry is considered a huge consumer of water in Egypt, the result is generating large amounts of dye-containing wastewater that is essential to be treated before the final disposal. However, searching for an efficient treatment is an important aspect for a sustainable environment. Advanced oxidation processes (AOPs) have been emerged as efficient techniques for industrial wastewater remediation. Among the AOPs, Fenton based reactions is considered a promising process for its simplicity in application and cost-efficient with high process efficiency. In this study, heterogeneous Fenton reaction using magnetite nanoparticles induced by ultraviolet radiation (UV) was applied as a green technology pathway for textile dying wastewater oxidation. Nanostructured magnetite was successfully synthesized by co-precipitation method that is used as the precursor of the Fenton's reaction process. The heterogeneous iron (Fe^{2+}/Fe^{3+}) supported catalyst with hydrogen peroxide (H_2O_2) was used as a coupled Fenton and Fenton-like oxidation system for methylene blue dye removal in aqueous media. The obtained results investigated that the dye oxidation rate increases with decreasing pH to 3.0. However, increasing H_2O_2 and magnetite nanoparticles catalyst results in an increase the dye oxidation rate and the optimum operating values were 80 and 1600 mg/L for Fe_3O_4 and H_2O_2, respectively. By optimizing the amount of reactants and process conditions, the results revealed that magnetite was considered an efficient Fenton-based catalyst for dye oxidation that is reached to 94% within 3 hr of oxidation time. Finally, magnetite catalyst could be easily recovered by magnetic separation to confirm the process sustainability.

Keywords:
Fenton's reagent, Methylene blue, Wastewater, Nanomaterial, Magnetite

[1]Faculty of Engineering, Menoufia University, Shebin El-Kom, Egypt
[2]Chemical Engineering Department, Faculty of Engineering, Cairo University, Egypt
[3]Cyclotron project, Nuclear Physics Department, Nuclear Research Center, Atomic Energy Authority, Egypt
* corrosponding author: Tel: +2048-2221549, Fax: +20482235695, Egypt, E-mail: **dr.maha.tony@gmail.com**

IOP Conf. Series: Materials Science and Engineering **975** (2020) 011001 doi:10.1088/1757-899X/975/1/011001

Military Technical College
Kobry El-Kobbah,
Cairo, Egypt

I.C.E.E.2020

10th International Conference on Chemical & Environmental Engineering
7-9 April 2020

Session No 2

Energetic Materials and its Applications (EMA)

Room: N
Date : Tuesday, 7-7-2020
Time : **13**:00 – 14:45
Board: Prof. Dr./ M. Amin El-Sahir BUE
Assoc.Prof. Dr. / M. Hussein Samour EAF
Lt.Col. Dr. / Mahmoud Abd El-Hafize MTC

005-ICEE-10	**Superior spectral performance of decoy flares compositions via inclusion of graphite as a black body emitter** Amr A. Elmotaz[1], Sherif Elbasuney[1], Hesham Tantawy[1], M. A. Sadek[2]
006-ICEE-10	**Assessment of antioxidant resistance to thermal- oxidative degradation of stabilized poly-butadiene binder** E.Elsaka, Sh. Elbasuney, A. Elbeih, H.Mostafa, M.E. Awad , T. M. Elhedery
007-ICEE-10	**A study of the detonation wave shape inside confined explosive with a metallic liner** Tamer Elshenawy, Ahmed Elbeih, Mohamed G. Zaki, Ahmed K. Hussein
008-ICEE-10	**The Potentials of Aluminum Nanoparticles: Novel High Energy Density Material for Underwater Explosions** Sh.Elbasuney, M. Gaber, M.Bennaya, Sh.M. Abdelkhalek, M.Radwan.
009-ICEE-10	**Aluminum Nanoparticles: a Novel High Energy Density Material for Metalized Explosives with Combined Destructive Effect (Combustion/Detonation)** Sherif Elbasuney[a,b*] , M. Gaber Zaky[a], Mostafa Radwan[c], Ahmed Maraden[a], Hesham Ramzy[a], Sherif M. Abdelkhalek[d], Sherif F. Mostafa[a]

IOP Conf. Series: Materials Science and Engineering **975** (2020) 011001 doi:10.1088/1757-899X/975/1/011001

Military Technical College
Kobry El-Kobbah,
Cairo, Egypt

10th International Conference on Chemical & Environmental Engineering
7-9 April 2020

005-ICEE-10

Superior spectral performance of decoy flares compositions via inclusion of graphite as a black body emitter

Amr A. Elmotaz[1], Sherif Elbasuney[1], Hesham Tantawy[1], M. A. Sadek[2]

Abstract

Graphite particles can offer high interfacial surface area as well as high emissivity as black body; therefore it can find wide application in advanced decoy flares. In the current study graphite particles of 100 μm size, were employed with different reactive metal fuels including Mg, Al, and Mg-Al alloy. Different formulations based on graphite ratio up to 8 wt % were developed via granulation with subsequent pressing. Thermal signature was measured using IR spectrometer (1-6 μm) Arc-Optics. Nanocomposite flare based on Mg-Al fuel, and graphite demonstrated superior spectral intensity; this formulation offered an increase in average intensity by 83 % to reference. In the mean time, nanocomposite flare based on Al and graphite offered the highest relative intensity value Θ of 0.6. Graphite, as an allotrope of carbon, can act as an excellent source of carbonaceous materials; that can strengthen incandescence emission. Furthermore Mg-Al reactive metal alloy con offer novel synergism as Mg could vaporize at 1000 0C offering efficient combustion process. It can be concluded that tailored flare can be developed according to the required spectral data.

Keywords:
Thermal signature; Infrared missile; Countermeasure; Decoy flare; Pyrotechnics

[1] School of Chemical Engineering, Military Technical College, Cairo, Egypt
[2] Chemical engineering department,. British University in Egypt (BUE)
Email: amr_elmotaz@hotmail.com

Military Technical College
Kobry El-Kobbah,
Cairo, Egypt

10th International Conference on Chemical & Environmental Engineering
7-9 April 2020

006-ICEE-10

Assessment of antioxidant resistance to thermal- oxidative degradation of stabilized poly-butadiene binder

E.Elsaka, Sh. Elbasuney, A. Elbeih, H.Mostafa, M.E. Awad, T. M. Elhedery

Abstract

Poly-butadiene is the most common polymeric binder for composite solid propellants. This polymer suffers from oxidative degradation reactions on storing with deterioration in mechanical properties. Certain anti-oxidant materials can retard such reaction offering enhanced aging characteristics. In this study, two different primary antioxidants including phenol-based (AO2246) and amine-based (Flexzone 6H) were employed to stabilize polybutadiene polymer. Oxidation induction time (OIT) was evaluated using isothermal DSC. High OIT value means high resistance to oxidative degradation. The impact of anti-oxidant on mechanical properties was evaluated using shore A. Flexzone 6H significantly improved binder oxidative stability with an increase in OIT value by 20 folds with significant decrease in oxidation enthalpy. Flexzone 6H can offer enhanced resistance toward oxidative reactions; as it can act as efficient H atom donor to break down chain degradation reaction. Additionally, the three stabilized aromatic rings can capture free radicals. AO2246 enhanced the binder mechanical characteristics during aging at 80 0C using Shore A test. In the meantime, AO2246 could enhance the mechanical properties due to its integration into the polymeric matrix due to –OH functional groups. The two anti-oxidants demonstrated controversy effects. Whereas Flexzone 6H offered enhanced resistance to oxidative degradation; AO2246 offered enhanced mechanical properties.

Keywords:
Antioxidants; Thermal-oxidative stability; Ageing; Composite propellants

[1]School of chemical Engineering, Military Technical College, Cairo, Egypt

IOP Conf. Series: Materials Science and Engineering **975** (2020) 011001 doi:10.1088/1757-899X/975/1/011001

Military Technical College
Kobry El-Kobbah,
Cairo, Egypt

I.C.E.E.2020

10th International Conference on Chemical & Environmental Engineering
7-9 April 2020

007-ICEE-10

A study of the detonation wave shape inside confined explosive with a metallic liner

Tamer Elshenawy[1*], Ahmed Elbeih[2], Mohamed G. Zaki[2], Ahmed K. Hussein[2]

Abstract

The behaviour of shock wave inside a shaped charge is so important that the detonation wave nature characterizes (determines) the degree of jet coherency at the liner interface. The detonation wave shape inside the shaped charge depends on the casing thickness, material, the liner material, geometry and the applied wave shaper geometry. In this paper, a series of numerical simulation models were performed to study the incident and reflected waves and to characterize the resulting shock wave front to evaluate its effect on the produced (generated) jet characteristics. The used code was Autodyn 3D based on non-linear dynamic analysis with jetting analysis model. Inert wave shapers (Lead, Aluminum and steel) and active ones (TNT and PETN) behaviour were studied. The liner type used in this study was copper with in which the apex angle of the liner cone was 40o. It was concluded that the wave front shape has an obvious influence on the characteristics of the jet formation and its properties such as mass of jet and velocities of its tip. Also, the bulk sound speed of the liner itself in addition to the collapse characteristics of the liner was discussed in this paper successfully.

Keywords:
Liner, wave shaper, explosive shock wave, jetting, Autodyn code

[1] Technical Research Center, Cairo, Egypt
[2] Military Technical College, Kobry Elkobbah, Cairo, Egypt

IOP Conf. Series: Materials Science and Engineering **975** (2020) 011001 doi:10.1088/1757-899X/975/1/011001

Military Technical College
Kobry El-Kobbah,
Cairo, Egypt

10th International Conference on Chemical & Environmental Engineering
7-9 April 2020

008-ICEE-10

The Potentials of Aluminum Nanoparticles: Novel High Energy Density Material for Underwater Explosions

Sh.Elbasuney [a*], M.Gaber[b], M.Bennaya [c], Sh.M.Abdelkhalek [b], M.Radwan [d]

Abstract

The destructive parameters of underwater explosives (i.e. shock wave energy, maximum pressure, and bubble radius) are limited to explosion heat; that is comparatively low. One approach for enhanced heat output can be accomplished by integrating reactive metal particles (i. e. Aluminium). However conventional aluminium particles (µm size) would contribute only with combustion gaseous products behind detonation wave front. Underwater, there is no oxygen for such contribution to take place. Furthermore, conventional Al particles could decrease the detonation velocity. So far, full exploitation of aluminium particles in underwater explosions couldn't be achieved. Aluminium nanoparticles would combust more efficiently within detonation wave front, offering smaller critical diameter, high reaction rate, and high heat release rate. Consequently, Al nanoparticles could be ideal high energy density material for underwater explosion. Ship model with positive metacentric height, GMT =4.7 cm for ship transverse stability, and GML = 19.3 for ship longitudinal stability was designed. Ship model offers large angle stability (heeling angles= 0-70 deg.). 2 g of explosive charge was detonated underneath the developed naval structure. Upon explosion, the acceleration of the naval structure was measured using shock accelerometer VC tri-axial, high frequency, 5000 ground acceleration, Dytran, Inc. While, Al particles (10 µm) offered an increase in mono-hull acceleration by 16 % compared to TNT; Al nanoparticles offered an acceleration increase by 49 %. This novel finding can be ascribed to the efficient combustion of Al nanoparticles within detonation wave front offering ideal detonation reaction with enhanced destructive effect.

Keywords: Underwater explosion (UNDEX); Metalized explosives; Shock wave.

[a] Head on nanotechnology research centre, Military technical college, Cairo, Egypt.
[b] Military Technical College, Cairo, Egypt.
[c] School of Ships and submarines engineering, Military Technical college, Cairo, Egypt.
[d] British University in Egypt, Elshorouk City, Cairo, Egypt.

IOP Conf. Series: Materials Science and Engineering **975** (2020) 011001 doi:10.1088/1757-899X/975/1/011001

Military Technical College
Kobry El-Kobbah,
Cairo, Egypt

I.C.E.E.2020

10th International Conferenc
on
Chemical & Environmental
Engineering
7-9 April 2020

009-ICEE-10

Aluminum Nanoparticles: a Novel High Energy Density Material for Metalized Explosives with Combined Destructive Effect (Combustion/Detonation)

Sherif Elbasuney [a,b*], M. Gaber Zaky [a], Mostafa Radwan [c], Ahmed Maraden [a], Hesham Ramzy[a], Sherif M. Abdelkhalek [d], Sherif F. Mostafa [a]

Abstract

Even though, reactive metal particles can boost the energy density with an increase in total impulse of explosive materials. Such particles will react behind the detonation wave front with decrease in detonation velocity and brisance (destructive effect). This study reports on the effective development of aluminium nanoparticles of 100 nm particle size. Al nanoparticles offered superior performance compared micron scale particles. Whereas conventional Al particles (10 µm) offered increase shock wave strength of TNT by 17 %, nanoparticles offered an increase in shock wave strength by 48 %. While micron-Al decreased the destructive effect of TNT by -6.5 %; Al nanoparticles offered an increase in destructive effect by 21 %. The main outcome of this study is that Al nanoparticles offered an enhanced detonation velocity 6330 m/s compared with 5650 m/s for TNT It can be concluded that while conventional Al particles could act as desensitizer; Al nanoparticles could act as sensitizer and could combust more effectively within detonation wave front. Additionally Al nanoparticles offered decrease in TNT critical diameter from 40 mm to 20 mm. This study reports on the real development of metalized nanocomposite explosives with superior performance (combustion/detonation).

Keywords:
Metalized explosives; Combustion, Destructive effect; Workability; Brisance; Thermal behavior

[a] School of Chemical Engineering, Military Technical College, Kobry Elkoba, Cairo, Egypt.
[b] Head of nanotechnology research centre, Military Technical College, Cairo Egypt.
[c] British University in Egypt, El-Shrouk City, Cairo, Egypt.
[d] Head of Engineering and Technology Research Center, Military Technical College, Cairo, Egypt
s.elbasuney@mtc.edu.eg, sherif_basuney2000@yahoo.com, +201112630789

IOP Conf. Series: Materials Science and Engineering **975** (2020) 011001 doi:10.1088/1757-899X/975/1/011001

Military Technical College
Kobry El-Kobbah,
Cairo, Egypt

I.C.E.E.2020

10th International Conference on Chemical & Environmental Engineering
7-9 April 2020

Session No. 3

Water and Waste Water Treatment by Synthetic Polymers and Adsorbents (WWWT)

Room : N
Date : Wednessday 8-7-2020
Time : 9:00 – 10:45
Board: Prof. Dr. /Omar El Farouk — Cairo Univ.
Brig. Gen. Prof. Dr. / M. Aboel-Fotoh — MTC
Col.Dr./ M. Mokhtar Kotb — MTC

010-ICEE-10	**Physico Chemical Studies for water at Rosetta Branch of River Nile, Egypt** Mamdouh S. Masoud[(a)], Amel M. Ismail[(a)], Hesham M. kamel[(b)], Ahmed S. El-Kholany*[(b)] and Ahmed A.M. ElKasas[(a)]
011-ICEE-10	**Rain Transmission Losses Assessment in Arid Environment, Egypt: Numerical and Experimental Study** Adel Bakheet[1], Ahmed Sefelnasr[2]
012-ICEE-10	**Synthesis and characterization of zinc-imidazolate coordination polymer and its use for photocatalytic degradation of methylene blue dye under visible light irradiation** Mostafa Mohsen[1, 2], Ibrahim Naeem[1], Mohamed I. Awaad[1] and Ahmad Baraka[1]
013-ICEE-10	**Synthesis , Characterization and application of analcime to control nitrate ions from the ground water samples from Wadi El-Assiuti – Egypt as a low-cost and locally available adsorbent** A.A. Abdelmoneim[1], M.Abdul-Moneim[2], A.A Geies[3], and Seham O. Farghaly[1]

IOP Conf. Series: Materials Science and Engineering **975** (2020) 011001 doi:10.1088/1757-899X/975/1/011001

Military Technical College
Kobry El-Kobbah,
Cairo, Egypt

I.C.E.E.2020

10th International Conferenc
on
Chemical & Environmental Engineering
7-9 April 2020

010-ICEE-10

Physico Chemical Studies for water at Rosetta Branch of River Nile, Egypt

Mamdouh S. Masoud[(a)], Amel M. Ismail[(a)], Hesham M. kamel[(b)], Ahmed S. El-Kholany*[(b)] and Ahmed A.M. ElKasas[(a)]

Abstract

This work was carried out to study the quality of water of Rosetta branch. Five sites were chosen for sampling during the period of January 2013 to December 2013. Physical parameters (Temperature, electrical conductivity (EC), turbidity and pH) and chemical parameters (concentration of Cl^-, Ca^{+2}, Mg^{+2}, NH_4^+, SO_4^{-2}, NO_2^-, and some of trace metals, alkalinity, total hardness, Dissolved Oxygen(DO), Chemical Oxygen Demand(COD) and Biological Oxygen Demand(BOD) were discussed. These results were compared with the results of another study that were carried out at Rosita Branch at 2018. Statistical analysis of data was computed to show the correlation between some parameters through the whole sampling sites during the period of the study, using simple correlation coefficient.

Keywords:

Rosetta branch - Physical and chemical parameters – Water Quality Index - Correlation Coefficients

a) Chemistry Department, Faculty of Science, Alexandria University, Egypt.
b) Medical Laboratory Department, Faculty of Allied Medical Sciences, Pharos University, Alexandria, Egypt

IOP Conf. Series: Materials Science and Engineering **975** (2020) 011001 doi:10.1088/1757-899X/975/1/011001

Military Technical College
Kobry El-Kobbah,
Cairo, Egypt

I.C.E.E.2020

10th International Conference on Chemical & Environmental Engineering
7-9 April 2020

011-ICEE-10

Rain Transmission Losses Assessment in Arid Environment, Egypt: Numerical and Experimental Study

Dr./Adel Bakheet[1], Assoc. Prof. Dr./ Ahmed Sefelnasr[2], and Mohamed Saber[3]

Abstract

In water-scarce regions, Transmission Losses Assessment from rainfall is required for the estimation of groundwater recharge. This Study is an attempt to assess Transmission Losses and the infiltration rate at Wadi El-Assiuti, Egypt, field infiltration experiments were conducted to measure the infiltration and porosity at the target area, such as using a double-ring infiltrometer. Quantifying the soil infiltration capacity is very important for determining components of the hydrological modeling, irrigation design and many other natural or man-made processes Steady state infiltration rates have been determined in selected sites of Wadi El-Assiuti. Accurate determination of infiltration rates is essential for reliable prediction of surface runoff and groundwater recharge. Double ring infiltrometer was used to measure infiltration rate at selected sites. An important part of the study was the accurate and consistent measurement of infiltration rates. .The infiltration rate was calculated according to the installation and operating instruction for the double ring in ASTM 2003. Final infiltration rates were 18 cm h^{-1}, for the first observation site, 24.2 cm $h^{-1}$for the second site and 59 cm $h^{-1}$for the third site. The analysis indicated that the soil for the three selected areas is sandy soil. Infiltration rates were taken at 0 to70 minutes of 10 minutes intervals. The infiltration rate at the third site is high compared to the other two selected sites. After performing the experiments, soil samples have been gathered and analyzed in the laboratory for porosity analysis. The results of this study can therefore be applied in the prediction of saturated hydraulic conductivity of the surface layers and groundwater recharge, and in developing or selecting the most efficient irrigation methods. The results from this work have been also involved in a numerical hydrological model to estimate the Transmission Losses form rainfall at Wadi El-Assiuti using Hydro-BEAM.

Keywords:
Transmission Losses, Double ring in-filtro-meters, hydrological modeling, Wadi El-Assiuti, Hydro-BEAM

[1] Ph D. Environmental Geology, water and microbiology lab, Chemistry Administration
[1]-Adel Bakheet Mostafa Tel.01007202821 Address- Assiut, bakheet_adel@yahoo.com
[2] Geology Department, Assiut University, ahmed.sefelnasr@science.au.edu.eg
3 Senior Researcher, Water Resources Research Center, DPRI, Kyoto University, japan

IOP Conf. Series: Materials Science and Engineering **975** (2020) 011001 doi:10.1088/1757-899X/975/1/011001

Military Technical College
Kobry El-Kobbah,
Cairo, Egypt

I.C.E.E.2020

10th International Conference on Chemical & Environmental Engineering
7-9 April 2020

012-ICEE-10

Synthesis and characterization of zinc-imidazolate coordination polymer and its use for photocatalytic degradation of methylene blue dye under visible light irradiation

Mostafa Mohsen[1,2], Ibrahim Naeem[1], Mohamed I. Awaad[1] and Ahmad Baraka[1]

Abstract

Zinc-Imidazole coordination polymer (Zn-Im CP) was facilely synthesized by hydrothermal method from zinc acetate and imidazole as ligand. The structure of the synthesized Zn-Im CP was analyzed by using scanning electron microscopy (SEM), EDX, FTIR, and powder X-ray diffraction (PXRD). Surface area of Zn-Im CP was determined by BET analysis of N_2 adsorption. The photocatalytic performance of Zn-Im CP using visible light irradiation source was investigated using UV–Vis spectrophotometer for the degradation of methylene blue. The pseudo first order reaction kinetic model was applied on the degradation data which gave the rate constant of 0.006 min^{-1}. The results indicate that Zn-Im CP exhibits good photocatalytic performance for the photodegradation of methylene blue following indirect mechanism pathway.

Keywords:
Photocatalytic degradation, Visible light, Coordination polymer, Dye, imidazole

[1] Military Technical College, Kobry Elkobbah, Cairo, Egypt
[2] The corresponding author, E-mail Moustafa.klm@gmail.com

IOP Conf. Series: Materials Science and Engineering **975** (2020) 011001 doi:10.1088/1757-899X/975/1/011001

Military Technical College
Kobry El-Kobbah,
Cairo, Egypt

I.C.E.E.2020

10th International Conference on Chemical & Environmental Engineering
7-9 April 2020

013-ICEE-10

Synthesis , Characterization and application of analcime to control nitrate ions from the ground water samples from Wadi El-Assiuti – Egypt as a low-cost and locally available adsorbent

Ahmed A. Abdelmoneim[1], M.Abdul-Moneim[2], A.A Geies[3], and Seham O. Farghaly[1]

Abstract

In this study, the sorption behavior of synthetic (Analcime) zeolites with respect to nitrate ions has been studied in order to consider its application to purity ground water. Analcime or also can be called analcite (from the Greek, means "weak") is a kind of grey, white or colorless tectosilicates minerals. It is hydrated sodium aluminum silicates which exist in cubic form in crystalline. Analcime was successively synthesized from kaolinite as a raw material using the fusion with NaOH method. The conditions of hydrothermal crystallization (zeolitization) were found to be at temperature of 170 C°, and time span between 36 h and 72 h for kaoline with the molar composition of $6Na_2O$: $0.75Al_2O_3$:$30SiO_2$. $780H_2O$.The synthetic materials have been characterized by X-ray diffraction (XRD), scanning electron microscope (SEM), Fourier transform infrared spectroscopy (FT-IR) and thermo gravimetric (DTA/TGA) analysis. The results indicate that the crystallization of analcime not affected by the hydrothermal reaction time. Also analcime was tested as adsorbents for the removal of nitrate ions from the ground water samples from Wadi El-Assiuti – Egypt .The adsorption capacities of nitrate ions by analcime, as a function of its concentration, were determined at room temperature by varying analcime concentration for each water samples. During the process, all the other parameters (pH and contact time) were kept constant with respect to the initial concentration of nitrate ions in the water samples. It was found analcime has good removal efficiency obtained at pH 6-7.6, adsorbent dosage 10-12 g/L, and contact time 60 min. The Langmuir constants model for NO_3^- ion sorption on the adsorption isotherms is fitted well. The R_L value in the present investigation was equal or less than one, indicating that the adsorption of NO_3^-ion by analcime is favorable.

Keywords:
Analcime, Ground water, Characterization, Element contents, Water treatment

[1]Geology Departments, Faculty of Science, Sohag University, Sohag, Egypt
[2]Geology Department, [3]ChemistryDepartment, Faculty of

IOP Conf. Series: Materials Science and Engineering **975** (2020) 011001 doi:10.1088/1757-899X/975/1/011001

Military Technical College
Kobry El-Kobbah,
Cairo, Egypt

I.C.E.E.2020

10th International Conference on Chemical & Environmental Engineering
7-9 April 2020

Session No 4

Chemical Engineering Applications (CEA)

Room: N
Date : Wednesday, 8/7/2020
Time : 11:00 – 12:45
Board: Prof. Dr. Hussein Sayed Hegazy NRC
Prof. Dr. / Zeinab Abd El Hamid CMRDI
Brig.Gen. Assoc. Prof. Dr. /M.M.Gobbara MTC

014-ICEE-10	**Ionic Liquids based on Different Chain Fatty Acids as Green Corrosion Inhibitors for C-steel in Produced Oilfield Water** Y Atef[1]* and A Ghanem[2]
015-ICEE-10	**Thermal and Potentiometric Studies for Complexes of Zr and Hf with Some Barbituric Acid Derivatives** Mamdouh S. Masoud [(a)], Azza S. Hassan[(a)], Nermeen M. Desouky[(a)],Hesham M. kamel[(b)], Ahmed S. El-Kholany* [(b)]
016-ICEE-10	**Thermodynamic Analysis for Tracking the Effect of the Egyptian Zirconium Silicate on the slag properties of low hydrogen Welding Electrodes** A.AbdElkarim[1], H.M.Abd El-Aziz[2], , Waleed.A.Mohrez[1] and H. Mashaal[1]

IOP Conf. Series: Materials Science and Engineering **975** (2020) 011001 doi:10.1088/1757-899X/975/1/011001

Military Technical College
Kobry El-Kobbah,
Cairo, Egypt

10th International Conference on Chemical & Environmental Engineering
7-9 April 2020

I.C.E.E.2020

014-ICEE-10

Ionic Liquids based on Different Chain Fatty Acids as Green Corrosion Inhibitors for C-steel in Produced Oilfield Water

Y Atef[1]* and A Ghanem[2]

Abstract

The work presents the investigation of fatty acids based ionic liquid as green corrosion inhibitors for C-steel in produced oilfield water. Different-chain length fatty acids C_6, C_{12} and C_{18-1} are mixed with di-ethanol amine (DEA) to produce some ionic liquids and evaluated the corrosion rates on C-steel surface using weight loss method and potential time measurements. The prepared ionic solutions were shown to be effective in reducing the corrosion rates as sustainable corrosion inhibitors. The polar unit is considered as the reaction center for the adsorption process. Generally, the inhibition efficiency increases with increase the inhibitors concentration.

Keywords:
Carbon steel, ionic liquids, fatty acids, corrosion inhibitors, adsorption

[1*] Egyptian Chemistry Administration, Cairo, Egypt, E-mail: jasy9@yahoo.com
[2] PVT Laboratory, Production Department, Egyptian Petroleum Research Institute (EPRI), Cairo, Egypt

Military Technical College
Kobry El-Kobbah,
Cairo, Egypt

I.C.E.E.2020

10th International Conference on Chemical & Environmental Engineering
7-9 April 2020

015-ICEE-10

Thermal and Potentiometric Studies for Complexes of Zr and Hf with Some Barbituric Acid Derivatives

Mamdouh S. Masoud [(a)], Azza S. Hassan[(a)], Nermeen M. Desouky[(a)],Hesham M. kamel[(b)], Ahmed S. El-Kholany* [(b)]

Abstract

The complexes of Barbital (Bal), Thiobarbituric acid thione (TBA), Thiobarbituric acid thiol (2-TBA), 2-Thiouracil (TU) with Zr (IV) and Hf (IV) were prepared. The structures and mode of bonding were characterized by elemental analysis, IR and electronic spectroscopy. Thermal analyses of the prepared complexes were done by DTA, TG, DSC techniques. The thermodynamic parameters were evaluated. The change of entropy values, $\Delta S^{\#}$, for all the complexes lay within the range -0.267,-0.299 $KjK^{-1}mol^{-1}$. The thermal processes proceed in complicated mechanisms with ordered transition states. In general, the complexes dissociated after losing small molecules such as H_2O, Cl_2 and CO. The decomposition is usually ended by the metal moiety. The DSC technique explained the effect of temperature on physical properties of some complexes. The glass transition phase doesn`t appear in all the complexes, while the crystallization temperature (Tc) was 78^0C-322.12^0C. Potentiometric titration studies calculated the dissociation constants of the ligands, the stability constants of the complexes and the distribution of species at different pH values.

Keywords:
IR, DTA, TGA, DSC, potentiometric titration, complexes, Zr, Hf, barbituric, acid derivatives

a) Chemistry Department, Faculty of Science, Alexandria University, Alexandria, Egypt.
b) Medical Laboratory Department, Faculty of Allied Medical Sciences, Pharos University in Alexandria, Egypt

IOP Conf. Series: Materials Science and Engineering **975** (2020) 011001 doi:10.1088/1757-899X/975/1/011001

Military Technical College
Kobry El-Kobbah,
Cairo, Egypt

10th International Conferenc
on
Chemical & Environmental
Engineering
7-9 April 2020

016-ICEE-10

Thermodynamic Analysis for Tracking the Effect of the Egyptian Zirconium Silicate on the slag properties of low hydrogen Welding Electrodes

A.AbdElkarim[1], H.M.Abd El-Aziz[2], Waleed.A.Mohrez[1] and H. Mashaal[1]

Abstract

One of the most common types of welding electrodes, used in welding processes; are the low hydrogen electrode classification; E 7018. This electrode is applied for fabrication most carbon steel alloys, where it used for welding plate works of ships, equipment, tanks, boilers and pressure vessels, steel structures of buildings, bridges as well.

One of the most common drawbacks resulting of using of this welding electrode is the difficulty of removing the slag between the deposited passes, which consumes time and money and hence decreasing the quality of welding process also one of the most drawbacks of this type of welding electrode is the rabid burning of coating flux at elevated amperes increase.

This study aimed to suggest rooted solution for these two problems by adding zirconium silicate to the flux recipe.

Different percentages of Egyptian zirconium silicate were added to the coating flux recipes of type E7018 where the changes at both weldability and slag removing between deposited passes were noticed at different amperes.

By using HSc simulation program of the thermodynamics data and expected phases in slag were studied. Microstructure samples of weld metal were imaged and analyzed, basicity index was calculated, XRF chemical analysis for all slag different samples, were determined. The study concluded that adding 5% of the Egyptian zirconium silicate to the flux clearly improved; the welding properties as well as the ease of slag removal.

Keywords:

[1]-Nuclear materials Authority, Cairo
[2]- Mining and petroleum Dep., faculty of Engineering, Al-Azhar University, Cairo

IOP Conf. Series: Materials Science and Engineering **975** (2020) 011001 doi:10.1088/1757-899X/975/1/011001

Military Technical College Kobry El-Kobbah, **Cairo, Egypt**	I.C.E.E.2020	**10th International Conferenc** **on** **Chemical & Environmental Engineering** 7-9 April 2020

Session No. 5

Environmental Studies and Pollution Control

(ESPC)

Room : **N**
Date : Wednessday 8-7-2020
Time : 1300 – 14:45
Board:

Prof. Dr. / Nabeel Abd El Monem	Cairo Univ.
Prof. Dr. / Said Abd El Magid Hassanein	MTC
Brig.Gen. Prof. Dr. / Ahmed Baraka	MTC

017-ICEE-10	**Radiological and environmental studies on the metamorphosed sandstones at Wadi Sikait area, Southeastern Desert, Egypt** A. H. El-Farrash[**], M. G. El-Feky[*], S. F. Hassan[*], S. A. Abu Elatta[*], and M. Refaat[*]

IOP Conf. Series: Materials Science and Engineering **975** (2020) 011001 doi:10.1088/1757-899X/975/1/011001

Military Technical College
Kobry El-Kobbah,
Cairo, Egypt

I.C.E.E. 2020

10th International Conference on Chemical & Environmental Engineering
7-9 April 2020

017-ICEE-10

Radiological and environmental studies on the metamorphosed sandstones at Wadi Sikait area, Southeastern Desert, Egypt

A. H. El-Farrash[**], M. G. El-Feky[*], S. F. Hassan[*], Soliman A. Abu Elatta[*], and M. Refaat[*]

Abstract

Wadi Sikait area lies in the southern part of the Eastern Desert of Egypt along the upper stream of Wadi Sikait. The exposed rock units in Wadi Sikait area are ophiolitic mélange roccks, metamorphosed sandstones, and porphyritic granites, lamprophyre dykes, in addition to fluorite and quartz veins. The activity concentrations of various radionuclides are measured by NaI-Tl activated detector and the radioactive mineralization were determined by Alpha track technique using CR-39 films. Radiometric investigations indicate that 238U activity concentrations range between 37.2 and 520.4 with average 148.8 BqKg-1, 232Th activity concentrations vary between 8.08 and 366.99 BqKg-1 with 76.75 BqKg-1 as an average. The studied metamorphosed sandstones have higher values relative the world concentration levels. High uranium content of the metamorphosed sandstone is attributed to presence radioactive minerals like uranophane and autonite, in addition to accessory minerals like allanite, zircon and monazite. Other U-bearing minerals are also recorded as biotite, muscovite, iron oxides and clays. Absorbed Dose Rate (D), annual effective dose equivalent (AEDE), radium equivalent activity (Raeq), external (Hex) and internal (Hin) hazard index, in addition to activity gamma index (Iγ) caused by gamma emitting natural radionuclide are determined from the obtained values of 226Ra, 232Th and 40K. Fairly, many of the studied metamorphosed sandstone do not satisfy the universal standards.

Keywords:
Radiological, Environment, Uranium, Thorium, Egypt

Nuclear materials Authority, Cairo

IOP Conf. Series: Materials Science and Engineering **975** (2020) 011001 doi:10.1088/1757-899X/975/1/011001

Military Technical College
Kobry El-Kobbah,
Cairo, Egypt

I.C.E.E.2020

10th International Conference on Chemical & Environmental Engineering
7-9 April 2020

Session No 6

Nuclear and Radiometric Applications (NRA)

Room: N
Date : Thrusday, 9/7/2020
Time : 9:00 – 10:45
Board: Prof. Dr / Amr M.I. Kany — Al Azhar Univ.
Prof. Dr. / Magdy Senna — AEA
Col. Assoc.Prof. Dr. / Mohamed Abd El Rahman — MTC

018-ICEE-10	**Mineralogical and Radiometric Studies of Granitic Rocks in Wadi Sabbagh Area, South Sinai, Egypt** Mira, H. I.[2]., El-Gharbawy, R.I.[1], Elmowafy, A.A.[2], Osman, A.F.[1] El Maadawy, W.M.[2]
019-ICEE-10	**Sub-channel thermal-hydraulic analysis for VVER- 1000 generation III PWR** Alsherief M Almessallmy[1], Mohamed K Shaat[2] and Saeed A Hassanien[1]
020-ICEE-10	**Effect of The Formation Parameters on The Thermal Properties of La2Zr2O7** Ahmed Abdelkareem

Military Technical College
Kobry El-Kobbah,
Cairo, Egypt

10th International Conferenc
on
Chemical & Environmental
Engineering
7-9 April 2020

I.C.E.E.2020

018-ICEE-10

Mineralogical and Radiometric Studies of Granitic Rocks in Wadi Sabbagh Area, South Sinai, Egypt

Mira, H. I.[2]., El-Gharbawy, R.I.[1], Elmowafy, A.A.[2], Osman, A.F.[1] El Maadawy, W.M.[2]

Abstract

Wadi Sabbagh area is located at southern part of Sinai Peninsula. It is constrained by Latitudes 28° 05′ and 28° 15′ N and Longitudes 34° 00′ and 34° 15′ E. Gneisses, migmatites, metasediments, syn- and post-orogenic granites in addition to pegmatites and post-granitic dykes are the main rock units encountered in this area. Geological, petrological and geochemical studies indicate that there are two distinct suites of granitoids: older tonalite to granodiorite assemblage and younger two phases (monzo- to syenogranites and alkaline granites). The field radiometric survey and lab investigations indicates that, the anomalous uranium contents are mainly restricted to the pegmatites and altered monzogranites. The eU contents of pegmatites reach up to 168 ppm with 117 ppm in the average, while the eTh reach up to 257 ppm with 175 ppm in the average. The eTh/eU ratios range from 1.1 to 2.4 with an average 1.6. In the altered monzogranites,eUreachs up to 305 ppm with an average of 126 ppm. The eThreachs up to 382 ppm with an average of 166 ppm. The average values of Ra and K are 60 ppm and 2.93 wt percentage, respectively. The average values eTh/eU ratios in the examined anomalies monzgranites (1.56) indicating addition of uranium. The high eU and eTh contents may be attributed to the secondary ascending hydrothermal solutions thatlead to the accumulationof radioactive mineralizations mostly along fractures, faults and shear districts. The recognized radioactive minerals in the studied pegmatites are thorite, uranothorite, zircon, fluorite, columbite, samarskite, monazite, xenotime and allanite, whereas thorite, uranothorite, zircon and fluorite are the foremostradioactive minerals in the altered younger granites. Additional accessory minerals such as ilmenite, magnetite, hematite, pyrite and rutile were found in the studied pegmatites. Hematite, magnetite and pyrite are the foremost accessories accompanying the radioactive minerals in the altered younger granites.

Keywords:
Sabbagh, alkaline granite, monzogranite, pegmatite, dykes, radioactive, hydrothermal solutions

[1] Geology Department, Faculty of Science, Ain Shams University, Egypt.
[2] Research Department, Nuclear Materials Authority, Cairo, Egypt

IOP Conf. Series: Materials Science and Engineering **975** (2020) 011001 doi:10.1088/1757-899X/975/1/011001

Military Technical College
Kobry El-Kobbah,
Cairo, Egypt

I.C.E.E.2020

10th International Conference on Chemical & Environmental Engineering
7-9 April 2020

019-ICEE-10

Sub-channel thermal-hydraulic analysis for VVER- 1000 generation III PWR

Alsherief M. Almessallmy[1], Mohamed K Shaat[2] and Saeed A Hassanien[1]

Abstract

The sub-channel thermal-hydraulic analysis of a nuclear reactor is essential for assessing of its safety aspects. In this paper, the VVER-1000 has been selected as an example of the third generation reactors since it meets most of the international safety standards and because it has been taken as a base for designing the VVER-1200 which is belonging to the III+ generation. A steady state mathematical model has been proposed and solved to validate and assure that the hottest channel temperature limits are satisfied. The various temperature distributions, the critical heat flux and the departure from nucleate boiling ratio (DNBR) for the hottest channel were evaluated. Also, a transient state model has also been presented and solved using the finite difference method with the aid of MATLAB algorithm. An exponential loss of flow rate of the reactor core coolant was triggered from the steady state conditions. We assumed that the neutron flux and the generated power were unchanged during the postulated event. The average core coolant flow time constant was treated as a single parameter expressing the rapidity of the event. A value of 250 seconds time constant was assumed for slow transient, whereas 10 seconds was assumed for fast one. The reactor core was assumed to be protected through the reactor control system and mitigated according to the regular emergency operating procedures. The time dependent temperature distributions were calculated for the cladding of the hottest coolant channel. For each value of the temperature, the response time required for reaching unsafe conditions was evaluated, discussed and presented.

Keywords:
Thermal-hydraulics, steady-state, transient state, hottest channel, MATLAB, DNBR, loss of flow, time constant, postulated event

[1]Egyptian Armed Forces, Cairo, Egypt, Email: shomedhat@gmail.com
[2]Egyptian Atomic Energy Authority, Cairo, Egypt

IOP Conf. Series: Materials Science and Engineering **975** (2020) 011001 doi:10.1088/1757-899X/975/1/011001

Military Technical College
Kobry El-Kobbah,
Cairo, Egypt

I.C.E.E. 2020

10th International Conference on Chemical & Environmental Engineering
7-9 April 2020

020-ICEE-10

Effect of the Formation Parameters on the Thermal Properties of $La_2Zr_2O_7$

Ahmed Abdelkareem

Abstract

Lanthanum zirconates has been suggested as a thermal barrier coating for many high temperature applications. This study is focusing on the effect of different preparation parameters on the thermal conductivity of $La_2Zr_2O_7$. We have used solid state reaction to form the zirconates powder was very good homogeneity. We have studied the effect of the intimal ZrO_2/La_2O_3 ratio, and found the 50% to be the optimum. The effect of the sintering temperature has also been studied and 1100°c was chosen as the optimum sintering temperature. The sintering time was also studied, where 3 hours was proved to be enough to obtain a homogeneous single phase compound with affordable method.

Keywords:
Ceramic composites, Sintering, Thermal properties, Thermodynamics and kinetics of processes in materials

*Nuclear Materials Authority, POBox 540 Madi, Cairo

IOP Conf. Series: Materials Science and Engineering **975** (2020) 011001 doi:10.1088/1757-899X/975/1/011001

Military Technical College
Kobry El-Kobbah,
Cairo, Egypt

I.C.E.E.2020

10th International Conference on Chemical & Environmental Engineering
7-9 April 2020

Session No 7

Extraction Techniques of Nuclear Materials

(ETNM)

Room: N
Date : Thrusday, 9/7/2020
Time : 11:00-12:45
Board: Prof. Dr / Hamid Mira NMA
Prof. Dr. / Gamal Othman NRC
Col. Dr./Ahmed Farghal Tawfik MTC

021-ICEE-10	**Bioleaching of uranium from El-Sella ore material using Epicoccum nigrum** Azhar M. Elsayad1, *Shimaa S. Hussien2, Mohamed G. Mahfouz2, Abdo A. El Mougith1 and Wesam A. Hassanien1
022-ICEE-10	**Commercial approach for highly pure thorium from Egyptian monazite mineral acid process** Abd El Fatah A.I.L
023-ICEE-10	**Biosorption OF Uranium BY Saragassum Detifolium from the Trachytic Sills OF Wadi-El Atshan Area, Central Eastern Desert, Egypt** Nagdi M. Farag and Nora . SH .Gad

IOP Conf. Series: Materials Science and Engineering **975** (2020) 011001 doi:10.1088/1757-899X/975/1/011001

Military Technical College
Kobry El-Kobbah,
Cairo, Egypt

I.C.E.E.2020

10th International Conference on Chemical & Environmental Engineering
7-9 April 2020

021-ICEE-10

Bioleaching of uranium from El-Sella ore material using Epicoccum nigrum

Azhar M. Elsayad[1], *Shimaa S. Hussien[2], Mohamed G. Mahfouz[2], Abdo A. El Mougith[1] and Wesam A. Hassanien[1]

Abstract

Uranium (U) is one of the strategic elements and essential for many applications as a fuel in nuclear power plants and nuclear weapons. Bioleaching of uranium was carried out with *Epicoccum nigrum* isolated from El-Sella ore material. El-Sella area is located in the southern eastern desert of Egypt. The maximum bioleaching of U(VI) was found to be 76.6 % at these optimum conditions: 9 days incubation period, 0.5% pulp density, 30 °C incubation temperature at shaking speed 175 rpm. Glucose and NH4Cl were considered the optimum carbon and nitrogen sources for fungal growth. *E. nigrum* exhibit good potential in generating varieties of organic acids which effective for bioleaching of uranium. It produced 112.3 and 23.5 µg/ml from gallic and ellagic acids in the culture filtrate under the optimum conditions. This work addresses the area of beneficiation of the used mineral to solubilize U(VI) through the biotechnological route in Egypt, where the bioleaching method is more effective than the chemical one using organic acids

Keywords:
Bioleaching; uranium; organic acids; Epicoccumnigrum

---

[1]Department of Botany, Faculty of Science, Zagazig University, Egypt
[2]Nuclear Materials Authority, P.O. Box 530 El-Maadi, Cairo, Egypt

Military Technical College
Kobry El-Kobbah,
Cairo, Egypt

I.C.E.E.2020

10th International Conference
on
Chemical & Environmental Engineering
7-9 April 2020

022-ICEE-10

Commercial approach for highly pure thorium from Egyptian monazite mineral acid process

Abd El Fatah A.I.L

Abstract

Thorium has a special interest in the future nuclear energy as a fuel especially the breeder reactors. The present research has been fulfilled throughout successive research schedule manner on separation and purification of thorium from Egyptian beach monazite mineral acid digtion. Monazite mineral was digested with sulfuric acid and dissolute with ice water, the clear solution precipitated to produce thorium pyrophosphate with 45% purity. The purification was carried out after the caustafication and leaching with hydrochloric acid with the addition of oxalic acid at pH 0.3-0.4 to produce 91.68 % purity of thorium oxalate. The high purification is done with the uses of thiosulphate salt to produce 99% purity.

Keywords:
Monazite mineral, thorium separation and production, sodium thiosulfate

Nuclear Materials Authority, Cairo, Egypt

IOP Conf. Series: Materials Science and Engineering **975** (2020) 011001 doi:10.1088/1757-899X/975/1/011001

Military Technical College
Kobry El-Kobbah,
Cairo, Egypt

I.C.E.E.2020

10th International Conference on Chemical & Environmental Engineering
7-9 April 2020

023-ICEE-10

Biosorption OF Uranium BY Saragassum Detifolium from the Trachytic Sills OF Wadi-El Atshan Area, Central Eastern Desert, Egypt

Nagdi M. Farag and Nora . SH .Gad

Abstract

In wadi El Atshan area, Central Eastern Desert of Egypt, the main rock types exposed in that area are the older granites, Hammamat sedimentary rocks, younger granites , trachytic sills and dikes. The Field studies and radiometric survey indicate that the trachytic sills are the main rock type in the investigated area which contain secondary uranium minerals. Sargassum detifolium is used to capture uranium from the trachytic sills. The analysis indicate that this algae absorb the uranium content by 100% in two samples where happened in the other samples as well as, it change the chemical composition of some minerals content, while some elements changed while other elements are decreased, Sargassum detifolium has the ability to biosorbe uranium content about 75% as in uranotile mineral $\{CaU_2(SiO_4)_2(OH)_6.3H_2O\}$.

Keywords:
Sargassum detifolium, trachytic rocks,

Nuclear Materials Authority, Cairo, Egypt, **norashenouda@yahoo.com**

IOP Conf. Series: Materials Science and Engineering **975** (2020) 011002 doi:10.1088/1757-899X/975/1/011002

Peer review declaration

- **Types of peer review : double-blind**
- **Conference submission management system : Online**
- **Number of submissions received : 35**
- **Number of submissions sent for review: 35**
- **Number of submissions accepted: 23**
- **Acceptance rate : 65.7 %**
- **Average number of reviews per paper : 2**
- **Total number of reviewers involved : 37**
- **Any additional info on review process : No**
- **Contact person for queries:**

- Dr. M.Kassem Abdel-Magid	**m.kassem @mtc.edu.eg,** **m.kassem.mtc @gmail.com**
- Dr. Mohamed Elkhatib	**melkhatib@mtc.edu.eg** **elkhatib.m@gmail.com**

IOP Conf. Series: Materials Science and Engineering **975** (2020) 012001 doi:10.1088/1757-899X/975/1/012001

Effect of pyrolysis temperature on the synthesis of high-quality MWCNTs by CVD method

Omar Abdel Aziz[1] , **Tamer Z.Wafy**[1], **Mahmoud Abdelhafiz**[1] **and Mohamed A. Elsayed** [1,2]

[1] Chemical Engineering Department, Military Technical College, Kobry Elkobbah, Cairo, Egypt
Email: m.abolfotoh@mtc.edu.eg[2]

Abstract. Production of high-quality Multi-Walled Carbon Nano-Tubes (MWCNTs) is essential in some field, especially in medical and mechanical applications. This work illustrates the synthesis of high-quality (MWCNTs) by Injection Chemical Vapor Deposition (Inj-CVD) method using Ferrocene / Toluene solution as a carbon precursor and a catalyst. The effect of synthesis pyrolysis temperature, which has a pivot impact on the synthesis process, was investigated. The structure defects, impurities, thermal stability and sample morphology, as well as the mean diameters of the MWCNTs, were analyzed using Raman spectroscopy, thermal gravimetric analysis (TGA) and scanning electron microscope (SEM), respectively. The results revealed that a change in quality, mean diameter and purity were observed when the pyrolysis temperature varied between 700°C to 850°C. High-quality MWCNTs were observed by Raman spectroscopy at a pyrolysis temperature of 700°C with ID/IG = 0.2. While the outer and inner diameters were 25±6 nm and 7.8±1.5nm respectively, measured by Transmission Electron Microscope (TEM). The Inj-CVD shows an excellent control of the quality of the prepared MWCNTs by optimizing the synthesis temperature.
Keywords: High-quality Carbon Nano-tubs, Effect of temperature on CNTs, Characterization of CNTs.

1. Introduction

In recent history, Iijima's synthesized carbon nanotubes CNTs by arc evaporation method for C_{60} [1]. At the same time, a small company in the USA was already able to produce defected carbon nanotubes called carbon fibrils using a chemical vapor deposition [2]. These achievements attracted considerable attention in both the technological and scientific communities and initiated a wide range of research in several important directions [3,4]. Due to their unique properties such as mechanical[5] thermal[6], chemical [7], electrical[8], optical[9,10] and magnetic [11], carbon nanotubes are designated as one of the most attractive materials for many applications. These properties are directly affected not only by a unique structure such as (zigzag, arm-chair and chiral); but also by how the graphene sheets were rolled up. CNTs have different forms as single-walled carbon nanotubes (SWCNTs) which defined by only one rolled graphene sheet [12,13]. However, Double-walled carbon nanotubes (DWCNTs) have two rolled graphene sheets [13,14] and multi-walled carbon nanotubes (MWCNTs) have more than three rolled graphene sheets[15]. In principle, Injection Chemical Vapor deposition (Inj-CVD)[16] is considered a simple and inexpensive synthesis method for the production of MWCNTs, compared with other bottom-up methods[12,17] such as Arc-discharge [18,19] and laser ablation[20]. The evaporation of carbon atoms

from solid targets at a temperature above 3,000°C, as in Arc-discharge and laser ablation methods, is neither economical nor convenient [21]. In addition, Inj-CVD can be operated under atmospheric pressure, moderate temperatures and does not require any special substrates. Furthermore, several researchers found that CVD [22] technique gives a very low aspect ratio. (ID /IG) define this ratio, where ID is the intensity of D band, which is a characteristics band refers to disordering in carbon nanotubes structure and located in the range of 1350 cm^{-1} Raman shift. While, IG is the intensity of the G band, which is a characteristics band that refers to graphitization in carbon nanotubes structure and located around 1500 cm^{-1} Raman shift [23, 24]. This aspect ratio refers to the disorder in Carbon Nano-materials Structure (CNMs) [25,26]. The lower aspect ratio indicates the highest [24,25,27].
In this work, high-quality MWCNTs were synthesized using Inj-CVD technique using an optimum concentration of Ferrocene in the Toluene solution as a carbon precursor and a catalyst. The uniform injection feed rate of this solution was used, under Argon as a carrier gas at 1 atm (pressure) and the surface of Quartz tube as a substrate. The influence of temperature on synthesizing MWCNTs quality was investigated using Raman spectra in the frequency range of 500 –3500 cm^{-1}. The thermal stability and the purity of syntheses MWCNTs were studied by thermos-gravimetric analysis (TGA). The morphology, as well as the outer diameter of the samples, were investigated using a scanning electron microscopy (SEM). In addition, the inner and outer diameters of the prepared MWCNTs samples were visualized and confirmed with high-resolution transmitted electron microscopy (TEM).

2. Experimental

2.1. Materials and chemicals

Carbon nanotube was synthesized using materials and chemicals characterized and listed in the Table 1.

Table 1. The main Characteristics of the chemicals and material used in the synthesis of MWCNTS

No	Name	Type	Source	Melting Point°C	Boiling point °C	Chemical form	Density g/cm^3	Act as
1	Toluene	Colorless liquid	Fisher Scientific, UK	-95	110-111	C_7H_8	0.87 at 25°C	Carbon precursor
2	Ferroce ne	Light orange powder	Sigma-Aldrich, UK	173	249	$C_{10}H_{10}Fe$	1.49 At	Carbon precursor, Catalyst donor

2.2. Synthesis of MWCNTs experimental procedure

The injection chemical vapor deposition (InJ-CVD) method was employed to synthesize MWCNTs [28-30] in a typical experiment setup as shown in figure 1. The main tubular furnace temperature, synthesizing pyrolysis temperature, was varied from 700 to 850 ± 5 °C under the flow of argon gas at atmospheric pressure. The volume of Toluene (carbon source) and Ferrocene (carbon source and catalyst) was adjusted to 5% (w/w) by weight respectively. A 2.21 cc/h of feed solution was uniformly injected using syringe pump (Razel R99-E LF28V) to a horizontal preheated tube at a temperature of 200°C (the evaporation temperature of Ferrocene) for 60 min. The evaporated feed is then flushed away to the main furnace by 100 cc/hr of argon (carrier gas & pyrolysis medium).

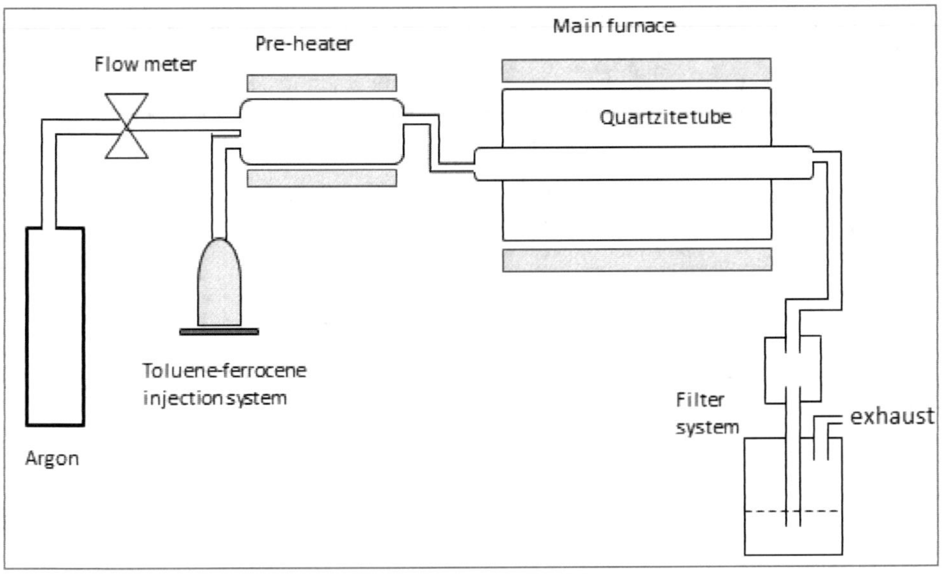

Figure 1. Schematic diagram of Injection CVD setup

2.3. Characterizations of synthesized MWCNTs

The effect of pyrolysis temperature on the type and quality of MWCNTs is investigated by measuring the Raman spectra of the samples. The results were recorded in the frequency range of 500 –3500 cm^{-1} using a Raman spectrometer (JY HR-800 type) with a laser excitation line at 532 nm and 12.5 nm power. The thermal stability of syntheses MWCNTs and the purity were studied by Thermal Gravimetric Analysis (TA55).The temperature profile starting from 30°C to 830°C under airflow rate 25 CC/hr with heating rate 20°C/min. The outer diameter, as well as morphology of the samples, were investigated by scanning electron microscopy (SEM), ZEISS SEM EVO 10 MA. To visualize the high quality of prepared MWCNTs; Transmission electron microscopy (TEM) was operated at 200 kV. For this purpose, sonication of the prepared samples in ethanol was conducted and then a small droplet of the dispersion was deposited on the TEM carbon grid and dried for 10 min.

3. Results and discussion

3.1. Quality investigation using RAMAN spectroscopy

Raman spectroscopy depends on induced dipole moment, So it's sensitive to the highly symmetric covalent bonds that have not dipole moment as found in CNMs [25,31]. Therefore, the Raman spectroscopy technique is usually used to detect the type and quality of Carbon Nano-materials (CNMs). In general, CNTs with all of its types, in addition to Graphene and fullerene were identified by three main Raman peaks [32]. These peaks called D band, G band and 2D band, which appear at Raman shift around 1350cm^{-1}, 1500 cm^{-1} and 2700 cm^{-1} respectively. Figure 2 shows the Raman spectra for the prepared samples at different synthesizing temperature. Raman spectra not only used to differentiated the types of CNTs as SWCNTs, DWCNTs or MWCNTs but also can be used to measure the quality of the prepared sample [31]. The quality of CNMs was investigated by dividing the intensity of the disordered peak (ID) over the intensity of the graphitization peak (IG) [33]. In this work, high-quality CNTs were recorded below 0.7 aspect ratio ID /IG [25] as shown in figure 3. The results revealed that the change in

IOP Conf. Series: Materials Science and Engineering **975** (2020) 012001 doi:10.1088/1757-899X/975/1/012001

synthesizing temperature affect directly the quality of the prepared samples, as the aspect ratio decreased from 0.7 to 0.2 with decreasing the synthesizing temperature from 800°C to 700 °C respectively, as shown in both figures 2 and 3.

The obtained lowest aspect ratio (0.22) by this work which achieved at a synthesizing temperature of 700°C, indicating the optimum temperature of synthesizing MWCNTs with a high-obtained quality of compared to other research work [23,25,31] .Note that under previous synthesized conditions, neither MWCNTs nor CNMs formed at 650 ° C, while CMSs formed at 850 °C as shown in figure 2.

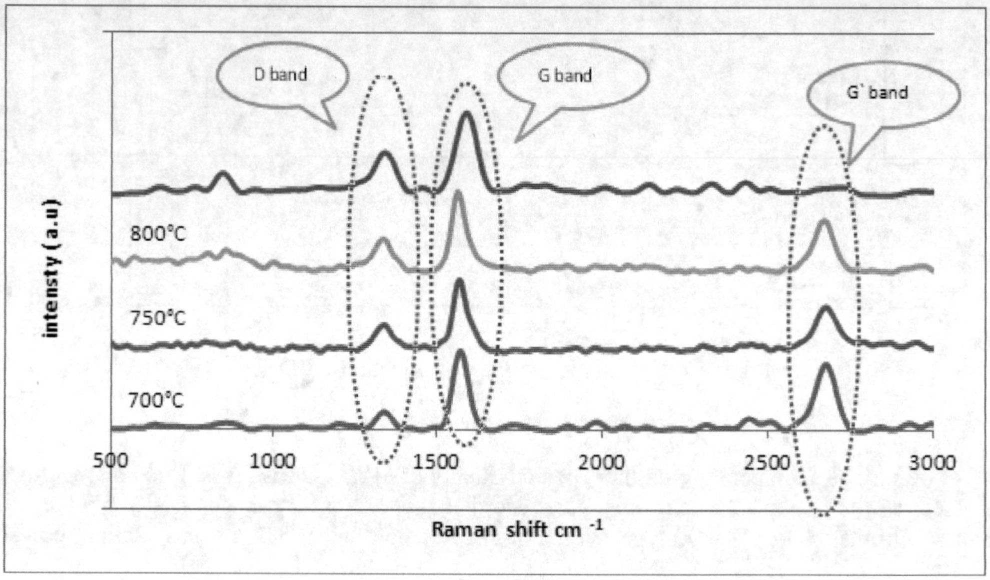

Figure 2. Raman spectra for the prepared CNMs at different pyrolysis temperature, by injection solution of 5%Wt Ferrocene in Toluene under 100cc/h argon atmosphere for 60 min.

3.2. Morphology and average outer diameter investigation using (SEM)

A scanning electron microscope (SEM) is a very important tool to show the material morphology and the outer diameter for MWCNTs. Where it can be used to estimate the average outer diameter of prepared MWCNTs. figure 4 shows the SEM images of MWCNTs samples synthesized at different pyrolysis temperatures with the same magnification which equals 35KX and tall bar equal 1μm. Under the studied experimental conditions there where no MWCNTs are formed at the temperatures range from 600 ˚C to 650 ˚C. While carbon microspheres structures are formed in the temperatures range from 850 ˚C to 900 ˚C. For the sample synthesized at 700˚C, the average outer diameter was 35±10 nm as shown in figure 4a. However, for the sample synthesized at 750˚C the average outer diameter equal to 50±16 nm figure 4b. Moreover, for the sample synthesized at 800˚C the average outer diameter equal to 75±12 nm figure 4c. While for the sample synthesized at 850˚C at the same experimental conditions, figure 4d, carbon microspheres (CMSs) was generated with average outer diameter equal 2±0.1 μm

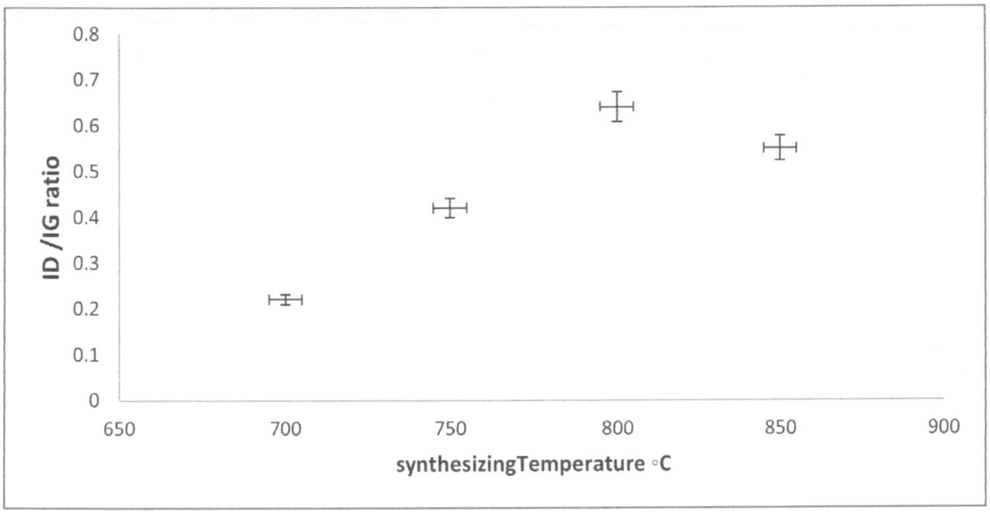

Figure 3. Raman spectra for the prepared CNMs at different synthesizing temperatures shown the quality (disordered ratio) vs. temperature

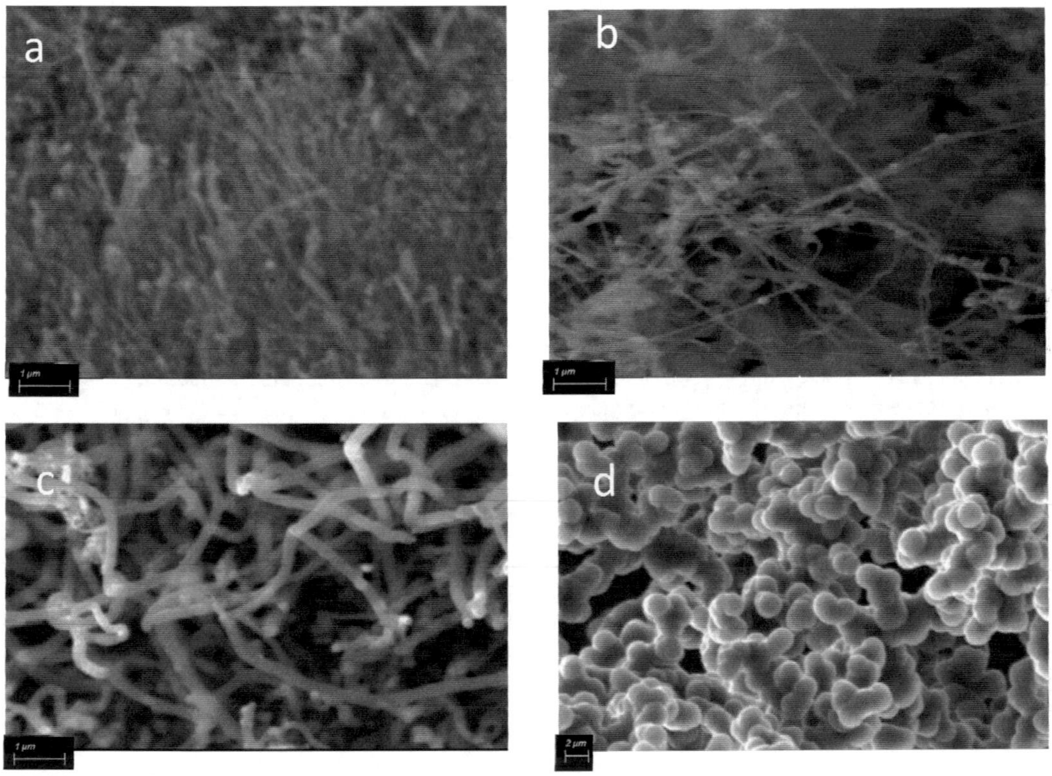

Figure 4. SEM images of MWCNTs samples synthesized at different pyrolysis temperatures: (a) 700˚C, (b) 750 ˚C, and (c) 800˚C while (d) SEM image of CMSs synthesized at 850 ˚C.

IOP Conf. Series: Materials Science and Engineering **975** (2020) 012001 doi:10.1088/1757-899X/975/1/012001

Figure 5 shows the relation between the outer diameter of the prepared samples and pyrolysis temperatures. The results confirm that the outer diameter depends strongly on the pyrolysis temperature. As the outer diameter increase with increasing the pyrolysis temperature. This could be attributed to, the increases of pyrolysis temperatures leads to increasing the catalyst particles (Iron) mobility on the quartz substrate. The lager catalyst clusters formed so the lager outer diameter synthesized.

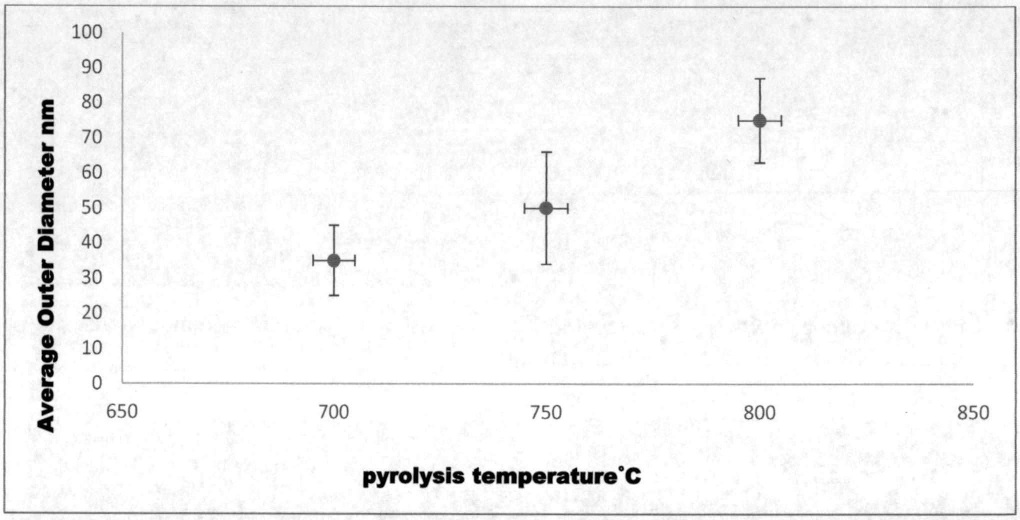

Figure 5. Average of outer Diameter for MWCNTs at 700±5˚C, 750±5˚C and 800±5˚C pyrolysis temperature

3.3. *Evaluation of dimensions of high-quality samples using (HITEM)*
High-resolution transmission electron microscope (HITEM) using to characterize the outer and inner diameter and calculate the number of tubes for MWCNTs. Here HITEM used especially to confirm that the sample synthesized at 700˚C, which had average ID/IG aspect ratio 0.22, is MWCNTs as shown in figure 6a. The averages of outer and inner diameter were 26±9 nm and 7.9±1.8 nm respectively, as shown in figure 7b. Because of the inner spacing between tunes around 0.36nm [30,34], the average number of tubes for this high-quality MWCNTs was 23±7 tube. The dark spots in both figures 6 a and b were the iron metal catalyst.

3.4. *Thermal Oxidation Stability of MWCNTs study by TGA*
In this part of the work, not only the thermal stability of syntheses MWCNTs was studied but also the purity was calculated by using Thermal Gravimetric Analysis (TGA)[35]. It has been known that the oxidation temperatures for amorphous carbon contaminants, SWCNTs, and MWCNTs are typically in the range of 200 – 300 °C, 350 – 500 °C and 400–650 °C respectively [36,37]. However, the DWCNTS bundles and bucky papers (a thin sheet made from DWCNTs) can be stable in air in the range of 500-800 °C [14,38].

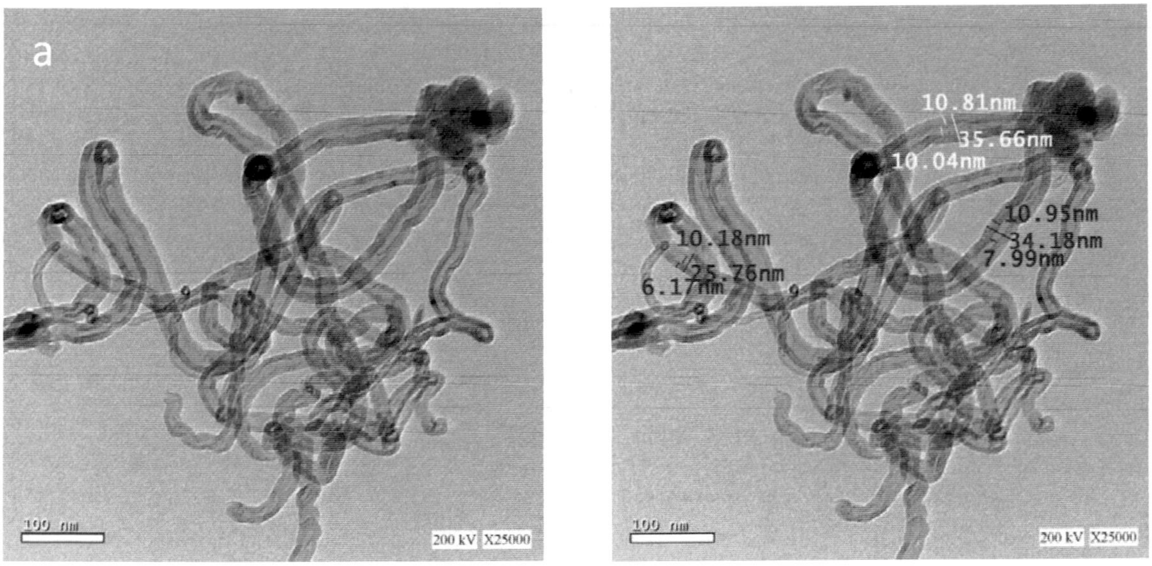

Figure 6. HITEM of MWCNTs at 700°C with Mag=25XK (a) without diameters information and (b)with outer and inner diameters.

Figure 7 shows the thermos-gram for MWCNTs synthesized at different pyrolysis temperatures. The results revealed that in the temperature range from 200°C to 400°C very low weight loss was observed. This could be attributed to the prepared samples are free of amorphous carbon.

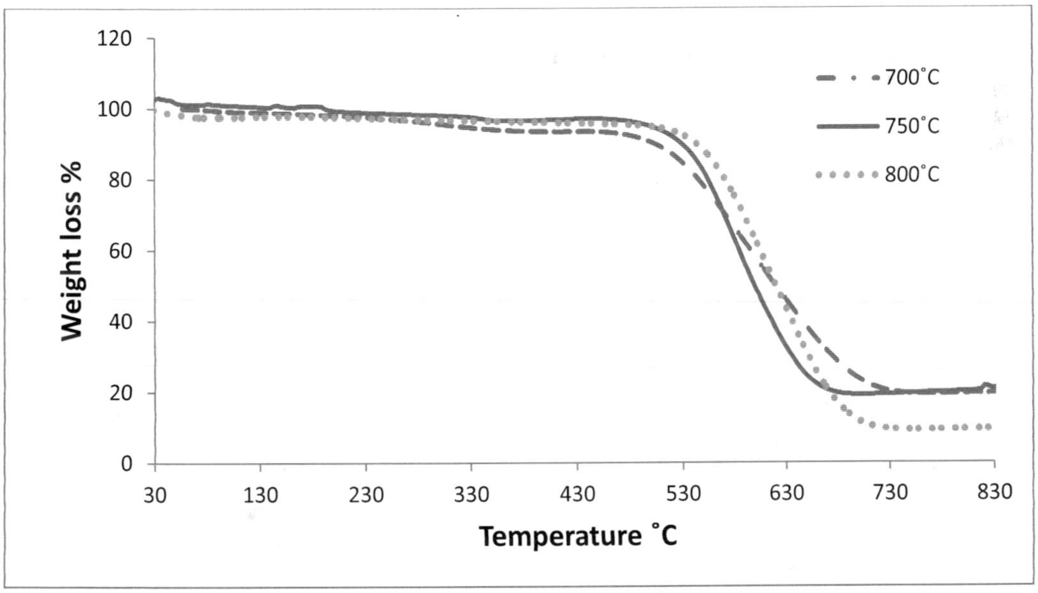

Figure 7. Thermo-gram for MWCNTs at 700°C, 750°C, and 800°C pyrolysis temperature

IOP Conf. Series: Materials Science and Engineering **975** (2020) 012001 doi:10.1088/1757-899X/975/1/012001

Figure 8 shows that there is a gradually increasing in oxidation temperature from 591˚C, 594.5˚C to 613.6˚C for the sample prepared at synthesizing temperature of 700 ˚C, 750 ˚C, and 800 ˚C respectively ; which confirm that there was a relation between the synthesizing temperature and the thermal stability of the prepared samples. It indicates also, the presence of unreacted catalyst, (Fe) nanoparticles, within the prepared samples which affect its thermal stability and also its oxidation temperature value [39]. Dash line in figure 8 refers to oxidation temperature which increases with decreasing ash content (solid line). After complete oxidation of the samples, the residual mass percentage of iron oxides (ash) were 19.8±0.1, 19.5±0.1 and 10 ± 0.1 wt% for the samples prepared at synthesizing temperature 700˚C, 750˚C, and 800˚C respectively.

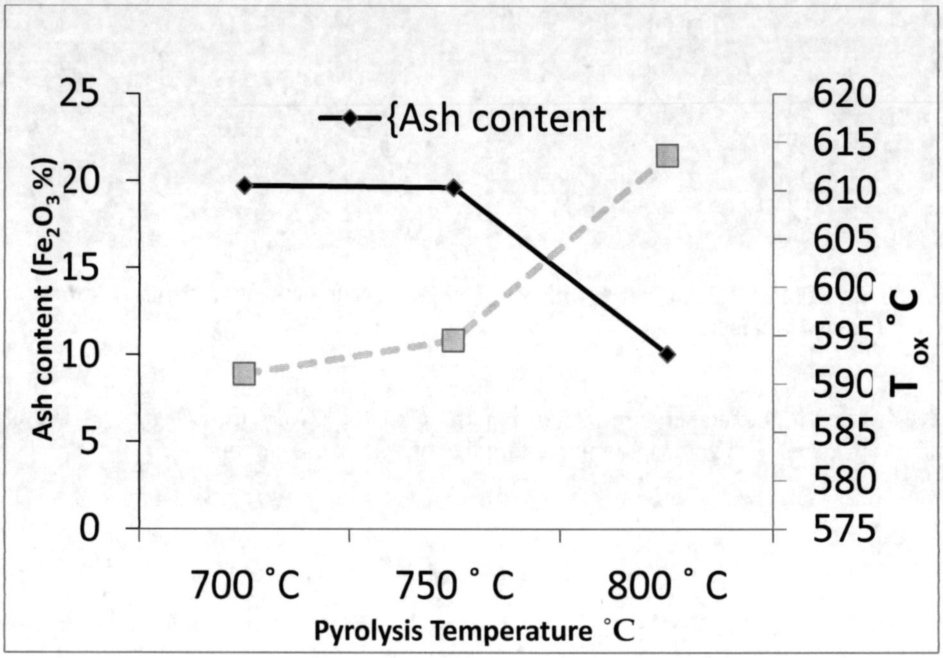

Figure 8. Thermal stability (TOX) and Purity (Ash content) vs. pyrolysis Temperature at 700˚C, 750˚C, and 800˚C

4. Conclusion

In this work, it has been confirmed that the variation of synthesizing temperature has a great influence on the purity and thermal stability of MWCNTs preparation by Injection Chemical Vapor Deposition of Ferrocene/toluene solution 5%Wt over the surface of the quartz tube substrate. Under the studied experimental conditions there where no MWCNTs are formed at the temperatures range from 600 ˚C to 650 ˚C. While carbon microspheres structures are formed in the temperatures range from 850 ˚C to 900 ˚C. The optimum pyrolysis temperature was 700 ˚C which has the lowest (ID/IG) aspect ratio equal to 0.22 recorded by Raman. HRTEM result confirms the high quality of the prepared sample. The high thermal stability of the prepared samples which were observed by TGA indicating that the prepared samples are free of amorphous carbon and consequently have high purity. SEM shows that the average outer diameter increase from 25±6 nm, 50±16 nm to 75±12 nm by increasing the pyrolysis temperature from 700 ˚C, 750 ˚C and 800 ˚C respectively.

5. References

[1] Y. J. Tang, Z.R., Han, B., Han, C., and Xu, "Helical microtubules of graphitic carbon," Nature, vol. 354, no. November, pp. 56–58, 1991.

[2] A. R. Dalila, A. B. Suriani, M. S. Rosmi, R. Rosazley, J. Rosli, and M. Rusop, "Carbon nanotubes: A brief outlook on history, synthesis methods and various bio-hydrocarbon sources," Adv. Mater. Res., vol. 832, pp. 792–797, 2014.

[3] S. Kumar, R. Rani, N. Dilbaghi, K. Tankeshwar, and K. H. Kim, "Carbon nanotubes: A novel material for multifaceted applications in human healthcare," Chem. Soc. Rev., vol. 46, no. 1, pp. 158–196, 2017.

[4] K. S. Ibrahim, "Carbon nanotubes-properties and applications: a review," Carbon Lett., vol. 14, no. 3, pp. 131–144, 2013.

[5] M. Felisberto et al., "Carbon nanotubes grown on carbon fiber yarns by a low-temperature CVD method: A significant enhancement of the interfacial adhesion between carbon fiber/epoxy matrix hierarchical composites," Compos. Commun., vol. 3, no. March 2016, pp. 33–37, 2017.

[6] J. Hansson et al., "Synthesis of a graphene carbon nanotube hybrid film by joule self-heating CVD for thermal applications," Proc. - Electron. Components Technol. Conf., vol. 2018-May, pp. 2450–2456, 2018.

[7] J. Xu et al., "A review of functionalized carbon nanotubes and graphene for heavy metal adsorption from water: Preparation, application, and mechanism," Chemosphere, vol. 195, pp. 351–364, 2018.

[8] T. W. Odom, J. L. Huang, P. Kim, and C. M. Lieber, "Atomic structure and electronic properties of single-walled carbon nanotubes," Nature, vol. 391, no. 6662, pp. 62–64, 1998.

[9] S.-J. Kwon, H.-K. Seo, S. Ahn, and T.-W. Lee, "Value-Added Recycling of Inexpensive Carbon Sources to Graphene and Carbon Nanotubes," Adv. Sustain. Syst., vol. 3, no. 1, p. 1800016, 2019.

[10] A. Hannon, Y. Lu, J. Li, and M. Meyyappan, "Room-temperature carbon nanotube-based sensor for carbon monoxide detection," J. Sensors Sens. Syst., vol. 3, no. 2, pp. 349–354, 2014.

[11] F. C. Dillon, A. Bajpai, A. Koós, S. Downes, Z. Aslam, and N. Grobert, "Tuning the magnetic properties of iron-filled carbon nanotubes," Carbon N. Y., vol. 50, no. 10, pp. 3674–3681, 2012.

[12] A. Eatemadi et al., "Carbon nanotubes: Properties, synthesis, purification, and medical applications," Nanoscale Res. Lett., vol. 9, no. 1, pp. 1–13, 2014.

[13] J. Prasek et al., "Methods for carbon nanotubes synthesis - Review," J. Mater. Chem., vol. 21, no. 40, pp. 15872–15884, 2011.

[14] H. Muramatsu et al., "Pore structure and oxidation stability of double-walled carbon nanotube-derived bucky paper," Chem. Phys. Lett., vol. 414, no. 4–6, pp. 444–448, 2005.

[15] W. Wang et al., "Three-dimensional few-layer graphene and carbon nanotube foam architectures for high fidelity supercapacitors," Nano Energy, vol. 2, no. 2, pp. 294–303, 2013.

[16] R. Andrews et al., "Continuous production of aligned carbon nanotubes: A step closer to commercial realization," Chem. Phys. Lett., vol. 303, no. 5–6, pp. 467–474, 1999.

[17] R. Chavan, U. Desai, P. Mhatre, and R. Chinchole, "Review Article A REVIEW : CARBON NANOTUBES," Int. J. Pharm. Sci. Rev. Res., vol. 13, no. 1, pp. 125–134, 2012.

[18] N. Arora and N. N. Sharma, "Arc discharge synthesis of carbon nanotubes: Comprehensive review," Diam. Relat. Mater., vol. 50, pp. 135–150, 2014.

[19] O. Volotskova, I. Levchenko, A. Shashurin, Y. Raitses, K. Ostrikov, and M. Keidar, "Single-step synthesis and magnetic separation of graphene and carbon nanotubes in arc discharge plasmas," Nanoscale, vol. 2, no. 10, pp. 2281–2285, 2010.

[20] J. Chrzanowska et al., "Synthesis of carbon nanotubes by the laser ablation method: Effect of laser wavelength," Phys. Status Solidi Basic Res., vol. 252, no. 8, pp. 1860–1867, 2015.

[21] S. A. Bakar, A. A. Aziz, P. Marwoto, and S. Sakrani, "Laser Welding of Structural Aluminum,"

Carbon N. Y., vol. 77, no. May, pp. 311–320, 1987.

[22] I. Gases, Thermodynamics of physical and chemical vapour deposition, vol. 60. 2016.

[23] C. S. S. R. Kumar, "Raman spectroscopy for nanomaterials characterization," Raman Spectrosc. Nanomater. Charact., vol. 9783642206, pp. 1–645, 2012.

[24] S. Suzuki and H. Hibino, "Characterization of doped single-wall carbon nanotubes by Raman spectroscopy," Carbon N. Y., vol. 49, no. 7, pp. 2264–2272, 2011.

[25] M. Endo et al., "Comparison study of semi-crystalline and highly crystalline multiwalled carbon nanotubes," Appl. Phys. Lett., vol. 79, no. 10, pp. 1531–1533, 2001.

[26] Q. L. Yan, M. Gozin, F. Q. Zhao, A. Cohen, and S. P. Pang, "Highly energetic compositions based on functionalized carbon nanomaterials," Nanoscale, vol. 8, no. 9, pp. 4799–4851, 2016.

[27] C. Wu, Z. Wang, L. Wang, P. T. Williams, and J. Huang, "Sustainable processing of waste plastics to produce high yield hydrogen-rich synthesis gas and high quality carbon nanotubes," RSC Adv., vol. 2, no. 10, pp. 4045–4047, 2012.

[28] S. W. Hong, T. Banks, and J. A. Rogers, "Improved density in aligned arrays of single-walled carbon nanotubes by sequential chemical vapor deposition on quartz," Adv. Mater., vol. 22, no. 16, pp. 1826–1830, 2010.

[29] A. Annu, B. Bhattacharya, P. K. Singh, P. K. Shukla, and H. W. Rhee, "Carbon nanotube using spray pyrolysis: Recent scenario," J. Alloys Compd., vol. 691, pp. 970–982, 2017.

[30] T. Wafy, "Characterization of Chemical and Mechanical Properties of Polymer Based Nanocomposites," pp. 1–183, 2013.

[31] A. A. Ezz, M. M. Kamel, and G. R. Saad, "Synthesis and characterization of nanocarbon having different morphological structures by chemical vapor deposition over Fe-Ni-Co-Mo/MgO catalyst," J. Saudi Chem. Soc., vol. 23, no. 6, pp. 666–677, 2019.

[32] M. S. Dresselhaus, A. Jorio, and R. Saito, "Characterizing Graphene, Graphite, and Carbon Nanotubes by Raman Spectroscopy," Annu. Rev. Condens. Matter Phys., vol. 1, no. 1, pp. 89–108, 2010.

[33] S. Musso, J. M. Tulliani, G. Ferro, and A. Tagliaferro, "Influence of carbon nanotubes structure on the mechanical behavior of cement composites," Compos. Sci. Technol., vol. 69, no. 11–12, pp. 1985–1990, 2009.

[34] E. Ertekin and D. C. Chrzan, "Ideal torsional strengths and stiffnesses of carbon nanotubes," Phys. Rev. B - Condens. Matter Mater. Phys., vol. 72, no. 4, pp. 1–5, 2005.

[35] Y. C. Hsieh, Y. C. Chou, C. P. Lin, T. F. Hsieh, and C. M. Shu, "Thermal analysis of multi-walled carbon nanotubes by Kissinger's corrected kinetic equation," Aerosol Air Qual. Res., vol. 10, no. 3, pp. 212–218, 2010.

[36] J. Wei et al., "Preparation of highly pure double-walled carbon nanotubes," J. Mater. Chem., vol. 13, no. 6, pp. 1340–1344, 2003.

[37] N. Hong, W. Yang, C. Bao, S. Jiang, L. Song, and Y. Hu, "Facile synthesis of graphene by pyrolysis of poly(methyl methacrylate) on nickel particles in the confined microzones," Mater. Res. Bull., vol. 47, no. 12, pp. 4082–4088, 2012.

[38] O. Liu, W. Ren, F. Li, H. Cong, and H. M. Cheng, "Erratum: Synthesis and high thermal stability of double-walled carbon nanotubes using nickel formate dihydrate as catalyst precursor (Journal of Physical Chemistry C (2007) 111C (5007)," J. Phys. Chem. C, vol. 111, no. 40, p. 14932, 2007.

[39] J. H. Lehman, M. Terrones, E. Mansfield, K. E. Hurst, and V. Meunier, "Evaluating the characteristics of multiwall carbon nanotubes," Carbon N. Y., vol. 49, no. 8, pp. 2581–2602, 2011.

Synthesis and characterization of barium hexaferrite ($BaFe_{12}O_{19}$) nanoparticles from iron ore waste and their possible application in water treatment

Nehal A. Erfan[*a], Salwa A. M. Abdel-Hameed[b], Asma A. Mohammed[a]

[a] Chemical engineering department, faculty of engineering, Minia university, Minia, Egypt.
[b] Glass Research Department, National Research Centre, Dokki, Cairo, Egypt
Email: n.erfan1@mu.edu.eg

Abstract. In this work hard magnetic glass ceramics has been fabricated using Egyptian Bahariya oasis ore; which contains not less than 51% of its weight as pure iron. High energy planetary ball mill was used to convert the prepared magnetic sample into nanoparticles. Different mechanical milling times were applied to study its effect on the properties of the produced magnetic glass ceramic. Scanning electron microscope (SEM), transmission electron microscope (TEM), X-ray florescence analysis (XRF), differential thermal analysis (DTA), X-ray diffraction analysis (XRD), Raman spectroscopy and Vibrating sample magnetometer (VSM) analysis techniques were performed to characterize the prepared glass ceramic samples before and after exposing to mechanical milling process. XRD revealed crystallization of barium hexaferrite as major phase with average crystallite size ~26 nm. Saturation magnetization Ms depicted 21.2 emu/g after 5 hours milling. The produces nanoparticles after 5 hours milling showed a good performance in removal of pollutants like methylene blue and Congo red from water by adsorption.
Key words
Magnetic glass ceramics; Ball milling; Barium hexaferrite; nanoparticles;Iron ore, Waste; Recycling

1. Introduction

East Aswan, Eastern desert, Bahariya oasis and Western desert are localities rich with iron ore mines in Egypt. In these localities, the iron ore deposits widely and vary in their chemical and mineralogical composition. In this work Bahariya oasis ore which contains not less than 51% of its weight as pure iron has been used for preparing magnetic glass ceramics. Huge amount of iron ore is wasted during ore extraction, tailing removal, crushing, grinding and screening processes; as pretreatment steps to reach the suitable size for blast furnace use. Not less than 20% of iron oxide from iron ore is wasted. This iron oxide waste could be used as raw material for magnetic glass ceramics production [1]. Hard ferrites or permanent magnets are viewed as important material for different engineering applications. It characterized by its high coercivity, remanence after magnetization, high magnetic permeability and the ability to conduct magnetic flux. Ferrite magnets are widely used compared to other magnets as metallic alloys because of their low cost. Hard ferrite magnets have a wide variety of applications such as refrigerator magnets,electric motors, magnetic separators for ferrous materials, automotive sensors, MRI, microwave devices, wireless communications, data storage and air crafts as well as many other novel applications.

An increasing interest in hexaferrite nano-fibers and the effect of fiberous structure on magnetic properties is grown recently. On the other hands, ferro-fluid technology is one of the important applications for magnetic nanoparticles. Nano-ferro-fluids have many biomedical applications, for example, hyperthermia treatment for cancer [1], MRI contrast agent [2], drug delivery [3], DNA hyberdization [4] and cell separation [5].Nano-ferrites and a different iron-containing minerals (akaganeite, feroxyhyte, ferrihydrite, goethite, hematite, lepidocrocite, maghemite, and magnetite [6]were used for waste water treatment.The produced magnetic nanoparticles can be added to a polymer solution prior to electrospinning for synthesis magnetic nanofibers by ex-situ approach [7]. Magnetic nanofibers have a wide range of applications in biomedical engineering [8], tissue engineering [9] and waste water treatment [10, 11].

Since the very early development of nanoscience, creating nanomaterials using simple and low cost techniques with high yield was a great challenge. High energy ball milling technique has been used widely for nanomaterials, nanograins, nanoalloy, nanocomposites and nano-quasicrystalline materials synthesis. During milling process the energy transferred from the balls to the powder which control the kinetics of alloying [12].Type of mill, kind of powder, milling speed, size of balls, wet or dry milling, milling temperature and milling duration are the main parameters govern the energy transfer from balls to milled powder [13]. Dense materials (steel or tungsten carbide) are preferable to ceramic balls because the kinetic energy of the balls is a function of their mass and velocity, in addition, the size and size distribution are optimized for the given mill [14]. Dilute distribution of the milling balls minimizes the collision frequency, while too dense packing of balls reduces the mean free path of the ball motion [15].
Material characteristic of the powder and milling media and the kinetic energy of the ball were the factors affecting temperature during milling process.Also, defect concentration and diffusivity of the milled powder are very sensitive to temperature which influences the phase transformations induced by milling [16]. The formation of phases which need higher atomic mobility (intermetallics) is induced by higher temperatures while, amorphous and nanocrystalline phases formation take place at lower temperatures [17, 18].

Particle size reduction, mixing, particle shape changes and synthesis of nanocomposites are the main objectives of mechanical milling. High energy ball mill is the typical mill used for these purposes such as vibratory mills, tumbler ball mills, planetary mills, and attritor mills etc. The internal mechanics of the specific mill, the power supplied to drive the milling chamber; the composition, size, and size distribution of the balls influence the energy of the milling media. Ball mass-to-powder mass ratios of 5 to 10 are typically used and are effective. Various industrial processes use synthetic chemical dyes over the last few years. Paper and pulp manufacturing, plastics, dyeing of cloth, leather treatment, printing are examples of the areas where these chemicals are frequently used. Industrial waste waters containing such dyes are generally considered as effluents. Because of the toxicity of some of these dyes, their removal from industrial effluents can be a major environmental problem [19, 20]. The highly colored waste water in the presence of various dyes block sunlight and oxygen penetration which are both vital for various forms of aquatic life. Moreover, carcinogenic compounds which end up in food chain can be formed as the dye solution undergo anaerobic degradation [21]. From an environmental and economic point of view, removal of dyes from waste water is of significant importance. To remove dyes from aqueous solutions, several methods have been used [22-25]. Among these methods, adsorption is efficient and economical for the removal of dyes from aqueous solutions. Traditional approaches for water treatment, such as chemical precipitation, electrochemical degradation, acidification, biological degradation, activated carbon adsorption and ozone oxidation are generally restricted because of high operating cost, sensitivity to treatment conditions and low efficiency. Using the characteristic of high separation efficiency of magnetic

nanoparticles, the detection and separation of the toxic components of waste water can effectively shorten the working hours and reduce the workload. The novelty of this study is two-fold: first,
Using Baharaiya oasis iron ore for hard ferrimagnetic glass ceramics nanoparticles preparation; second, it provides an assessment of the adsorption efficiency of the proposed nanoparticles.

2. Experimental methods

2.1 Preparation of the glass

Table 1 represents the X-ray fluorescence analysis (XRF) of IOW (iron ore waste) sample. In this works, the prepared samples composition based on $BaFe_{12}O_{19}$ as hard magnet. Table 2 illustrates the composition of the prepared samples. IOH is referred to hard magnetic glass ceramics with composition of ~37% IOW beside Na_2CO_3, $NH_4H_2PO_4$, $BaCO_3$, H_3BO_3 and TiO_2as a source for Na_2O,P_2O_5, BaO, B_2O_3 and TiO_2 respectively.

Platinum crucibles were used to melt the required amounts of chemicals with composition shown in Table 2 at 1200°C. An electrically heated furnace was used for the melting process which lasted for 2 h with occasional swirling every 30 min. to ensure homogenization.The melts were poured onto a stainless steel plate and pressed into 1-2 mm thick strips by another cold steel plate at room temperature. In order to investigate the effect of heat treatment on the phase transformations and sample properties, the produced strips were heat treated at 900°C for 2 h.

Table 1. Quantitative XRF analysis of iron ore waste (IOW)

Constituents	**SiO_2**	**TiO_2**	**Al_2O_3**	**Fe_2O_3tot**	**MgO**	**CaO**	**Na_2O**	**K_2O**	**P_2O_5**	**CuO**	**Cr_2O_3**	MnO	
IOW (wt%)	7.96	0.28	2.79	67.15	0.31	3.96	1.53	0.28	0.48	-	0.028	1.2	
Constituents	**PbO**	**ZnO**	**NiO**	**As_2O_3**	**BaO**	**V_2O_5**	**SrO**	**ZrO_2**	**Y_2O_3**	**Co_3O_4**	**SO_3**	LOI	Cl
IOW (wt%)	0.009	0.041	0.065	0.005	0.293	0.07	0.014	0.011	-	-	1.16	11.63	0.692

IOP Conf. Series: Materials Science and Engineering **975** (2020) 012002 doi:10.1088/1757-899X/975/1/012002

Table 2. Chemical composition of the prepared magnetic glass ceramics

Chemical Composition wt%	Fe_2O_{3tot}	SiO_2	TiO_2	ZnO	Al_2O_3	MgO	CaO	Na_2O	K2O	ZrO_2	PbO
IOH	74.71	16.92	0.311	0.046	3.1	0.3444	4.41	1.7	0.3114	0.0122	0.01
Chemical Composition wt%	**P_2O_5**	**CuO**	**Cr_2O_3**	**MnO**	**B_2O_5**	**BaO**	**V_2O_5**	**Ni_2O**	**SrO**		
IOH	0.54	0.012	0.031	1.39	27.33	72.4	0.078	0.073	0.0156		

2. 2 Preparation of the magnetic glass ceramic nanoparticles (MGNPS)

To produce magnetic nanoparticles, the heat treated IOH sample was used. Desktop high speed vibrating ball mill was used to carry out the high energy ball milling process at room temperature. One end of the principal axis of the mill connected with an eccentric wheel that has some certain angle degree with the principal axis, and the eccentric wheel is connected with the foundation of the ball mill through a tensile spring. When the electric machinery is in high speed operation, the eccentric wheel will swing at a high speed due to the tension of the spring. With the powerful vibration generated by the high speed swinging of the eccentric wheel, the ball milling jars fixed on the eccentric wheel also rotate and vibrate in a high – speed manner of 1200 rpm. Zirconium oxide jar was used and a vacuum stainless steel cover was used to operate the milling process in vacuum environment. Granularity of the material fed to the jar was less than 1mm and the maximum material loading was two thirds of the cubage of the ball milling jar (including milling balls). Two Zirconium oxide balls of 20 mm diameter were used in milling process. After the selected times (1, 3 and 5 h), the milling was stopped in order to prevent overheating and rapid engine wear, in addition, to collect the samples

2.3 water treatment experiments

The MB and CR solution was made up in stock solution of concentration 1000 (ppm) and was subsequently diluted to required concentrations using distilled water. Standard solutions of MB and CR were made up using distilled water as a solvent. Dilutions of concentrated samples were undertaken in order that the reduced optical density as found from the Spectrophotometer was in the range 0.001 to 0.6; it was found that more accurate results were obtained in this range. Then by multiplying these reduced optical densities by their dilution factor, the optical density of the original stock solution was obtained. All tests were carried out at room temperature (25±2°C) to eliminate any temperature effects, unless otherwise stated.

The calibration curves for both MB and CR were prepared by recording the absorbance (optical density) values for a range of known concentrations of dye solution at the wave length for maximum optical density (λ_{max}). This value, λ_{max}, was found from a full scan of the MB and CR spectrum. The values of λ_{max}, were determined using a double beam UV-visible Spectrophotometer (T80 UV-VIS Spectrophotometer PG Instruments ltd), UK, and this value is (665 nm) for MB and (556nm) for CR was used in all subsequent investigations using a Spectrophotometer.

Optical density readings for various samples (Tables 3.1& 3.2) were compared with the calibration curve (Figures 3.2&3.3) and transformed into concentration (ppm) terms.

In accordance with the Beer-Lambert's law and using least-squares method applied to the straight lines the optical densities follows a linear relationships to concentration as follows:

1) For MB Dye Conc. = 4.405*abs
2) For CD Dye Conc. =34.483*abs
The correlation coefficients are0.9902 and0.9986 for CR and MB, respectively

Table 3. Calibration curve for MB

Conc. (ppm)	0	1	1.5	2	2.5	3
Abs	0	0.224	0.322	0.503	0.565	0.666

Table 4. Calibration curve for CR

Conc.(ppm)	0	1	2	5	10	15
Abs	0	0.052	0.062	0.144	0.303	0.453

2.4 Characterization

The morphological study for the precursor hard magnetic glass ceramic sample and for the nanoparticles was carried out by using scanning electron microscope (SEM, FEI Quanta 200 Environmental Scanning Electron Microscope) and transmission electron microscope (TEM 2010).For TEM imaging, applying usual preparation methodology on the ball milled samples under investigation resulted in not evenly distributed particles on the form var grid but rather were on the grid bars therefore unobservable in the TEM. For successful imaging under the TEM, freeze-drying technique has been used to improve the long-term stability of the colloidal nanoparticles. The process of water removal from frozen samples by sublimation and desorption under vacuum is known as freeze-drying or lyophilization, this process can be controlled to reach a shelf life of several years.

SDTQ600 differential thermal analysis (DTA) instrument under inert gas was used to determine the thermal properties of the glass samples. Alumina was the inert reference material and 10°C/min was the used heating rate. DTA results were the guide to determine the heat-treatment temperatures needed to induce samples crystallization. To determine samples content and precipitated crystalline phases, X-ray diffraction using Ni-filled Cu Kα radiation was performed to samples before and after heat treatment. Bruker D8, an advanced X-ray diffraction instrument was used. ASTM x-ray diffraction card files were used for the interpretation of the XRD patterns of the reference data. Raman spectroscope (Model: Horiba Jobin Yvon Lab Ram ARAMIS) has been used to characterize the magnetic nanoparticles. Vibrating sample magnetometer (Quantum Design(R), Magnetic Property Measurement System (MPMS), SQUID VSM) was used to determine the magnetic properties of the magnetic nanoparticles and the precursor hard magnetic glass ceramic samples.

A double beam UV-visible Spectrophotometer (T80 UV-VIS Spectrophotometer PG Instruments ltd), was used for water treatment determination.

3 Results and discussions

Figure 1 illustrates the differential thermal analysis (DTA) of the prepared IOH sample before heat treatment. It revealed one endothermic peak at 554°C followed by one exothermic peak at 716°C. The endothermic effect mean increase in heat capacity, which cased due to transformation of glass from rigid to plastic structure resulted in the endothermic effect at ~ 554°C **[26]**. However, the exothermic peak is due to the crystallization effects, which is accompanied by releasing heat due to its lower free energy, reaches its maximum rate near the summit of the exothermal peak. The higher the intensity and sharpness of the exothermic peak is, the higher the propensity, velocity and bulk (internal) crystallization. In our case the exothermic peak is corresponding to the crystallization of Ba-hexaferrite crystallization as will be certified later by XRD. The broadness of the exothermic peak is due to surface crystallization or hardness to crystallize.

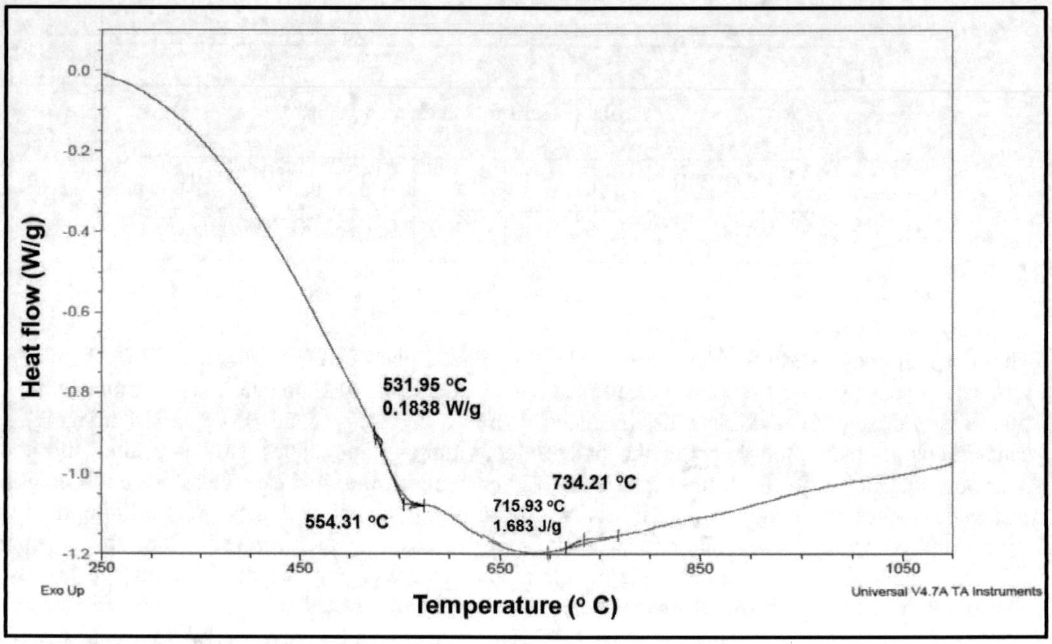

Figure 1. DTA scan for the as prepared IOH sample.

Figure 2. depict XRD analysis for unmodified iron ore waste (IOW) and the modified glass sample (IOH) before and after heat treatment. Heat treatment at 900°C / 2 h was applied to the as quenched sample to convert it from glassy to crystalline structure. Un modified IOW revealed presence of hematite (Fe_2O_3), goethite FeO (OH) and wuestite ($Fe_{0.925}O$) phases as indexed in ASTM cards Nos89-0599,81-0464 and 89-686 respectively. On the other hands glass sample before heat treatment revealed amorphous structure, where hump at 2-theta 20 and 40° where appeared, no narrow XRD peaks were detected. Applying heat treatment at 900°C/2 h revealed crystallization of pure Ba-hexaferrite ($BaFe_{12}O_{19}$); where XRD peaks were matched well with ASTM card No84-0757. This results evidence the success of the design composition to get pure Ba- hexaferrite.

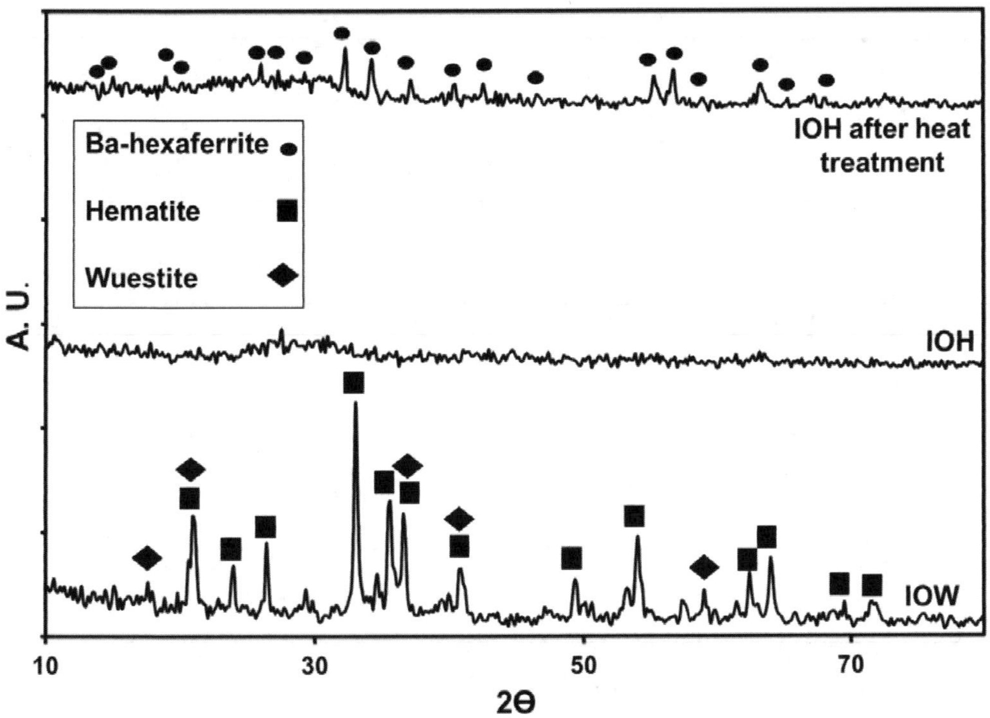

Figure 2. XRD spectra of IOW and IOH before and after heat treatment.

The comparative SEM images of the hand milled and high energy milled for 5h, at the same magnification for hard magnetic glass ceramics were shown in figure 3. (a, b). The hand milled samples are irregular with a rough surface, and the particles have a non-uniform distribution figure.3a. On the other hands, the mechanical milled revealed uniform distribution of very fine particles figure.3b; this confirms the effective conversion to magnetic nanoparticles during the 5 h high energy ball milling process. TEM of sample after high energy milled for 5h figure.3c revealed distribution of highly uniform agglomerated nanoparticles. The inherently large surface energy (> 100 dyn. cm-1) caused by the high surface area to volume ratio [27], in addition to strong magnetic dipole-dipole interactions and van der waals between particles are the reasons for agglomeration tendency of iron oxide nanoparticles [28]. After 5h milling duration the powder particles average size reached 26 nm. At lower temperature, the high reactivity of the 5 h-milled powders can facilitate densification and homogenization of the glass ceramics. This clarifies the reason of apparent agglomeration of the 5-h milled powder.

Figure 4. demonstrates the Raman spectra of heat treated IOH sample hand milled before and after 5 hours milling time. The fundamental Raman scattering peaks for $BaFe_{12}O_{19}$ observed at 686cm^{-1}, 615cm^{-1} refering to A_{1g} vibrations of Fe-O bonds at the bipyramidal 2b, octahedral 4f$_2$, 2a+12k, 12k dominated and 2a sites respectively.The peak at 507cm^{-1} is due to E_{1g} vibrations, whereas the peak observed at 339 cm^{-1} is due to the E_{2g} vibration modes. The peak at 186 cm^{-1} originates from E_{1g} vibrations of the entire spinel block [29]. The particles after 5 hours milling time have the same characteristic peaks as the hand milled

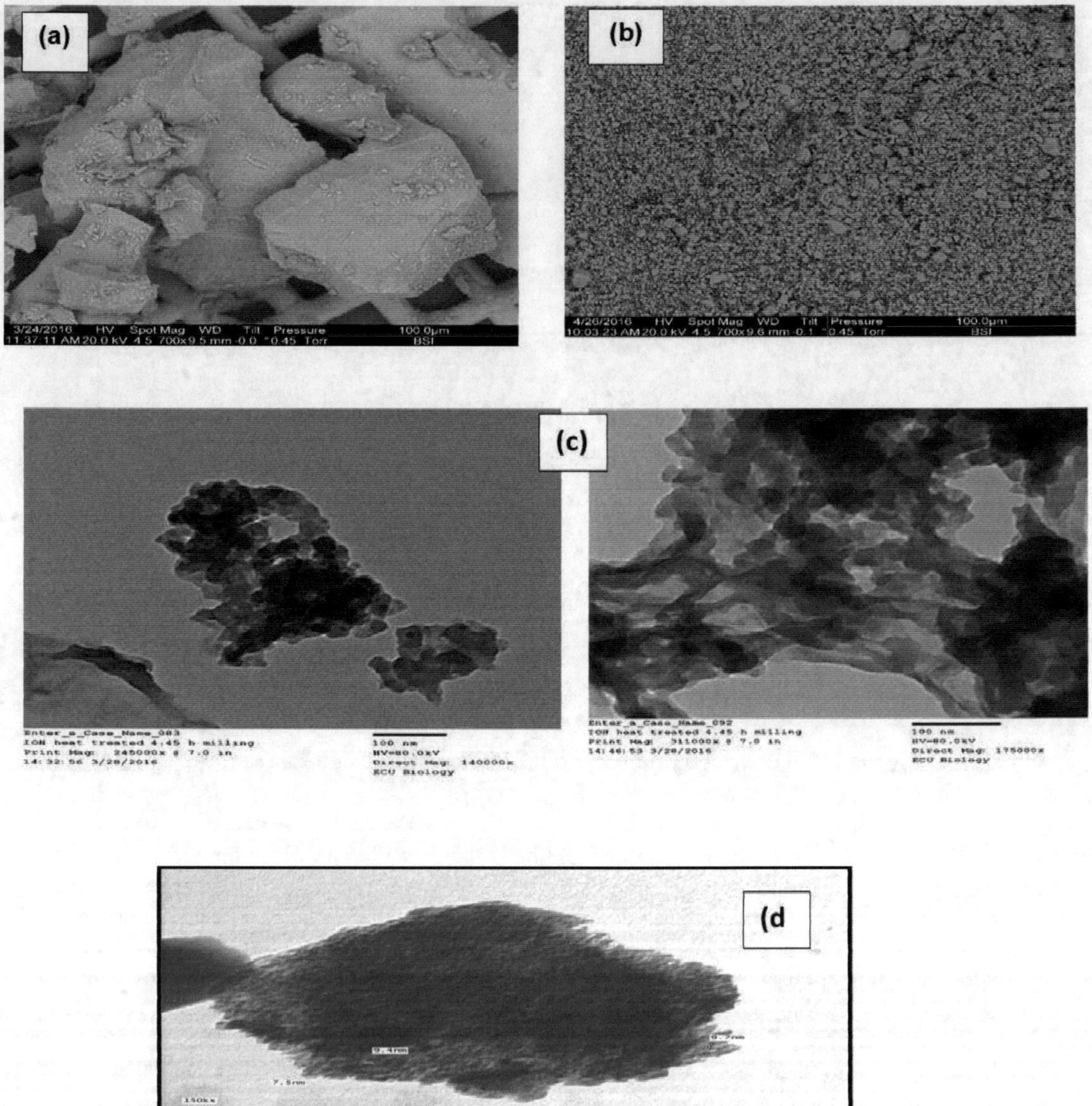

Figure 3. SEM images of (a) Hand milled and (b) 5 h milled IOH (700 x) magnification, (c, d) TEM images of 5 h milled IOH with 14000 X magnification compared to TEM image of (d) the hand milled heat treated IOH sample (150000x).

magnetic glass ceramic.The distortion of Ba-hexaferrite crystals due to milling process can be the reason of peaks intensity variation [30]. Intense peaks can be considered an indication of high degree of crystallinity of nanoparticles, in addition, defects such as spin glass structures formation, BaO presence, cation redistribution and surface spins existence [31]. Increasing fractured particles percentage, higher aspect ratios for longer milling time or the formation of most of the nanoparticles rather than the disordered boundary regions with increasing milling time can lead to decrease in peaks intensity [32].

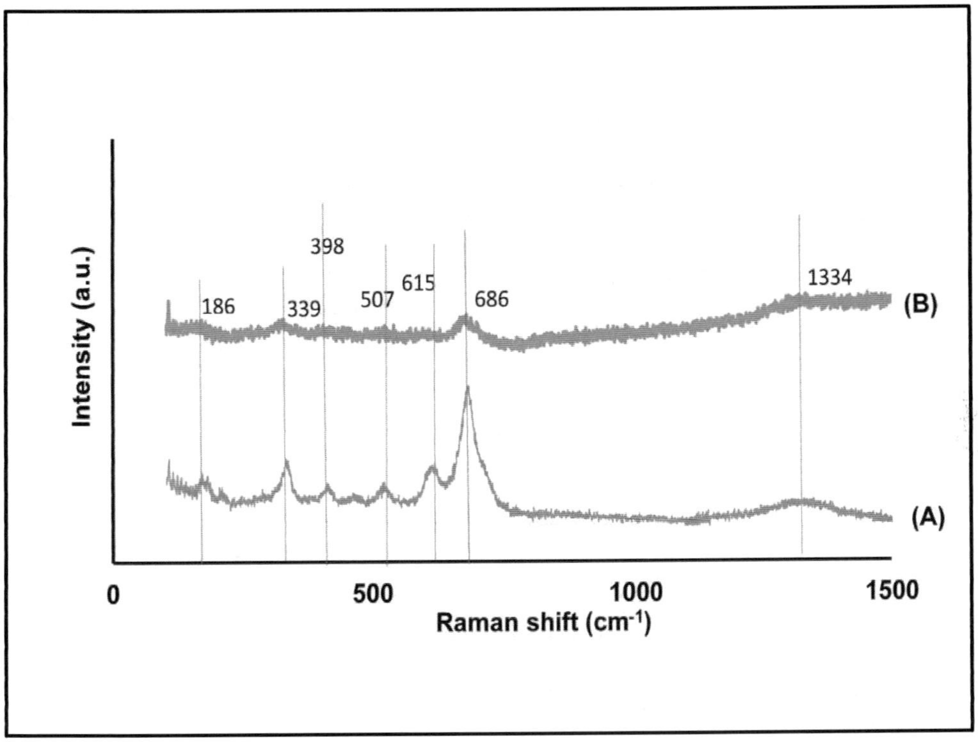

Figure 4. Raman spectra of (A) Hand milled and (B) 5hrs milled particles

Figure 5. Shows the room temperature magnetic hysteresis (*M-H*) loops of hand milled IOH and after 1, 3 and 5 hours milling time under a magnetic field strength of 20 KOe. Table 5 displays the relevant magnetic parameters: saturation magnetization (Ms), remanence (*Mr*), Coercivity (*Hc*) and *Mr/Ms* ratio obtained from M-H loops. In general, the magnetic field necessary to saturate the samples increases as the Ba-hexaferrite crystallization increases. As expected, saturation magnetization is due to crystallization of Ba-hexaferrite. It should be considered that Ba-hexaferrite is a ferromagnetic phase and amorphous glass matrix is an antiferromagnetic material with a very low saturation magnetization. As a consequence, the quantity of Ba-hexaferrite crystals present in the glass ceramic samples can be followed up by the variation of the saturation magnetization [30]. It is noticed that, *Ms* decreased by increasing milling time to reach 21.18 emu/g after 5 hours milling. The coercivity decreased from 1306Oe for hand milled sample to 83Oe after 1 hour milling and after that kept increasing by increasing milling time to reach the maximum value 1550Oe after 5 hours milling. Grain size, defects presence, strains, magneto-crystalline anisotropy, shape anisotropy and surface disorder usually affect the coercivity [31], [33].

IOP Conf. Series: Materials Science and Engineering **975** (2020) 012002 doi:10.1088/1757-899X/975/1/012002

The continual ball collisions with high energy in ball milling process is responsible primarily for the defects and strain resulted as well as the difference in magnetic properties of IOH powder after different milling time .

In the hand milling process, the reactants were in the interaction region of the mortar and pestle, can be ground to a limited extent. In the planetary ball milling process, almost all of the reactants are in the interaction region of the balls so that they can be simultaneously ground. The strains and defects generated by the repeated high energy ball collisions are still abound in the products. These findings explain the increase in H_c by increasing milling time (table 5).It can be observed from the loops that *Mr/Ms* values increased from 8.89 for the hand milled to 9.32 after 5 h milling. The remanence ratio enhancement were due to the increasing percentage of fractured particles and the formation of most of the nanoparticles with increasing milling time [32].These results consistent with Raman spectroscopy results. The magnetic properties results confirm that after milling process, the material is still strongly magnetic and enable magnetic manipulation and recovery of the sorbent in the water treatment as shown in inset in figure 5.

Table 5. Magnetic properties of 1- hand milled, 2- 1hr, 3- 3hrs, and 4- 5 hrs milling particles.

Sample	Ms (emu/g)	Mr (emu/g)	Hc (Oe)	Mr/Ms
Hand milled	22.56	8.89	1306	0.39
5 hrs milled particles	21.18	9.32	1550	0.44

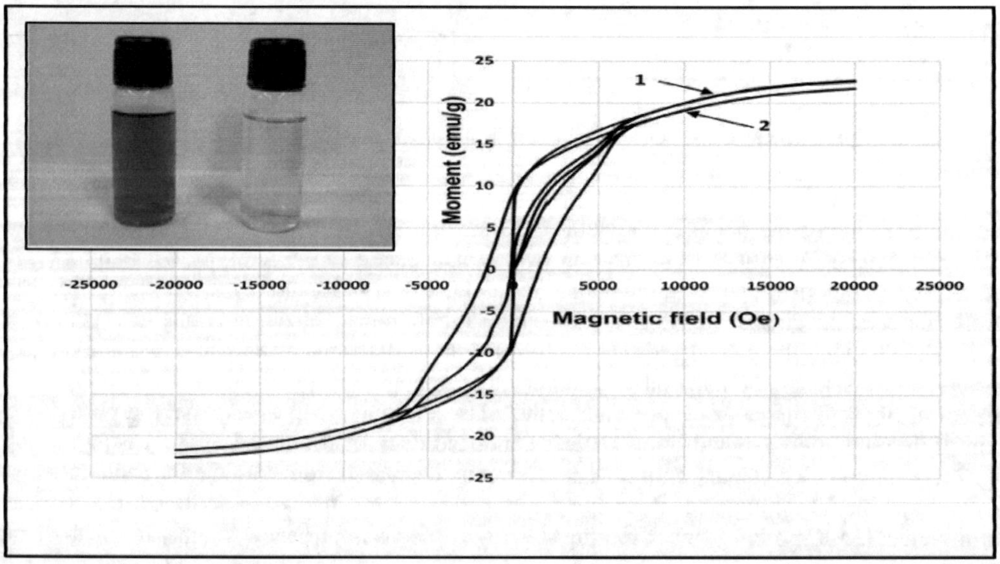

Figure 5. Room temperature M-H hysteresis loops for 1- Hand milled 2-5hrs milled IOH particles.

The presence of organic pollutants in the water resources is a major risk to the human health and ecosystem. Therefore, an effective low cost adsorbent with high adsorption capacity for organic pollutants removal is desirable. Many researchers studied the organic contaminants adsorption using iron oxide nano-material as, particularly for the efficient treatment of large-volume water samples and fast separation by employing a strong external magnetic field.

In this study, Methylene blue (MB) and Congored (CR) as common dyes in many industries were selected. These dyes were used for studying the removal capacity of the hand milled and the 5 hrs. milled magnetic nanoparticles. A constant mass of the hand milled and 5 hours milled particles were added to 100 ml (cm^3) bottles containing 50 ml of solution. The bottles were sealed together with appropriate blanks and mechanically shacked in a constant temperature, constant agitation shaker bath for a period of 3-h was used. The speed of agitation was kept constant for each run throughout the experiments to ensure equal mixing. After that time, the samples were centrifuged at 5000 rpm for 30 min. to separate the magnetic particles. The adsorption isotherm calculation was important to describe the interaction mechanism between adsorbate and adsorbent and to show the adsorption capacity of adsorbent. The amount of dye adsorbed is calculated from the following equation

$$qe = \frac{(Co - Ce)V}{m}$$

Where qe is the equilibrium adsorbent capacity (mg/g), Co and Ce are the initial and final dye concentration (mg/l) respectively, V is the volume of dye solution and m is the mass of adsorbent (g).

Figures 6. shows the effect of contact time on the adsorption of Methylene blue (MB) dye by 5 hrs. milled nanoparticles compared to the hand milled particles. The equilibrium concentration for MB on the hand milled particles reached after 25 min. Adsorption of dye was quit rapid initially and the rate of adsorption became slower with time and reached constant value (equilibrium time). The initial faster rate may be due to the availability of uncovered surface area of the adsorbent [37]. For MB dye, the adsorption capacity of the 5 hours milled magnetic particles is 200 mg/g while it is 166.5 mg/g for the hand milled particles. Similarly, for CR dye, the adsorption capacity of the 5 hours milled nanoparticles is 124.5 mg/g while it's 104 mg/g for the hand milled magnetic particles. The fast removal rate is mainly attributed to the electrostatic attraction between the surface of the particles and the MB and CR dyes in the solution and the nanostructure with high surface area.

Table 6. A comparison of the adsorption capacities of the nano-sorbent used by present study and selected studies.

Group	Nanosorbents	Dye	Adsorption capacity	Reference
Ying Qi Wang et al.	$MnFe_2O_4$	Congo red	39.3 mg/g	[34]
Kah Aik Tan et al.	Fe_3O_4-MCP	Methylene blue	24.4 mg/g	[35]
Saksornchai et al.	CTAB coated Fe_3O_4	Congo red	93.5 mg/g	[36]
This study	$BaFe_{12}O_{19}$	Methylene blue Congo red	200 mg/g 124.5 mg/g	

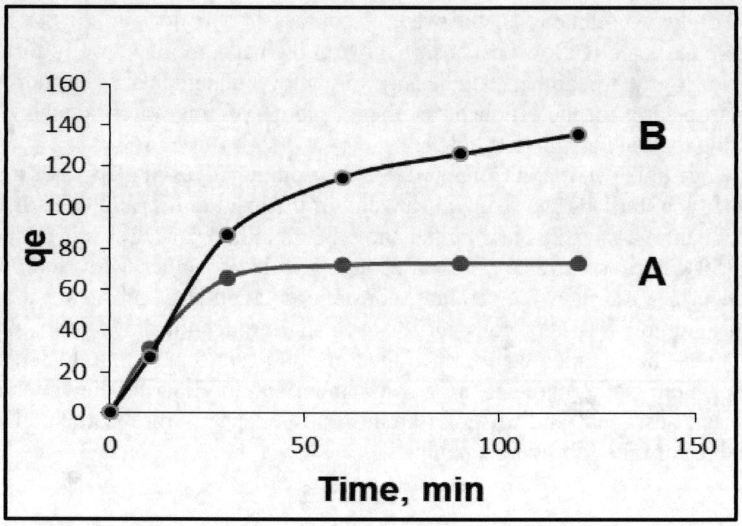

Figure 6. Effect of contact time on removal of methylene blue dye using A, Hand milled and B, 5 hrs milled particles.

Similar to heavy metal adsorption, the adsorption of organic contaminants take place by surface exchange reactions until the surface functional sites are fully occupied then contaminants could diffuse into the adsorbent for further interactions with the functional groups [38-40]. The nanostructured materials with high surface area increase the contact between the dye and the adsorbent surface. It is apparent that the adsorption capacity of the $BaFe_{12}O_{19}$ nanoparticles prepared in this study is higher than that obtained previously by other workers (table 6). This confirms that the recycled magnetic nanoparticles could be a promising adsorbent for waste water treatment.

4 Conclusions

1-Hard ferromagnetic glass-ceramics were prepared by recycling Baharaya oasis iron ore waste.

2-XRD data confirmed that, the heat treatment schedule (900oC/ 2h) chosen according to DTA results had a positive effect on Ba-hexaferrite crystallization.

3-By using mechanical milling technique, uniformly distributed hard ferromagnetic glass-ceramics with 26 nm particle size has been obtained after 5 h milling time.

4-Raman spectroscopy analysis confirmed that, the produced nanoparticles after 5 hours milling time have the same characteristic peaks like the hand milled magnetic glass ceramic precursor.

5-Saturation magnetization decreased by increasing milling time to reach 21.18 emu/g after 5 hours milling. The coercivity decreased from 1306 Oe for hand milled sample to 83 Oe after 1 hour milling and after that kept increasing by increasing milling time to reach the maximum value 1550 Oe after 5 hours.

6-Increasing percentage of fractured particles and higher aspect ratios for longer milling time explain the remanence ratio enhancement after 5 hours milling.

7- Reducing particle size increased the adsorption capacity of the magnetic particles. For MB dye, the adsorption capacity of the 5 hrs milled magnetic particles is 200 mg/g while it is 166.5 mg/g for the hand milled particles. For CR dye, the adsorption capacity of the 5 hrs milled nanoparticles is 124.5 mg/g while it's 104 mg/g for the hand milled magnetic particles.

5. References

[1] Aderhold J, Davydov V Yu, Fedler F, Klausing H, Mistele D, Rotter T, Semchinova O, Stemmer and Graul J 2001 J. Cryst. Growth 222 701
[2] S. A. M. Abdel-Hameed, A. M. El Kady, 2012. J. Adv. Mater. 3 167.
[3] Jason Hill, Erik Nelson, David Tilman, Stephen Polasky and Douglas Tiffany, 2006 Proceedings of the national academy of sciences of the United States of America 103 30.
[4] J. Smit and H.P.J. Wijn, 1959 New York, Wiley.
[5] G.Schmid,2004 Wiley-VCH, Wein-heim.
[6] K.J.Klabunde, 2001 Wiley-Interscience,New York.
[7] S.Sun, H.Zeng, D.B.Robinson, S.Raoux, P.M.Rice,S.X.Wang, 2003 Journal of the American chemical society, 126 273.
[8] Xiaoyan Zhang, Qiaoyu Dai, Xiaobin Huang and Xiaozhen Tang, 2009 Solid state sciences, 11 1861.
[9] Mani Diba, Qurania-MentiGoudouri, Felipe Tapia and Aldo R. Boccaccini, 2014 Journal of current openion in solid state and materials science 18 147.
[10] SeungyounLyu, Chunlan Huang, Hong Yang and Xinping Zhang, 2013 Journal of orthopedics research 31 1382.
[11] C. Feng, K. C. Khulbe, T. Matsuura, S.Taba and A. F. Ismail, 2013 Separation and purification technology 102 118.
[12] JeehyeByun, Hasmukh A. Patel and Cafer T. Yavuz, 2014 Journal of nanoparticles research 16 2787.
[13] S.C. Tjong, and H. Chen, 2004 Materials Science and Engineering, 45 1.
[14] J.S. Benjamin, 1990 Metal Powder Rep. 45 122.
[15] D.L. Zhang, 2004 Progress in Materials Science 49 537.
[16] T. S. Ward, W. Chen, M. Schoenitz, R. N. Dave, and E. Dreizin, 2005 ActaMaterialia, 53 2909
[17] Q.S. Mei, and K. Lu, 2007 Progress in Materials Science 52 1175.
[18] M. Sherif El-Eskandarany, K. Sumiyama, and K. Suzuki, 1997 ActaMateialia, 45 1175.
[19] S. Zghal, R. Twesten, F. Wu, and P. Bellon, 2002 ActaMaterialia, 50 4711.
[20] O. Ligrini, E. Oliveros, A. Braun, 1993 chem. Rev. 93 671.
[21] C. T. Helmes, C. C. Sigman, Z. A. Fund, M. K. Thompson, M. K. Voeltz, M. Makie, 1984 J. Environ. Sci. Health A. 19 97.
[22] M. Boeningo, 1994 Washington, DC: US Gov. printing off., DNHS (NIOSH), 80.
[23] R. Han, D.Ding, Y. Xu, W. Zou, 2008 Bioresour. Technol. 99 2938.
[24] V. S. Mane, P.V. Vijay Babu, 2011 Desalination 273 321.
[25] M. Ghaedi, M. N. Biyareh, S. N. Kokhdan, S. Shamsaldini, R. Sahraei, A. Daneshfar, S. Shahriyar, 2012 Mater. Sci. Eng., C 32 725.
[26] A. Afkhami, R. Moosavi, 2010 J. Hazard. Mater. 174 398.
[27] Salwa A.M. Abdel-Hameed , Mohamed A. Marzouk , Mohamed M. Farag, 2014 Journal of Advanced Research 5 543.
[28] C.Buzea, I.Pacheco, K.Robbie, 2007 Biointerphases 8630 MR 17
[29] DonyaRamimoghadam, Samira Bagheri and Sherifah Bee AbdHamid, 2015 Colloids and surfaces B:Biointerfaces 133 388.
[30] V. Rane, S. Meena, S. Gokhale, S. M. Yusuf, G. Phatak and S. Date, 2013 J. Electron. Mater.42 761.
[31] M. K. Roy, B. Haldar, H. C. Verma, 2006 Nanotechnology 17 232.
[32] S. Chkoundali, S. Ammar, N. Jouini, F. Fiévet, P. Molinié, M. Danot, F. Vallain, J. M. Grenèche, 2004 J. Phys.: Condens. Matter 16 4357.
[33] V. M. Chakka, B. Altuncevahir, Z. Q. Jin, Y. Li, and J. P. Liu, 2006 Journal of applied physics 99

08E912.
[34] A.B. Nawale, N. S. Kanhe, K. R. Patil, S. V. Bhoraskar, V. L. Mathe, A. K. Das, 2011 J. Alloy. Compd. 509 4404.
[35] Ying Qi Wng, RongMin Cheng, Zi Wen, and LiJun Zhao, 2011 Eur. J. Inorg.Chem. 2942.
[36] Kah Aik Tan, Norhashimah Morad, Tjoon Tow Teng, Ismail Norli and P. Panneerselvam, 2012 APCBEE Procedia 1 83
[37] Saksornchai E, Kavinchan J, Thongtem S, Thongtem T., 2018 Mat. Lett. 142 213
[38] Tayyebeh Madrakian, Abbas Afkhami, MazaherAhmadi, 2012 Spectrochimicaacta part a: molecular and biomolecular spectroscopy 99 102.
[39] Hu J, Shao DD, Chen CL, Sheng GD, Ren XM, Wang XK., 2011 J Hazard Mater 185 463.
[40] Ma ZY, Guan YP, Liu XQ, Liu HZ., 2005 J Appl Polym Sci 96 2174.
[41] Zhao XL, Wang JM, Wu FC, Wang T, Cai YQ, Shi YL, 2010 J Hazard Mater 173 102.

Thermal Properties of Nano Tungsten - Ethylene Vinyl Acetate (EVA) Composites

M. Sobhy[*], Ahmed Saleh[*], Alaa I Eid[**], W. M. Khairaldien[***]

[*]Science and Technology Center of Excellence (STCE), El- Salam City, Cairo, Egypt
[**]Center of Metallurgy Research and Development Institute
[***]Asiut University, Faculty of Engineering, Mechanical Engineering Department
Correspondence to: Maged Sobhy; maged@stce-egypt.org or msobhy2009@live.com

Abstract. Ethylene vinyl acetate (EVA) with different percentage of nano tungsten powder (0wt%, 30/wt %, 45/wt %, 60/wt%, and 70/wt %) composites (EVA/W nano-composites) were prepared by melt blending in a twin-screw brabender. The structural, morphological, thermal and mechanical properties of the prepared nano-composites were investigated using powder X-ray diffraction (PXRD), Scanning Electron Microscopy (SEM), thermogravimetric analysis (TGA), and dynamic mechanical thermal analysis (DMTA). It was found that the EVA composites show a higher thermal stability at high levels of tungsten loading.
Keywords. EVA, Nano tungsten powder, Higher thermal stability.

1 Introduction:

Polymers in general have low stiffness and strength, poor thermal properties, good dielectric properties and they are good insulators of electricity and heat. In addition, they are non-magnetic. Many attempts have been made over the years to change some of these properties for use in more applications. One of these attempts is reinforcing polymers with metal nanoparticles to enhance the polymer properties in different applications. The overall composites reinforced with nano-fillers was found to vary widely depending on type and content of reinforcing filler, dispersion, filler size and shape, and type of coupling agent. For polymeric based composites type and percentage of reinforced nano-composites are selected to impart some degree of metallic behavior to the polymer matrix. [1]

Ethylene vinyl acetate (EVA) is the copolymer of ethylene and vinyl acetate segments characterized by different acetate contents, and typically formed via free radical polymerization [2, 3]. Due to its desirable physico-chemical properties and its ability to accept additives, EVA is widely used in many applications: it can be used in electronic devices, electrical engineering, wire and cables, buildings, and transportation (aircraft, cars) [4]. This reflects the extent of its industrial importance and potential use. However, because of its low tensile strength, dense smoke generation as well as release of toxic gases during combustion its practical application has been strictly limited [5–7]. To overcome these problems and to improve its strength, and thermal stability, several studies have been conducted [8–11].

In this study, EVA with different percentage of Nano Tungsten powder -composites (0%, 30%, 45%, 60%, and 70%) was prepared, to improve its dynamic mechanical and thermal properties.

2 Experimental

2.1. Materials and Methods

IOP Conf. Series: Materials Science and Engineering **975** (2020) 012003 doi:10.1088/1757-899X/975/1/012003

EVA copolymer with a content of 18% vinyl acetate by weight, a melt flow index of 1.83 g/10 min and density of 0.98 g/cm3 was supplied by Tane Elf Atofina, China. Tungsten powder with purity 99.9%, particle size 0.5-2 µm, density 19.3 g/cm3, and melting point 3410°C was continuously ball milled for particle size reduction until reaching nano scale. Milling was carried out using a SPEX 8000M MIXER/MILL. The ball milling process was performed using balls and tank made from chromium steel with a ball to powder mass ratio of 10:1. Scherrer equation, was used to calculate size of sub-micrometer crystallites in a solid to the broadening of a peak in a diffraction pattern using X-ray diffraction and crystallography, and the results showed that the tungsten particles after milling was in the Nano scale range.

2.2. Sample Preparation (Preparation of EVA Nano-composites)

EVA/W nano-composites, with different amounts of W nano particles (0, 30, 45, 60, and 70 wt. %) with respect to EVA content, were extruded. The EVA/W nano-composites were prepared via melt blending. Melt blending was performed in a twin-screw extruder (10 mm twin-screw extruder, Rondol, England) at 160 °C and screw speed of 60 rpm for 10 min. The extrudates were cooled down through passing in a water bath and then pelletized (10mm Diameter * 2 ± 0.05 mm thickness).

2.3. The phase structure

2.3.1. Powder X-ray Diffracto-meter.

The crystallite phases of sheet samples with dimensions 10×10×1 mm3 of the prepared EVA composites were identified by Powder X-ray diffraction (PXRD) using a step size of 0.04◦ and at a scanning rate down to 1.2 ◦/min. with continuous scan mode using ARL™ X'TRA powder diffracto-meter. Thermo-Fisher Scientific Inc. The diffraction data was recorded for 2θ values between 10◦ and 7◦.

2.3.2. Scanning Electron Microscopy (SEM)

The microstructure of the prepared EVA composite samples was also analyzed using Quanta FEG 250 scanning electron microscope (FEI Company, USA). Samples were mounted onto SEM stubs. Applied SEM conditions were: a 10.1 mm working distance.

2.3.3. Thermal Analysis

TGA instrument model Q500 (TA instruments), characterized by thermo balance with sensitivity 0.1µ g, was used to evaluate the thermal stability of the EVA composite sheets. Samples were heated in nitrogen, flow rate at 50 ml min−1, from 25 to 500 °C at a heating rate of 10 °C min−1. Sample weights ranging from 5 mg to 10 mg were used. The storage modulus of EVA and their nano-composites were conducted by DMA Q800 (TA instruments) from RT to 190 °C with heating rate 10 °C min−1 using dual cantilever mode.

3 Results and discussion

3.1. Powder X-ray Diffractometer

Our data in figure 1 shows that the main peaks of tungsten powder are identified at angles 40.41◦, 73.32◦ and 58.35◦ as per the standard card no. 00-001-1204 which indicates that tungsten powder did not react with the polymer but physically dispersed.

The diffraction patterns displayed in figure 2, show that adding tungsten nano powder with different weight ratio from 0% (pure polymer EVA) up to 70% changed the degree of crystallinity, which measured by area under the peaks to the total area, of the prepared samples.

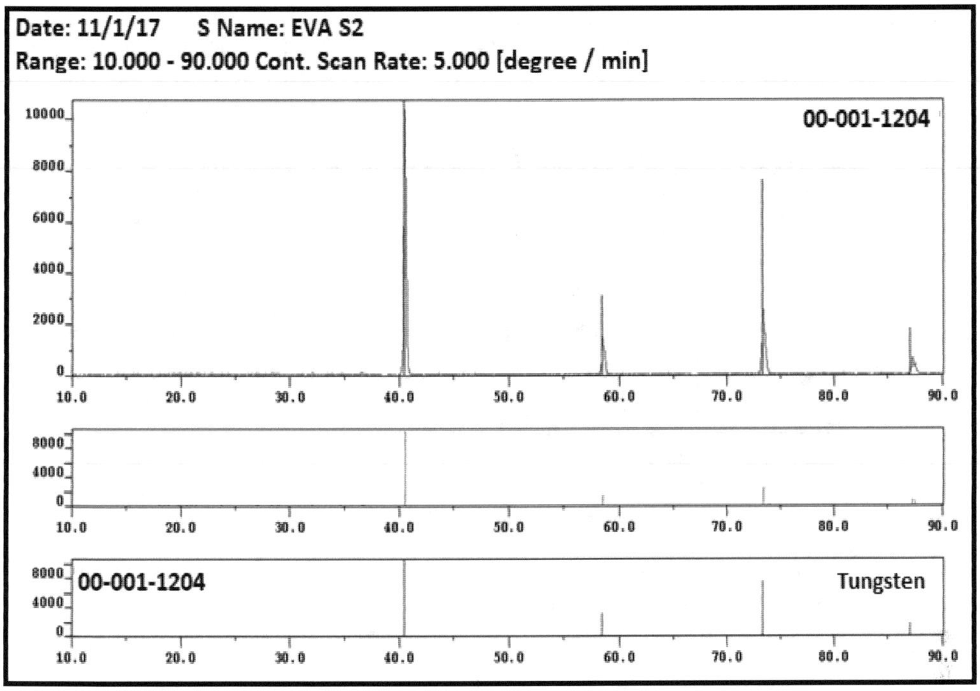

Figure 1. EVA-30%W sample with respect to W XRD card

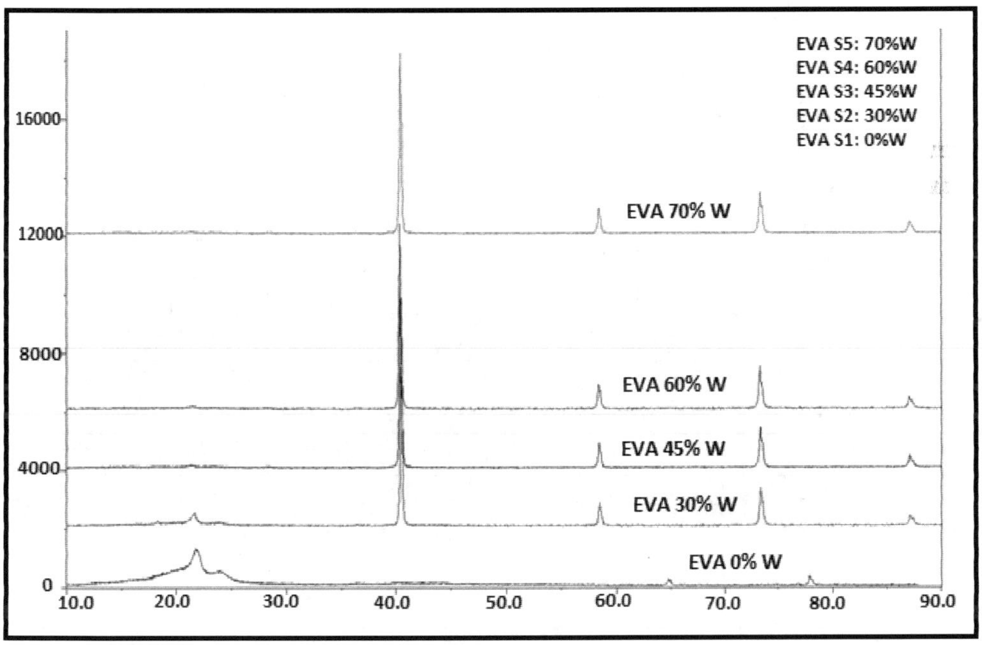

Figure 2. The PXRD profile of EVA samples with 0 wt% W,30 wt% W,45 wt% W,60 wt% W,70 wt% W

IOP Conf. Series: Materials Science and Engineering **975** (2020) 012003 doi:10.1088/1757-899X/975/1/012003

3-2 Scanning Electron Microscopy (SEM)

Figure 3 a, b, c, d and e, show the scanning electron microscope (SEM) images of the microstructure of EVA-0%W, EVA-30%W, EVA-45%W, EVA-60%W, and EVA-70%W nanocomposites, respectively using liquid nitrogen fractured samples.

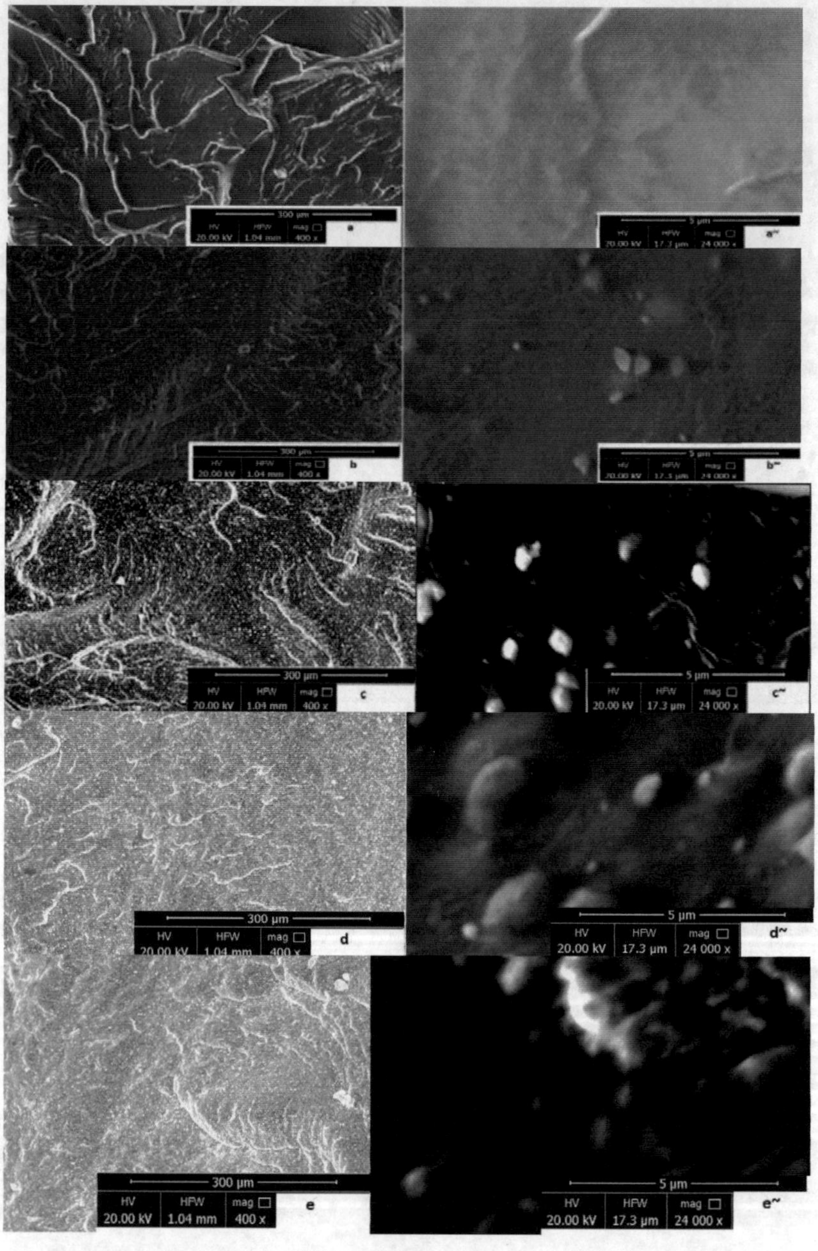

Figure 3. SEM images of EVA composites: (a) 0 wt% W, (b) 30 wt% W, (c) 45 wt% W, (d) 60 wt% W and (e) 70 wt% W.

The images show that the nano-tungsten particles are well dispersed and adhered in/with the EVA matrix and the tungsten powder spreads homogeneously within the polymer matrix. The examination also shows that the matrix/ nanoparticles adhesion and dispersion are very efficient since the fractured cracks tend to develop through the matrix itself, not at the interface between the particles and polymer matrix, which confirm the strong mechanical bond (Particle mechanical interlock with in polymer matrix) of the reinforced composite.

3.3. Thermo-gravimetric analysis (TGA)

The mass loss profile of EVA in figure 4 shows a two stage of degradation pathway with the first step attributed to the elimination of acetate side groups (de-acetylation) leaving behind an unsaturated polymer backbone or polyene [2], while the second step is due to the allylic chain scission of the polyene to reach full thermal degradation at high temperatures [12], Scheme 1.

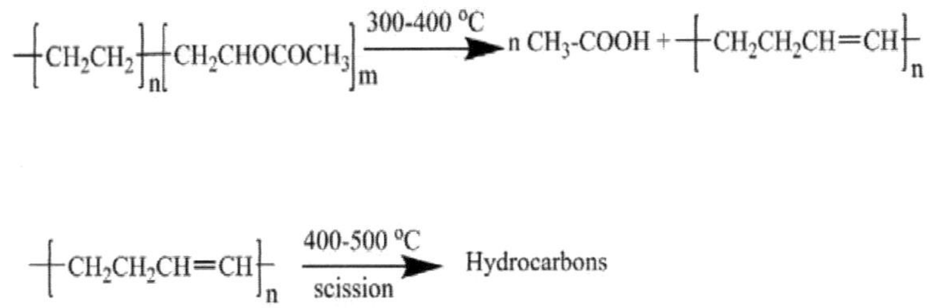

Scheme 1. Degradation of EVA

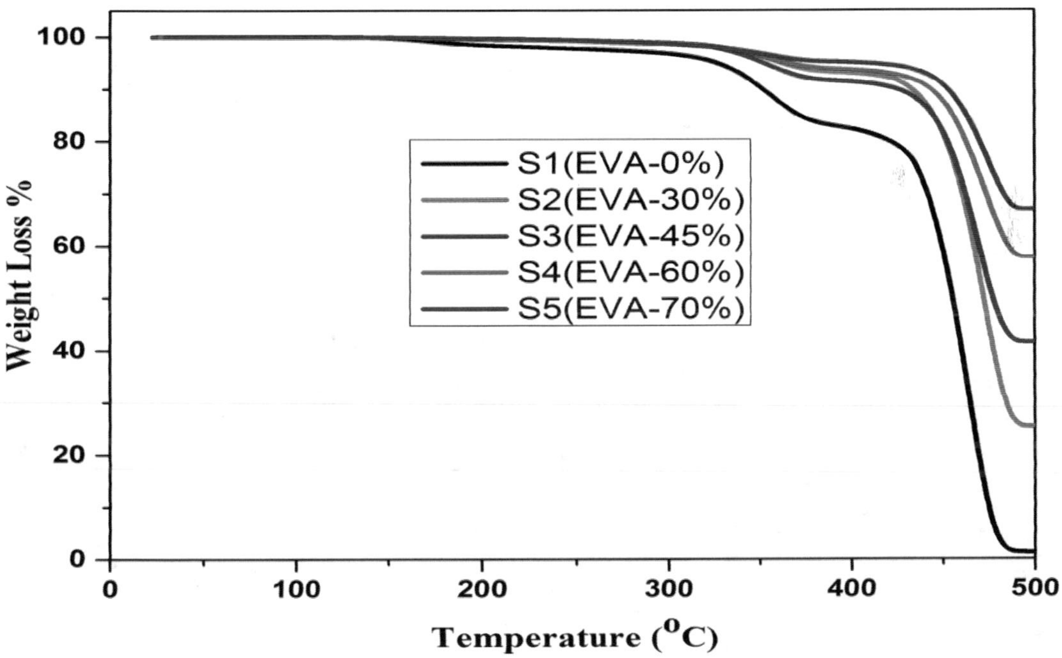

Figure 4. TG curves of EVA and its nano-composites

Introducing tungsten into the matrix of EVA has enhanced its thermal stability as shown in the previous figure. The TGA profiles of pure EVA and its composites are similar however, the curves shift to a higher degradation temperatures relative to pure EVA with the increase of reinforcing filler contents. The total weight loss of pure EVA, which was 98.97%, decreased by introducing of tungsten as shown in table 1, where the residue at 500 °C reaches its maximum value (66.97%) in sample S5 (70%W).

Table 1. Data derived from TGA curves of EVA composites

sample	Mass fraction of total weight loss (%) at 500°C	Residue % at 500°C
	EVA samples	
S1 (EVA-0%W)	98.79	1.24
S2 (EVA-30%W)	74.63	25.37
S3 (EVA-45%W)	58.48	41.53
S4 (EVA-60%W)	42.23	57.77
S5 (EVA-70%W)	33.11	66.89

3.4. Dynamic Mechanical Thermal Analysis (DMTA)

DMTA is a technique in which the storage modulus (elastic response) of the sample, under oscillating load, is monitored against temperature, time or frequency of oscillation. During the test, a fraction of energy is absorbed (E", viscous component), whereas the other fraction is returned elastically) E', elastic component).

Figure 5 shows the DMA traces (storage modulus versus temperature) of EVA and EVA nano-composites. Pure EVA sample (S1) showed lower storage modulus than other samples. By incorporation of tungsten nano into the matrix of EVA sample (S2), a remarkable increase in the value of storage modulus was observed in the rubbery region.

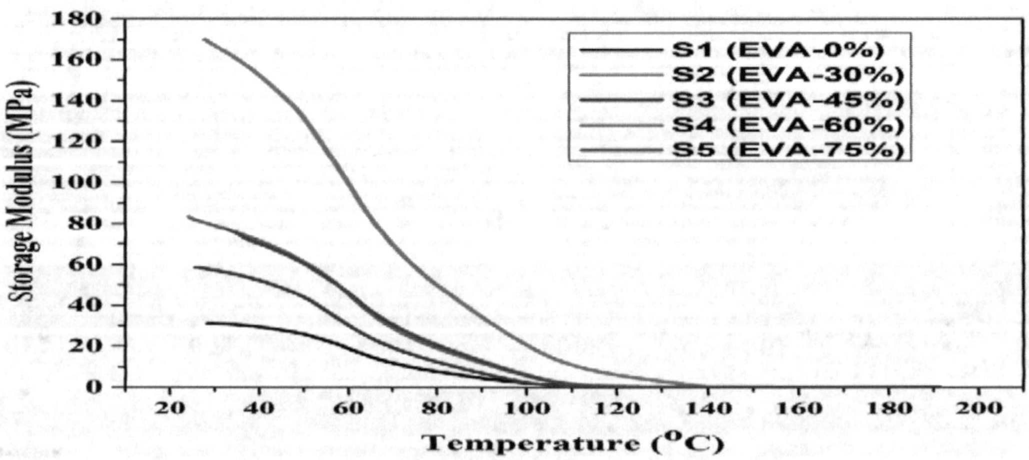

Figure 5. Storage modulus of EVA and its nano-composites

More increase in the filling content leads to decrease in the storage but still higher values than virgin polymer, this means that 30% nano-filler is sufficient to give the best optimum condition in the case of EVA. The high storage modulus of EVA nano-composite samples could be explained by the reinforcing role of the tungsten nanoparticles and strong established link. More increase of tungsten filler leads to failure in the storage moduli. This is may be attributed to the aggregation and agglomeration of nano-filler in the matrix of EVA.

4 Conclusions

EVA composite samples with different percentage of Tungsten nano Powder (0 wt%, 30/wt%, 45/wt%, 60/wt%, and 70/wt%) were prepared by melt blending in a twin-screw brabender mixer. The structural, morphological, thermal and mechanical properties of the prepared nanocomposites were investigated. It was found that the prepared composite have homogeneous dispersion of the reinforcement particles within the composite matrix. The thermal and dynamic properties of the composite were enhanced with the increase of tungsten concentration due to the high melting point of tungsten.

References

[1] Metal–polymer nanocomposites G. HENESS, University of Technology, Sydney. Australia
[2] R. Wilson, S.M. George, H.J. Maria, T.S. Plivelic, S.A. Kumar, and S. Thomas 2012, Phys. Chem. C, 116, 20002.
[3] R. Wilson, T.S. Plivelic, A.S. Aprem, C. Ranganathaiagh, S.A. Kumar, and S. Thomas 2012, Appl. Polym. Sci., 123, 3806.
[4] C. Hoffendahl, G. Fontaine, S. Duquesne, F. Taschner, M. Mezger, and S. Bourbigot 2015, Polym. Degrad. Stab., 115, 77.
[5] D.S. Chaudhary, R. Prasad, R.K. Gupta, and S.N. Bhattacharya 2005, Polym. Eng. Sci., 45, 889.
[6] S. Kim and L.T. Drzal 2009, Adhes. Sci. Tech., 23, 1623.
[7] Y. Tian, H. Yu, S. Wu, G. Ji, and J. Shen 2004, Mater. Sci., 39, 4301.
[8] B. Ramaraj and K.R. Yoon 2008, Appl. Polym. Sci., 108, 4090.
[9] Y. Yu, S. Qi, J. Zhan, Z. Wu, X. Yang, and D. Wu 2011, Mater. Res. Bull., 46, 1593.
[10] M. Liu, M. Pu, and H. Ma 2012, Compos. Sci. Technol., 72, 1508.
[11] N. Garcıa, M. Hoyos, J. Guzman, and P. Tiemblo 2009, Polym. Degrad. Stab., 94, 39.
[12] Marius C. Costache, David D. Jiang, Charles, WI 53201, USA, May 2005.

IOP Conf. Series: Materials Science and Engineering **975** (2020) 012004 doi:10.1088/1757-899X/975/1/012004

Catalytic oxidation over nanostructured heterogeneous process as an effective tool for environmental remediation

Rahma H. Thabet[1], Maha A. Tony[1*], Shakinaz A. El Sherbiny[2], I. A. Ali[3], Mai K. Fouad[2]

[1]Advanced Materials/Solar Energy and Environmental Sustainability (AMSEES) Laboratory, Basic Engineering Science Department, Faculty of Engineering, Menoufia University, Shebin El-Kom, Egypt.
[2]Chemical Engineering Department, Faculty of Engineering, Cairo University, Egypt
[3]Cyclotron project, Nuclear Physics Department, Nuclear Research Center, Atomic Energy Authority, Egypt
E-mail: dr.maha.tony@gmail.com

Abstract. Industrialization has led to a severe deterioration in water quality. Textile industry is considered a huge consumer of water in Egypt; the result is generating large amounts of dye-containing wastewater that is essential to be treated before the final disposal. However, searching for efficient treatment is an important aspect for a sustainable environment. Advanced oxidation processes (AOPs) have been emerged as efficient techniques for industrial wastewater remediation. Among the AOPs, Fenton based reactions is considered a promising process for its simplicity in application and cost-efficient with high process efficiency. In this study, heterogeneous Fenton reaction using magnetite nanoparticles induced by ultraviolet radiation (UV) was applied as a green technology pathway for textile dyeing wastewater oxidation. Nanostructured magnetite was successfully obtained by co-precipitation technique that is used as the precursor of the Fenton's reaction process. The heterogeneous iron (Fe^{2+}/Fe^{3+}) supported catalyst with hydrogen peroxide (H_2O_2) was used as a coupled Fenton and Fenton-like oxidation system for methylene blue dye removal in aqueous media. The obtained results investigated that the dye oxidation rate increases with decreasing pH to 3.0. However, increasing H_2O_2 and magnetite (Fe_3O_4) nanoparticles catalyst results in an increase in the dye oxidation rate and the optimum operating values were 80 and 1600 mg/L for Fe_3O_4 and H_2O_2, respectively. By optimizing the amount of reagents, process conditions as well, the results revealed that magnetite was considered an efficient Fenton-based catalyst for dye oxidation that is reached to 94% within 3 hr of oxidation time. Finally, magnetite catalyst could be easily recovered by magnetic separation to confirm the process sustainability.

Keywords: Fenton's reagent, Methylene blue, Wastewater, Nanomaterial, Magnetite

1. Introduction

Egypt is listed as a country with a rapid population growth and is known as one of those countries that have a gap between water supply and demand [1]. In addition, a large quantity of water is consumed as a result of industrialization and a massive amount of wastewater is produced by processing and the result is a danger to the environment [2]. Therefore, the major problem affecting water scarcity and human health is wastewater generation from industrial sectors [3], domestic[4] and agricultural[5] effluents. It is also in the interests of many researchers to search for a viable solution [6].

Textile industry is considered one of the industrial sectors responsible of discharging a highly contaminated colored wastewater. Such effluents are released into the environment causing the aquatic system to severely deteriorate [7, 8]. Thus, the treatment of fabric effluents is important for a clean environment and also for the protection of water supplies.

Dyes could be classified according to their application to c dyes. However, according to the chemical structure they could classified into azo, triarylmethanes, diphenyl-methanes, anthraquinones, stilbenes, methines, polymethines, xanthenes, phthalocyanines, sulfurs,..etc [9]. Therefore, the presence of such chemicals decreases the transparency of the water and inhibits the penetration of solar radiation. Thus, the result is decreasing the coloured photosynthetic activity and cause disruptions in the gases solubility, causing damage to the gills of the aquatic organisms. These chemicals can remain for several decades in the aquatic environments causing severe damage to the ecosystems. Also, causing danger to human beings and other living organisms. Based on various studies by different researches, there are numerous literature cited dealing with Fenton's reagent for treating several types of wastewater that contaminated dye effluents [10]. Such as Hammami et al. [11] treated the wastewater polluted with Direct Orange 61, Wang et al. [12]studied wastewater contaminated with Acid Red 14, Ali et al. [13] removed Basic Blue 3 dye from waste effluents and Bocos et al. [14] treated Reactive Black 5 contaminated wastewater. Moreover, Gulkaya et al. [15] treated wastewater effluent with real industrial dyestuff wastewater. Methylene blue (MB) that is a cationic dye is widely used as a coloring agent in coloring, painting and wool dyeing. The wastewater effluent containing this dye is causing hazard to the surrounding environment [16-18].

Several conventional wastewater techniques have been widely applied for wastewater treatments, such as flocculation [19], adsorption [20], ion exchange [21], filtration [22] and reverse osmosis [23]. However, those techniques may be signified by the incomplete mineralization of pollutants since they are only transferring the pollutant from one phase to another without destroying them. Therefore, a secondary treatment is needed [24, 25]. In recent two decades advanced oxidation processes (AOPs) have recognized as an effective method for treatment of wastewater. Such AOPs were first suggested for treatment of potable water in the 1980s [26, 27]. AOPs are characterized by the formation of highly reactive hydroxyl radicals (•OH) in appropriate amounts to oxidize wastewater contaminants and purify water [26]. When AOPs are introduced for the treatment of wastewater, (•OH) radicals are needed to effectively oxidize the contaminants as an efficient oxidizing agent for transformation into less and even non-toxic substances (CO_2 and H_2O). Thus, this is considered the best solution for wastewater treatment [28, 29]. The main processes of producing these radicals are including: UV [30], O_3/H_2O_2 [29], O_3/UV, [29, 30] TiO_2 photo-catalysis [31, 32], Fe^{3+}/UV [33] besides Fenton and photo-Fenton processes [34].

Fenton's reagent, i.e. Fe^{2+} and H_2O_2 is known as an efficient treatment technique as it implies high efficiency in the oxidation of organic compounds at low cost, in addition to the lack of toxicity with reagents. [35].

The photo-Fenton process is an oxidation mechanism that incorporates the Fenton reaction and ultraviolet light to oxidize ferrous ions Fe(II) to ferric ions Fe(III), as well as the decomposition of H_2O_2 into •OH radicals as in Eq. (1). Ferric ions are the source of iron in the photo-Fenton process; UV irradiation helps to regenerate the ferrous ions by equations (2-4) called Fenton-like process, which are responsible for the continuation of the Fenton reaction [36].

$$H_2O_2 + \mathrm{Fe}^{2+} \rightarrow \mathrm{Fe}^{3+} + OH^* + OH^- \tag{1}$$

Fe^{3+} can react with H_2O_2 in the so-called Fenton-like reaction:

$$Fe^{3+} + H_2O_2 \rightarrow FeOOH^{2+} + H^+ \tag{2}$$
$$FeOOH^{2+} \rightarrow Fe^{2+} + OH^* \tag{3}$$
$$Fe^{3+} + OOH^* \rightarrow Fe^{2+} + O_2 + H^+ \tag{4}$$

The higher concentration of Fe^{2+} ions can improve the rate of decomposition of organic pollutants since more hydroxyl radicals are produced according to Eq. (1) [37].

Magnetite (Fe_3O_4) is a semimetal semiconductor with a band gap 0.14 eV. It contains both iron ions, i.e. Fe^{2+} and Fe^{3+} cations. Consequently, magnetite could be considered an efficient photocatalyst. Various studies have been undertaken to treat wastewater using magnetite to remove organic and inorganic pollutants [38].

Recently, nano-compounds have been introduced to the field of photocatalysis owing to its efficient removals. Nano-catalysts is applied to the Fenton systems in the form of iron nanoparticles to treat various contaminants such as emerging pollutants and in dyes oxidation [39]. Their addition to the system showed a significant enhancement to the photocatalytic oxidation reaction [40].

Magnetite, (Fe_3O_4), nanoparticles have got a great attention not only for their super paramagnetic properties, but also for their low toxicity to human body. However, based on the authors' knowledge, it is not applied so far for treating wastewater polluted with the textile effluents. Therefore, in the present investigation, magnetite (Fe_3O_4) nano-crystalline powder was synthesized using low-cost simple co-precipitation technique. Based on such nano-powder as photo-catalysis, ultraviolet illumination (UV) augmented with nano-magnetite Fenton's reagent (NMFR) was used to treat Methylene Blue dye-containing wastewater. The NMFR parameters were investigated, and the process was compared with other treatment processes.

2. Materials and Methods

2.1. Chemicals and reagents

Methylene Blue (MB) dye was obtained from Sigma-Aldrich Co, (Germany). The nano-Magnetite was prepared from the precursors of ferrous sulfate ($FeSO_4 \cdot 7H_2O$) and Ferric Sulfate $Fe_2(SO_4)_3$; which were all obtained from Qualikems Fine Chem Pvt. Ltd (India). Hydrogen peroxide (H_2O_2) 40% w/w. Sulfuric acid and sodium hydroxide were used for pH adjustment. All chemicals were of analytical grade and were used as received without further purification or treatment. Distilled water was used throughout this study.

2.2. Preparation of dye solution

The dye solution was prepared by dissolving 1.0 g of (MB) in 1.0 L of distilled water to produce a the stock solution of 1000 mg/L of the dye solution, which is further diluted to attain the required concentration for treatment.

2.3. Experimental procedure

2.3.1. Synthesization of Fe_3O_4 nanoparticles

The co-precipitation method has been used to synthesize Fe_3O_4 nanoparticles samples. This technique is reported elsewhere [41]. A 2 mol of $Fe_2(SO_4)_3$ and 1 mol of $Fe(SO_4)$ were mixed according their stoichiometric ratios in double-distilled water of 1.5 L volume. More than 8 mol of sodium hydroxide (NaOH) solution was then added dropwise with stirring until the pH of the solution reached 11.0. The precipitate solution was heated to 80 ^{o}C with continuous stirring. The prepared nanoparticle was washed many times afterwards with distilled water to remove the sodium hydroxide and sodium sulfate until the pH reached 7.0. Finally, the produce Fe_3O_4 was dried in an electrical oven at 60 °C and manually grounded.

2.3.2. Fe_3O_4 NPs Characterization

For the X-ray diffraction analysis, the Fe_3O_4 precipitate was dried to be a powder and analyzed using an X-ray diffractometer. The instrument used for the analysis was X-lab Shimadzu X-6000), and identified with Cu-Kα radiation.

2.3.3. Photocatalytic methodology

Methylene Blue (MB) photo-oxidation process was studied using 100 mL of 10 mg/L dye-wastewater solution. The synthesized magnetite nano-powder under UV illumination (254 nm wavelength), as illustrated in figure 1 was subjected to the system and the reaction is initiated by the hydrogen peroxide addition.

Firstly, the pH of the solution was adjusted if needed by adding sulfuric acid and/or sodium hydroxide using a digital pH-meter (AD1030, Adwa instrument, Hungary). Adding different dosage of hydrogen peroxide H_2O_2 to the solution to initiate the Fenton reaction. Prior to the UV illumination, the desired amount of the nano-Fenton's reagent was dispersed in the MB aqueous solution using sonication for 5 min. Thereafter, the reaction is subjected to the Ultraviolet illumination using 12-W UV lamp. Then, within a specific time interval, samples are filtered using a micro filter then subjected to spectrophotometer (UV-1601, Shimadzu, Model TCC-240A, Japan) to monitor the dye concentrations remaining in solution.

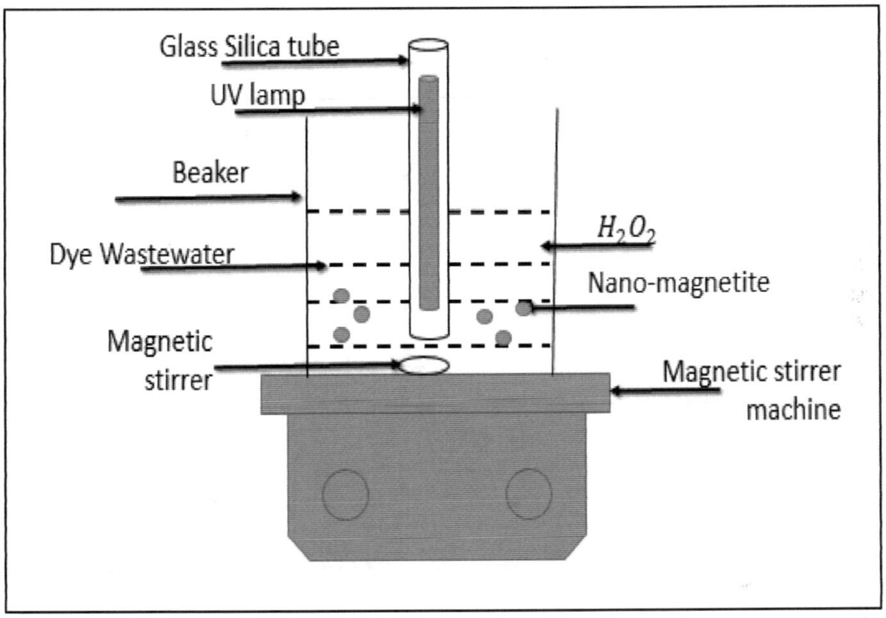

Figure 1. Schematic diagram of lab-scale photo-Fenton for MB removal

3. *Results and discussions*:

3.1 XRD analysis and Morphological characterization of the synthesized Fe_3O_4 nano-crystalline powder

Firstly, it is worthy to note that according to the preliminary results, the XRD pattern for the Fe_3O_4 shows diffraction peaks at 2θ (degrees) of 30.15°, 35.61°, 37.27°, 43.25°, 53.73°, 57.37°, 62.81°, 71.33°, 74.57°, 75.51° and 79.31 which correspond to the miller indices (220), (311), (222), (400), (422), (511), (440), (620), (533), (622) and (444), respectively (graph is not shown) indicate a single phase of spinel structure for nano-magnetite.

A high-resolution Transmission Electron Microscope (TEM) has been used to study the particle size of the Fe_3O_4.

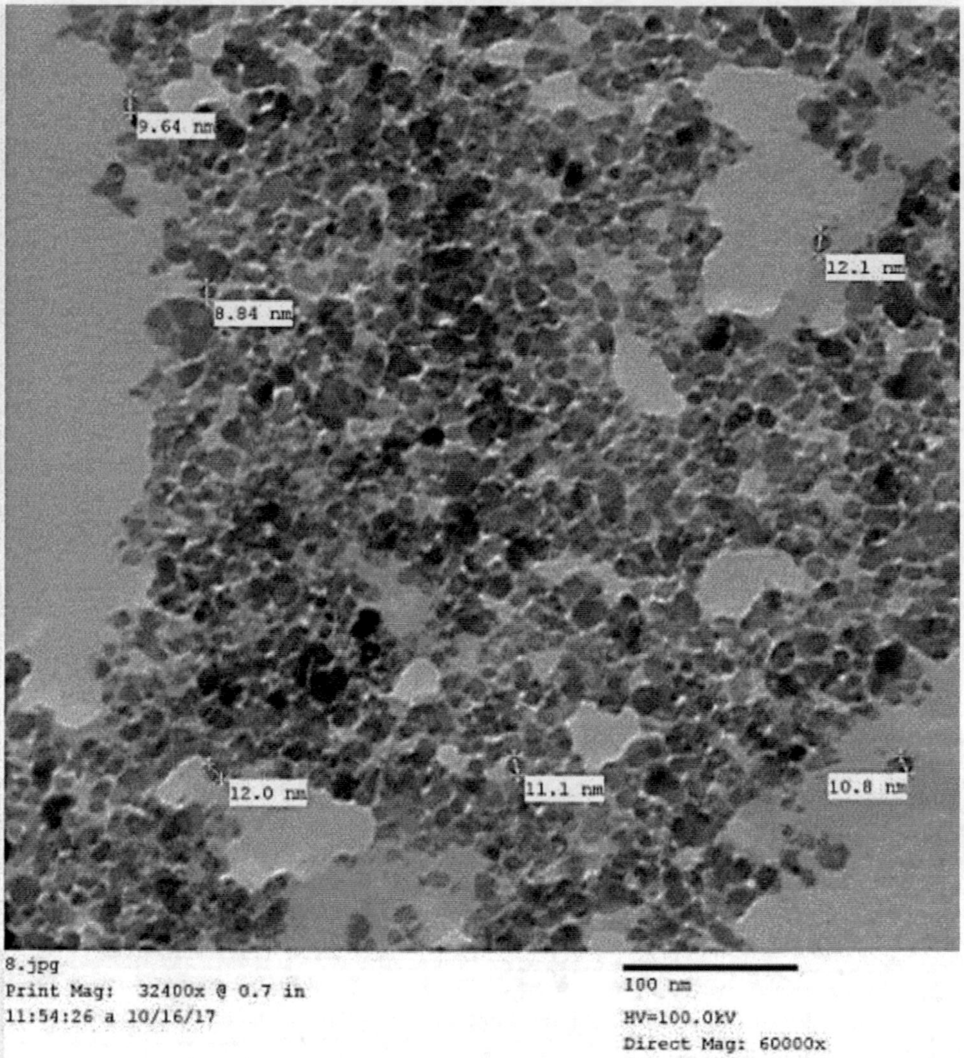

Figure 2. Tem Image of the synthesized Fe_3O_4 nanoparticles

Figure 2. shows the TEM micrographs of the synthesized Fe_3O_4 nanoparticles demonstrated that the shape of most of the nanoparticles are in spherical shape with except for small edges attached to certain nanoparticles. The average particle size estimated from the TEM images is in the range of 9-12 nm. This size gives a reasonable high surface area for the nanoparticles to perform the photocatalytic oxidation reaction.

3.2. Comparison of different oxidation systems

Firstly, three different experiments were conducted separately for eliminating dye-contaminated wastewater in order to investigate and compare the performance of the systems with the photo-Fenton system.

The results in figure 3 show the comparison of different oxidation systems with the performance of nano-magnetite Fenton reagent using the UV illumination. Analysis of the results reveals that the UV photolysis (UV-vis), for 10 ppm MB dye solution concentration and pH 3.0, removes only a 7% of MB within 60 min of reaction time. Nevertheless, the sole use of 40 mg/L of magnetite with UV radiation without H_2O_2 addition, which is called iron photo-assisted system Fe^{3+}/UV-vis achieved 68% of MB oxidation. This is due to the hydroxyl radicals which are formed via an intermolecular photo-redox process in Fe^{3+} excited hydroxy-complexes ($Fe(OH)^{2+}$) which is the most active complex in terms of $HO^{\bullet}$ formation. Such radicals are the responsible of attacking aromatic rings consisted on MB molecule . Therefore, opening the rings and creating reaction intermediates, that are ultimately converted to harmless end products, i.e. CO_2 and H_2O [2, 7]. However, when Fenton's reagent doses were used at: magnetite 60 mg/L; hydrogen peroxide 1600 mg/L at a starting pH of 3.0 under the UV illumination (photo-Fenton process), the dye removal rate reached 95.3%. This could be illustrated by the presence of UV illumination with the nanocrystals as the source of Fenton's reagent gives higher oxidation rate, since higher hydroxyl radicals were produced that are considered the main responsible of the dye oxidation [19, 34].

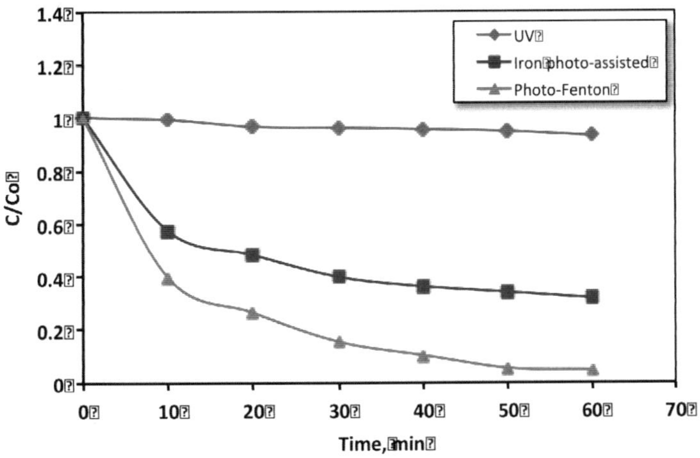

Figure 3. Comparison of different oxidation systems on the removal of MB

3.3 Effect of photo-Fenton's parameters

3.3.1. Effect of catalyst dose.

In order to investigate the role of initial concentration of iron ions on the oxidation of MB, series of experiments were conducted with different catalyst does, i.e. 40, 60, 80 and 200 mg/l and their results are observed in figure 4. The results recorded in figure4. revealed that the oxidation of MB is remarkably dependent on the iron ions concentrations at fixed initial concentration of hydrogen peroxide (1600 mg/l) and pH 3.0.

At low iron concentration (40 mg/L), the oxidation efficiency was 92% after 60 min of reaction time. Whereas, for 60 mg/L, 80 mg/L and 200 mg/L, the oxidation efficiency are 95.29, 93.97 and 95.52% during 60 min of reaction time, respectively.

The results showed that 200 mg/L of the nano powder ions gives a maximum removal of MB dye 95.52%, Furthermore, 60 mg/L reveals an almost removal rate. Therefore, considering the need to remove iron ions from the solution before discharge and with economic point of view, low catalyst consumption is preferred, the optimal iron ions concentration was selected to be 60 mg/L. This result is in accordance with, the work of Hariani et al. [40] who revealed that the high concentration of Fe_3O_4 competing the hydraoxyl radicals for attaching the dye molecules . Thus, the overall reaction rate is scavenged [42, 43] and the organics oxidation is declined. Additionally, with increasing the iron concentration, the catalytic efficiency increases. This could be attributed by the presence of Fe^{3+} in high concentration in the solution, higher amounts of UV illumination is absorbed since it has a strong absorbent band. However, higher amounts of the reagent rinders the oxidation rate since it forms $Fe(OH)^{2+}$ [44]. Furthermore, Fe^{3+} ions may react with hydrogen peroxide reagent to form hydroperoxyl radicals that has a less oxidizing efficiency than •OH radicals [33].

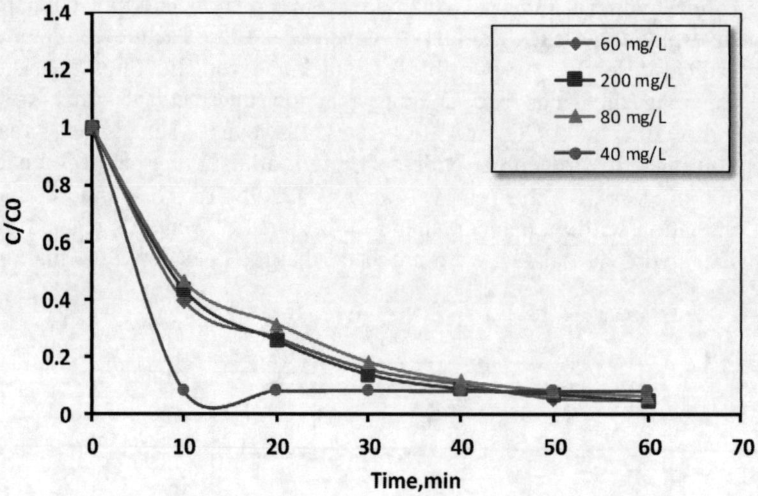

Figure 4. Effect of the concentration of Nano-magnetite on the removal of MB, [MB]o= 10 mg/L, H_2O_2= 1600 mg/L, pH 3.0.

3.3.2. Effect of hydrogen peroxide dose.

According to the literature cited [4, 5], hydrogen peroxide is described as an important parameter in the Fenton's reaction system. figure 5 shows the effect of increasing hydrogen peroxide reagent on the MB removal rate while keeping the other parameters fixed (pH 3.0 and iron dose of 40 mg/L). Various initial H_2O_2 concentrations ranged from 100 to 3000 mg/L was studied. The results show that by increasing the hydrogen peroxide concentration from 100 to 1600 mg/L, MB removal increased from 40.3 to 91.9% after 60 minutes of illumination time. It might be due to the increasing of •OH with the addition of hydrogen peroxide. The effect of high dose of H_2O_2 on Fenton's reaction is previously reported by several authors [45, 46].

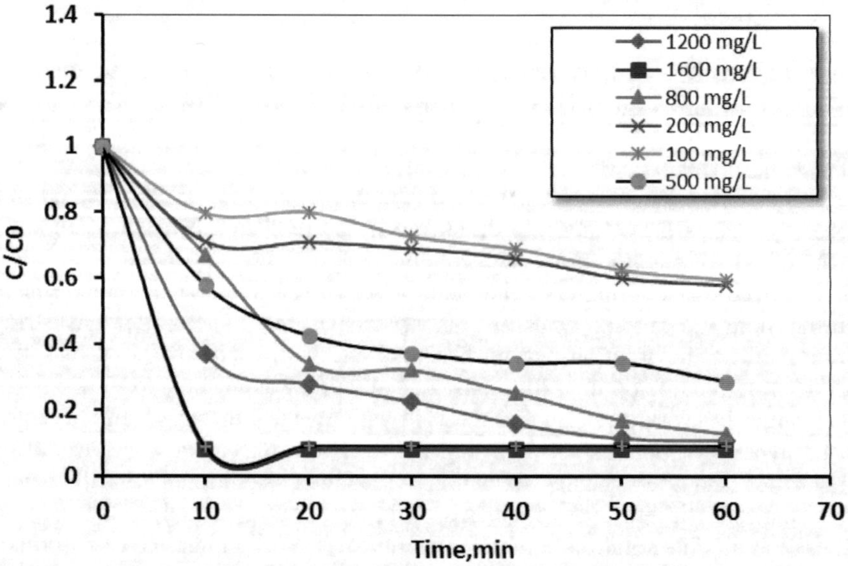

Figure 5. Effect of the concentration of H_2O_2 on the removal of MB, [MB] = 10 mg/L, iron = 40 mg/L, pH 3.0

IOP Conf. Series: Materials Science and Engineering **975** (2020) 012004 doi:10.1088/1757-899X/975/1/012004

Additionally, further increase in H_2O_2 concentration (3000 mg/L) rather than the optimal value (1600 mg/L), the oxidation rate is decreased. This may be attributed by the excess hydrogen peroxide reagent will react with •OH radicals instead of generating them equation 6, Also perhydroxyl radicals (•OOH) are produced in the solution. This radicals have much less potential for oxidation than •OH. Additionally •OH radical was consumed to H_2O and O_2 according to Equations (7, 8). Thus, the overall reaction rate is declined according to the excess initial hydrogen peroxide concentration rather than the optimal dose that is called hydrogen peroxide scavenging effect [4, 19, 21].

$$H_2O_2 + h\upsilon \rightarrow 2OH^* \quad (5)$$
$$OH^* + OH^* \rightarrow H_2O_2 \quad (6)$$
$$H_2O_2 + OH^* \rightarrow H_2O + OOH^* \quad (7)$$
$$OOH^* + OH^* \rightarrow H_2O + O_2 \quad (8)$$

3.3.3. Effect of pH

The pH of solutions is a primary parameter of the treatment efficiency of Fenton's reagent, since it influences the production rate of hydroxyl radical and the concentration of Fe^{2+} [47]. The effect of initial pH value of solutions on the degradation of MB was obtained in the pH range of 3.0- 8.0 and the results are shown in figure 6.

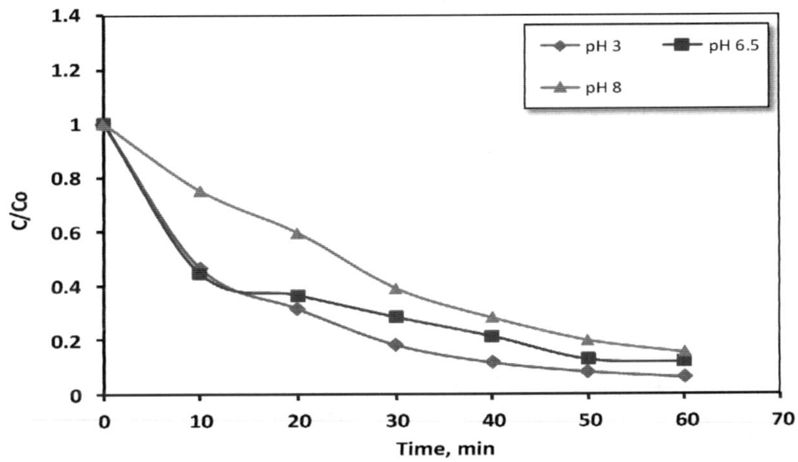

Figure 6. Effect of pH on the removal of MB, $[MB]_o$= 10 mg/L, iron= 60 mg/L, $[H_2O_2]$=1600 mg /L

The results in figure 6 showed that the dye oxidation was significantly influenced by the solution pH. The optimal pH-value is 3.0. At this pH value and within 60 minutes of reaction time, there is a 93.9% removal of MB compared to the removal rate of 84.8%, which obtained at pH 8.0. This could be illustrated by, at low pH (2.0-4.0), the oxidation reaction may be reduced because hydrogen peroxide will remain stable, potentially by dissolving a proton in the presence of a high concentration of H^+ forming an ion of oxonium (H_3O^+). An oxonium (H_3O^+) ion makes H_2O_2 electrophilic to be stable and likely greatly to reduce the reactivity of Fe (II) ion. The results reported in literature [49-50], displayed that low activity of Fenton's reagent is attained at high pH values since undesired radicals are produced rather highly reactive •OH radicals. Thus, the reaction rate is scavenged.

Moreover, Fe^{3+} complexes are formed in the reaction media, which decrease the dissolved iron and also declines the oxidation process.

4. Conclusion

Nano-Magnetite powder can be efficiently prepared using co-precipitation method and it is attained a good photo reactivity as a Fenton's reagent Photo-Fenton catalyst for the oxidizing Methylene Blue dye (MB) dye in aqueous media. MB oxidation via Fenton process was investigated in the current work and the results revealed that the solution pH, initial H_2O_2 concentration and initial iron concentration are the main influencing parameters in the dye oxidation through the Fenton system. The optimal operating parameters for the Fenton oxidation of MB are 1600 mg/L and 60 mg/L for H_2O_2 and iron nanoparticles, respectively, at an initial pH of 3.0. Under such conditions, 95.3% oxidation of the MB dye in the aqueous solution is attained within 3 hrs. of reaction time. Hence, this study is recommending the use of nano-magnetite for its low toxicity effect and simplicity in removal from aqueous media. The study suggests that the catalyst could be reused for successive cycles; however, further research is needed to signify this investigation.

5. References

[1] Ashour E, Tony MA, Purcell PJ 2014 Use of agriculture-based waste for basic dye sorption from aqueous solution Kinetics and isotherm studies. Amer. J. Chem. Eng. 2 (6) 92-98.
[2] Tony M. A. 2019 An Industrial Ecology Approach: Green Cellulose Based Bio-Adsorbent From Sugar Industry Residue For Treating Textile Industry Wastewater Effluent. Inter. J. Environ. Anal. Chem.1-18
[3] Tony M. A. 2020 Central composite design optimization of Bismarck Dye oxidation from textile effluent with Fenton's reagent. Appl. water Sci. 10(5), 108.
[4] Tony MA, Parker HL, Clark JH 2016 Treatment of laundrette wastewater using Starbon and Fenton's reagent. J. Environ. Sci. Health, Part A 51(11) 974-979.
[5] Tamimi M, Qourzal S, Assabbane A, Chovelon JM, Ferronato C, Ait-Ichou Y 2006 Photocatalytic degradation of pesticide methomyl: determination of the reaction pathway and identi- fication of intermediate products. Photochem. Photobiol. Sci. 5 477–482
[6] Tony M A, Zhao Y Q and El-Sherbiny M F 2011 Fenton and Fenton-like AOPs for alum sludge conditioning: Effectiveness comparison with different Fe2+ and Fe3+ salts Chem. Eng. Commun. 198(3) 442-452
[7] Rahman M A, Amin S M R and Alam A M S 2012 Removal of Methylene Blue from Waste Water Using Activated Carbon Prepared from Rice Husk Dhaka Univ. J. Sci. 60(2) 185-189.
[8] Ashour E, Tony MA 2017 Equilibrium and Kinetic Studies on Biosorption of Iron (II) and Iron (III) Ions onto Eggshell Powder from Aqueous Solution. Applied Eng 1(3) 65-73.
[9] Booth G., Zollinger H., McLaren K., Sharples W.G., Westwell, A. 2000 Dyes, General Survey. Ullmann's Encycl. Ind. Chem.
[10] Pinheiro, H. M., Touraud E., and Thomas O., 2004 Aromatic amines from azo dye reduction: Status review with emphasis on direct UV spectrophotometric detection in textile industry wastewaters. Dye. Pigment., 61, 121–139
[11] Hammami, S., Oturan N., Bellakhal N., Dachraoui M., and Oturan M. A., 2007: Oxidative degradation of direct orange 61 by electro-Fenton process using a carbon felt electrode: Application of the experimental design methodology. J. Electroanal. Chem., 610, 75–84
[12] Wang A., Qu, J., Ru J., Liu H., Ge J., 2005 Mineralization of an azo dye Acid Red 14 by electro-Fenton's reagent using an activated carbon fiber cathode. Dye. Pigment., 65, 227–233
[13] Ali O. ¸Sahin Y., Koparal A.S., Oturan M.A., 2008 Carbon sponge as a new cathode material for the electro-Fenton process: Comparison with carbon felt cathode and application to degradation of synthetic dye basic blue 3 in aqueous medium. J. Electroanal. Chem., 616, 71–78
[14] Bocos E., Pazos M., Sanromán M.A., 2014: Electro-Fenton decolourization of dyes in batch mode by the use of catalytic activity of iron loaded hydrogels. J. Chem. Technol

[15] Gulkaya I, Surucu GA, Dilek FB, 2006 Importance of H2O2/Fe2+ ratio in Fenton's treatment of a carpet dyeing wastewater. J. Hazard. Mater., 136, 763–769.
[16] Sareen D, Garg R and Grover N 2014 A Study on Removal of Methylene blue dye from waste water by Adsorption Technique using fly ash briquette Int. J. Chem. Environ. Eng. 5(7) 610-613
[17] Ehrampoush M H, Moussavi G R, Ghaneian M T, Rahimi S and Ahmadian M 2011 Removal of methylene blue dye from textile simulated sample using tubular reactor and TiO 2/UV-C photocatalytic process Iran. J. Environ. Heal. Sci. Eng. 8(1) 35–40
[18] Tony M A, Purcell P J, Zhao Y, Tayeb A M and El-Sherbiny M F 2018 Kinetic Modeling Of Diesel Oil Wastewater Degradation Using Photo-Fenton Process Environ. Eng. Manag. J. 14(1) 11-16.
[19] Tony MA, Purcell PJ, Zhao YQ 2012 Oil refinery wastewater treatment using physicochemical, Fenton and Photo-Fenton oxidation processes. J. Environ. Sci. Health, Part A, 47 435–440.
[20] Tony MA 2020 Zeolite-based adsorbent from alum sludge residue for textile wastewater treatment. Inter. J. Sci. Technol. 17 2485–2498
[21] Chakraborty R., Asthana A., Singh A. K., Jain B., and Susan A. B. H. 2020 Adsorption of heavy metal ions by various low-cost adsorbents: a review. Int. J. Environ. Anal. Chem., 1–38,
[22] Tayeb AM, Farouq R, Mohamed O, Tony MA 2019 Oil spill clean-up using combined sorbents: a comparative investigation and design aspects, Inter. J. Environ. Anal. Chem. 1-13
[23] Abid M.F., Zablouk M.A., Abid-Alameer A.M., 2012 Experimental Study Of Dye Removal From Industrial Wastewater by Membrane Technologies Of Reverse Osmosis and Nanofiltration. J. Environ. Health Sci. Engineer. 9(17) 1-9
[24] Samira S and Raja P A 2012 Photocatalytic Degradation of Crystal Violet (C.I. Basic Violet 3) on Nano TiO2 Containing Anatase and Rutile Phases (3:1) J. Thermodyn. Catal. 3(5) 117–119
[25] Glaze W H 1987 Drinking-water treatment with ozone. Environ. Sci. Technol. 21(3) 24-30
[26] Zhao YQ, Keogh C, Tony MA 2009 On the Necessity of Sludge Conditioning with Non-organic Polymer: AOP Approach. J. Res. Sci. Technol. 6(3) 151-155.
[27] Coelho A, Castro A V., Dezotti M and Sant&apos Anna G L 2006 Treatment of petroleum refinery sourwater by advanced oxidation processes J. Hazard. Mater. 137(1) 78–184
[28] Andreozzi R, Caprio V, Insola A, Marotta R and Sanchirico R 2000 Advanced oxidation processes for the treatment of mineral oil-contaminated wastewaters Water Res. 34(2) 620–628.
[29] Coelho A, Castro A V., Dezotti M and Sant&apos Anna G L 2006 Treatment of petroleum refinery sourwater by advanced oxidation processes J. Hazard. Mater. 137(1) 78–184
[30] Kositzi M, Poulios I, Malato S, Caceres J and Campos A 2004 Solar photocatalytic treatment of synthetic municipal wastewater Water Res. 38(5) 1147-1154
[31] Gernjak W, Maldonado M L, Malato S, Cáceres J, Krutzler T, Glaser A and Bauer R 2004 Pilot-plant treatment of olive mill wastewater (OMW) by solar TiO_2 photocatalysis and solar photo-Fenton Sol. Energy 77(5) 567-572.
[32] Tony MA, Zhao YQ, Purcell PJ, EL-Sherbiney MF 2009 Evaluating the photo-catalytic application of Fenton's reagent augmented with TiO2 and ZnO for the mineralization of an oil-water emulsion, J. Environ. Sci. Health Part A 44 488–493.
[33] Tony MA, Lin LS 2020 Iron recovery form acid mine drain sludge as a Fenton source for municipal wastewater treatment, Inter. J. Environ. Anal. Chem 1-15
[34] Aplin R, Waite T D 2000 Comparison of Three Advanced Oxidation Processes For Degradation of Textile Dyes Water Sci. Technol. 42(5-6) 345-354.
[35] Babuponnusami A, Muthukumar K 2014 A review on Fenton and improvements to the Fenton process for wastewater treatment J. Environ. Chem. Eng. 2(1) 557-572.
[36] Pereira M C, Oliveira L C A and Murad E 2012 Iron oxide catalysts: Fenton and Fenton like reactions – a review Clay Miner. 47(3) 285-302.
[37] Reza K M, Kurny A and Gulshan F 2016 Photocatalytic Degradation of Methylene Blue by Magnetite+H2O2+UV Process Int. J. Environ. Sci. Dev. 7(5) 325-329.
[38] Sirdeshpande K D, Sridhar A, Cholkar K M and Selvaraj R 2018 Structural Characterization Of

Mesoporous Magnetite Nanoparticles Synthesized Using the Leaf Extract of Calliandra Haematocephala and their Photocatalytic Degradation of Malachite Green Dye Appl. Nanosci. 8(4) 675-683.

[39] Tony M A, Parker H L and Clark J H 2019 Evaluating Algibon Adsorbent and Adsorption Kinetics for Launderette Water Treatment: Towards Sustainable Water Management Water Environ. J. 33(3) 401-408.

[40] Hariani P L, Faizal M, Ridwan R, Marsi M and Setiabudidaya D 2013 Synthesis and Properties of Fe3O4 Nanoparticles by Co-precipitation Method to Removal Procion Dye Int. J. Environ. Sci. Dev. 4(3) 336-340

[41] Mazellier P and Bolte M 2001 3-chlorophenol elimination upon excitation of dilute iron(III) solution: Evidence for the only involvement of Fe(OH)2+ Chemosphere 42(4) 361-366.

[42] Chenini H, Djebbar K, Zendaoui S M, Sehili T and Zouchoune B 2011 Removal of an Azo Dye (Orange G) By Various Methods in Homogenious Phase. Comparative Study Jordan J. Chem. 6(3) 307–19

[43] Modirshahla N, Behnajady M A and Ghanbary F 2007 Decolorization and mineralization of C.I. Acid Yellow 23 by Fenton and photo-Fenton processes Dye. Pigment. 73(3) 305-310.

[44] Neyens E and Baeyens J 2003 A review of classic Fenton's peroxidation as an advanced oxidation technique J. Hazard. Mater. 98(1-3) 33-50.

[45] Gulkaya I, Surucu G A and Dilek F B 2006 Importance of H2O2/Fe2+ ratio in Fenton's treatment of a carpet dyeing wastewater J. Hazard. Mater. 136(3) 763–769

[46] Neyens E and Baeyens J 2003 A review of classic Fenton's peroxidation as an advanced oxidation technique J. Hazard. Mater. 98(1-3) 33-50.

[47] Kumar A, Rana A, Sharma G, Naushad M, Dhiman P, Kumari A and Stadler F J 2019 Recent advances in nano-Fenton catalytic degradation of emerging pharmaceutical contaminants J. Mol. Liq. 290 111177.

[48] Kato S and Walling C 1971 The Oxidation of Alcohols by Fenton's Reagent. Effect of Copper Ion J. Am. Chem. Soc. 93(17) 4275–4281.

[49] Tayeb AM, Tony MA, Ismaeel EK 2019 Engineered nanostructured ZnO for water remediation: operational parameters effect, Box–Behnken design optimization and kinetic determinations. Appl Water Sci 9 (3): 1-11.

[50] Tony MA, Lin LS 2020 Attenuation of organics contamination in polymers processing effluent using iron-based sludge: Process optimization and oxidation mechanism Environ. Technol. DOI: 10.1080/09593330.2020.1803417

Superior spectral performance of decoy flares compositions via inclusion of graphite as a black body emitter

Sherif Elbasuney[1], **Amr A. Elmotaz**[1], **M. A. Sadek**[2], **M. Yehia**[1] and **Hesham Tantawy** [1]

[1] School of Chemical Engineering, Military Technical College, Cairo, Egypt
[2] Chemical engineering department, British University in Egypt (BUE)
Email: sherif_basuney2000@yahoo.com

Abstract. Graphite particles can offer high interfacial surface area as well as high emissivity as black body; therefore it can find wide applications in advanced decoy flares. In the current study, graphite particles of 100 µm size, were employed with different reactive metal fuels including Mg, Al, and Mg-Al alloy. Different formulations based on graphite ratio up to 8 wt % were developed via granulation with subsequent pressing. Thermal signature was measured using IR spectrometer (1-6 µm) Arc-Optics. Composite flare based on Mg-Al fuel, and graphite demonstrated superior spectral intensity; this formulation offered an increase in average intensity by 83 % to reference MTV flare. In the mean time, composite flare based on Al and graphite offered the highest relative intensity value ϴ of 0.6. Graphite, as an allotrope of carbon, can act as an excellent source of carbonaceous materials; that can strengthen incandescence emission. Furthermore Mg-Al reactive metal alloy can offer novel synergism as Mg could vaporize at 1000 ^{0}C offering efficient combustion process. It can be concluded that tailored flare with customized thermal signature can be developed.

1. Introduction

Decoy flares are effective countermeasures that are ejected from aircraft and separates away to divert away infrared (IR) missile [1]. Decoy flare must provide thermal signature similar to aircraft for effective countermeasure of IR missile. During the last decades, several attempts were directed to enhance performance of traditional Magnesium/Teflon/Viton (MTV) decoy flare, in an attempt to keep up with recent developed IR-missiles. Enhanced decoy flare formulation should secure thermal signature with higher intensity over α band (2-3µm) and β band (3-5µm) [2]. Consequently, IR missile could favor the decoy flare over the aircraft. Another vital factor for IR countermeasure is the average intensity over α band / average intensity over β band ($\Theta = I_\alpha / I_\beta$). Modern decoy flares should offer thermal signature with high intensity and in the mean time should maintain ϴ value similar to aircraft. It has been reported that recommended ϴ value should be in the range from 0.5 to 0.8 according to the aircraft signature. Tailored spectral performance of the traditional MTV composition can be achieved via introducing novel materials for instance silicon [3-7], phosphorous [8-12], boron[2, 6], ytterbium [13], deuterium [14], ferric oxide [1, 7, 15-18], and graphite [2, 19-20]. Graphite is an attractive material as it can act as black body emitter [2, 15, 19]. Black body radiation is the major source of decoy flare emissivity. According to Plank's law, the emissivity of a radiating body can be described by Equation 1:

$$W_\lambda = \frac{2\pi hc^2}{\lambda^5} \frac{1}{e^{\frac{hc}{\lambda KT}}-1} \quad (1)$$

Where: W_λ, λ, h, k, c, and T are the spectral radiance in W cm^{-2} μm^{-1}, wavelength in µm, Plank´s constant 6.626 x 10^{-34} W S^2, Boltzmann constant, velocity of light 2.998 x 10^{10} cm/s, and is absolute temperature in Kelvin respectively. The maximum emission wavelength (λ_{max}) shifts to shorter wavelengths with increase in radiator temperature equation 2.

$$\lambda_{max} T = 2898 \ \mu m.K \quad (2)$$

Traditional metal fuel such Mg can suffer from lack of chemical stability and low heat output. Another metal fuels can offer substantial heat output with an increase in reactive emitting species signature [21-22].

In this study graphite was employed in standard MTV composition with content ranging from 4-8 wt %. Graphite was further employed with different reactive metal fuels such as Al, and Mg-Al alloy. MTV, ATV, and MATV refer to flare based on Mg, Al, Mg-Al alloy respectively. The impact of graphite content on thermal signature was measured using (FT-MIR 1–6 µm) spectrophotometer. Spectral performance was evaluated to standard MTV flare (reference). Whereas MTV based on 4 wt % graphite demonstrated an increase in thermal signature average intensity by 67%; MATV based on 8 wt % graphite offered an increase in average intensity by 83 %. In both cases good average relative intensity ratio (ϴ) in the order of 0.5 was achieved. Superior performances of MATV flare based on 8 wt % graphite can be ascribed to the synergism between the two components (Mg, Al) of bi-metal alloys. Where Mg could act as tinder; and could combust efficiently with substantial heat output. Surplus magnesium could vaporization at 1000 ^{0}C. These features could secure fine environment for efficient combustion of aluminum. High temperature is required for graphite to act as black body emitter; in the mean time the resulted black carbon can further strengthen the developed thermal signature.

2. Experimental

2.1 Materials

Different metal powders including Mg, Al, and Mg-Al alloy were employed as fuels. An energetic fluorocarbon polymer (Teflon) was employed as energetic binder; Viton A was employed as plasticizer. Graphite was employed as black body emitter. The structure, function, and supplier of all employed chemicals are listed in table 1.

Table 1. The function and structure of different used chemicals.

Chemicals	Function	Formula	Grade	Supplier
Teflon	Oxidizer	$[C_2F_4]_n$	n= 20000 Fine powder (Reagent Grade)	Alpha chemika
Magnesium	Fuel	Mg	98%, ribbon (Reagent Grade)	Alpha chemika
Aluminum	Fuel	Al	98%, powder (Reagent Grade), Average particle size (4–7 µm)	Alpha chemika
Viton A	Binder	$C_5H_2F_9X$	Fluorine: 65 % (Reagent Grade), 1.85 g/cm³	Alpha chemika
Acetone	Solvent	C_3H_6O	100% (Reagent Grade)	Sigma Aldrich
Graphite	Enhancer	C	99.5%, flakes (Commercial Grade)	NICE (India)

2.2 Decoy flare formulation

Decoy flare development should secure intimate mixing of different ingredients. Decoy flare grain was developed via granulation and subsequent pressing; this technology includes the following processes.

- Sieving of solid particles to fine powder less than 300 μm
- Teflon as oxidizer and binder along with Viton A® as a plasticizer were dissolved in acetone.
- Granulation was conducted to assure composition homogeneity.
- The final grain was developed by pressing 25 g composition in aluminum mold.

All formulations were developed with the same methodology to avoid any variability that could affect performance. Chemical composition of investigated formulations is tabulated in table 2.

Table 2. Chemical composition of developed formulations.

Sample Number	MTV (Teflon 30% ,Viton 5%)		ATV (Teflon 30% ,Viton 5%)		MATV (Teflon 30% ,Viton 5%)	
	Mg	*Graphite*	*Al*	*Graphite*	*MA*	*Graphite*
1	65 %	---	65 %	---	65%	---
2	61 %	4 %	61 %	4 %	61 %	4 %
3	59 %	6 %	59 %	6%	59 %	6 %
4	57 %	8 %	57 %	8 %	57 %	8 %

2.3 Spectral measurements

Experimental setup for spectral performance measurements includes combustion chamber, optical fibers, IR spectrometer (1-6 μm) Arc-Optics, and data receiving and a recording system (Figure 1).

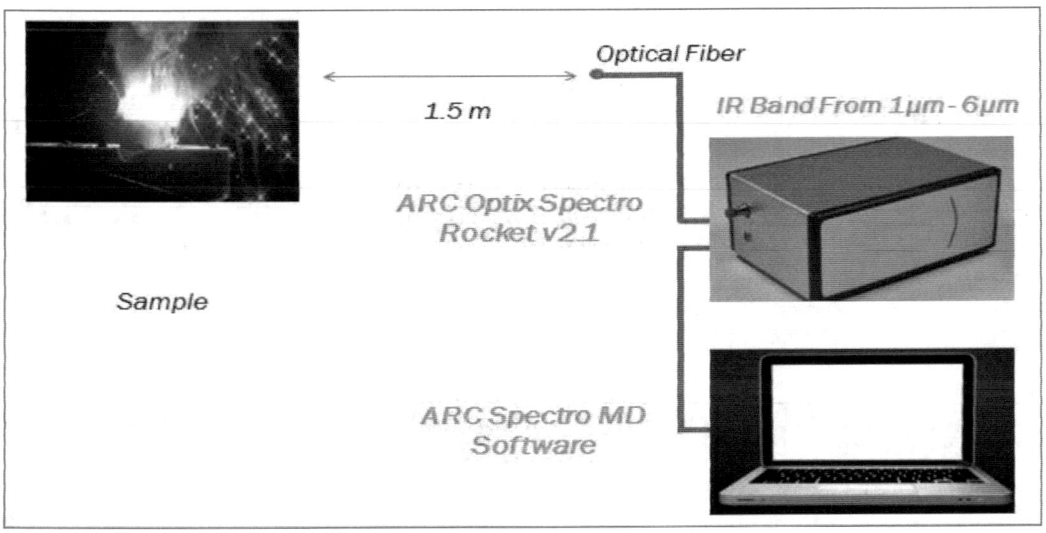

Figure 1. Schematic for spectral measurements of decoy flares.

IOP Conf. Series: Materials Science and Engineering **975** (2020) 012005 doi:10.1088/1757-899X/975/1/012005

IR spectrometer recorded thermal signature with interval time of 6 seconds. The spectrometer response was recorded in counts/s over IR band 1-6 µm. After acquiring the spectral data, the average intensity value was calculated. Additionally the intensity over α and β bands was calculated; consequently the relative intensity ratio $\theta = I_\alpha / I_\beta$ was calculated.

3. Results and discussion

3.1 Characterization of graphite

Graphite was characterized by scanning electron microscopy (SEM) ZEISS, EVO-MA10. Morphological characterization of employed graphite confirms its flake nature figure 2. SEM micrograph demonstrates multiple layers structure; each layer of one µm thickness. Graphite demonstrated flakes of 150 µm lateral dimensions.

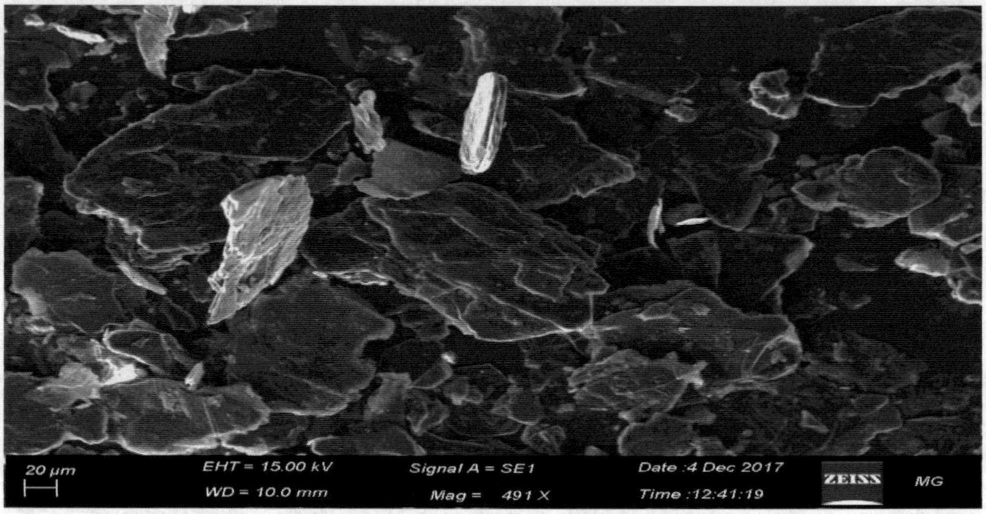

Figure 2. SEM micrograph of the employed graphite.

Graphite powder was examined using X-ray powder diffracto-meter, Panalytical XPERT PRO MPD. XRD diffracto-gram of as obtained graphite flakes is presented in figure 3.

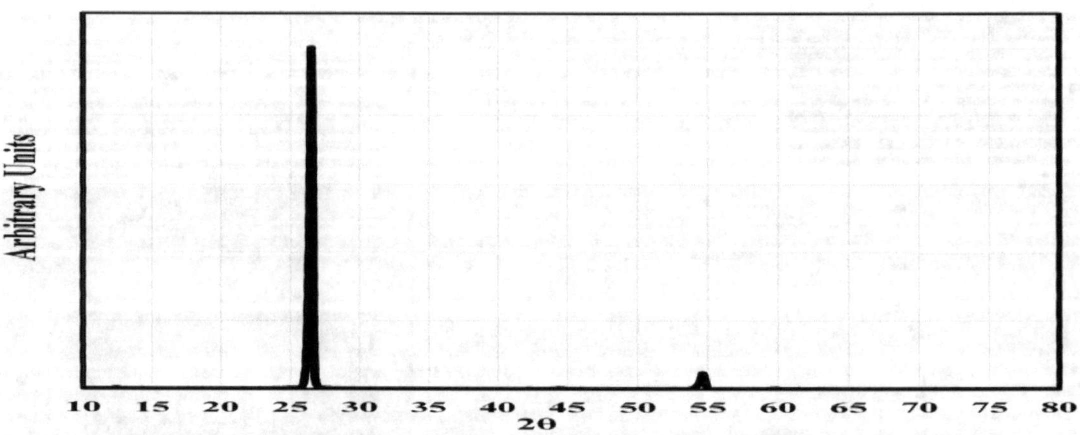

Figure 3. XRD diffracto-gram of graphite.

The prominent peak at ~ 26°, was attributed to the (002) plane of ordered hexagonal graphite structure. This is an evident of crystalline carbon structure. Raman spectra were acquired using dispersive

Raman microscope (model Sentera, Bruker, Germany). A Nikon 20x objective was used to focus the Raman excitation source (10 mW, 532 nm neodymium-doped yttrium aluminum garnet (Nd:YAG) laser. The most prominent features in the Raman spectra of graphitic materials are G band appearing at 1582 cm^{-1} (graphite), the D band at about 1350 cm^{-1}, and the G′-band at about 2700 cm^{-1} figure 4.

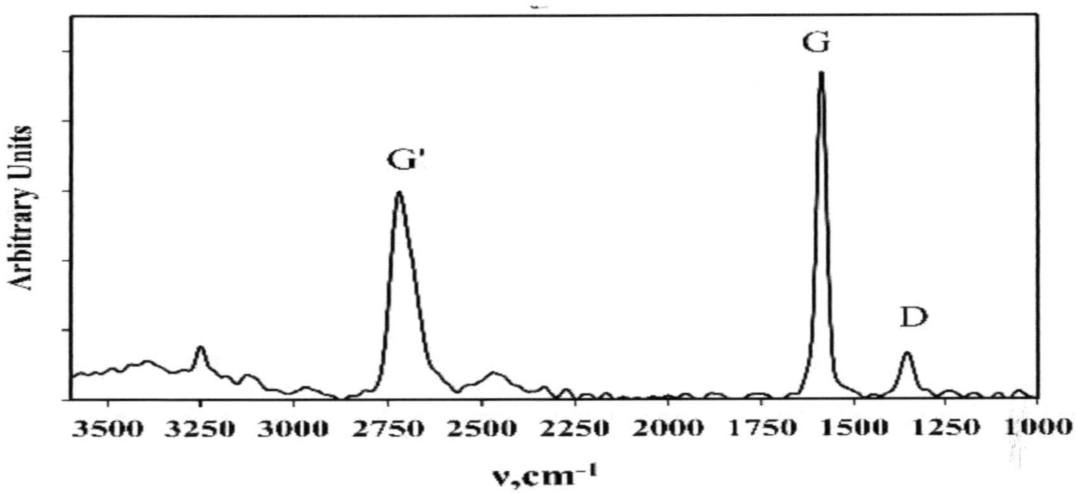

Figure 4. Raman spectra of graphite.

The integrated intensity ratio D band and G band is widely used for characterizing the defect quantity in graphitic materials. It is obvious the employed graphite has relatively low D-band defect intensity, which is localized where the crystalline structure is not perfect, mostly at the edges of the crystallite. Raman results confirmed high crystalline structure of employed graphite material.

3.2 Impact of graphite on spectral performance of MTV flare

Graphite was found to improve total ignition time (TIT) of standard MTV flare. The optimum wt % of graphite was found to be 4 wt %. This ratio offered an increase in TIT by 75 % to reference formulation figure 5.

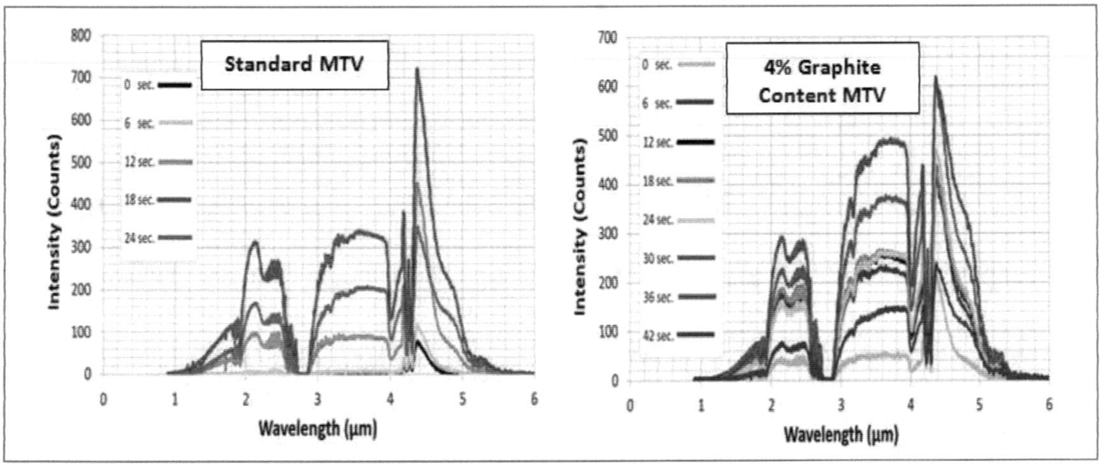

Figure 5. Impact of graphite content on total ignition time of MTV flare.

IOP Conf. Series: Materials Science and Engineering **975** (2020) 012005 doi:10.1088/1757-899X/975/1/012005

The impact of graphite content on spectral performance of MTV flare was conducted. Graphite content of 4 wt % offered the highest intensity to reference MTV flare figure 6.

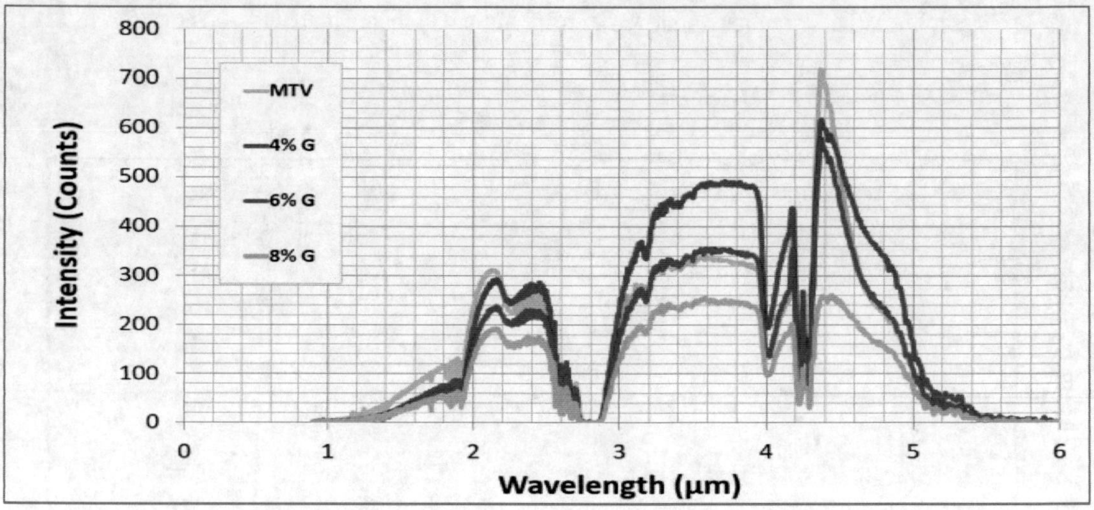

Figure 6. Impact of graphite content of spectral performance of standard MTV flare.

The received thermal signature was transformed into EXCEL data. Average total intensity was calculated using matlab software. Average intensity over α (2-3 μm) and β (3-5 μm) bands were calculated; consequently θ value was retrieved. The impact of graphite content on average intensity and θ value is demonstrated in Figure 7.

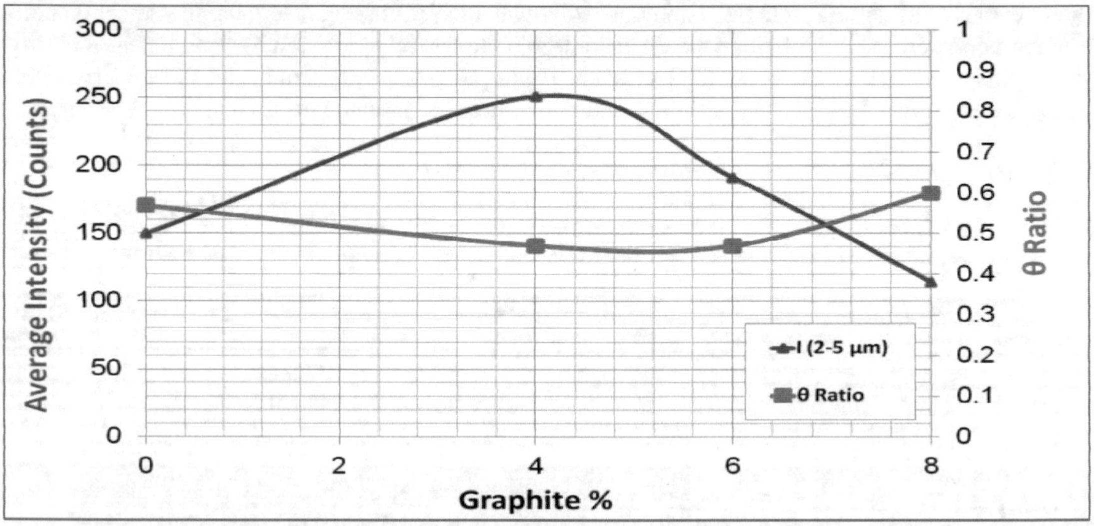

Figure 7. Impact of graphite content of spectral performance of standard MTV flare.

Figure 8 confirmed that graphite content 4 wt % offered superior spectral performance compared to reference formulation. This ratio offered the highest total intensity as well as low θ value. This enhanced spectral performance can be ascribed to graphite as an effective source of carbon particles which absorbs heat and emits nearly as a black body emitter offering enhanced total emission. Also evolved carbon at high temperature would combust with atmospheric oxygen at the tertiary zone producing CO_2 which results in further increase in overall intensity.

3.3 Impact of graphite on spectral performance of ATV flare

ATV with 6 wt % graphite content secured optimum performance; this formulation offered an increase in TIT by 100 % figure 8. 6 wt % graphite offered dramatic increase in ATV thermal signature average intensity figure 9.

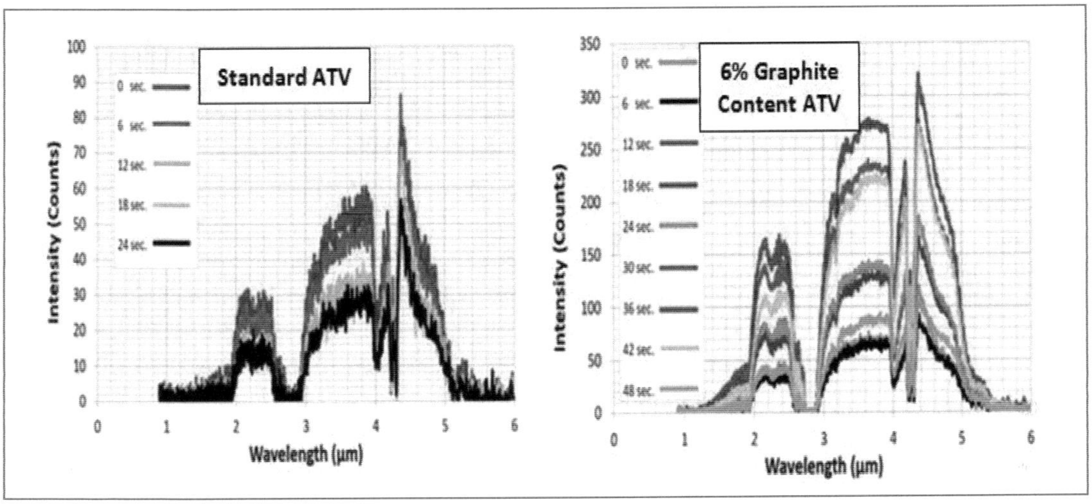

Figure 8. Impact of graphite content on total ignition time of ATV flare.

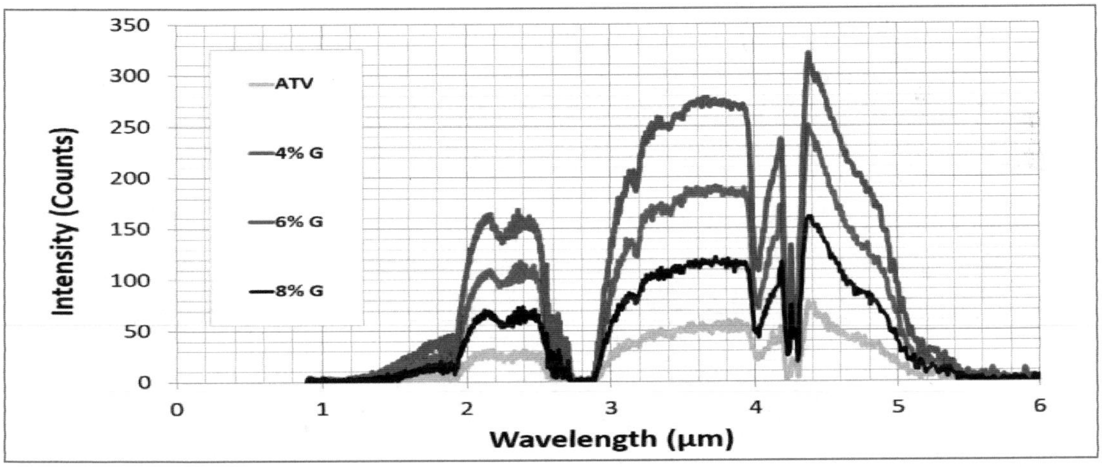

Figure 9. Impact of graphite content of spectral performance of standard ATV flare.

It is undeniable that the inclusion of 6 wt % graphite greatly enhanced the spectral performance to reference formulation. The impact of graphite content on both average intensity and Θ value is represented in figure 10.

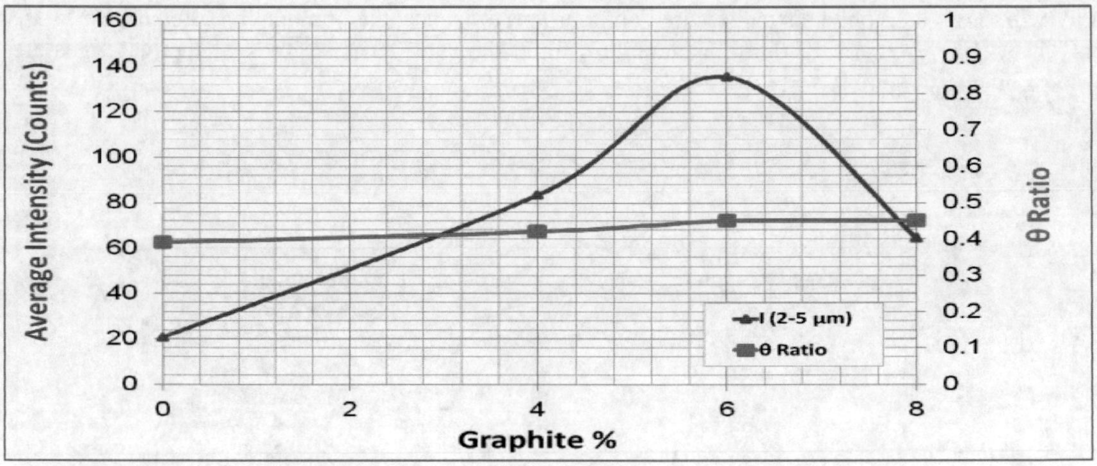

Figure 10. Impact of graphite content of spectral performance of standard ATV flare.

It is obvious that sample containing 6 wt % graphite offered enhanced spectral performance compared to reference ATV formulation. It offered dramatic increase in average total intensity by 590 %. Additionally θ value was found be within the accepted range. This can only be credited to the graphite role in amplifying the spectral output.

3.4 Impact of graphite on spectral performance of MATV flare

Graphite improved TIT of standard MATV flare, 6 wt % graphite was reported to be the optimum loading level. This ratio offered an increase in TIT by 20 % figure 11.

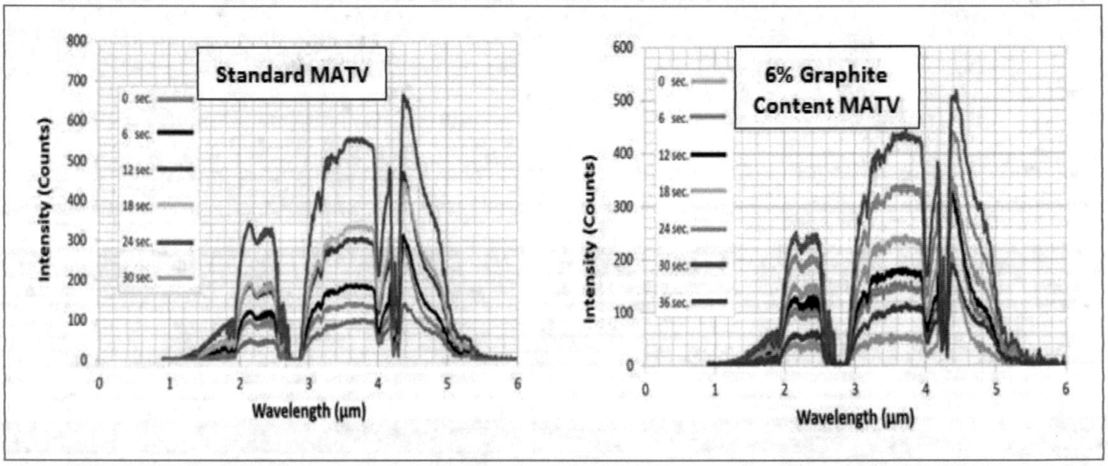

Figure 11. Impact of graphite content on total ignition time.

Spectral performance of MATV formulations with different graphite content revealed that 8 wt % graphite offered that highest spectral performance figure 12.

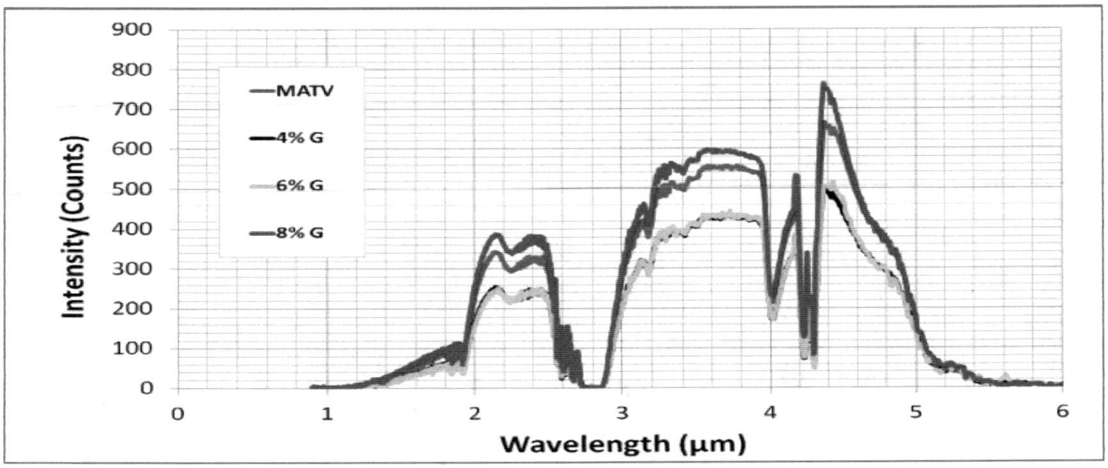

Figure 12. Impact of graphite content of spectral performance of standard MATV flare.

The impact of graphite content on average intensity and θ value for MATV fare is represented in figure 13. MATV demonstrated an increase in average intensity with graphite content.

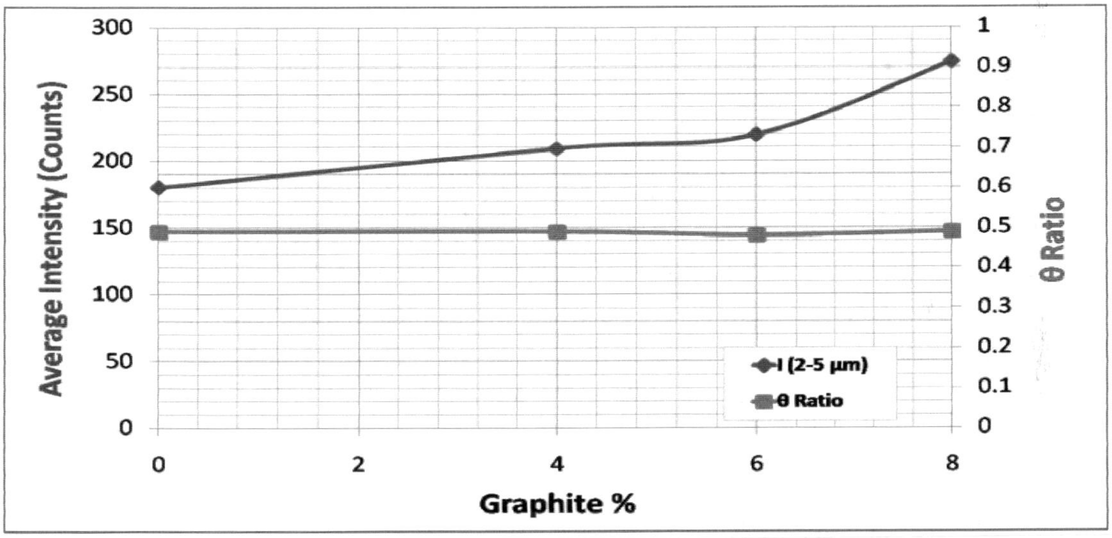

Figure 13. Impact of graphite content of spectral performance of standard MTV flare.

Optimum graphite content (8 wt %), demonstrated the highest average intensity. In the mean time, θ value was almost constant.

3.5 Comparative spectral performance assessment

Comparative investigation for graphite impact on spectral performance of different flare formulations (based on different metal fuels) were conducted to reference ATV flare figure 14.

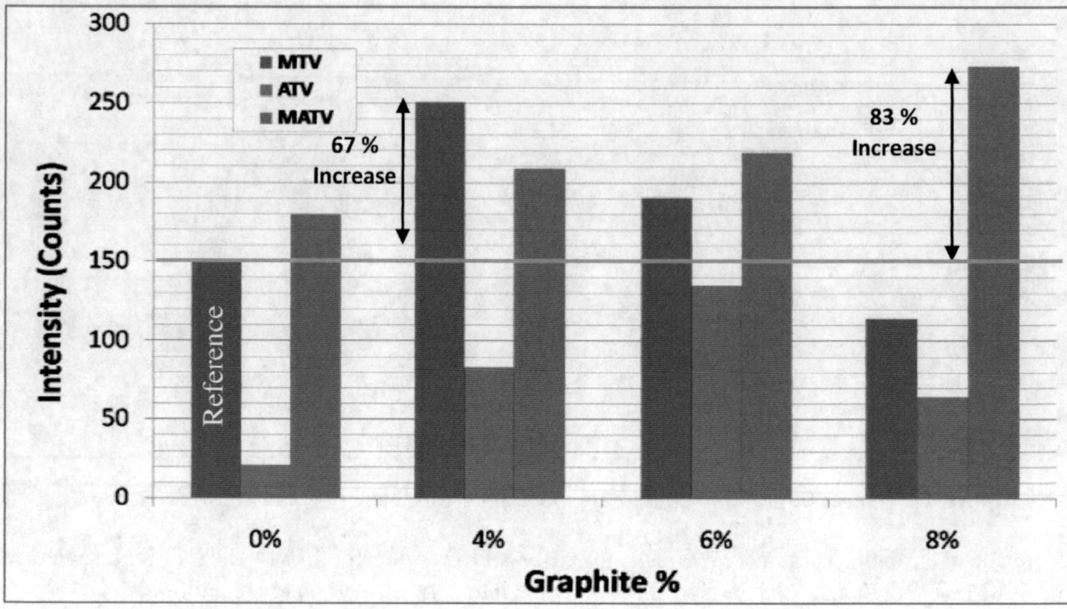

Figure 14. Comparison between average intensity values for all formulations.

Whereas MTV flare with 4 wt % graphite offered an increase in average intensity by 67 %; MATV flare with 8 wt % graphite offered an increase in average intensity by 83 %. The explanation to graphite's impact on spectral performance can be further clarified through understanding the combustion reaction mechanisms within each of the three flame zones separately.

3.6 Combustion wave of decoy flare

It is widely accepted that combustion wave of decoy flare consists mainly of three reaction zones. These zones can be varied in size and associated chemical reactions according to chemical composition.

Primary Combustion Zone

Magnesium and Teflon react within the primary anaerobic zone close to the pyrolant surface to yield condensed magnesium fluoride, carbon (soot), excess vaporized magnesium and heat as in eq. (3) [23].

$$m\,Mg + (C_2F_4) \longrightarrow 2\,MgF_2 + m\text{-}2\,Mg + 2\,C + h\upsilon \quad (3)$$

ATV follows the same trend as MTV, and yield aluminum products instead of magnesium (Equation 4). At high temperature, the formed AlF_3 reacts with metallic aluminum to produce AlF (active transient species) equation 5 [24-26].

$$n\,Al + 3\,C_2F_4 \longrightarrow 4\,AlF_3 + n\text{-}4\,Al + 6\,C + h\upsilon \quad (4)$$

For MATV, magnesium could acts as a tender for aluminum. Aluminum would react with teflon to yield aluminum fluoride. Bimetal fuel (Mg & Al) would induce exothermic reactions via oxidation with teflon equation 5

$$m\,Mg + n\,Al + 10\,(C_2F_4) \longrightarrow 8\,MgF_2 + 8\,AlF_3 + m\text{-}8\,Mg + n\text{-}8\,Al + 20\,C + h\upsilon \quad (5)$$

Secondary combustion zone

Oxidation of primary products will occur in this zone. Excess metal fuel will be combusted in secondary reaction zone, according to the type of employed metal (Equations 6-8).

$$Mg + O_2 \longrightarrow MgO + h\upsilon \quad (6)$$

$$4\,Al + 3\,O_2 \longrightarrow 2\,Al_2O_3 \quad (7)$$

$$Mg + 2\,Al + 2\,O_2 \longrightarrow MgO + Al_2O_3 + h\upsilon \quad (8)$$

MATV formulation, surplus magnesium could vaporize and oxidized with air oxygen offering substantial heat output; as it has low boiling point of 1000 ^{0}C. Therefore excess Al could combust more efficiently with additional heat. This action could stimulate the active emitting species in the primary reaction zone offering extended primary reaction zone at higher temperature.

Tertiary combustion zone

Upon mixing with air in the outer boundaries of the third zone, the remaining carbon would be oxidized with air developing CO_2 (selective radiator) offering enhanced flare intensity equation 9 [6].

$$2\,C + 2\,O_2 \longrightarrow 2\,CO_2 + h\upsilon \quad (10)$$

The enhanced average intensity for all formulations can be attributed to the fact that graphite, is a carbonaceous material, supplies the flame with carbon particles which absorbs heat and emit in IR region. The impact of graphite content of Θ value for investigated flare formulations based on different metal fuels was evaluated figure 15.

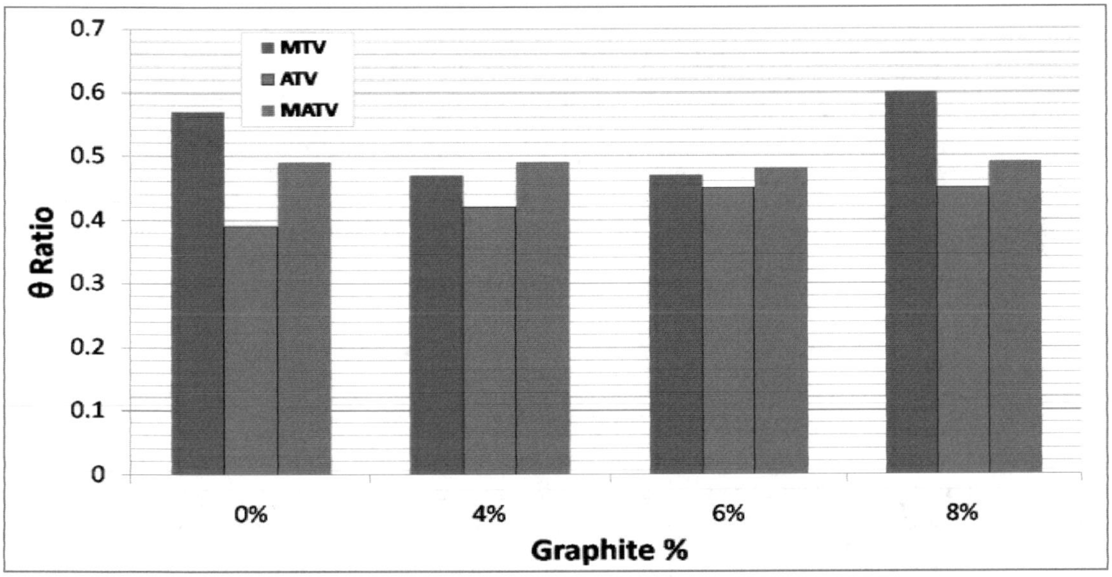

Figure 15. Comparison between theta ratio behaviors for all formulations.

Graphite content did not significantly impact Θ value for different flare formulation. It is apparent that MTV flare with 8 wt % graphite offered enhanced Θ value of 0.6. It can be concluded that customized flare can be developed according to spectral intensity as well as relative intensity ratio.

4. Conclusion

Graphite can find an application in decoy flares field as it improved thermal signature overall intensity. MATV as multi-component flares based on 8 wt % graphite demonstrated superior spectral intensity. This flare formulation offered an increase in average intensity by 83 % to reference MTV formulation. In the meantime MTV flare based on 8 wt % offered the highest relative intensity value ϴ of 0.6. Graphite, as an allotrope of carbon, can act as an excellent source of carbonaceous materials. That can strengthen incandescence emission as ideal black body. Mg-Al alloy con offer novel synergism as Mg can act as tinder for Al offering efficient combustion process. Mg could vaporize at 1000 ^{0}C offering efficient combustion process. It can be concluded that tailored flare can be developed according to the required spectral data.

5. References

[1] E.-C. Koch, Metal-fluorocarbon based energetic materials: John Wiley & Sons, 2012.
[2] E. C. Koch, "Pyrotechnic countermeasures: II. Advanced aerial infrared countermeasures," Propellants, Explosives, Pyrotechnics, vol. 31, pp. 3-19, 2006.
[3] J. Andreotti and A. Hirschman, "Infrared decoy method using polydimethylsiloxane fuel," ed: Google Patents, 1994.
[4] S. Erwin, "Anti-Missile Program for Airliners on a Fast TrackSmart'Flares Being Designed to Defeat Heat-Seeking Missiles," NATIONAL DEFENSE, pp. 14-16, 2003.
[5] T. M. Klapotke and C. M. Rienacker, "Drophammer test investigations on some inorganic and organic azides," Propellants Explosives Pyrotechnics, vol. 26, pp. 43-47, 2001.
[6] E. C. Koch, et al., "Metal-Fluorocarbon Pyrolants. XIII: High Performance Infrared Decoy Flare Compositions Based on MgB2 and Mg2Si and Polytetrafluoroethylene/Viton®," Propellants, Explosives, Pyrotechnics, vol. 37, pp. 432-438, 2012.
[7] V. Weiser, et al., "Pyroorganic flares–a new approach for aircraft protection," in Europyro 2007–34th International Pyrotechnics Seminar, 2007, pp. 8-11.
[8] M. Comet, et al., "Phosphorus-based nanothermites: a new generation of energetic materials," Journal of Physics and Chemistry of Solids, vol. 71, pp. 64-68, 2010.
[9] L. Klusáček and P. Navratil, "The Use and Application of Red-Phosphorous Pyrotechnic Composition for camouflage in the infrared region of radiation," Propellants, explosives, pyrotechnics, vol. 22, pp. 74-77, 1997.
[10] E. C. Koch and S. Cudziło, "Safer pyrotechnic obscurants based on phosphorus (V) nitride," Angewandte Chemie International Edition, vol. 55, pp. 15439-15442, 2016.
[11] E.-C. Koch and A. Dochnahl, "Pyrotechnic active mass for producing an aerosol highly emissive in the infrared spectrum and inpenetrable in the visible spectrum," ed: Google Patents, 2003.
[12] N. Wardecki, et al., "Camouflage and decoy munitions for protecting objects against guided missiles," ed: Google Patents, 2006.
[13] E. C. Koch and A. Hahma, "Metal-Fluorocarbon Pyrolants. XIV: High Density-High Performance Decoy Flare Compositions Based on Ytterbium/Polytetrafluoroethylene/Viton®," Zeitschrift für anorganische und allgemeine Chemie, vol. 638, pp. 721-724, 2012.
[14] E.-C. Koch, "Pyrotechnic charge," ed: Google Patents, 2009.
[15] E. C. Koch, "2006–2008 Annual Review on Aerial Infrared Decoy Flares," Propellants, Explosives, Pyrotechnics, vol. 34, pp. 6-12, 2009.
[16] E. C. Koch, "Metal–Fluorocarbon Pyrolants: X. Influence of Ferric Oxide/Silicon Additive on Burn Rate and Radiometric Performance of Magnesium/Teflon/Viton®(MTV)," Propellants, Explosives, Pyrotechnics, vol. 34, pp. 472-474, 2009.
[17] S. Elbasuney, et al., "Infrared signature of novel super-thermite (Fe 2 O 3/Mg) fluorocarbon nanocomposite for effective countermeasures of infrared seekers," Journal of Inorganic and Organometallic Polymers and Materials, pp. 1-10, 2018.
[18] S. Elbasuney, et al., "Super-thermite (Al/Fe 2 O 3) fluorocarbon nanocomposite with stimulated infrared thermal signature via extended primary combustion zones for effective countermeasures of infrared seekers," Journal of Inorganic and Organometallic Polymers and Materials, vol. 28, pp. 2231-2240, 2018.
[19] E. Koch, "Metal-fluorocarbon pyrolants: VIII. Behaviour of burn rate and radiometric performance of two magnesium/Teflon/Viton (MTV) formulations upon addition of graphite," Journal of Pyrotechnics, vol. 27, p. 38, 2008.

[20] M. P. Nadler, "Pyrotechnic pellet decoy method," ed: Google Patents, 2004.
[21] S. Elbasuney, et al., "Infrared Signature of Novel Super-Thermite (Fe 2 O 3/Mg) Fluorocarbon Nanocomposite for Effective Countermeasures of Infrared Seekers," Journal of Inorganic and Organometallic Polymers and Materials, pp. 1-10.
[22] S. Elbasuney, et al., "Super-Thermite (Al/Fe 2 O 3) Fluorocarbon Nanocomposite with Stimulated Infrared Thermal Signature via Extended Primary Combustion Zones for Effective Countermeasures of Infrared Seekers," Journal of Inorganic and Organometallic Polymers and Materials, pp. 1-10.
[23] E.-C. Koch and A. Dochnahl, "Investigation of combustion process in Magnesium/Teflon/Viton (MTV) compositions by emission spectroscopy," in INTERNATIONAL ANNUAL CONFERENCE-FRAUNHOFER INSTITUT FUR CHEMISCHE TECHNOLOGIE, 1999, pp. P 87-P 87.
[24] W. Witt and R. Barrow, "The heat of sublimation of aluminium trifluoride and the heat of formation of aluminium monofluoride," Transactions of the Faraday Society, vol. 55, pp. 730-735, 1959.
[25] J. Dyke, et al., "A study of aluminium monofluoride and aluminium trifluoride by high-temperature photoelectron spectroscopy," Chemical physics, vol. 88, pp. 289-298, 1984.
[26] R. W. Conner and D. D. Dlott, "Comparing boron and aluminum nanoparticle combustion in Teflon using ultrafast emission spectroscopy," The Journal of Physical Chemistry C, vol. 116, pp. 2751-2760, 2012.

IOP Conf. Series: Materials Science and Engineering **975** (2020) 012006 doi:10.1088/1757-899X/975/1/012006

Assessment of antioxidant resistance to thermal - oxidative degradation of stabilized polybutadiene binder

Eslam Elsaka[1,3], Sherif Elbasuney[2], Ahmed Elbeih[1], Hosam E. Mostafa[1], Mahmoud E. Awad[1], Tarek M. Elhedery[1]

[1]School of chemical Engineering, Military Technical College, Cairo, Egypt
[2]Head of Nanotechnology Research Center, Military Technical College (MTC), Egyptian Armed Forces, Cairo, Egypt
[3]Email:eng.elsaka12@gmail.com

Abstract. polybutadiene is the most common polymeric binder for composite solid propellants. This polymer suffers from oxidative degradation reactions on storing with deterioration in mechanical properties. Certain anti-oxidant materials can retard such reaction offering enhanced aging characteristics. In this study, two different primary antioxidants including phenol-based (AO2246) and amine-based (Flexzone 6H) were employed to stabilize polybutadiene polymer. Oxidation induction time (OIT) was evaluated using isothermal DSC. High OIT value means high resistance to oxidative degradation. The impact of anti-oxidant on mechanical properties was evaluated using shore A. Flexzone 6H significantly improved binder oxidative stability with an increase in OIT value by 20 folds with significant decrease in oxidation enthalpy. Flexzone 6H can offer enhanced resistance toward oxidative reactions; as it can act as efficient H atom donor to break down chain degradation reaction. Additionally, the three stabilized aromatic rings can capture free radicals. AO2246 enhanced the binder mechanical characteristics during aging at 80 0C using Shore A test. In the meantime, AO2246 could enhance the mechanical properties due to its integration into the polymeric matrix due to –OH functional groups. The two anti-oxidants demonstrated controversy effects. Whereas Flexzone 6H offered enhanced resistance to oxidative degradation; AO2246 offered enhanced mechanical properties.

1. Introduction

Composite propellant is a heterogeneous mixture composed mainly of inorganic oxidizer, metallic fuel, and polymeric binder. Polymeric binder is the main ingredient responsible for holding all ingredients together and rendering the required mechanical properties to the polymeric matrix. Hydroxy terminated polybutadiene (HTPB) is the most common polymer in use for composite solid propellants. HTPB has several advantages including: low glass transition temperature, high tensile strength, good chemical resistance, high compatibility with propellant solid ingredients [1, 2]. HTPB has high ability to accommodate solid loading up to 88 %. Additionally, HTPB can act as fuel producing large volume of gaseous products. HTPB is the backbone of polymeric matrix; it constitutes about 93 % of polymeric binder [3, 4]. High mechanical properties are mandatory for safe storing, transportation, and firing. During firing the propellant grain will be exposed to severe thermal and pressure stresses [5, 6].

HTPB can offer high degree of un-saturation that is required to provide the composite propellant with its viscoelastic nature and flexibility. On the other hand, these un-saturated bonds are very sensitive to oxidation [7, 8]. Various slow chemical oxidative reactions (aging reactions) can take place on storing. These oxidative reactions include peroxidation, hydro-peroxidation, cross linking by peroxide linkage figure 1; further details can be found in the following reference [9].

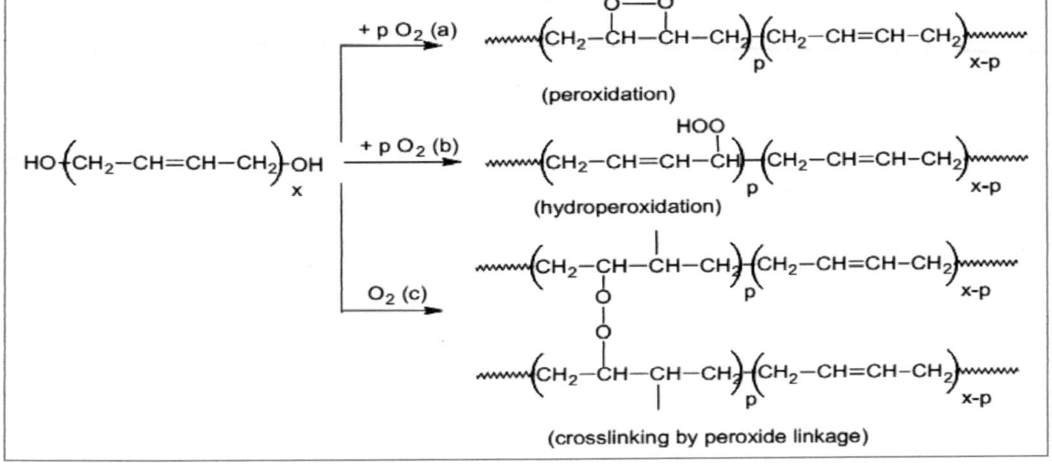

Figure 1. Thermal oxidative degradation reactions of HTPB prepolymer.

As the binder is exposed to thermal loading during processing and storage, free radicals ($R^{\cdot}$) are formed. Such radicals can react with oxygen forming peroxy radicals ($ROO^{\cdot}$). These radicals further react with labile hydrogen atoms of the binder to form unstable hydroperoxides (ROOH) which immediately decompose via hemolytic cleavage to form alkoxy and hydroxyl radicals figure 2.

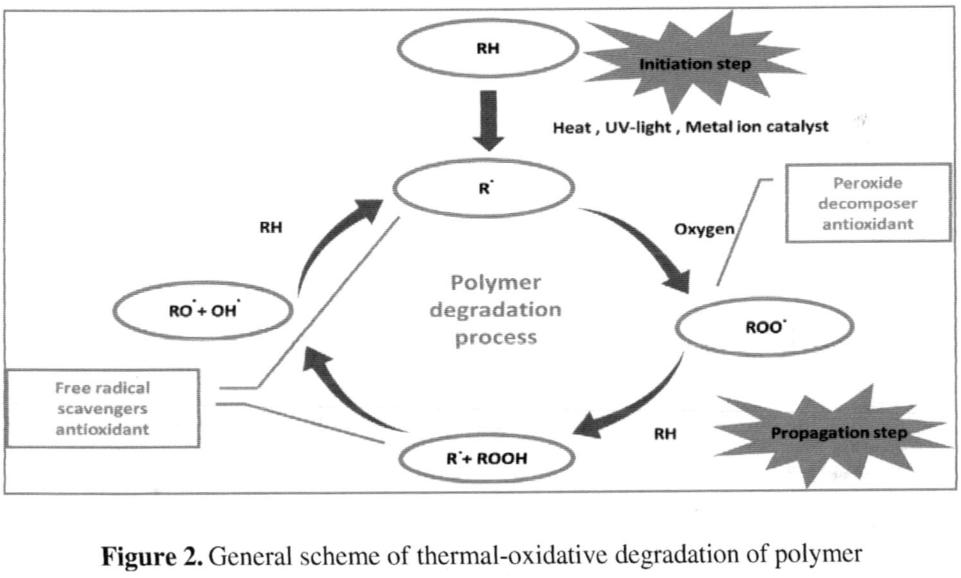

Figure 2. General scheme of thermal-oxidative degradation of polymer

These oxidative reactions have unfavorable effect on the characteristic viscoelastic nature of the composite propellants and could result in deterioration of propellant mechanical properties by propagating through the whole propellant grain and causing integral system failure. Antioxidant materials can stabilize the binder against these oxidative reactions [10, 11]. Generally, antioxidants can be classified into two main classes including:

- Primary antioxidants: these antioxidants break the degradation chain by donating H- atoms to free radicals. Therefore, it could prevent those radicals from propagating the chain reaction. The chemical structure of such anti-oxidant includes pendant phenols or aromatic amines.
- Secondary antioxidants: these antioxidants can act as hydroperoxide decomposers like phosphites and thioester [12, 13].

Primary antioxidant is the most common for composite propellant as it suppresses the oxidative reactions from the start point. Antioxidant depletion is strongly related to an oxidative crosslinking process as once the antioxidant is depleted, thermal oxidative degradation will proceed rapidly [12]. Such a complex oxidative crosslinking reaction could be investigated by monitoring the increase of mass variation as a result of oxygen uptake, double bond disappearance, carbonyl build-up, and change of mechanical property (hardening) as a result of further crosslinking [13]. Kinetic parameters such as activation energy could be considered as the key parameter for service life assessment of composite propellant by analysis of the provided data about binder thermal-oxidative degradation [14, 15].

2. Experimental

2. 1. Chemicals and materials
The employed polymer was hydroxyl terminated polybutadiene (HTPB) with 0.8495 mg. eqv. OH/g (Aldrich). Curing agent was hexamethylene diisocyanate (HMDI) with 11.86 mg. eqv. NCO/g (Aldrich). Different anti-oxidants including 2.2'-methylene bis 6-t-butyl-4methylphenol (AO 2246) and N, N'-diphenyl-p-phenylenediamine (Flexzone 6 H) (Alpha Chemica) were employed. The chemical structure of employed anti-oxidants is demonstrated in figure 3. All chemicals were used as received without further treatment

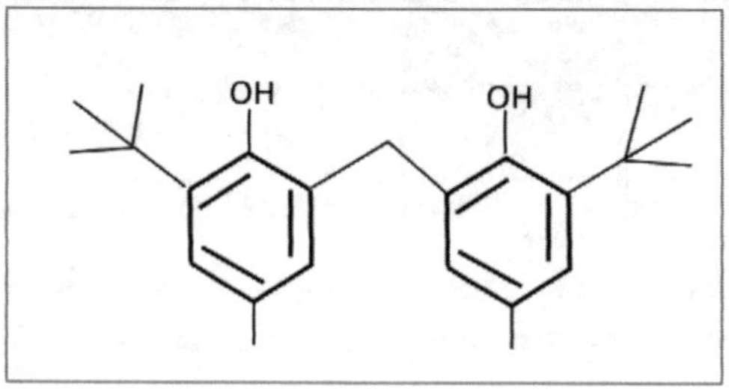

(AO 2246)

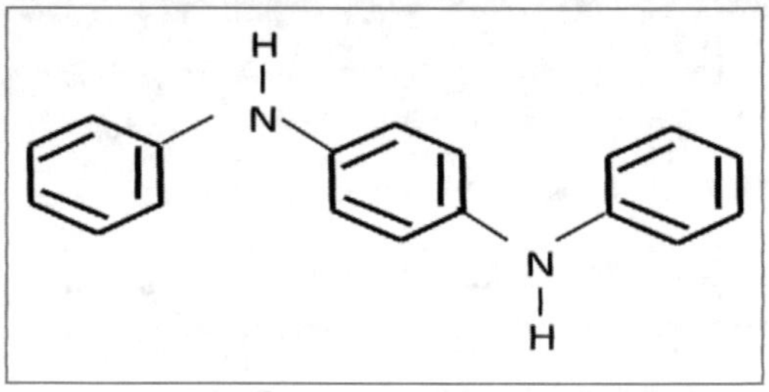

(Flexzone 6 H)

Figure 3. Chemical structure of antioxidants: AO-2246, and Flexzone 6H.

2.2. Binder stabilization
Poly-butadiene polymer was mixed with representative amount (2 wt %) of different anti-oxidants. The mixing process was conducted in a vertical mixer; the employed mixing speed was 30 rpm. The mixing temperature was 45 ^{0}C, controlled with hot circulated water. Cross linking agent (HMDI) was introduced by the end of mixing process. Consequently, the cross-linked binder was cast and cured at 40 ^{0}C, under vacuum for 7 days. The employed curing ratio (NCO/OH) was 0.7.

2.3. Impact of anti-oxidant on thermal behaviour
TGA measurement can provide valuable information about degradation temperature, degradation rate, and total weight loss. The impact of anti-oxidant on thermal decomposition was evaluated by

thermal gravimetric analysis TGA, Q500, TA instruments, USA. The tested samples were heated from 25 to 500 ^{0}C at 5 ^{0}C/min, under N_2 flow of 30 mL/min. DSC measurement can provide valuable data on initial oxidation temperature (T_i), onset oxidation temperature (OOT), oxidation peak temperature (OP), final oxidation temperature (T_f), as well as oxidation enthalpy (ΔH).
Anti-oxidant efficiency for oxidative degradation was evaluated via precise measurement of oxidation induction time (OIT). High OIT value means high resistance to oxidative degradation. This key parameter was evaluated using an isothermal DSC, Q1000 TA Instrument, USA according to the standard method (ASTM D 3895) [16]. The tested sample was placed in an aluminum crucible and heated to 180 ^{0}C at 5 ^{0}C/min, under N_2 flow of 50 ml/min. The temperature was hold at 180 ^{0}C for 5 min. Consequently, the flowing gas was switched from nitrogen to pure oxygen gas. The time interval between flowing gas change to oxygen and the onset point of exothermic oxidation peak (catastrophic degradation or oxidation) was recorded as OIT time [17].

2.4. Impact of anti-oxidant on hardness

Hardness is defined as the material resistance to penetration. Hardness measurements can offer an indication about polymer cross linking efficiency. Higher hardness value is associated with higher degree of crosslinking. The impact of different anti-oxidants on HTPB hardness was measured using Shore A Durometer.

3. Results and discussion

3.1. Initial decomposition temperature using TGA

Thermal behavior of reference binder to binder stabilized with different antioxidants was investigated using TGA (Figure 4). It is obvious that the thermal behavior of stabilized binder is similar to non-stabilized binder; no degradation could take place up to 210 ^{0}C. It can be assumed binder stabilized with AO2246 is more resistant to degradation as it could be integrated into the polymer chain via bonding with its – OH functional group [18].

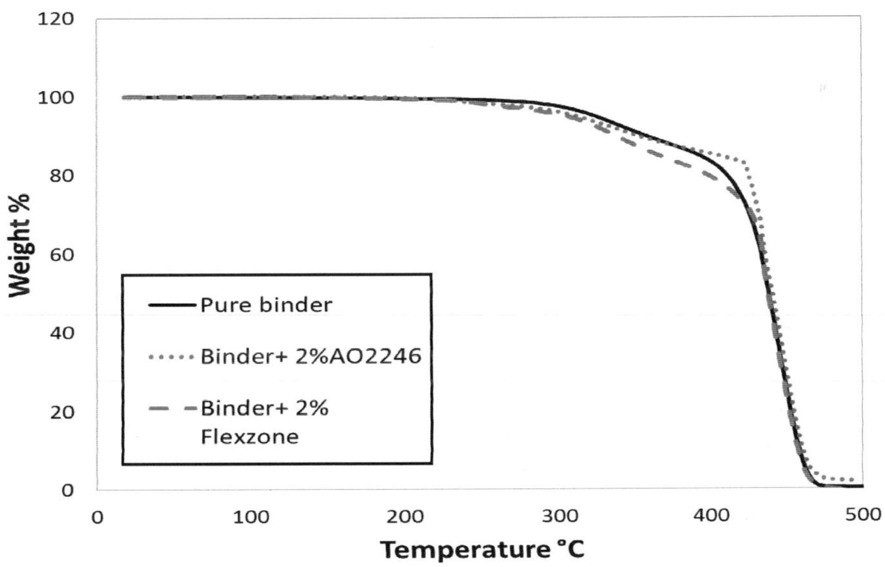

Figure 4. Thermo-gravimetric (TGA) results.

3.2. Thermal behaviour using dynamic DSC

IOP Conf. Series: Materials Science and Engineering **975** (2020) 012006 doi:10.1088/1757-899X/975/1/012006

It apparent that thermal decomposition of HTPB take place in two separated stages including: thermal oxidation peak, and exothermic heat involved in cyclization process figure 5 [19, 20].

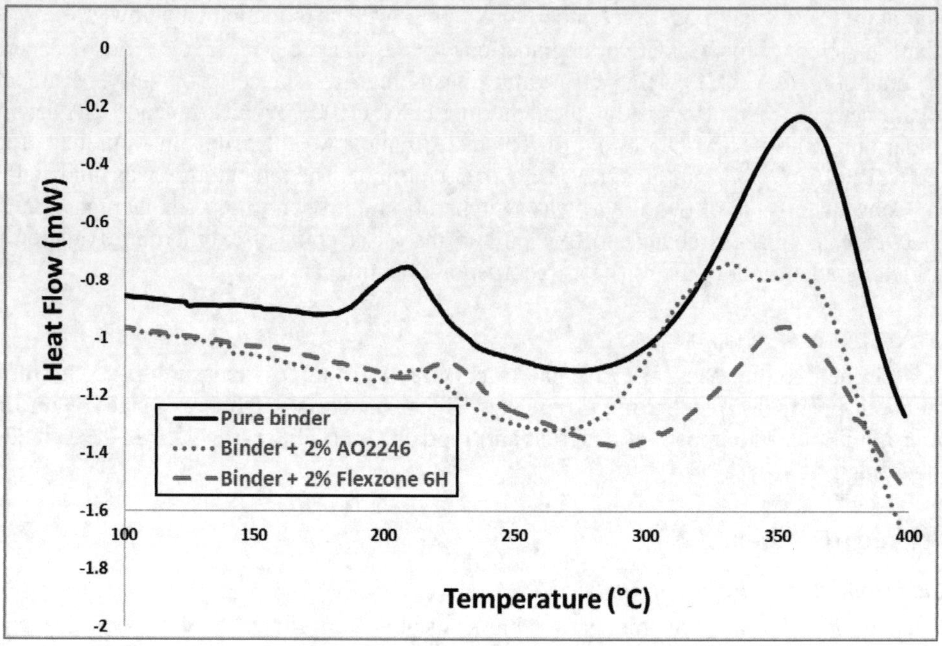

Figure 5. Dynamic DSC thermogram.

Figure 5 demonstrated that there is a dramatic impact of anti-oxidant on oxidation temperature and oxidation enthalpy. A reduction in oxidation enthalpy (energy) and fractional decomposition (relative oxidation energy) were related to improved stability via reduced oxidation effect [19].
Oxidation onset temperature (OOT), and oxidation peak temperature (OP) was evaluated for stabilized binder to non-stabilized binder. Whereas, AO2246 offered an increase in OP and OOT by 4 ^{0}C and 5 ^{0}C respectively; Flexzone 6H offered increase in OP, and OOT by 13 ^{0}C and 17 ^{0}C respectively table .1.

Table 1. Thermal-oxidative critical parameters.

	Ti	**Tf**	**OP**	**OOT**	**ΔHox**	**α**
	(◦ C)	**(◦ C)**	**(◦ C)**	**(◦ C)**	**(J/g)**	**(-)**
Pure binder	177.3	237.2	208.4	190.4	37.2	-
Binder + 2 % AO2246	188.2	233.0	213.5	196.5	18.3	0.49
Binder + 2 % Flexzone 6 H	197.1	242.3	221.7	207.8	16.4	0.44

Keys: T_i and T_f: Initial and final oxidation temperature, OP: Oxidation peak temperature OOT: Oxidation onset temperature, ΔHox: Oxidation enthalpy (energy) and α: fractional decomposition ($\Delta H/\Delta H^0$).

It is apparent that Flexzone 6 H offered the lowest oxidation enthalpy; this oxidation could take place at higher temperature compared with AO2246 and reference polymer.

3.3. Oxidation induction time (OIT) using isothermal DSC

Oxidation induction time (OIT) is an effective parameter to characterize the isothermal oxidation reaction. It evaluates the efficiency of different stabilizers to evaluate the relative resistance of various polymeric hydrocarbon material to oxidation. Figure 6 demonstrates the impact of anti-oxidants on oxidation induction time (OIT).

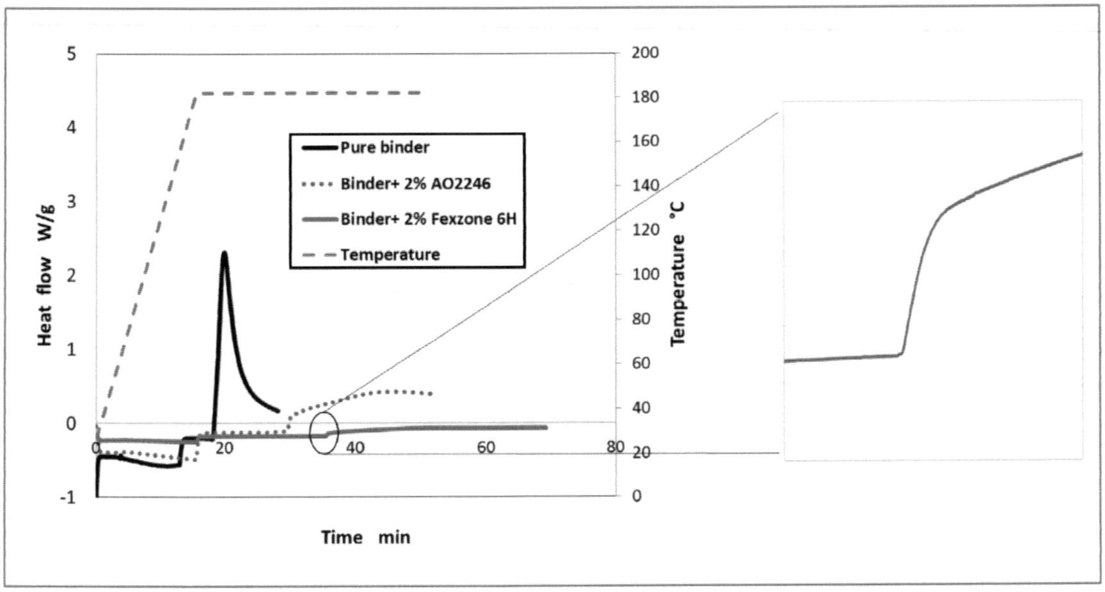

Figure 6. Comparative isothermal DSC thermo-gram.

It is apparent that pure binder has low OIT value of 0.9 min; this means that it is extremely sensitive polymer to oxidation. The two anti-oxidants offered dramatic increase in OIT value. While AO2246 offered an increase in OIT by 13.2 min.; flexzone 6H offered an increase in OIT by 18.9 min table .2. Multifold was defined as the capability to stabilized polymer to resist oxidation with respect to pure binder.

Table 2. OIT values of stabilized and un-stabilized binder

	OIT	**Effectiveness**
	(min)	**(Multi-fold)**
Pure binder	0.9	-
Binder + AO2246	13.2	13.7
Binder + Flexzone 6H	18.9	20

Flexzone 6H offered an increase in polymer resistance to oxidation by 20 times. The stabilization mechanism of both AO2246 and Flexzone 6H depend on H atom donor to break down HTPB chain degradation reaction. Whereas stabilization with AO2246 depend on –OH group; the stabilization mechanism of Flexzone 6H depends on –NH group. It can be concluded that part of AO2246 –OH stabilization groups could be involved in cross linking reactions. Therefore less stabilization could

IOP Conf. Series: Materials Science and Engineering **975** (2020) 012006 doi:10.1088/1757-899X/975/1/012006

be achieved compared with enhanced mechanical properties due to integration into the polymeric matrix [20 - 24] .

3.4 Hardness monitoring

Hardness testing is a useful method for determination of polymer mechanical changes [25]. Hardness of reference polymer to stabilized polymer was measured; the change of hardness with aging time was further investigated table .3.

Table 3. Hardness variation after aging at 80 °C.

	After curing	After aging (days)		
		12	24	35
Pure binder	68	93	94	95
Binder + 2 % AO2246	42	44	50	58
Binder + 2 % Flexzone 6 H	37	38	43	49

Pure binder demonstrated dramatic increase in hardness with aging time. This gives an indication that it is extremely sensitive polymer to oxidative cross linking. AO2246 offered minimum change in mechanical properties with aging time. This can be ascribed to its integration into the polymeric matrix via OH group [26, 27].

During aging periods visual changes in binder color were observed figure 7. Yellow color is a qualitative indication to investigate the occurrence of oxidative aging reactions. Yellowing could be quantitively determined by using special spectrophotometer [28].

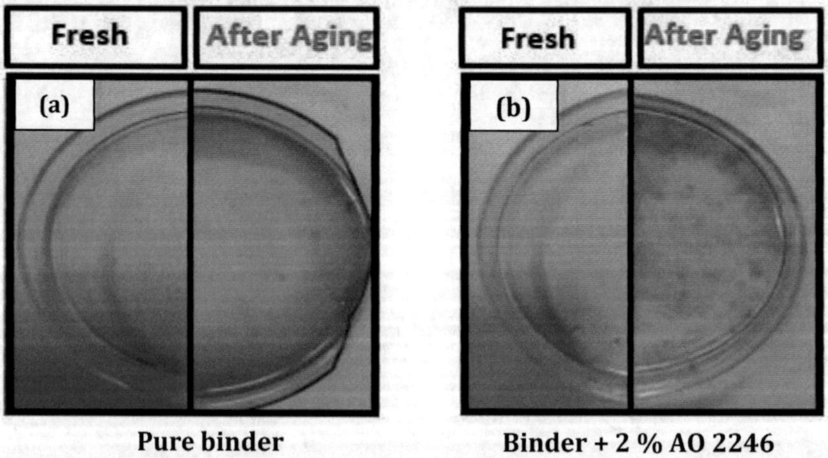

Figure 7. Visual change of non-stabilized and stabilized polymer with 2 percentages of AO2246 antioxidant after curing and after aging at 80 °C for 35 days

Whereas virgin polymer demonstrated dramatic change in color with aging figure 7-a; polymer stabilized with AO2246 demonstrated slight change in yellow color figure 7-b. On the other hand, this test couldn't be performed for Flexzone 6H as it has a dark base color.

4. Conclusions

Antioxidant plays a vital role to increase the thermal stability of the base polymeric matrix. Flexzone 6H offered significant increase in OIT value and enhanced binder resistance to oxidative degradation by 20 folds. This can be ascribed to the availability of H donor groups to break down the oxidative degradation process. In the meantime, AO2246 was found to have the ability to integrate itself into the polymeric matrix (via – OH groups) offering enhanced mechanical properties with aging. This was verified via hardness measurement. It could minimize the availability of reactive OH – groups for oxidative stabilization. It can be summarized that the two antioxidants have controversy stabilization mechanisms that could complete each other.

References

[1] S. Cerri, L. Galfetti, M. A. Bohn, K. Menke, P. Milano, and I. Aerospaziale, 2009, Chemik, **162(2)**, pp. 1–12.
[2] H. L. J. Keizers, G. C. R. Brouwer, J. Weijl, and F. P. Weterings, 2002, **12(1)**, pp. 23–26.
[3] S. Pf. Bm. A, 2016, Chemik **48 (2)**, pp. 21–25.
[4] M. F. Lemos, G. Mussbach, and M. A. Bohn, 2015 J. Aerosp. Technol. Manag. **15(1)**, pp. 379–388.
[5] N. Gligorijević et al., 2014, Sci. Tech. Rev., **64(4)**, pp. 21–29.
[6] A. Göçmez, C. Erişken, Ü. Yilmazer, F. Pekel, and S. Özkar, 2002, J. Appl. Polym. Sci., **67(8)**, pp. 1457–1464.
[7] P. Bunyan, A. V Cunliffe, A. Davist, and F. A. Kirby, 1993, Polym. Degrad. Stab **40(3)**, pp. 239–250.
[8] L. F. C. Sbegue and L. D. Villar, 2016, J. Aerosp. Technol. Manag., **8(2)**, pp. 122–129.
[9] A. K. and D. Dalie, 2012, Propellants, Explos. Pyrotech., **14(4)**, pp. 25-31.
[10] Y. S. Duh, T. C. Ho, J. R. Chen, and C. S. Kao, 2010 J. Aerosp. Technol. Manag. **45(3)**, pp. 174–187.
[11] M. Celina, J. M. Skutnik Elliott, S. T. Winters, R. A. Assink, and L. M. Minier, 2006 Polym. Degrad. Stab., **91(8)**, pp. 1870–1879.
[12] I. Vulic, G. Vitarelli, and J. M. Zenner, 2002 Chinese J. Aeronaut. **78(6)**, pp. 27–34.
[13] M. Lokander, B. Stenberg, and R. Sandén, 1998 Propellants, Explos. Pyrotech., **23(5)**, pp. 272–278.
[14] M. Coquillat, J. Verdu, X. Colin, L. Audouin, and M. Celina, 2008 Polym. Degrad. Stab **93(1)**, pp. 1689–1694.
[15] W. M. Adel and G. Liang, 2018 Chinese J. Aeronaut., **14(2)**, pp. 25-31.
[16] O. Okamba-diogo et al. 2016 Polym. Test. **60(4)**, pp. 21–29.
[17] M. Schmid, A. Ritter, and S. Affolter, 2006 Propellants, Explos. Pyrotech. **83**(1), pp. 367–371.
[18] J. A. F. F. Rocco, J. E. S. Lima, A. G. Frutuoso, K. Iha, M. Ionashiro, and J. R. Matos, 2015 Asian journal of nanoscience, **77**, pp. 803–813, 2004.
[19] E. M. lRichard A. Pgard and E. Zaikov, 2015 physical chemistry research for engineering and applied sciences. **68(1)**, pp. 21–29
[20] P. Pla, A. Masek, and M. Latos-brozio, 2018 Polym. Degrad. Stab **40(5)**, pp. 239–250.
[21] T. Gupta and B. Adhikari, 2003 Propellants, Explos. Pyrotech.**402**, pp. 169–181.
[22] I. Thear, pp. 20–23, 2012 Polym. Test. **11(4)**.
[23] W. Wu, X. Zeng, H. Li, X. Lai, and H. Xie, 2015 Polym. Degrad. Stab., **111(2)**, pp. 232–238.
[24] J. A. F. F. Rocco et al., 2004 J. Therm. Anal. Calorim.,**75(2)**, pp. 551–557.
[25] L. D. Villar, R. F. da Silva, M. F. Diniz, M. F. K. Takahashi, and L. C. Rezende, 2010 J. Aerosp. Technol. Manag., **2(2)**, pp. 163–168.
[26] G. Li and J. L. Koenig, 2010 J. Aerosp. Technol. Manag. **81(2)**, pp. 377–385.
[27] L. Dias, T. Cicaglioni, M. Faria, D. D. E. Química, and E. De Lorena, 2011 physical chemistry research for engineering and applied sciences, **14(3)**, pp. 372–375.

[28] M. Celina, A. B. Trujillo, K. T. Gillen, and L. M. Minier, 2006 Polym. Degrad. Stab., **91**(**2**), pp. 552–558.

A Study of the detonation wave shape inside confined explosive with a metallic copper liner

Tamer Elshenawy[1], Ahmed Elbeih[2*], Mohamed G. Zaki[2], Ahmed K. Hussein[2]

[1] Technical Research Center, Cairo, Egypt
[2] Military Technical College, Kobry Elkobbah, Cairo, Egypt

Abstract: The behaviour of shock wave inside a shaped charge is so important that the detonation wave nature characterizes (determines) the degree of jet coherency at the liner interface. The detonation wave shape inside the shaped charge depends on the casing thickness, material, the liner material, geometry and the applied wave shaper geometry. In this paper, a series of numerical simulation models were performed to study the incident and reflected waves and to characterize the resulting shock wave front to evaluate its effect on the produced (generated) jet characteristics. The used code was Autodyn 3D based on non-linear dynamic analysis with jetting analysis model. Inert wave shapers (Lead, Aluminium and steel) and active ones (TNT and PETN) behaviour were studied. The liner type used in this study was copper with in which the apex angle of the liner cone was 40°. It was concluded that the wave front shape has an obvious influence on the characteristics of the jet formation and its properties such as mass of jet and velocities of its tip. Also, the bulk sound speed of the liner itself in addition to the collapse characteristics of the liner was discussed in this paper successfully.
Keywords: Liner, wave shaper, explosive shock wave, jetting, Autodyn code.

1. Introduction

Regarding to the theory of shaped charges, the explosive types and its energy affect the performance of the jet formed and its penetration-ability in addition to the type of the liner and its characteristics [1-3]. Several different energetic materials could be used for the charge filling [4, 5]. Trinitrotoluene (TNT) is one of the most useable explosive for filling of charges by casting method [6]. In addition, it might be used in mixture with other explosives to increase the penetration performance of the charges [7]. Different high energy materials can also be used such as pentaerythritol tetranitrate (PETN) and other advanced high performance explosives [8-10]. Therefore, it is essential to do the best use of the explosive mass by modifying the detonation wave at the interface of the liner wall [1]. The detonation wave head could be modified by placing of inert or active space part in the explosive itself [11]. This spacer can be placed in the explosive charge itself under the initiation point of the charge where it has the ability to delay the detonation wave head. This space part is named as wave shaper or explosive lenses [11]. The wave shaper is also used to decrease the head height of the hollow cavity by modifying the shock wave over short distance travel. Changing the incident angle of the DW could be performed and cause obvious change of the jet properties [12]. The successful and effective wave shapers that have been used were plastics, ceramics, metals and concrete [12].

The detonation wave moves in hemisphere shape inside the explosive charge where these hemispheres smoothly propagate through the whole explosive charge without any deflection away from the explosive boundaries and a metallic confinement. The coherent jet could be enhanced by decreasing the collapse angle of the liner which depends mainly on the angle lies between the detonation wave front and the wall of the liner [13].

In this paper, several models of shaped charge models were studied and analyzed applying a selected hydro-code named Autodyn. The geometry of wave shaper in addition to the type of its material was studied. The effect of different parameters; tip velocity, collapse angle and the angle of its collapse in addition to the material of the confinement on the jet formed were investigated.

The used shaped charge in this study is a small caliber of 40mm outer diameter and a copper liner of uniformed thickness of 1.4mm and HMX explosive material of density of 1.8 g cm^{-3}. Steel was used as casing for the charge with 6mm thickness.

2. Description of the Studies Parameters

The influence of the shape of the detonation wave on the characteristics of jet produced is studied by changing the following parameters:

2.1. The wave shaper (W.S.) geometry

The time delay of detonation wave arrival until touching the liner material is so important that the produced jet velocity and its coherency is dependent upon the applied wave shaper configuration.

In this paper, four different wave shaper geometries applied inside the high explosive charge were tested as a time delay and so as the produced jetting characteristics such as jet tip velocity, flow velocity and collapse angle. These wave former shapes are cylindrical disc of thickness 3mm and 5mm radius, the hemispherical disc is of radius 5mm, arc disc of radius 7.5mm and inclined plate are used. The shaped charge configuration without wave shaper and its interior detonation wave is illustrated in figure 1; while other W.S. of different geometries and their detonation wave shape are illustrated in figures 2-5. The material of these wave lenses is Aluminium of density 2.7g cm^{-3}.

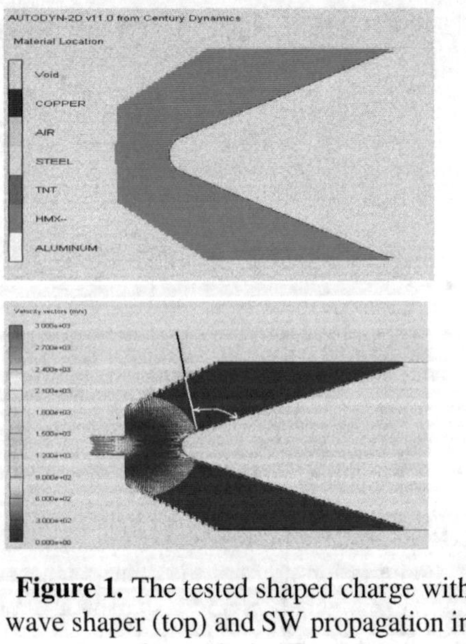

Figure 1. The tested shaped charge without wave shaper (top) and SW propagation inside the explosive (bottom)

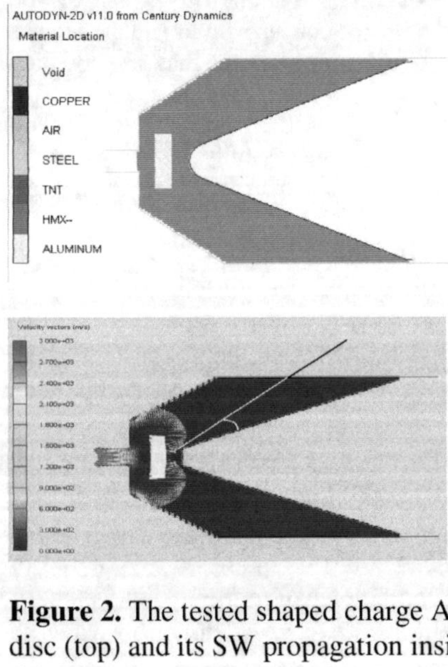

Figure 2. The tested shaped charge Al disc (top) and its SW propagation inside the explosive (bottom)

2.2. The wave shaper material (Al, lead steel, TNT and PETN)

One design of wave shaper is used to investigate the effect of different materials on the performance of this shaper on the jet characteristics. The used shaper is the inclined one using the different inert materials (Al, lead and steel) and Active explosive materials (TNT and PETN). The post detonation jet characteristics are considered to compare between the produced jet properties for these different materials.

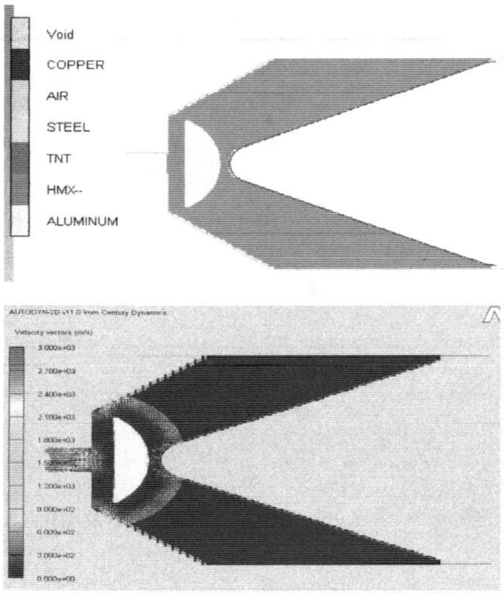

Figure 3. The tested shaped charge Al hemisphere (top) and its SW propagation inside the explosive (bottom)

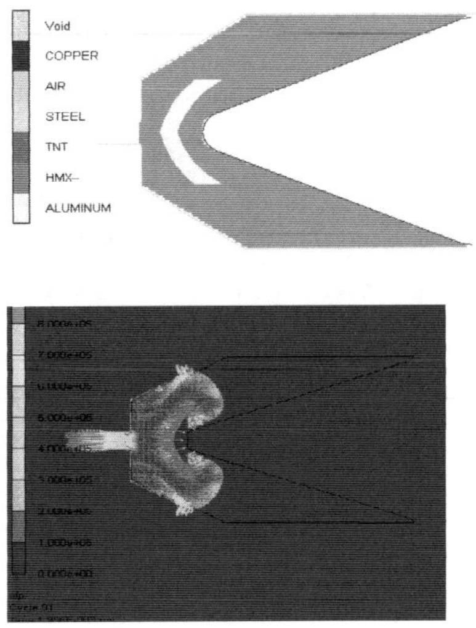

Figure 4. The tested shaped charge Al Arc disc (top) and its SW propagation inside the explosive (bottom)

2.3. The thickness of the shaped charge confinement

The shaped charge should have confinement of thin thickness (nearly 10% caliber of the charge[1]) made from light materials such as Aluminum or composite materials; but in case of oil perforation; the casing material should be thick and made from heavy materials such as steel in order to endure the downhole severe pressure and temperature conditions [14]. In this study, different models of jetting were studied using shaped charge of uniform liner thickness of 1.2mm with a cone apex angle of 40mm to determine the jet formation characteristics.

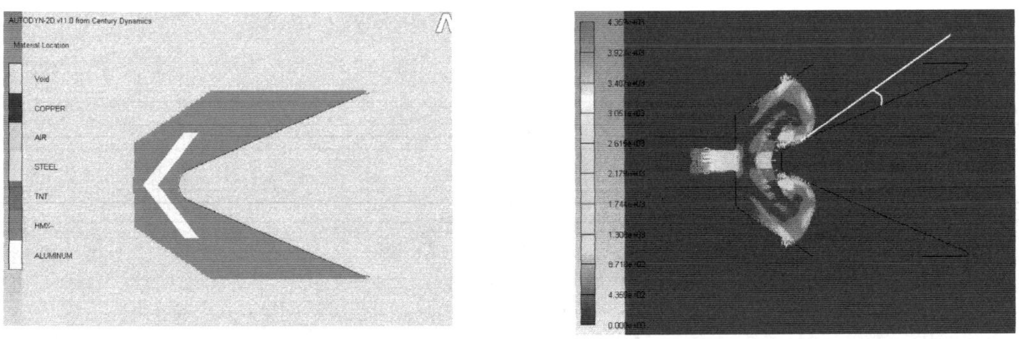

Figure 5. The tested shaped charge with inclined Al disc (left) and its SW propagation inside the explosive (right)

3. AUTODYN-3D code

The AUTODYN-3D program is non-linear dynamic analysis software developed by Century Dynamics. The program is general-purpose engineering software package that uses finite difference

IOP Conf. Series: Materials Science and Engineering **975** (2020) 012007 doi:10.1088/1757-899X/975/1/012007

technique for solving a wide variety of non-linear problems in solid, fluid and gas dynamics. This code is used to simulate the jetting phenomenon of designed shaped charge. It can predict velocity, cumulative jet mass, collapse angles, time of jet formation, particulization time, and jet mass characteristics.

Autodyn-2d-3d version of the numerical finite difference code Autodyn (AD), has a built in jetting routine which is coupled to the code in which PER –type calculation is performed when the collapse of the liner is complete [15, 16]. Lagrangian or Eulerian grids were applied to model the explosive charge while the liner was modeled as a shell in which specified mass points are used "jetting" points. A Lagrange grid was used to model the charge and wave shaper and shell segments were used to model the steel encasement. After the jetting of the specified points on the liner is complete, a history of the standard jetting analysis data is obtained together with the well known fitting curves such as cumulative jet mass and collapse angle with liner elements and liner initial position.

4. Results and discussions:

4.1. Wave shape geometry:

The jet characteristics data for the four different wave shaper configurations are listed in table 1. It was found that the jet mass for the five shaped charges is nearly constant but there is a significant difference in the velocity of the jet tip of these configurations. The shaped charge with an inclined W.S. configuration shows the highest tip velocity because the detonation wave around this barrier has been shaped to meet the liner wall with a very narrow angle between the normal to wave front and the liner wall. This angle is about 20° when the wave starts to meet the liner. The W.S. effect also appears when the collapse angle of the jet is considered. The minimum collapse angle in this configuration is 24° while it is 27.5^o in the case when Al disc W.S. is present. Therefore, a more coherent jet with higher tip velocity is expected when the inclined W.S. element is used.

Table 1. The different W.S. configurations jetting results of conical liner wall thickness of 1.2mm and cone apex angle 46° with HMX main explosive charge

Wave shaper Geometry	Al-Disc	Hemisphere	Arc disc	Inclined
Jet tip velocity (m/s)	6587.2	6703.4	6601.6	8808
Jet mass (g)	3.6157	3.5789	3.586	3.591
Jet K.E (kJ)	39.979	38.541	40.346	41.027
Min. Collapse angle B (deg.)	27.5	26.1	26.5	24
Max flow velocity (m/s)	3083	3151	3095	4190

Figure 6 illustrates an abnormal jump in the jet flow velocity near the apex angle of the cone, this may be attributed to the very narrow angle between the normal to wave front and liner wall in case of inclined W.S., which in turn accounts for the increase in the jet tip velocity in figure 7. The collapse angle of the inclined W.S. Figure 8 also indicates small angles close to the apex after which the collapse angle of the different shapers nearly have the same pattern.

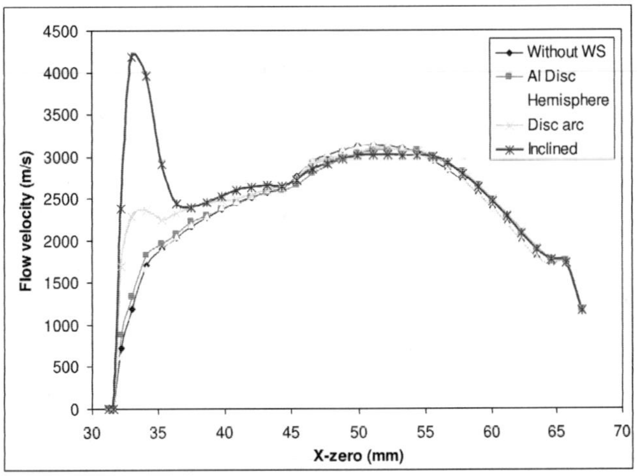

Figure 6. The flow velocity of different W.S. with initial x-position of the liner

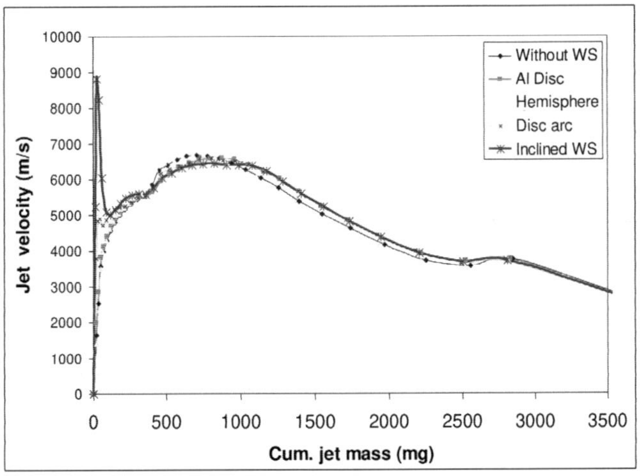

Figure 7. The jet velocity of different W.S. with cumulative jet mass

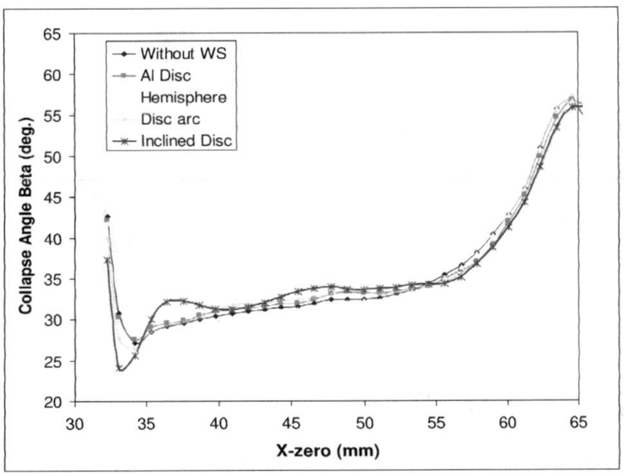

Figure 8. The collapse angle of the jet of different W.S with initial x-position of the liner

4.2. Wave shaper material

Table 2 lists the jetting output data for different W.S. materials. The inclined WS configuration is used with both active and inert W.S. tested shaped charges. It was found that Al W.S. is better than Lead and steel in the produced jet velocity but the collapse angle is greater than those of the Lead and steel. This means that the coherence of the jet in case of steel and Lead W.S. is better than that of Al W.S. but the relevant kinetic energy in case of Al is much higher than that of the others. Hence the predicted penetration depth is higher in case of Al W.S. The reason why the Al WS exhibits the best performance may be attributed to shock parameters where the detonation wave speed in case of Al is much lower than that of Lead and Steel, therefore the elapsed time for a shock wave to pass through Al barrier may be smaller than that of the two others.

The effect of explosive is also obvious in both shapers TNT and PETN. Both of these shapers resulted in much faster jet than without WS. The reason is the difference in the detonation velocity between these explosives and main charge HMX explosive. The detonation velocity of the three high brisant explosives TNT, PETN and HMX (ρ=1.89 g cm^{-3}) at the previously mentioned densities are 6930, 5170 and 9110 m s^{-1} respectively.

Table 2. The jetting output data for different W.S. materials of inclined configuration of conical liner wall thickness of 1.2mm and cone apex angle 46° with HMX main explosive charge

Wave shaper Material	Inert			Active (Explosive)	
	Al (ρ=2.8 g cm^{-3})	Lead (ρ=11.35 g cm^{-3})	Steel (ρ=7.8 g cm^{-3})	TNT [17] (ρ=1.63 g cm^{-3})	PETN [18] (ρ=1.7 g cm^{-3})
Jet tip velocity (m s^{-1})	8808	6570	6535	9872	9408
Jet mass (g)	3.5914	3.481	3.53	3.631	3.583
Jet K.E (kJ)	41.027	39.74	40.144	41.52	40.94
Min. Collapse angle B (deg.)	24	22.57	23.73	24.3	23.244
Max flow velocity (m s^{-1})	4190	3083.7	3068.8	4680	4485

4.3. Casing thickness effect

The jetting output data for the studied casing thicknesses effect on the jet characteristics are listed in table 3. This effect of casing material on the jet characteristics is attributed to the reflected detonation waves on the casing surface inside the explosive itself, which meet the liner with different incidence angles between the detonation front and liner wall axis.

The subsequent reflected waves from the casing surface can produce regions of high pressure on the liner surface resulting in a jet with higher tip velocity [19]. This was verified by adding 7 gauge points to the shaped charge, shown in figure 9. It can be observed that both the obtained pressure-time histories shown in figure 10-a,b for the two cases (i.e. 1 and 8mm casing thicknesses) have nearly the same pattern However, the impulse-time history in figure 11-a,b explained the reason why the jet tip velocity for 8mm case is higher than that for 1mm case. This velocity is preferred to be as much as

IOP Conf. Series: Materials Science and Engineering **975** (2020) 012007 doi:10.1088/1757-899X/975/1/012007

possible provided being under a certain critical value defined by Mach limit. The Mach number limit for the coherent jet is calculated according to the equation [20]:

$$Ma = \frac{v_f}{c_o} < 1.23$$

where v_f is the flow velocity of the jet and c_o is the sound speed in the liner material.

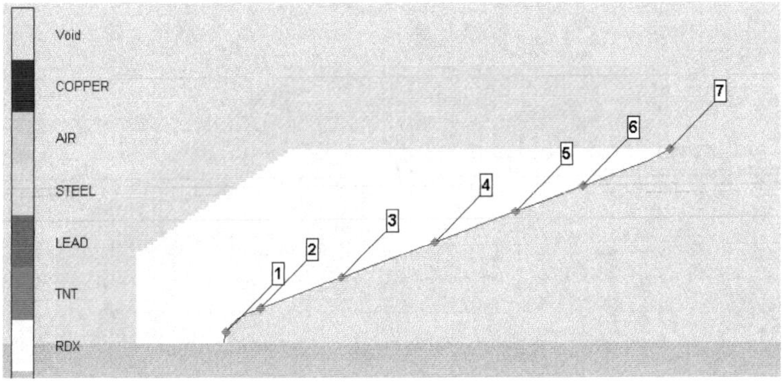

Figure 9. Different fixed target points along the liner axis to predict the p-t history of the DW inside the explosive charge using 8mm casing wall thickness

The Mach limit velocity of the copper material (v_f) is 4676 m s^{-1}. So, all the resulting jets are coherent because their collapse velocities are lower than the Mach number limit for the coherent jet calculated from bulk speed of sound of the copper material which is 3940 m s^{-1} [20] while the maximum collapse velocity for the highest velocity tip of casing 8mm model was 3748 m s^{-1}.

Table 3. The jetting analysis data obtained from the jetting analysis of OWP using RDX main charge with 1.4mm liner wall thickness and 40° cone apex angle and different confining thickness

Wave shaper	Casing thickness				
Casing thickness	1mm	2mm	4mm	6mm	8mm
Jet tip velocity (m s^{-1})	6489.4	6540.3	6790.5	7035.5	7232.8
Jet mass (g)	5.6449	5.6601	5.693	5.718	5.7567
Jet K.E (kJ)	35.6	39.1	43.3	49.2	54.1
Min. Collapse angle B (deg.)	22.42	22.42	22.42	22.42	22.42
Flow velocity (m s^{-1})	3115.4	3213.8	3549.3	3671	3748.1

Figure 12 illustrates the profile of flow velocity with the initial liner element position. This means that all the selected casing thicknesses are suitable to produce a coherent jet. But, an optimization for this casing should be done in accordance with the lowest casing thickness that is capable of confining the oil well perforators (OWP) explosive charge and protecting it against premature explosion. Both the jet velocity with cumulative jet mass; figure 13 and collapse angle with the initial liner position; Figure 14 exhibit the same pattern for the different casing thicknesses although a significant difference between them illustrates how much the casing thickness will affect the jet characteristics.

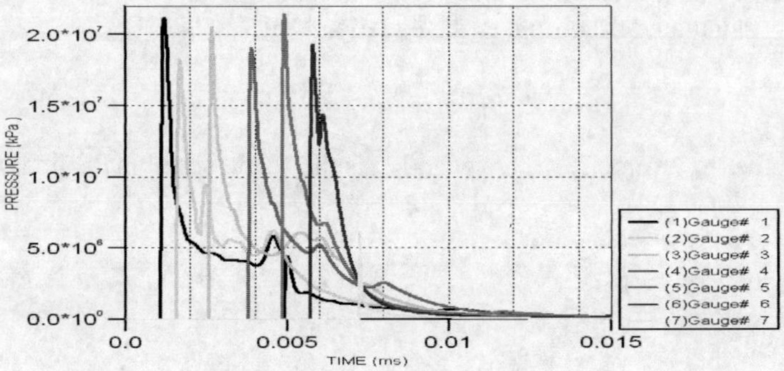

Figure 10 (a) P-t history of OWP with 1mm casing thickness

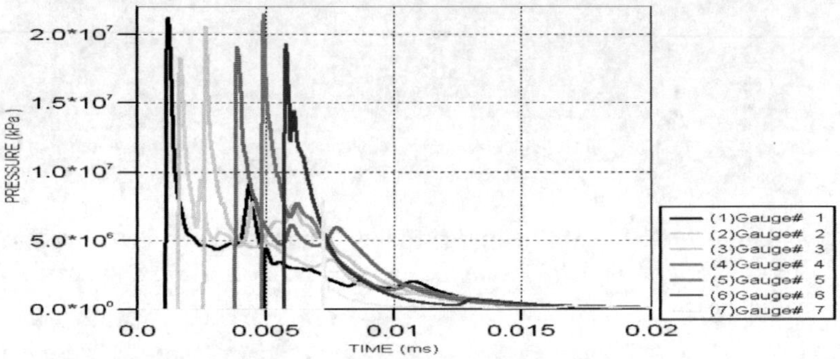

Figure 10 (b) P-t history of OWP with 8mm casing thickness

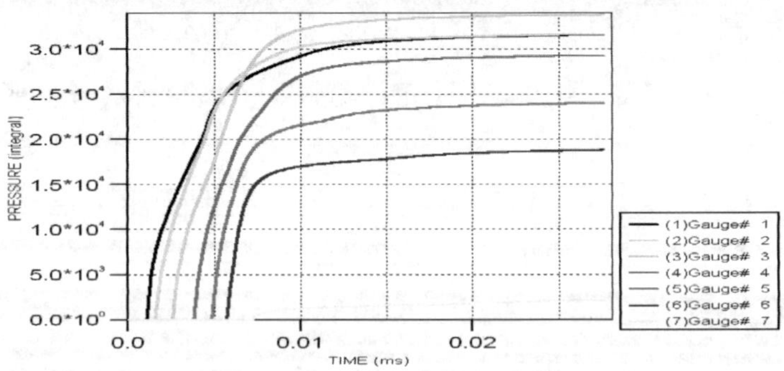

Figure 11 (a) Impulse –t history of OWP with 1mm casing thickness

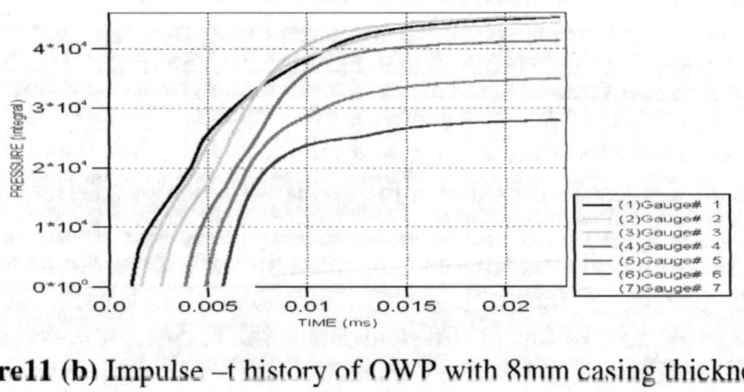

Figure11 (b) Impulse –t history of OWP with 8mm casing thickness

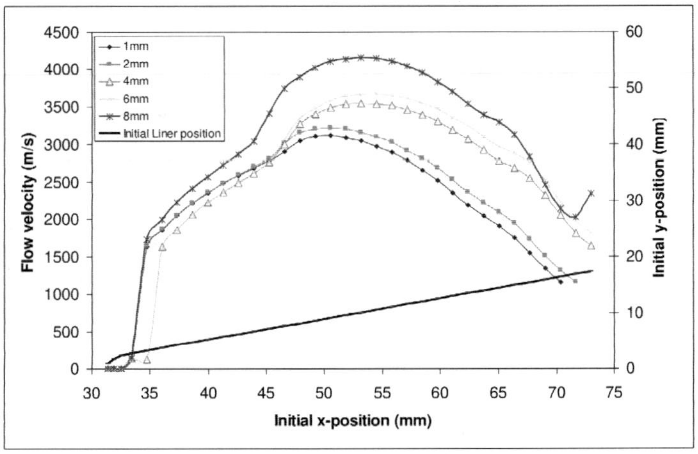

Figure 12 The flow velocity of different casing thicknesses with initial x-position of the liner

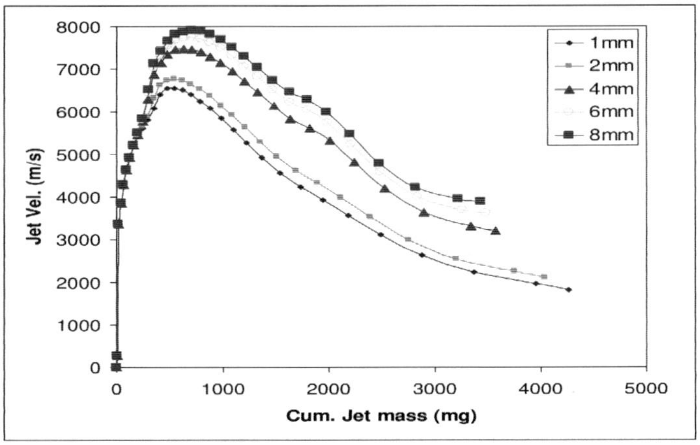

Figure 13 The jet velocity of different casing thicknesses with cumulative jet mass

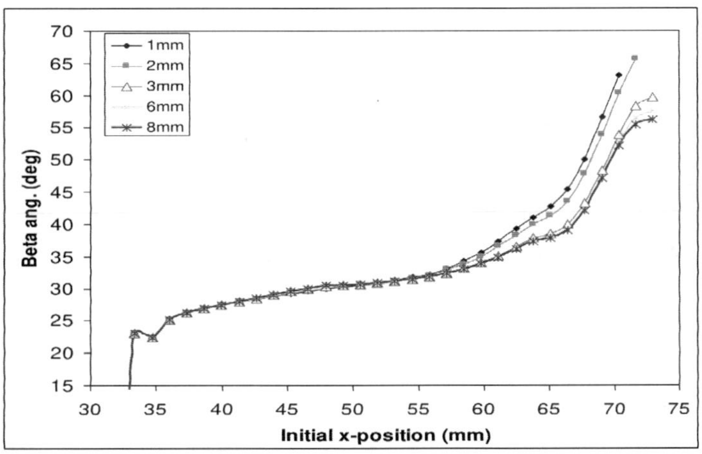

Figure 14 The collapse angle of the jet of different casing thicknesses with initial x-position of the liner

IOP Conf. Series: Materials Science and Engineering **975** (2020) 012007 doi:10.1088/1757-899X/975/1/012007

5. Conclusion

From the above work analysis including the W.S. geometry, material, and the shaped charge confinement thickness, the following conclusions have been drawn:

- The effect of the applied waver on the produced jet is not only attributed to the time delay; but also to the shape of the detonation wave where a narrower angle between the wave front normal and the liner wall will produce a faster jet, hence increasing the jet velocity.
- The casing material thickness has a significant effect on the formed jet characteristics such as jet tip velocity and its kinetic energy, therefore the predicted penetration capability of this jet with a thick confinement is predicted to be much higher than that with a thin confinement.
- The jet collapse angles for the tested casing thicknesses charges have nearly the same pattern and the same minimum collapse angle at the apex but at the region close to the liner base, a significant difference appears where the thick confinement produce narrower angles than that with thin confinement. Hence a better coherent behavior for the thick confinement shaped charge is observed.
- The W.S. effect is improved by imploding an explosive barrier such as TNT and PETN. The difference in the main charge and W.S. detonation velocity will create a two velocity gradients jet which could be effective to open holes in rock to connect the well bore to oil reservoir when the shaped charge is used as oil well perforator.

6. References

[1] Walters P and Zukas J 1989 Fundamentals of shaped charge: Wiley Interscience Publication, New York.

[2] Klapötke T M 2015 Chemistry of high-energy materials, Walter de Gruyter GmbH & Co KG

[3] Licht H-H 2000 *Propellants Explosives, Pyrotechnics* **25** 126-132

[4] Pelikán V, Zeman S, Yan Q L, Erben M, Elbeih A and Akštein Z 2014 *Central European Journal of Energetic Materials*, **11(2)** 219-235.

[5] Zeman S, Elbeih A and Yan Q L 2013 *Journal of Thermal Analysis and Calorimetry,* **111(2)** 1503-1506

[6] Elbeih A and Zeman S 2014 *Central European Journal of Energetic Materials* **11** 501-514

[7] Arnolda W and Rottenkolber E 2013 *Procedia Engineering* **58** 184-193.

[8] Elbeih A, Husarova, A and Zeman S 2011 *Central European Journal of Energetic Materials* **8(3)** 173-182.

[9] Yan Q L, Zeman S, Sánchez Jiménez P E, Zhang T L, Pérez-Maqueda L A and Elbeih A 2014 *J. Phys. Chem. C* **118** 22881-22895.

[10] Elbeih, A, Zeman, S and Pachman, J 2013 *Central European Journal of Energetic Materials* **10(3)** 339-350

[11] Elshenawy T, Elbeih A and Li Q M 2018 *International Journal of Mechanical Sciences*, **136** 234-242.

[12] Courtney-Green P R 1991 *Ammunition for the land battle*, Potomac Books Inc, UK.

[13] Zeman S, Yan Q.-L and Elbeih A 2014 *Central European Journal of Energetic Materials*, **11(3)** 395-404.

[14] Gui-Xi L 1995 *Propellants, Explosives, Pyrotechnics*, **20** 279-282.

[15] Grove B, Heiland J and Walton I 2008 *International Journal of Impact Engineering* **35** 1563–1566.

[16] AUTODYN Team 1997 *Theory Manual*, Revision 3.0.

[17] Zeman S and Elbeih A 2011 *Hanneng Cailiao/Chinese Journal of Energetic Materials*, **19(1)** 8-12.

[18] Elbeih A, Abd-Elghany M and Klapötke T M and Zeman S 2017 *Journal of Analytical and Applied Pyrolysis*, **126** 267-274.

[19] Elbeih A, Abd-Elghany M and Elshenawy T 2017 *Acta Astronautica*, **132** 124-130.

[20] Hasenberg D 2010 *naval postgraduate school*, Monterey, California, Report no. 93943-500.

The Potentials of Aluminium Nanoparticles: Novel High Energy Density Material for Underwater Explosions

Sherif Elbasuney [a,b*], M. Gaber Zaky [b], Mostafa Radwan c, Mohamed Bennaya [d], Sherif M. Abdelkhalek [e].

[a] Head on nanotechnology research centre, Military technical college, Cairo, Egypt.
[b] School of Chemical Engineering, Military Technical College, Cairo, Egypt.
[c] British University in Egypt, Elshorouk City, Cairo, Egypt.
[d] School of Ships and submarines engineering, Military Technical college, Cairo, Egypt.
[e] Head of Engineering and Technology Research Center, Military Technical College, Cairo, Egypt.
s.elbasuney@mtc.edu.eg, sherif_basuney2000@yahoo.com

Abstract. The destructive parameters of underwater explosives (i.e. shock wave energy, maximum pressure, and bubble radius) are limited to explosion heat; that is comparatively low. One approach for enhanced heat output can be accomplished by integrating reactive metal particles (i. e. Aluminium). However conventional aluminium particles (μm size) would contribute only with combustion gaseous products behind detonation wave front. Underwater, there is no oxygen for such contribution to take place. Furthermore, conventional Al particles could decrease the detonation velocity. So far, full exploitation of aluminium particles in underwater explosions has not been accomplished. Aluminium nanoparticles would combust more efficiently within detonation wave front, offering smaller critical diameter, high reaction rate, and high heat release rate. Consequently, Al nanoparticles could be ideal high energy density material for underwater explosion. Ship model with positive metacentric height, GMT =4.7 cm for ship transverse stability, and GML = 19.3 for ship longitudinal stability was designed. Ship model offers large angle stability (heeling angles= 0-70 deg.). 2 g of explosive charge was detonated underneath the developed naval structure. Upon explosion, the acceleration of the naval structure was measured using shock accelerometer VC tri-axial, high frequency, 5000 ground acceleration, Dytran, Inc. While, Al particles (10 μm) offered an increase in mono-hull acceleration by 16 % compared to TNT; Al nanoparticles offered an acceleration increase by 49 %. This novel finding can be ascribed to the efficient combustion of Al nanoparticles within detonation wave front offering ideal detonation reaction with enhanced destructive effect.

Keywords: Underwater explosion (UNDEX); Metalized explosives; Shock wave.

Introduction

Underwater explosion (UNDEX) is defined as the violent balance disturbance owing to detonation of explosives in water environment. Different weapons can be employed in naval warfare such as naval mines. Such weapons are submerged enough, a passing ship will initiate its firing mechanism [1-2]. Torpedoes are self-propelled guided projectiles that operate underwater [3]. The influence of underwater explosion does not constitute a single impulse but a few large energy pulsations of gas bubbles [4].

IOP Conf. Series: Materials Science and Engineering **975** (2020) 012008 doi:10.1088/1757-899X/975/1/012008

1.1 Underwater explosives

Many trials have been performed to develop more powerful explosive compositions, for underwater explosions (UNDEX). TNT compositions are known to be less sensitive, with extended shelf life [5]. Enhanced performance of TNT compositions for UNDEX is under improvement. Aluminium reaction at the Chapman-Joguet (C-J) plane is relatively slow. Therefore, aluminium could not contribute efficiently to the destructive effect of explosive materials. Aluminium contribution would be limited primarily to gaseous products, offering sustained pressure wave [6-8]. In UNDEX there is no oxygen available for such contribution to take place.

1.2 Shock wave propagation

Cole reported that chemical reaction in UNDEX takes place in two parts: the initial chemical reaction and the detonation process. If initial chemical reaction generated a high pressure; a thermo-mechanical shock wave "detonation wave" will propagate at supersonic speed through the explosive material producing intense heat [9-10]. The detonation gaseous products can offer a very dense, superheated, and spherical gas bubble; at this instance the detonation process of underwater explosion is complete. Shock wave can be defined as the pressure wave created by the arrival of the detonation wave at the boundary between gas bubble and water. Shock wave propagates in water as a spherical wave at supersonic speed [11]. The shock wave energy (Esw) can be represented as equation 1.

$$E_{sw} = 98000\,(W^{1/3})\,(W^{1/3}/R)^{2.1} \quad (1)$$

Where: Esw, W, and R are shock wave energy in (J/m^2), equivalent weight expressed in kg of TNT (Q_{exp}/Q_{TNT}), and standoff distance in m. Furthermore the maximum peak pressure (P_m) can be calculated according to equation 2:

$$P_m = 52.16\,(W^{1/3}/\,R)^{1.13} \quad (2)$$

Where: P_m is the maximum peak pressure in MPa, W is the mass of charge expressed in equivalent value of TNT, and R is the standoff distance in meter. It can be concluded that the main UNDEX parameters including shock wave and bubble energy are limited by the explosion heat output that is comparatively low.

1.3 Gas bubble behavior and bubble pulse loading

The gas bubble begins to expand by explosion hot gases. The expansion continues until the internal gas bubble pressure is at its minimum and the diameter is at its largest value [3, 11]. The bubble begins to contract until the compression of the gaseous products stop the contraction, force the bubble to expand again making an oscillating system [3]. Cole and Reid, displayed the shape of the bubble pulses as a gradual rise to the peak pressure followed by a concave shaped decay to hydrostatic pressure [12-13]. The migration of the bubble has a significant effect on the shape of the bubble pulse. It causes the bubble pulse to develop a triangular shape figure 1.

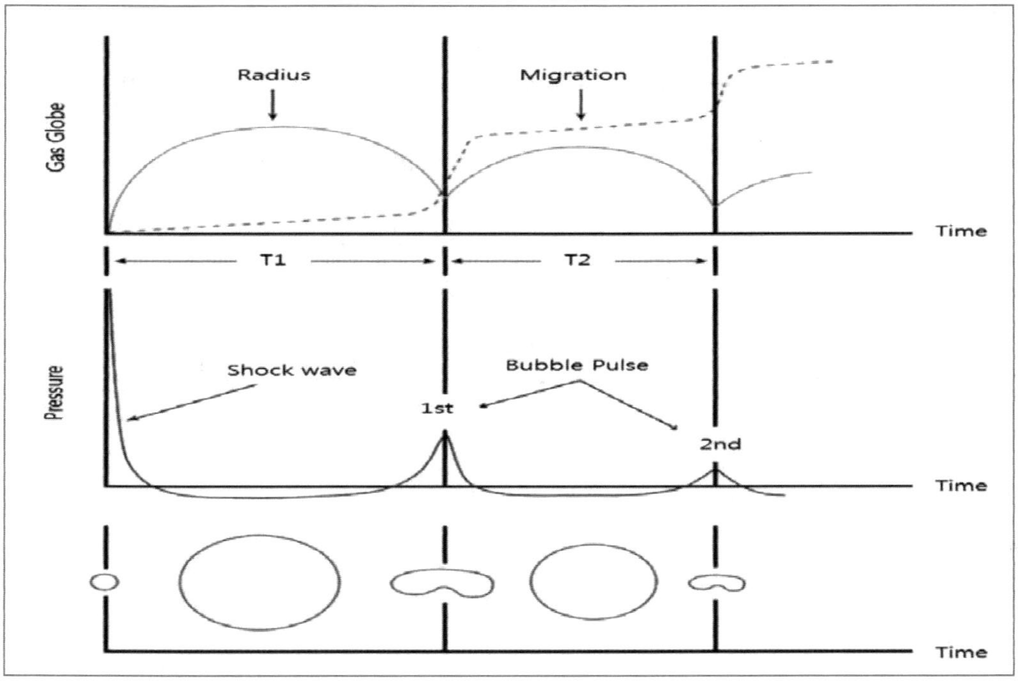

Figure 1. Bubble migration behavior versus bubble expansion.

The velocity of the bubble pulse is similar to the shock wave; it propagates at the speed of sound in water [14-15]. The decay time constant Θ in (μ sec) can be expressed by Equation 3.

$$\Theta = 96.5\,(W^{1/3})\,(W^{1/3}/\,R\,)^{\,0.22} \tag{3}$$

The total impulse can be expressed as equation 4.

$$I = 5760\,(w^{\,1/3})\,[w^{1/3}/s]^{\,0.891} \tag{4}$$

Where: I, S, and W are specific impulse in Ns/m^2, Stand of distance in m, TNT equivalent in Kg (Q_{exp}/Q_{TNT}) [16]. The maximum bubble radius can be estimated by Equation 5:

$$A_m = (13\,Q_v\,M\,/\,z)^{\,1/3} \tag{5}$$

Where; Q_v is the heat of detonation (MJ/kg), M is the mass of explosives (kg) [14]. When the frequency of the shock response and bubble pulse response matches the resonant frequency of the ship, whipping occurs. The whipping motion of the ship can break the ship's hull or cause other types of severe structural damage [12-13].

1.4 Damages due to UNDEX

The shock can develop a large hole in case of the small hull thickness and sufficient explosive charge [2]. The hull is heavily deformed; the level of deformation decreased with the increase in

IOP Conf. Series: Materials Science and Engineering **975** (2020) 012008 doi:10.1088/1757-899X/975/1/012008

standoff distances. However different portions of the ship will respond at different velocities depending upon the mass per unit area [2]. Another damage mechanism is the bubble collapse. If the oscillating gas bubble is close enough to a rigid body surface such as a submarine or ship hull. The differential pressure created due to bubble volume decrease could result in bubble collapsing onto the hull with the development of a high speed water jet [2].

1.4.1 Surface cut-off phenomena.
Surface cut-off occurs when a plane compressive wave hits a free surface. Compressive wave would be reflected off at that surface as a tensile wave. The reflected wave could interact with (cancels out) the compressive wave. Consequently a slightly negative pressure could be developed figure 2.

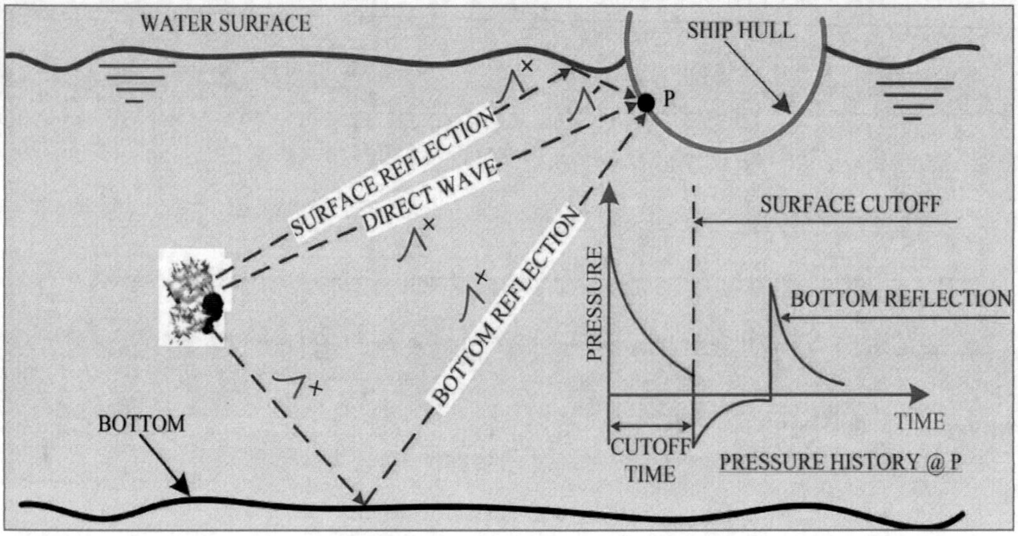

Figure 2. Shock wave pressure profiles with cut-off time [17]

Surface cut-off phenomena means that over pressure applied on the hull may be suddenly dropped to negative value ; this action could force the hull to move towards the water surface [2].

1.4.2 Cavitation damage.
Cavitation phenomenon occurs when there is a region of negative absolute pressure in water. The phenomenon of bulk cavitations occurs when a shock-wave is reflected off a free surface such as air/water interface. The compression shock-wave reflects off the free water surface as a tensile wave and since water can only sustain a very small level of tension it begins to cavitate figure 3.[18-19].

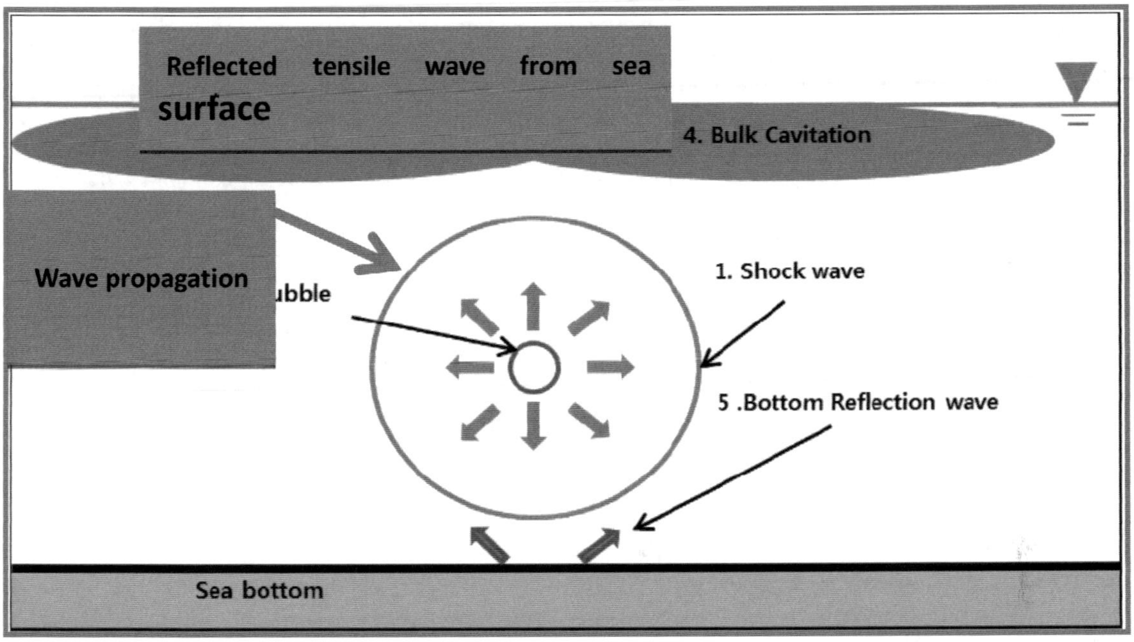

Figure 3. Bulk cavitation developed during underwater explosion [2]

Cavitations can be divided into two types local cavitations and bulk cavitations. Bulk cavitations is considered as large region of low pressure, at the free surface [18]. Fluid-structure interaction takes place at certain parts of the ship's hull. The overpressure turns out to be negative. Water cannot sustain tension; therefore local cavitations occurs [20].

1.5 Destructive effect measurement

1.5.1 Shock wave measurement.

Measurements for UNDEX are entirely different from air explosions. The main parameter to be experimentally determined is the pressure time period. Pressure sensors can offer measurements of pressure-time profile; therefore shock wave energy can be calculated.

1.5.2 Structure response subjected to underwater explosion.

Displacement and acceleration of naval structure caused by underwater explosion is of immense significance for performance evaluation and in determining the naval structure safety limits.

Explosion heat output is the key parameter affecting UNDEX performance i. e. shock wave and bubble energy. Heat output of common explosive is quite low, and is limited by their heat of formation. Reactive metal particles (i.e. Al) can act as effective high energy density material. Therefore reactive metal particles can propose extended pressure pulse. This improved performance can be correlated to high thermal loading developed behind C. J. plane [21]. The reaction rate of reactive metal particles i. e. aluminium (Al) is relatively slow. Therefore Al particles would act as inert component at the Chapman-Joguet (C-J) plane. Conventional Al particles would contribute with detonation products behind C-J plane [22-23]. Such contribution could not be achieved in UNDEX as there is no air oxygen to react with Al particles. So far, full exploitation of aluminium particles in UNDEX has not been achieved [9, 24-26].

Al nanoparticles can offer an increase in detonation heat and detonation velocity, presumably due to the efficient combustion of Al nanoparticles within the detonation zone. It is expected that Al nanoparticles could be consumed more rapidly in the CHNO reaction zone, as compared to larger particles [27].
Recently, aluminized explosive formulations employing small aluminium particles were developed to produce high initial detonation temperatures, pressures, and velocities. They can offer high expansion blast capabilities [28-29]. Effective development of novel cast metalized formulation based on TNT and Al nanoparticles for UNDEX has been developed. During explosive conversion, aluminum nanoparticles would combust more efficiently within detonation wave front. Consequently this approach could offer ideal combustion process. Naval structure with high floating stability was designed. The effectiveness of developed explosives for UNDX was evaluated via measuring acceleration of floating structure. Upon explosion, the acceleration of the naval structure (mono-hull) was measured using shock accelerometer VC tri-axial, high frequency, 5000 ground acceleration, Dytran,Inc. While, Al particles (10 µm) offered an increase in mono-hull acceleration by 16 % compared to TNT; Al nanoparticles (100 nm) offered an acceleration increase by 49 %. This novel finding can be ascribed to the reactivity and high interfacial surface area of Al nanoparticles. This manuscript would open the route for the effective development and testing of novel cast metalized formulation where full energy could be released and exploited within the detonation wave front.

Experimental Work

2.1 Development of ultra-fine aluminium particles
Commercial aluminium particles with particle size ≤ 32 µm were dispersed in 250 ml isopropyl alcohol (IPA). The slurry was ground using ball mill Retch 100 for 24 hours. IPA was employed to dissipate any developed heat during grinding. Morphology of developed aluminium particles was visualized TEM. Dry powder morphology was investigated with SEM , Zeiss EVO-10 by Carl Zeiss Corporation. The crystalline phase was investigated with XRD, D8 advance by Burker Corporation over the angle range 2θ from 20 to 90 degrees.

2.2 Integration of metallic fuels particles into energetic matrix
Metallic fuel was effectively dispersed in IPA using ultrasonic probe homogenizer, to break down any aggregates and to achieve even dispersed particles. Integration of colloidal particles into energetic matrix is an effective approach for enhanced dispersion characteristics [30-33]. The colloidal particles were integrated into molten TNT under stirring.

2.3 Design of Mono-Hull structure
2.3.1 Geometric model for the ship model.
A 3-D mono-hull model for the ship was developed using Inventor (3D CAD software) as demonstrated in figure 4.

After developing the 3D CAD model, naval architecture software MAXSURF had been employed to for floating stability study.

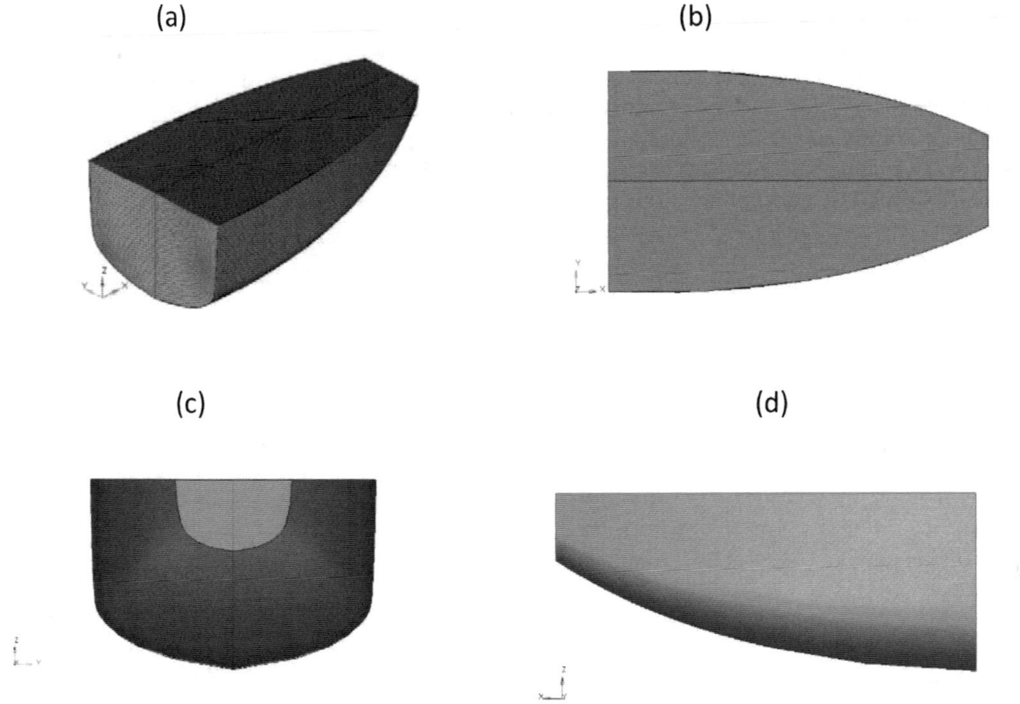

Figure 4. 3D Geometric model for the ship model: isometric view (a), plan view (b), profile view (c and d).

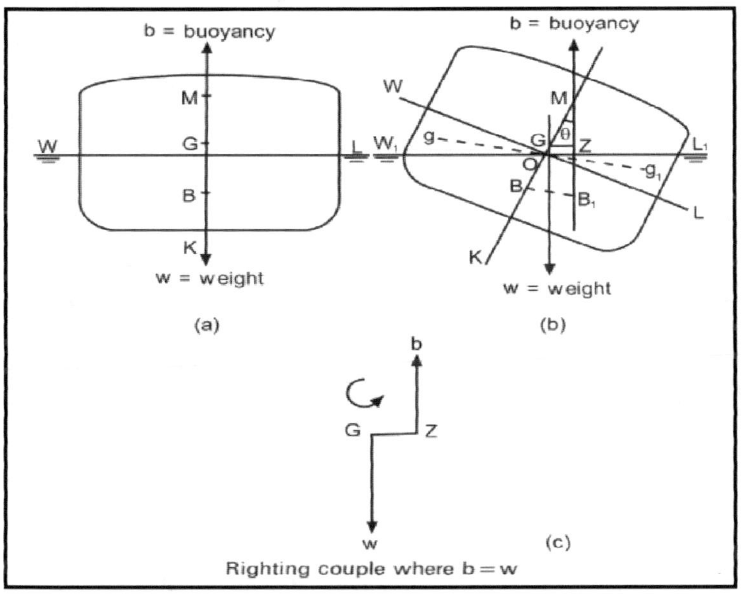

Figure 5. Stable equilibrium of a ship

IOP Conf. Series: Materials Science and Engineering **975** (2020) 012008 doi:10.1088/1757-899X/975/1/012008

2.3.2 Ship model stability

The naval structure can be considered stable if it has the ability to maintain to its equilibrium position after heeling or pitching. Floating stability was investigated to secure stability under sever working conditions during explosions. The ship is considered stable when vertical position of center of gravity (G) is lower than the position of transverse meta-center M figure 5.

Under these conditions, if the ship is heeled with an angle θ, an arm GZ will be developed. GZ arm will creates a righting moment that get equilibrium upright position.

2.3.3 Ship model simulation using MAXSURF

MAXSURF was employed to build ship model and to calculate the model hydrostatics and its stability under severe conditions. The 3D model of the ship was imported in the modeler module to develop the ship model figure 6.

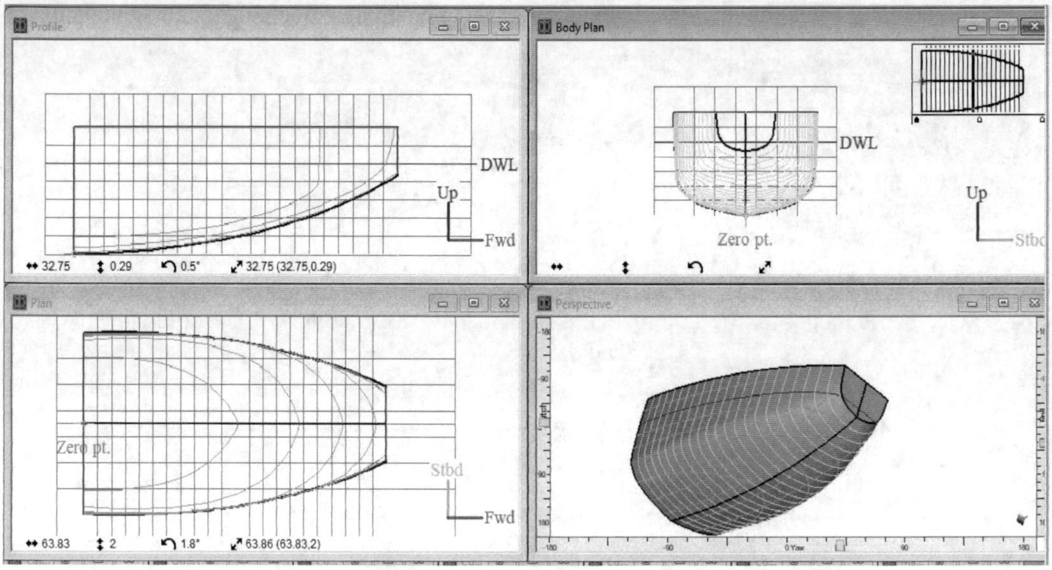

Figure 6. Ship model in modeler module under MAXSURF

a) Ship model initial stability

The hydrostatics was calculated using the MAXSURF Modeler module at design water line (DWL). The vertical center of gravity (VCG) is assigned at 7.4 cm figure 7.

The main hydrostatic results for the ship model are presented in table 1.

IOP Conf. Series: Materials Science and Engineering **975** (2020) 012008 doi:10.1088/1757-899X/975/1/012008

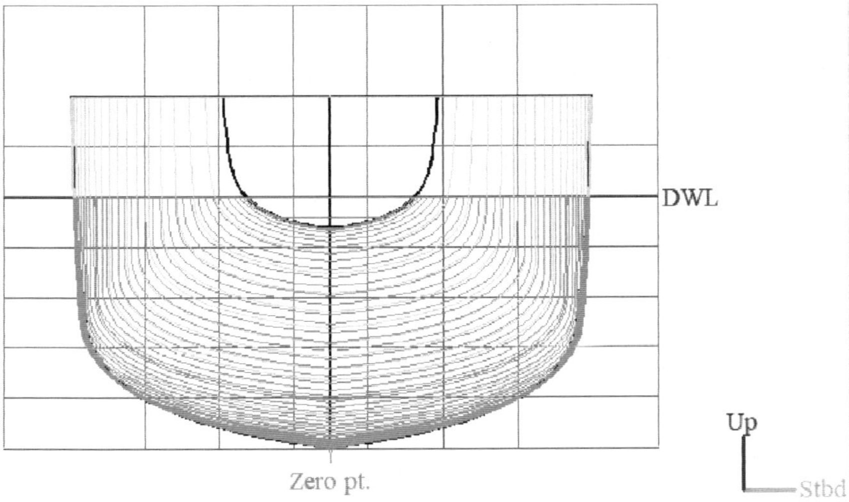

Figure 7. Body plan submerged volume at DWL

Table 1. Ship model hydrostatic calculations

Displacement	7.147	kg
Volume (displaced)	6972.7	cm^3
Draft Amidships	11.36	cm
Immersed depth	11.35	cm
Prismatic coeff. (Cp)	0.666	
Block coeff. (Cb)	0.563	
LCB length	16.6	from zero pt. (+ve fwd) cm
LCF length	19.92	from zero pt. (+ve fwd) cm
KB	7.12	cm
KG fluid	7.4	cm
GMT	4.8	cm
GML	19.3	cm

The ship model displacement =7.147 kg. This model is initially stability, as GM > 0. The main stability criteria include:

The model is transversely stable as GMT= 4.8 cm

The Model is longitudinally stable as GML= 19.3 cm

IOP Conf. Series: Materials Science and Engineering **975** (2020) 012008 doi:10.1088/1757-899X/975/1/012008

b) *Ship model upright hydrostatics*

As the ship will be to an explosion, it will be subjected to transient motions as follows:

Heaving motion (motion in vertical direction)

Rolling motion (motion about longitudinal direction)

Pitching motion (motion about transverse direction)

Upright stability analysis was conducted. The draft was assigned to be 8-13 cm. The calculated upright hydrostatics at draft=10 cm is represented in Figure 8.

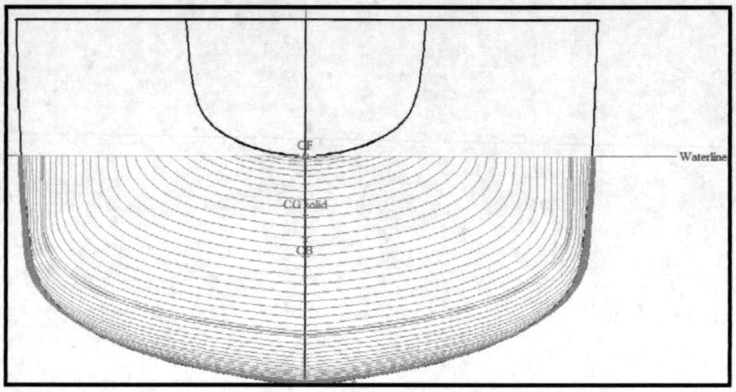

Figure 8. Upright hydrostatics

If the ship is inclined at any draft it tends to maintain its initial position; as the centre of gravity (CG) is below the metacenter (M) figure 9.

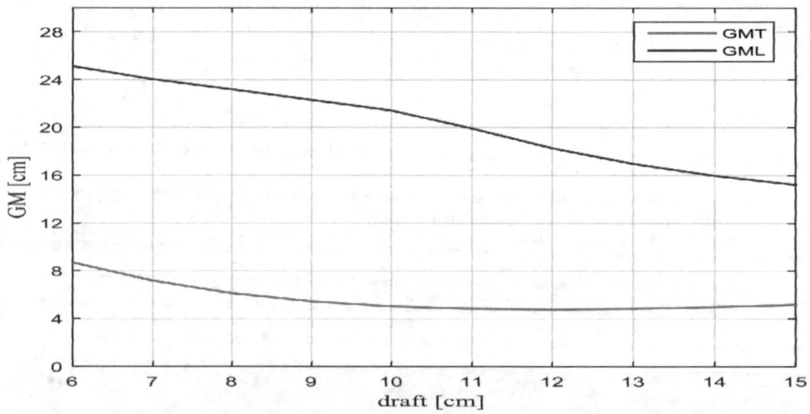

Figure 9. Change of GM with draft

The transverse and longitudinal stabilities are conserved as the GMT and GML have positive values. The ship model has upright stability for large draft range (6-15 cm). These severe drafts are expected during underwater explosions.

c) Ship model large angle stability

During explosions, the ship might be subjected to sever heeling (Rolling) conditions. Consequently, large stability angle must be secured. Large angle stability was conducted using stability module for heeling angle range=0 - 70 deg. (Figure 10).

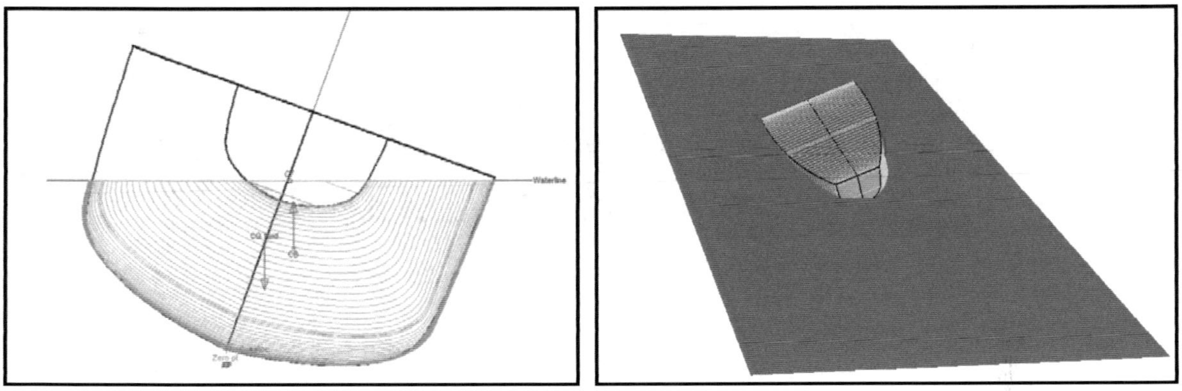

Figure 10. Large angle stability for heeling 20 deg.

Body plan and 3-D isometric view for the ship model simulation during large angle stability calculations at rolling angle =20 deg. Calculations for the large angle stability are represented in figure 11.

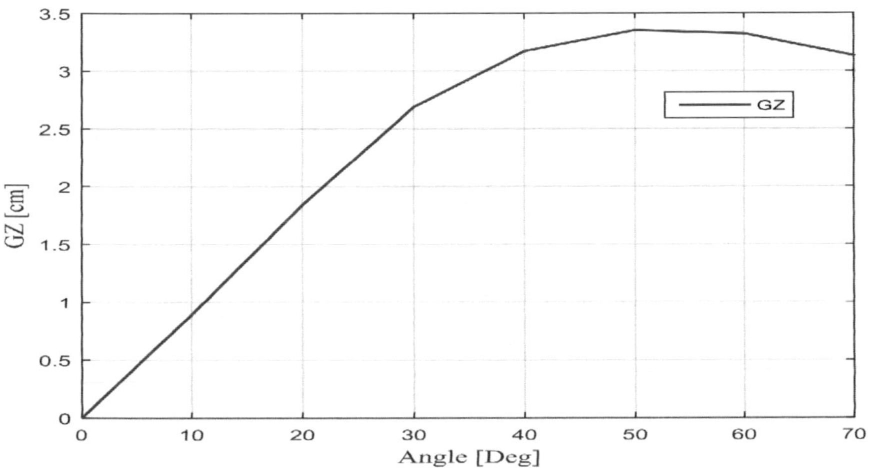

Figure 11. GZ curve

Ship model demonstrated positive righting arm GZ for large heeling angles (GZMax at 50 deg.). GZ arm will bring the ship to its initial position after severe rolling during the explosion activities. The designed model was manufactured and accelerometers were mounded on the ship as represented in Figure 12.

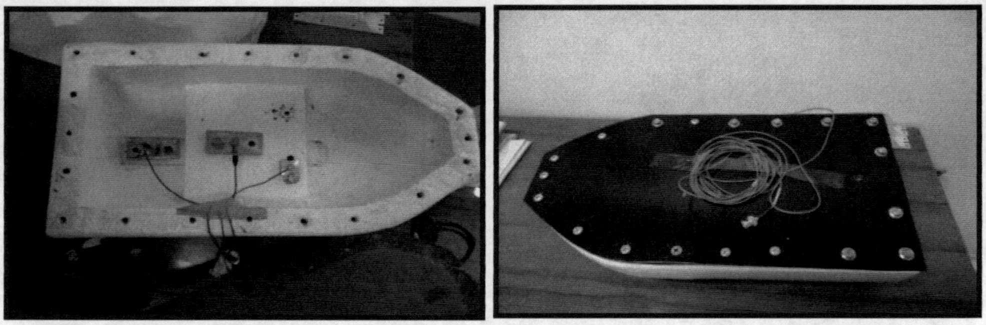

Figure 12. Manufactured ship model

2.4 Underwater explosion strength

UNDEX will produce shock wave in the water as well as bubbles of gaseous products. The explosion strength of different exploding charges was evaluated through the measurement of the acceleration of naval structure subjected to explosion. The acceleration of the ship model was employed as an indication for the strength of explosive charge. Steel tank 1 m (L) ×1 m (W) ×1.5 m (H) was filled with water for conducting the experimental measurements. Floating ship model was assembled with interior accelerometer. 2 g explosive charge was exploded underneath the ship at 50 cm. The target acceleration due to explosion was measured.

3. Results and Discussions

3.1 Characterization of Al particles

The morphology of employed commercial Al particles was investigated with SEM. Al particles in the shape of spheres with 10 μm average particle size were reported from SEM micrograph figure 13-a. aluminium nanoparticles in the shape of plates of 100 nm were visualized using TEM figure 13-b.

3.2 Theoretical performance evaluation

Detonation heat and volume of gases products represent the ability of explosive charge to develop gas bubble and effective incident shock damage. Al particles could enhance the heat output offering enhanced UNDEX parameters including shock wave energy (Esh), totl impulse (I), decay time constant (ϴ) equations 1-3. Table 1 summarize the impact of Al particles on main UNDEX parameters.

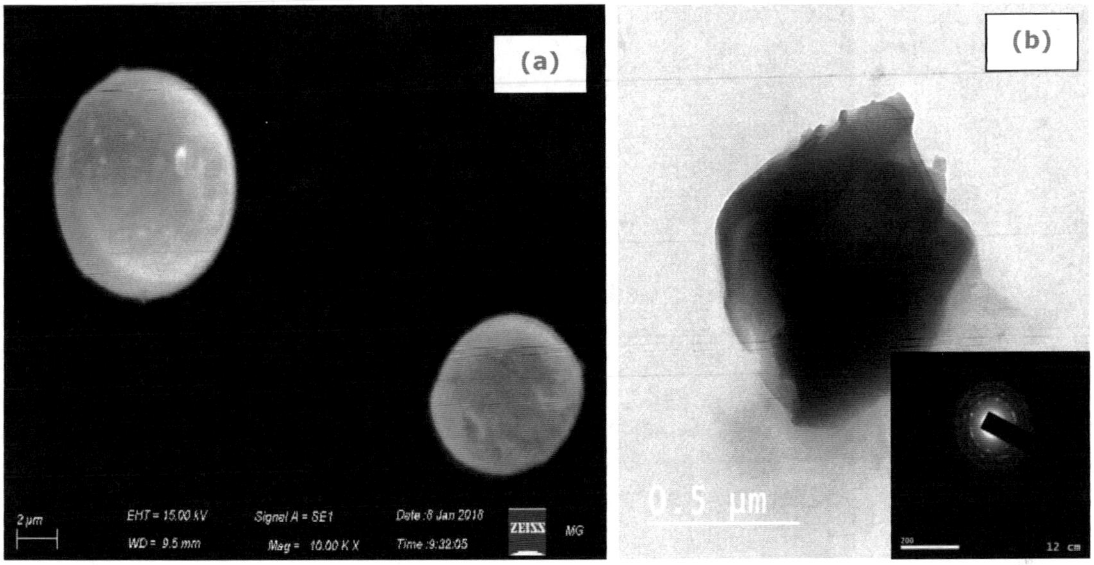

Figure 13. Morphology of employed conventional Al particles (10 µm) (a), Al nanoparticles (100 nm) (b) using SEM and TEM respectively.

Table 2. Impact of Al particles on UNDEX shock wave parameters

Shock wave parameters	TNT	TNT+ Al	Increment (%)
Esh (J/m)	559.0	762.5	36.5%
I (Ns/m)	195.2	236.0	21%
Θ (µ sec)	16.8	18.2	8.2%

However the employed empirical equations do not take into account the particle size. Al nanoparticles could contribute more efficiently within detonation wave front compared with conventional Al particles.

In this study, the impact of reactive metal particles on destructive effect of TNT for UNDEX was evaluated. 2 g of different explosives including pure TNT, Al (10 µm) / TNT, and Al (100 nm) /TNT were exploded, 50 cm underneath the floating naval structure. The solid loading level of reactive metal particles was limited to 12 w t%. Shock accelerometer was employed to measure the structure dynamic response. Practical measurement of floating structure acceleration is demonstrated in figure 14.

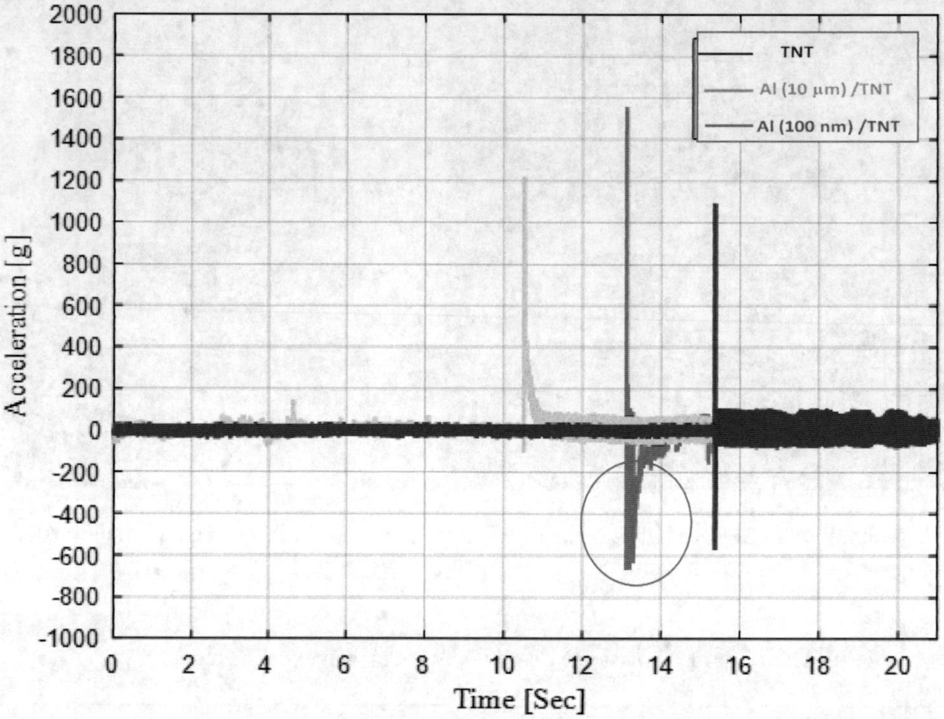

Figure 14. The impact of Al particles on TNT strength.

Whereas conventional Al particles (10 μm) offered enhanced ship acceleration by 16 % compared to TNT; Al nanoparticles (100 nm) offered an acceleration increase by 49 %. Furthermore Al nanoparticles offered a broad deceleration peak (circle in red).

This novel finding was ascribed to the reactivity of aluminium nano-particles; as well as efficient combustion within detonation wave front. This could offer higher detonation temperature and could speed up the detonation reaction compared with conventional Al particles. This approach could offer full exploitation of Al particles with UNDEX. It has been reported that Al nanoparticles have tendency to react with not only with oxygen but also with the inert decomposition gases adding additional exothermic reactions [34-35].

$$2Al_{(s)} + 3/2\ O_2 \longrightarrow Al_2O_{3\,(s)} + 1700\ KJ/mol \quad (6)$$

$$2Al_{(s)} + 3CO_{(g)} \longrightarrow Al_2O_3 + 3C + 1251\ kJ/mol \quad (7)$$

$$2Al_{(s)} + 3H_2O_{(g)} \longrightarrow Al_2O_{3\,(s)} + 3H_{2\,(g)} + 866\ kJ/mol \quad (8)$$

$$2Al_{(s)} + 3CO_{2\,(g)} \longrightarrow Al_2O_3 + 3CO + 741\ kJ/mol \quad (9)$$

$$2Al\ (s) + N2\ (g) \longrightarrow 2AlN + 346\ kJ/mol \quad (10)$$

IOP Conf. Series: Materials Science and Engineering **975** (2020) 012008 doi:10.1088/1757-899X/975/1/012008

Al nanoparticles could efficiently initiate this series of secondary reaction. This series of exothermic reactions could take place more rapidly within the detonation wave [36-38]. Therefore, a large amount of heat would be liberated within detonation wave front; excess metal fuel could combust more effectively behind C. J. point [37, 39].

Aluminium nanoparticles would combust more efficiently within detonation wave front, offering high reaction rate, and high heat release rate

3.3 Thermal behaviour of metalized compositions

Differential scanning calorimetry (DSC) can monitor any physical or chemical change which involves the evolution/absorption of heat. The impact of Al particles on TNT thermal behaviour was evaluated using DSC. While micron-Al offered an increase in total heat release by 3 %, nanometric Al offered an increase in total heat release with 78 % figure 15.

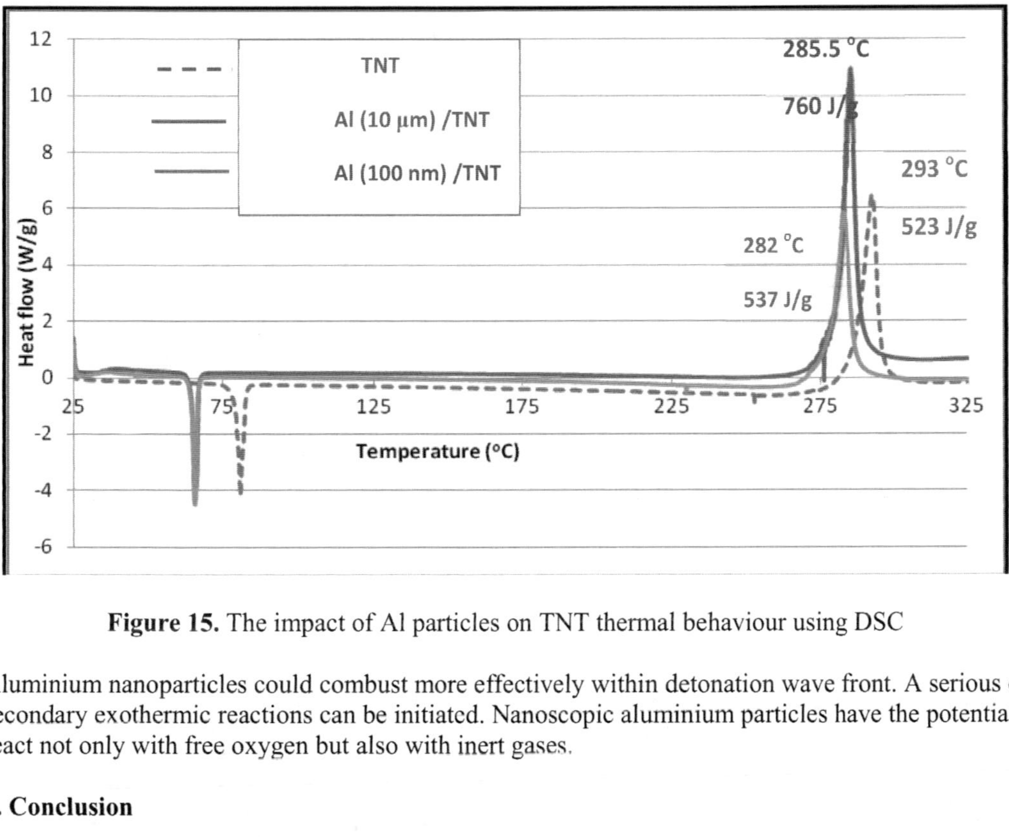

Figure 15. The impact of Al particles on TNT thermal behaviour using DSC

Aluminium nanoparticles could combust more effectively within detonation wave front. A serious of secondary exothermic reactions can be initiated. Nanoscopic aluminium particles have the potential to react not only with free oxygen but also with inert gases.

4. Conclusion

Novel approach for UNDEX testing is represented; this was accomplished via measurement of target acceleration upon explosion. A novel design for naval structure with superior stability was represented. The effectiveness of reactive Al nanoparticles as a novel high energy density material to conventional Al particles in TNT for UNDEX was evaluated. Al nanoparticles (100 nm) offered enhanced performance by 49 % compared with 16 % for conventional Al particles (10 μm). Al nanoparticles could react more efficiently within detonation wave front offering enhanced heat output.

IOP Conf. Series: Materials Science and Engineering **975** (2020) 012008 doi:10.1088/1757-899X/975/1/012008

5. References

[1] Yarom, T., et al., DESIGN AND TESTING OF UNDERWATER IM EXPLOSIVE CHARGE.

[2] DeRuntz Jr, J., The Underwater Shock Analysis (USA) Manual. Unique Software Applications, Colorado Spring, CO, 1996.

[3] Arons, A., Underwater explosion shock wave parameters at large distances from the charge. The journal of the acoustical society of america, 1954. 26(3): p. 343-346.

[4] Grządziela, A., Model of impact underwater detonation. Journal of KONES, 2012. 19: p. 191-199.

[5] LYNCH, R. Development of insensitive high explosives using propellant technology. in 26th Joint Propulsion Conference. 1990.

[6] Kumar, A.S., et al., Evaluation of plastic bonded explosive (PBX) formulations based on RDX, aluminum, and HTPB for underwater applications. Propellants, Explosives, Pyrotechnics, 2010. 35(4): p. 359-364.

[7] HAWASS, A., et al., REVIEW ON UNDERWATER EXPLOSION.

[8] Yen, N.H. and L.Y. Wang, Reactive metals in explosives. Propellants, Explosives, Pyrotechnics, 2012. 37(2): p. 143-155.

[9] Misovec, A. and W. David, Explosion phenomena. Taylor Naval Ship Research and Development Center, 1976.

[10] Kalavalapally, R., R. Penmetsa, and R. Grandhi, Multidisciplinary optimization of a lightweight torpedo structure subjected to an underwater explosion. Finite Elements in Analysis and Design, 2006. 43(2): p. 103-111.

[11] Reid, W.D., The Response of Surface Ships to Underwater Explosions. 1996, DEFENCE SCIENCE AND TECHNOLOGY ORGANIZATION CANBERRA (AUSTRALIA).

[12] Filler, W.S., Propagation of shock waves in a hydrodynamic conical shock tube. The Physics of Fluids, 1964. 7(5): p. 664-667.

[13] Shin, Y., Naval Ship Shock and Design Analysis. Course Notes for Underwater Shock Analysis, Naval Postgraduate School, Monterey, CA, 1996. 87.

[14] Mair, H.U., et al. Lagrangian hydrocode modeling of underwater explosive/target interaction. in 61st Shock and Vibration Symposium. 1990.

[15] Keil, A., The response of ships to underwater explosions. 1961, DAVID TAYLOR MODEL BASIN WASHINGTON DC.

[16] Kowsarnia, E., Y. Alizadeh, and H.S. Pour, Experimental evaluation of blast wave parameters in underwater explosion of hexogen charges. International Journal of Engineering-Transactions B: Applications, 2011. 25(1): p. 65-72.

[17] 박종완, Underwater explosion testing of catamaran-like structure vs. simulation and feasibility of using scaling law= 수중폭발에 의한 구조물의 충격응답 실험과 시뮬레이션 및 상사법칙 적용에 관한 연구. 2012.

[18] Costanzo, F. and J. Gordon, A procedure to calculate the axisymmetric bulk cavitation boundaries and closure parameters. David Taylor Research Center, Ship Structures and Protection Department, Underwater Explosions Research Division, Bethesda Maryland. Report SSPD-89-177-78 August, 1989.

[19] Yennie, D., A. Arons, and T.J. Cotter, Long range shock propagation in underwater explosion phenomena II. 1949: Navy Department, Bureau of Ordnance.

[20] DeRuntz, J. The underwater shock analysis code and its applications. in 60^< th> Shock and Vibration Symposium Proceedings. 1989.

[21] Vadhe, P.P., et al., Cast aluminized explosives (review). Combustion, Explosion, and Shock Waves, 2008. 44(4): p. 461-477.
[22] Strømsøe, E. and S. Eriksen, Performance of high explosives in underwater applications. Part 2: Aluminized explosives. Propellants, Explosives, Pyrotechnics, 1990. 15(2): p. 52-53.
[23] Keicher, T., et al., Influence of aluminium/ammonium perchlorate on the performance of underwater explosives. Propellants, Explosives, Pyrotechnics, 1999. 24(3): p. 140-143.
[24] V. Dolezal, a.V.J., "Theory Of Explosives". MTC, Egypt. 1974.
[25] Brinkmann, J. The behaviour of different explosive types and the effects on blast results. in ISRM International Symposium. 1990: International Society for Rock Mechanics.
[26] Kittell, D.E., N.R. Cummock, and S.F. Son, Reactive flow modeling of small scale detonation failure experiments for a baseline non-ideal explosive. Journal of Applied Physics, 2016. 120(6): p. 064901.
[27] Brousseau, P. and C.J. Anderson, Nanometric aluminum in explosives. Propellants, Explosives, Pyrotechnics, 2002. 27(5): p. 300-306.
[28] Capellos, C., et al. Eigenvalue detonation of combined effects aluminized explosives. in SHOCK COMPRESSION OF CONDENSED MATTER- 2007: Proceedings of the Conference of the American Physical Society Topical Group on Shock Compression of Condensed Matter. 2007: AIP Publishing.
[29] Baker, E.L., et al., Combined effects aluminized explosives. 2010, DTIC Document.
[30] Elbasuney, S., Dispersion characteristics of dry and colloidal nano-titania into epoxy resin. Powder Technology, 2014. 268(0): p. 158-164.
[31] Elbasuney, S., Sustainable steric stabilization of colloidal titania nanoparticles. Applied Surface Science, 2017. 409: p. 438-447.
[32] Elbasuney, S., Novel Colloidal Nanothermite Particles (MnO2/Al) for Advanced Highly Energetic Systems. Journal of Inorganic and Organometallic Polymers and Materials, 2018. 28(5): p. 1793-1800.
[33] Elbasuney, S., et al., Infrared Signature of Novel Super-Thermite (Fe_2O_3/Mg) Fluorocarbon Nanocomposite for Effective Countermeasures of Infrared Seekers. Journal of Inorganic and Organometallic Polymers and Materials, 2018. 28(5): p. 1718-1727.
[34] Lin, B.-Q., et al., Experimental investigation on explosion characteristics of nano-aluminum powder—air mixtures. Combustion, Explosion, and Shock Waves, 2010. 46(6): p. 678.
[35] Talawar, M., et al., Emerging trends in advanced high energy materials. Combustion, Explosion, and Shock Waves, 2007. 43(1): p. 62-72.
[36] CONKLING, J. and C. MOCELLA, eds. Chemistry of Pyrotechnics Basic Principles and Theory. Second ed. 2012, CRC: London.
[37] Krier, H.J., J.M. Peuker, and N. Glumac, Aluminum Combustion in Aluminized Explosives: Aerobic and Anaerobic Reaction, in 49th AIAA Aerospace Sciences Meeting including the New Horizons Forum and Aerospace Exposition. 2011: Orlando, Florida.
[38] McNesby, K.L., et al., Afterburn Ignition Delay and Shock Augmentation in Fuel Rich Solid Explosives. Propellants, Explosives, Pyrotechnics, 2010. 35(1): p. 57-65.
[39] Dreizin, E.L., Metal-based reactive nanomaterials. Progress in Energy and Combustion Science, 2009. 35(2): p. 141-167.

IOP Conf. Series: Materials Science and Engineering **975** (2020) 012009 doi:10.1088/1757-899X/975/1/012009

Aluminium Nanoparticles: The Potentials of Metalized Explosives with Combined Destructive Effect (Combustion/Detonation)

Sherif Elbasuney [a,b*], M. Gaber Zaky [a], Mostafa Radwan[c], Ahmed Maraden [a], Sherif M. Abdelkhalek[d]

[a] School of Chemical Engineering, Military Technical College, Kobry Elkoba, Cairo, Egypt.
[b] Head of nanotechnology research centre, Military Technical College, Cairo Egypt.
[C] British University in Egpyt, El-Shrouk City, Cairo, Egypt.
[d] Head of Engineering and Technology Research Center, Military Technical College, Cairo, Egypt
s.elbasuney@mtc.edu.eg, sherif_basuney2000@yahoo.com, +201112630789

Abstract. Even though, reactive metal particles can boost the energy density of explosive materials. Such particles will react behind the detonation wave front with decrease in detonation velocity and brisance (destructive effect). This study reports on the effective development of aluminium nanoparticles (NPS) of 100 nm particle size. Al NPs offered superior performance compared micron scale particles. Al NPs offered an increase in shock wave strength by 48 % compared with 17 % for micron scale particles. While micron-Al decreased the destructive effect of TNT by -6.5 %; Al NPs offered an increase in destructive effect by 21 %. The main outcome of this study is that Al NPs offered an enhanced detonation velocity of 6330 m/s compared with 5650 m/s for TNT. Additionally Al NPs offered decrease in TNT critical diameter from 40 mm to 20 mm. While conventional Al particles could act as desensitizer; Al NPs could act as sensitizer and could combust efficiently within detonation wave front. This study reports on the real development of metalized nano-composite explosives with combined destructive effect (combustion/detonation).

Keywords: Metalized explosives; Combustion, Destructive effect; Workability; Brisance; Thermal behaviour.

1. Introduction

The energy density of common energetic compounds is quite low; as it is limited by their enthalpy of formation. Higher energy density can be achieved through integration of reactive metal particles figure 1 [1].

Aluminium particles with combustion heat of 32000 J/g; is the common high energy density material [1-3]. Reactive metal fuels (i.e. Al) could offer a longer duration pressure pulse figure 2 [4].

In general, the long ignition delays and slow combustion rates could contribute behind detonation wave front. This action could decrease the detonation velocity [5]. Consequently reactive metal particles could transfer ideal detonation into non-ideal. This could limit the capability of producing sharp pressure pulses required for many applications. Furthermore, aluminium particles are covered with protective oxide layer; that could hinder the rapid combustion process [6].

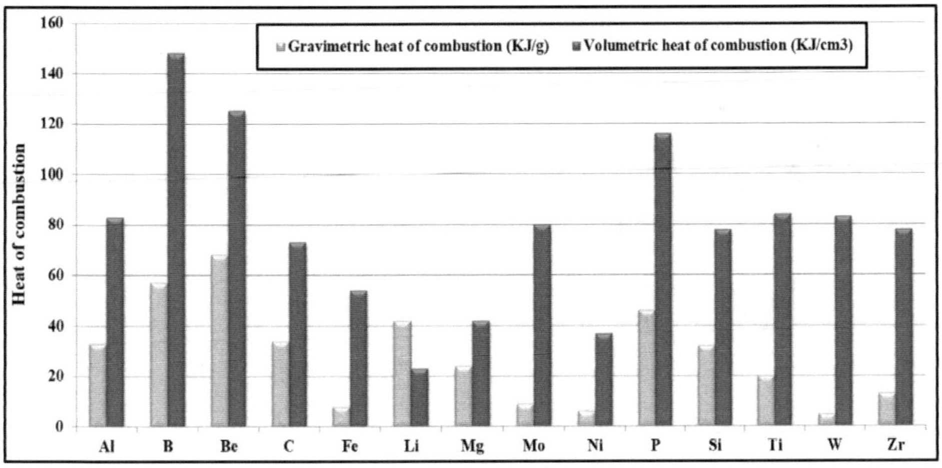

Figure 1. Heat output for common reactive fuels (Gravimetric and Volumetric).

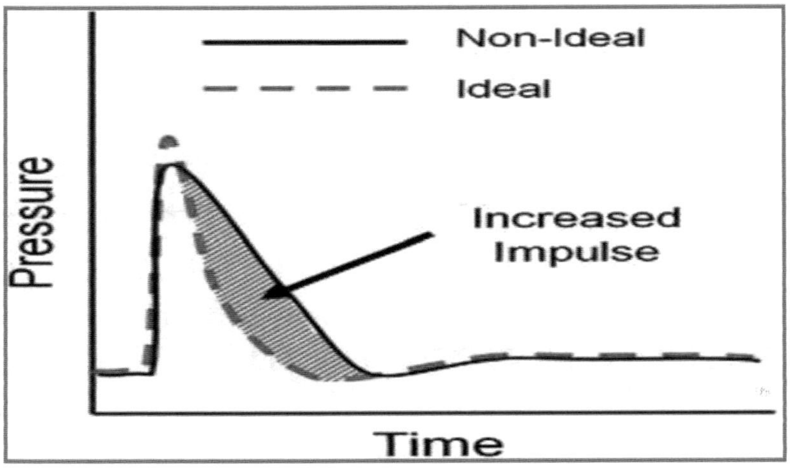

Figure 2. Pressure-time profile for an ideal vs. non-ideal detonation reaction.

Oxide layer should be pealed out for reactive metal core to combust with free oxygen. Aluminium combustion consists of two main stages including ignition and quasi-steady burning. The ignition of aluminium particles begins with a relatively short heterogeneous surface reaction (HSR) stage (Figure 3-a) and quickly transitions to a quasi-steady diffusion flame figure 3-b [7]. During the ignition phase, the particle heats up from convection and volumetric compression to its melting temperature; the short lived HSR stage consisting of solid metal (m_s) and liquid metal (m_l) surrounded by a layer of metal oxide (m_{mox}). Upon heating from convection and volumetric compression, the particle temperature increases until a melting phase transition occurs [5-8].

At sufficiently high temperature and mechanical loading, the particle oxide layer peels back allowing for molten aluminium to evaporate and form a diffusion flame [7-11]. This combustion mechanism confirmed that metal particles could react over longer timescale than the detonation of explosive material. Hence metal particles could induce non-ideal detonation [12-19].

For ideal detonation reaction, all potential energy of the explosive is liberated almost instantly in a thin reaction zone prior to C-J plane; behind the C-J plane are stable detonation products (Figure 4-a) [20-22]. In non-ideal detonation, the potential energy is not liberated completely within reaction zone; significant degree of chemical reactions take place behind the C-J plane figure 4-b [20-25].

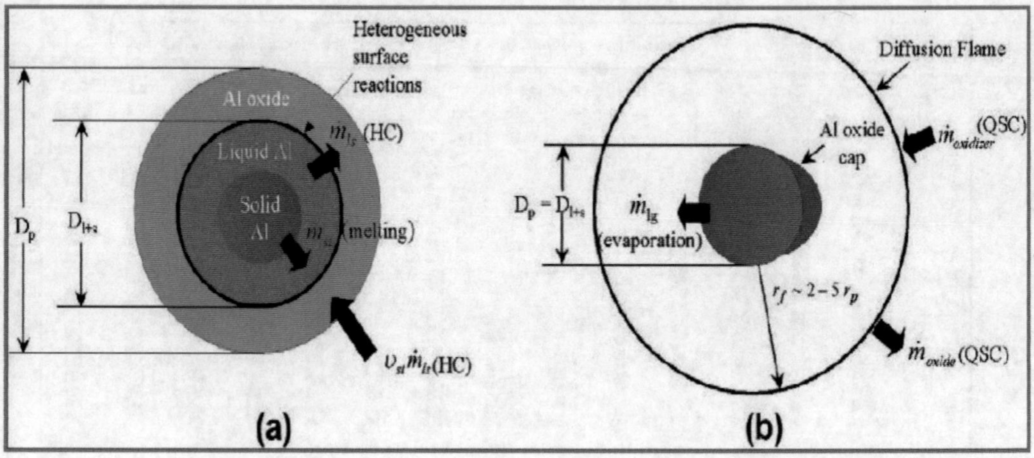

Figure 3. Sketch combustion model of aluminium particles.

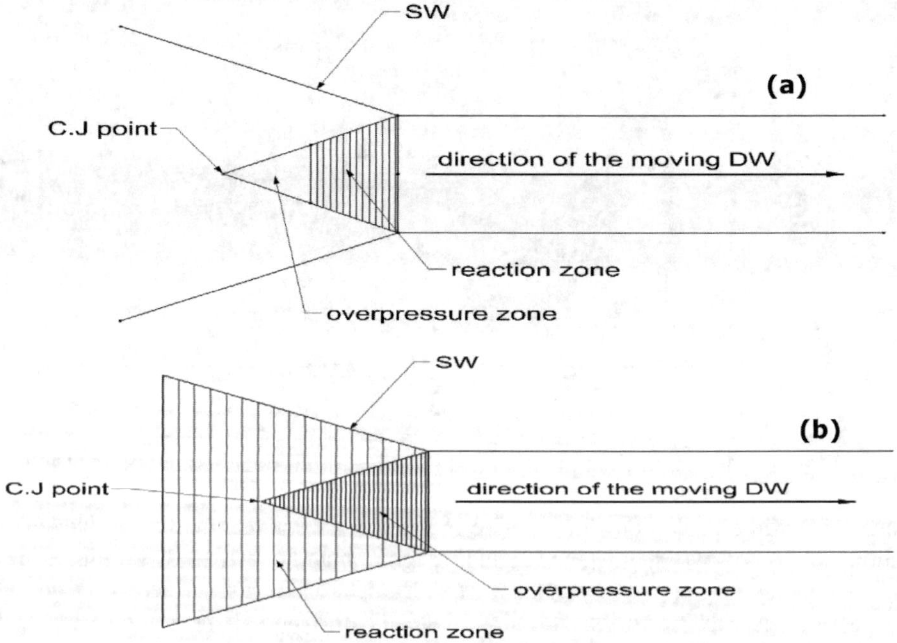

Figure 4. Schematic for ideal detonation reaction (a), and non-ideal detonation reaction (b).

Though aluminium could support the shock front via the energy released in the primary reaction zone; aluminized explosive offered non-ideal detonation with decrease in detonation velocity [26]. The reaction zone length for metalized explosives was reported to be more than two times that for organic explosives [26]. While conventional Al particles could contribute only to the work done by the expanding combustion products; they could reduce the destructive effect of explosive material [27-33].

Aluminized explosives are non-ideal, with long sequential reaction zones that depend on particle size [25]. It was presumed that fine Al particles are consumed more rapidly in the CHNO reaction

zone compared to larger particles [2]. Recently, aluminized explosive formulations employing fine aluminium particles were developed to produce high initial detonation temperatures, pressures, and velocities [5, 30]. They can offer both excellent early volume expansion metal pushing and high expansion blast capabilities [30-31]. The great impetus of nanotechnology is the massive increase of the surface area/weight [27-29].
Colloidal Al NPs were developed via wet milling. Al NPs offered superior performance compared with conventional Al particles. Al NPs offered an increase in workability of TNT by 48 % using ballistic mortar test, compared with 17 % for micron size particles. While micron-scale particles decreased the destructive effect of TNT by -6.5 %; Al NPs offered an increase in destructive effect of TNT by 21 % using kast test and Hess test. Nano-scale Al particles could combust more efficiently within the detonation wave front. Detonation velocity measurements confirmed that Al NPs offered enhanced detonation velocity and lower critical diameter.

2. Materials and methods

2.1 Chemicals and reagents
Industrial aluminium particles 10 μm were purchased from Alpha chemika.. Isopropyl alcohol (IPA) was employed as dispersing medium of reactive metal particles; IPA was purchased from Sigma-Aldrich. All chemicals were used as received.

2.2 Characterization of aluminium particles
The particle size of aluminium particles was visualized using scanning electron microscope SEM, Zeiss EVO-10 by Carl Zeiss Corporation. The crystalline phase was investigated with X-ray diffraction (XRD) D8 advance by Burker Corporation over the angle range 2θ from 5 to 65 degrees.

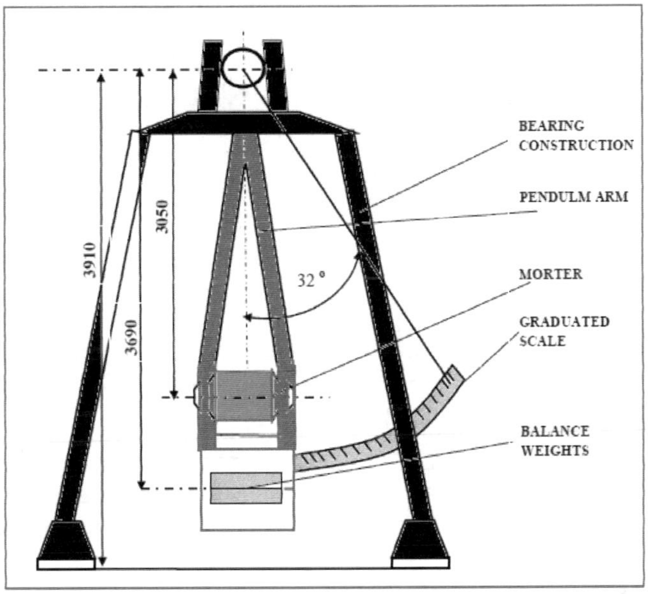

Figure 5. Schematic diagram of ballistic mortar [19]

2.3 Integration of metallic fuels particles into energetic matrix
It is widely accepted that enhanced dispersion characteristics can be accomplished via integration of colloidal particles into energetic. Ultrasonic probe homogenizer was employed to disperse Al NPs in IPA, to break down any aggregates. Colloidal Al particles were integrated into molten TNT.

2.4 Workability evaluation

Workability in terms of shock wave strength was evaluated using Ballistic mortar test. Ballistic mortar composed of suspended massive steel mortar figure 5. Upon exploding standard charge (10 g) within the mortar cavity, the maximum mortar displacement was recorded. Mortar displacement is function of detonation overpressure. Consequently, the equivalent overpressure value was evaluated using standard charts.

2.5 Explosive strength evaluation

The impact of Al particles on TNT brisance was evaluated using Kast test developed by PHYWE (Germany). Destructive effect evaluation was conducted via measurement of copper crusher compression figure 6.

Figure 6. Schematic for destructive effect using Kast test [19].

Moving steel cylinder (1) transmitted the overpressure to copper crusher (2). The copper crusher height was precisely measured, upon exploding explosive charge (2 g).

2.6 Destructive effect by Hess test

The brisance of an explosive was evaluated on the basis of the compression of a lead cylinder under the action of the shock wave of standard explosive charge. Lead cylinder compression is proportional to the destructive effect of an explosive charge figure 7.

The difference in the lead cylinder height would act as a direct measure of the brisance of the tested explosive formulation relative to TNT.

2.7 Detonation velocity measurement

The impact of Al particle size on detonation velocity of TNT was evaluated using Exploment-Fo-Multi channel, by Kontiniro A. G. (Swiss made). The time interval for detonation wave to travel between two points (start and end) was measured using an electronic counter figure 8.

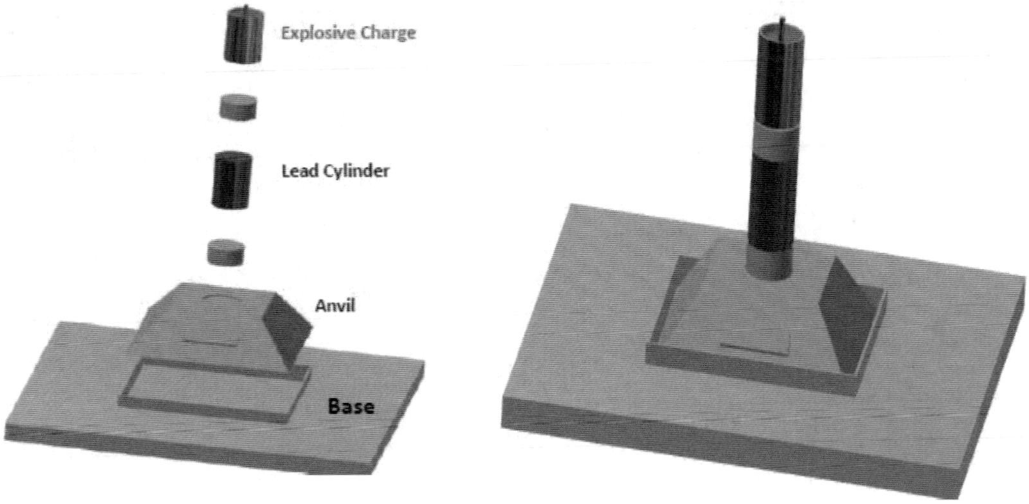

Figure 7. Setup for the determination of destructive effect by Hess test [9]

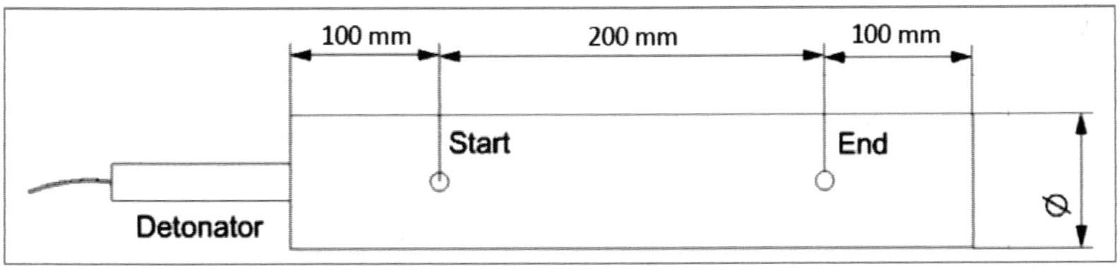

Figure 8. Detonation velocity measurement using an electronic timer

3. Results and discussions

3.1 Characterization of Al particles

The morphology of employed aluminium particles was investigated with SEM. Al particles in the shape of spheres with 10 μm average particle size were reported from SEM micrograph figure 9-a. The particles size of employed Al NPs was investigated with TEM. TEM micrographs demonstrated Al particles in the shape of flakes with average particle size of 100 nm figure 9-b.

3.2 Workability evaluation

Under the action of the detonation products, the mortar was moved from its equilibrium position. The maximum displacement was recorded. This movement can represent the workability of the tested explosive material figure 10.

Whereas micron-Al particles demonstrated an increase in TNT workability by 17 %; Al NPs offered an increase in TNT workability by 48 %. This enhanced performance was ascribed to the fact that Al NPs could vaporize easily and combust efficiently within the detonation wave front. Excess Al NPs could contribute to the shock wave strength and could combust more efficiently with oxygen in the air adding substantial heat output [25].

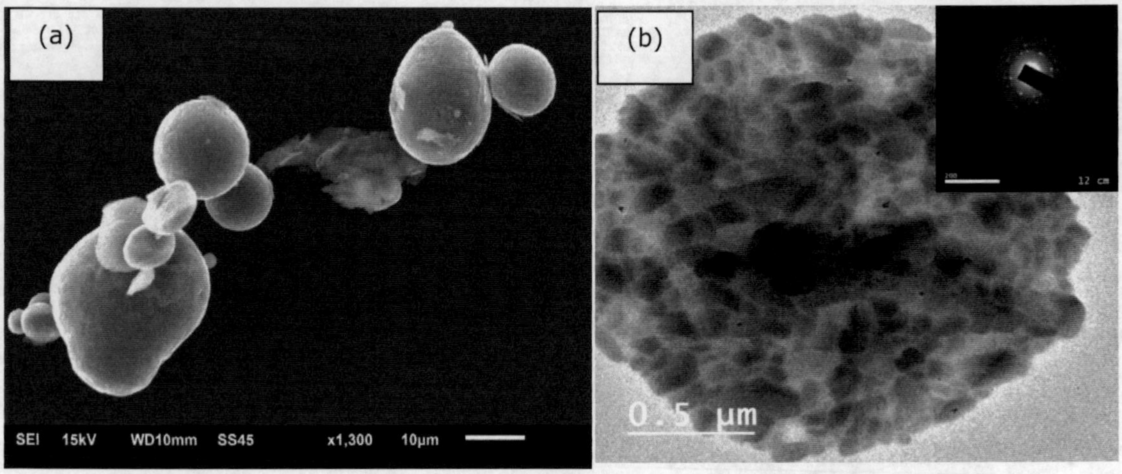

Figure 9. TEM micrograph of micron scale Al particles (a), nano-scale Al particles (b).

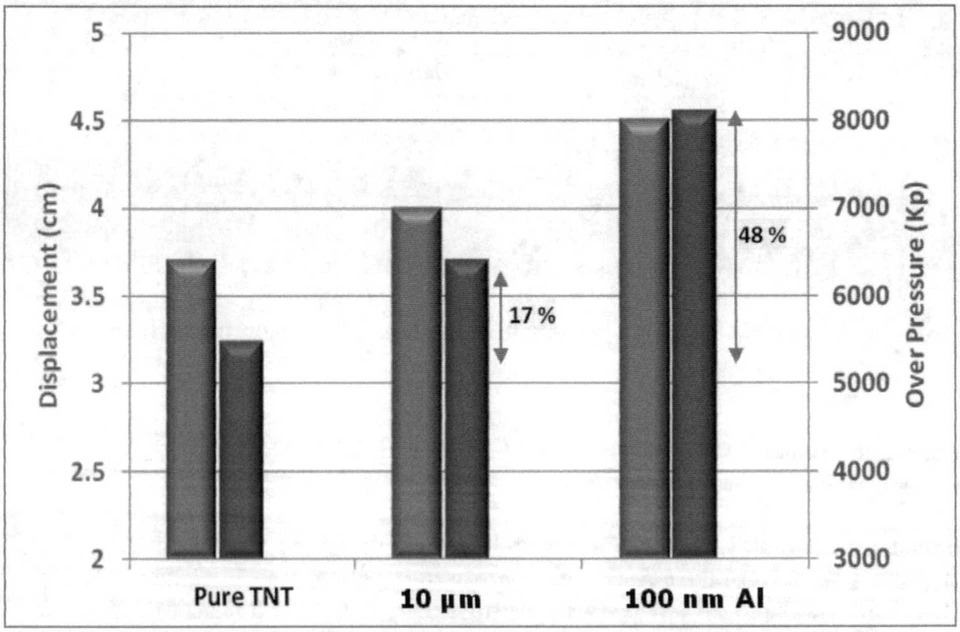

Figure 10. The impact of Al particle size on TNT workability

3.3 Destructive effect evaluation via Kast test

The impact of Al particle size on the destructive effect of TNT was evaluated using standard Kast test. The reduction of copper crusher height was precisely measured; the corresponding overpressure value was retrieved using standard charts. While, micron aluminium particles decreased the destructive effect of TNT by -6.5 %; Al NPs offered enhanced localized destructive effect of TNT by 21 % figure 11.

This novel finding could be ascribed to the fact that Al NPs could contribute more effectively within detonation wave front. Al NPs could offer enhanced heat output, combustion temperature, explosion force, and high detonation velocity. Consequently enhanced brisance can be achieved.

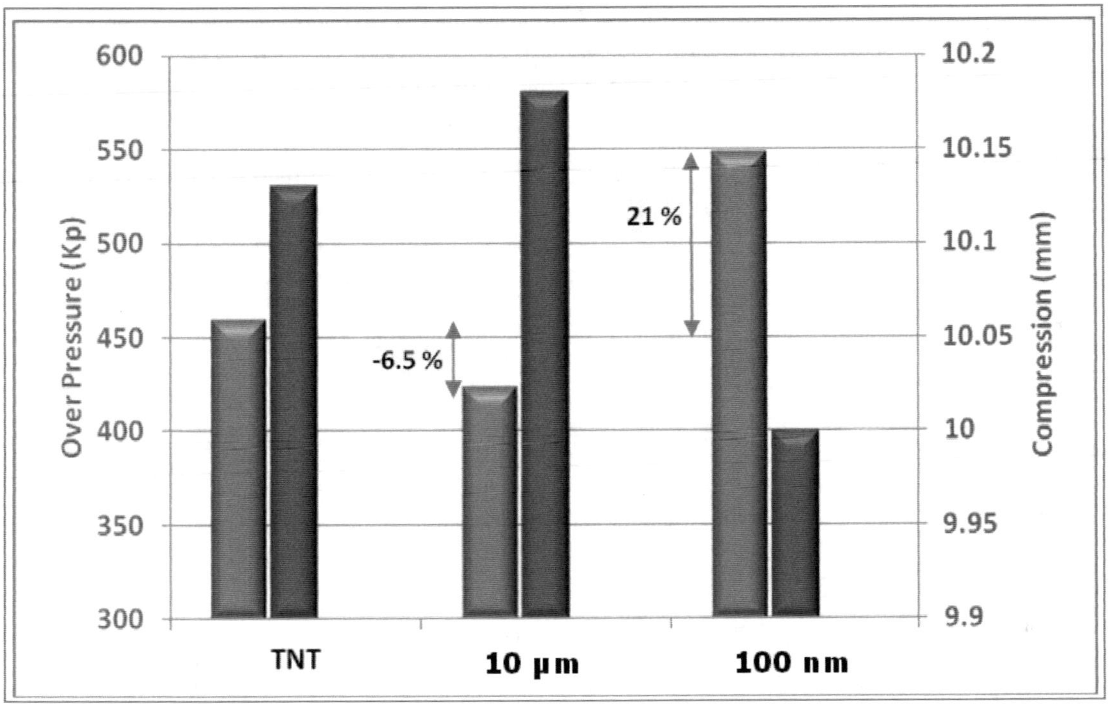

Figure 11. Impact of Al particle size on TNT brisance using Kast test.

3.4 Destructive effect evaluation via Hess test

The impact of micron and nano-scopic Al on destructive effect of TNT was further evaluated using Hess test. The compression in lead crusher height was employed as a measure of destructive effect relative to TNT. Explosive charge of 50 g was exploded; the height of lead crusher was precisely measured. Whereas micron Al particles decreased the destructive effect of TNT; nanoscopic Al particles offered an increase in TNT destructive effect figure 12.

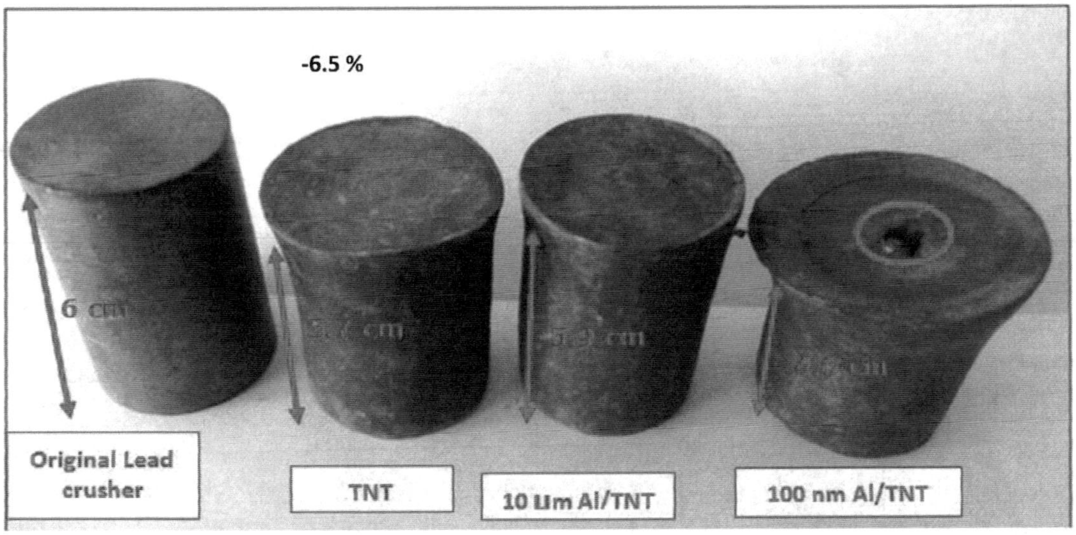

Figure 12. Effect of Al particle size on TNT brisance using Hess test

Whereas 10 μm Al particles demonstrated lead crusher height of 5.9 compared with 5.7 for TNT; Al NPs offered enhanced destructive effect as it offered lead crusher height of 4.6 cm compared with 5.7 cm of TNT. It can be concluded that micron-scale Al particles could contribute only to shock wave strength, with the expansion of gaseous products behind C. J. point. Micron-scale Al particles would act as inert material within detonation wave front prior to C.J. point.

These findings confirmed the findings by Kast test. It can be concluded that the effective integration of colloidal Al NPs could allow combustion to take place more efficiently within the detonation wave narrow zone. This zone lies in the order of few μm. This could withstand the enhanced destructive effect for Al NPs. The detonation action during Hess test was video recorded. Figure 13 demonstrates the moment of initiation (a), explosion of TNT (b), 10 μm Al/TNT (C), and 100 nm Al/TNT (d). It is apparently clear that Al NPs offered an enhanced fire ball action. This could withstand enhanced shock wave strength as well.

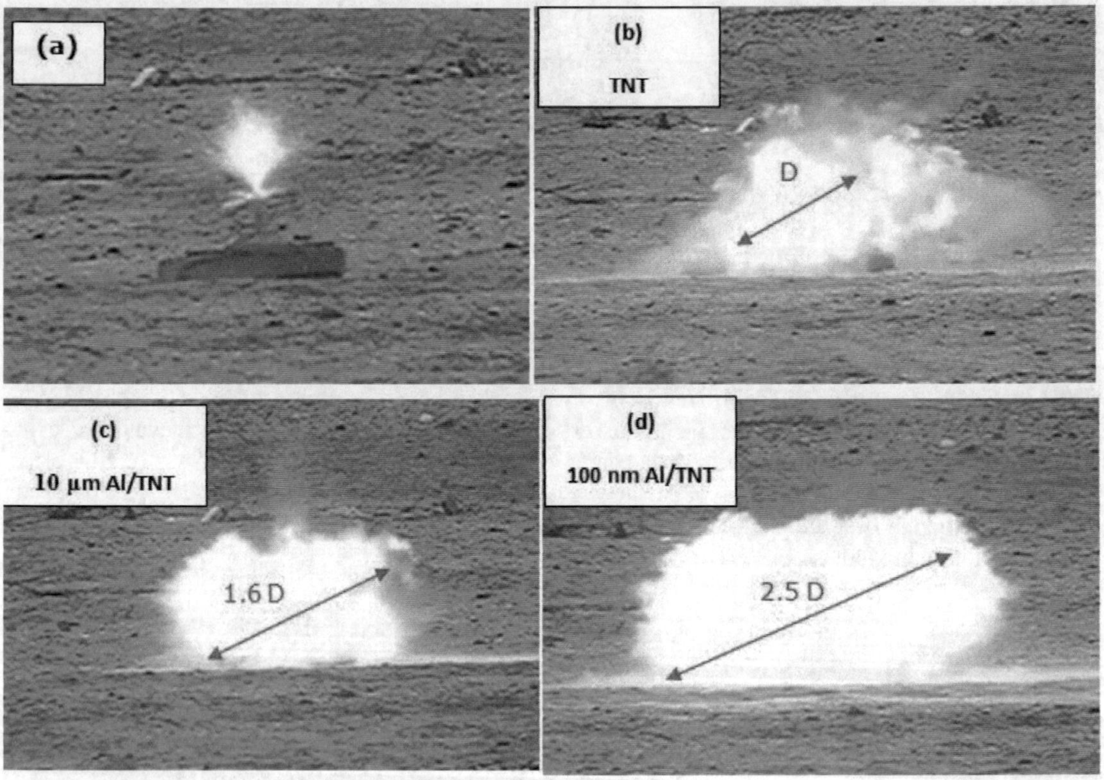

Figure 13. Hess test actions: moment of initiation (a), TNT (b), 10μm Al/TNT (C), 100 nm Al/TNT (d).

Al nanoparticles demonstrated extended fire ball with fireball diameter 2.5 times that of TNT, compared with 1.6 times for 10 μm Al/TNT. It is apparently clear that the Al nanoparticles not only contribute to the destructive effect within detonation wave front; but also excess fuel could contribute to the shock wave with efficient combustion. Therefore high thermal loading can be generated [1-2].

3.5 Detonation velocity measurement

Detonation velocity was determined by measuring the time required for the detonation wave to travel a measured distance longitudinally. Detonation velocity can provide information about detonation nature (ideal or non-ideal). 10 μm Al particles not only decreased the detonation velocity of TNT; but also it increased the critical diameter for detonation to take place. In contrast Al NPs offered ideal detonation reaction with detonation velocity higher than TNT. Furthermore Al NPs offered lower critical diameter figure 14.

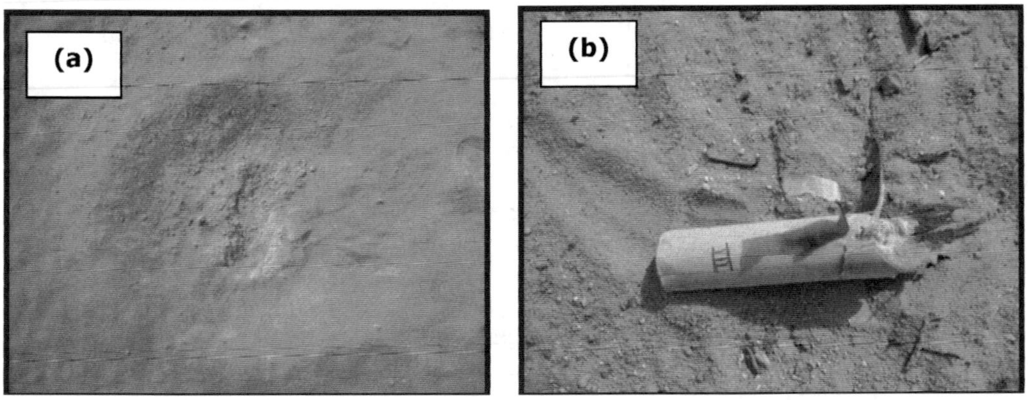

Figure 14. Ideal detonation for Al(100 nm)/TNT (a), no reaction for Al (10 μm) /TNT (b), at charge diameter of 40 mm.

It can be concluded that while commercial Al particles could act as a desensitizer, nanoscopic Al particles could act as a sensitizer. Therefore Al NPs could offer super-ideal detonation proceeding with superior detonation velocity [32-33]. Whereas TNT demonstrated detonation velocity of 5650 m/s; Al (100 nm) /TNT offered super-ideal detonation with detonation velocity of 6330 m/s.

4. Conclusion

This study shaded the light on Al NPs as a novel high energy density material that could offer not only enhanced shock wave strength but also brisance compared with conventional Al particles. It can be concluded that while conventional Al could contribute only to gaseous products behind C.J. point; Al NPs could contribute not only to shock wave strength but also with the detonation wave front with an increase in the brisance of explosive material. Al NPs offered lower critical diameter of TNT. Whereas Al NPs could act as a sensitizer at C.J. point; conventional Al particles could act as desensitizer. Enhanced dispersion characteristics could be achieved by integrating colloidal metal particles into TNT.

5. References

[1] N. H. Yen, L. Y. Wang, Reactive metals in explosives, Propellants Explosives Pyrotechnics, 37 (2) (2012) 143-155.

[2] E. L. Dreizin, Metal-based reactive nanomaterials, Progress in Energy and Combustion Science, 35 (2) (2009) 141-167.

[3] X. Zhou, D. Xu, J. Lu, K. Zhang, CuO/Mg/fluorocarbon sandwich-structure superhydrophobic nanoenergetic composite with anti-humidity property, Chemical Engineering Journal 266 (0) (2015) 163-170.

[4] P. P. Vadhe, R. B.Pawar, R. K. Sinha, S. N. Asthana, A. R. Subhananda, Cast aluminized explosives (review), Combustion Explosion and Shock Waves 44 (4) (2008) 461-477.

[5] V. E.ZARKO, A. A. GROMOV, ENERGETIC NANOMATERIALS Synthesis, Characterization, and Application. Elsevier, Amsterdam, Netherland, 2016.

[6] J. CONKLING, C. MOCELLA, Chemistry of Pyrotechnics Basic Principles and Theory. Second ed., CRC: London, U.K., 2012.

IOP Conf. Series: Materials Science and Engineering **975** (2020) 012009 doi:10.1088/1757-899X/975/1/012009

[7] K P.Ruggirello, P. E.DesJardin, M. R.Baer, M. J.Kaneshige, E. S.Hertel, A reaction progress variable modeling approach for non-ideal multiphase explosives, International Journal of Multiphase Flow 42 (0) (2012) 128-151.
[8] E.-C. Koch, T. M. Klapötke, Boron-Based High Explosives, Propellants Explosives Pyrotechnics 37 (3) (2012) 335-344.
[9] S. Elbasuney, Novel Colloidal Nanothermite Particles (MnO2/Al) for Advanced Highly Energetic Systems, Journal of Inorganic and Organometallic Polymers and Materials 28 (5) (2018) 1793-1800.
[10] S. Elbasuney, A. Elsaidy, M. Kassem,H. Tantawy, Infrared Signature of Novel Super-Thermite (Fe2O3/Mg) Fluorocarbon Nanocomposite for Effective Countermeasures of Infrared Seekers, Journal of Inorganic and Organometallic Polymers and Materials 28 (5) (2018) 1718-1727.
[11] S. Elbasuney, A. Elsaidy, M. Kassem, H. Tantawy, R. Sadek, A. Fahd, M. Gobara, Super-Thermite (Al/Fe2O3) Fluorocarbon Nanocomposite with Stimulated Infrared Thermal Signature via Extended Primary Combustion Zones for Effective Countermeasures of Infrared Seekers, Journal of Inorganic and Organometallic Polymers and Materials 28 (2018) 2231-2240.
[12] S. Elbasuney, A. Fahd, H. E. Mostafa, Combustion characteristics of extruded double base propellant based on ammonium perchlorate/aluminum binary mixture, Fuel 208 (Supplement C) (2017) 296-304.
[13] S. Elbasuney, A. Fahd, H. E. Mostafa, S. F. Mostafa, R. Sadek, Chemical stability, thermal behavior, and shelf life assessment of extruded modified double-base propellants, Defence Technology 14 (1) (2018) 70-76.
[14] S. Elbasuney, M. G. Zaky, M. Radwan, S. F. Mostafa, Stabilized super-thermite colloids: A new generation of advanced highly energetic materials, Applied Surface Science 419 (2017) 328-336.
[15] A. M. A. Elghafour, M. A. Radwan, H. E. Mostafa, A. Fahd, S. Elbasuney, Highly energetic nitramines: A novel platonizing agent for double-base propellants with superior combustion characteristics, Fuel 227 (2018) 478-484.
[16] A. K. Mohamed, H. E. Mostafa, S.Elbasuney, Nanoscopic fuel-rich thermobaric formulations: Chemical composition optimization and sustained secondary combustion shock wave modulation, Journal of Hazardous Materials 301 (2016) 492-503.
[17] R. Sadek, M. Kassem, M. Abdo, S. Elbasuney, Novel blue flare tracer with enhanced color quality and luminous intensity, Journal of Luminescence 195 (2018) 8-13.
[18] K. P. Ruggirello, P. E. DesJardin, M. R.Baer, M. J.Kaneshige, E. S. Hertel, A reaction progress variable modeling approach for non-ideal multiphase explosives, International Journal of Multiphase Flow 42 (2012) 128-151.
[19] M. Suceska, Test Methods for Explosives, Springer, New York, U. S. A., 1995.
[20] M.Sućeska, Calculation of the Detonation Properties of C-H-N-O explosives, Propellants Explosives Pyrotechnics , 16 (4) (1991) 197-202.
[21] M. Sućeska, Evaluation of detonation energy from EXPLO5 computer code results, Propellants Explosives Pyrotechnics 24 (5) (1999) 280-285.
[22] Jr. H. Bruckman, J. Guillet, Theoretical calculations of hot-spot initiation in explosives, Canadian Journal of Chemistry 46 (20) (1968)3221-3228.
[23] T. M.Klapϙtke, Chemistry of High-Energy Materials, Third ed., DE GRUYTER, Munich, Germany, 2015.
[24] C. ROSSI, Two Decades of Research on Nano-Energetic Materials, Propellants Explosives Pyrotechnics 39 (2014) 323-327.
[25] V. W. Manner, S. J.Pemberton, J. A. Gunderson, T. J. Herrera, J. M. Lloyd, P. J. Salazar, P. Rae, B. C. Tappan, The Role of Aluminum in the Detonation and Post-Detonation Expansion of Selected Cast HMX-Based Explosives, Propellants Explosives Pyrotechnics 37 (2) (2012) 198-206.

[26] P. Brousseau, C. J. Anderson, Nanometric aluminum in explosives, Propellants Explosives Pyrotechnics 27 (5) 2002 300-306.
[27] H. Abe, T. Shiina, Visualization of photoacoustic images in a limited-View measuring system using eigenvalues of a photoacoustic transmission matrix, Photoacoustics 8 (2017) 1-7.
[28] S. Elbasuney, Dispersion characteristics of dry and colloidal nano-titania into epoxy resin, Powder Technology 268 (0) (2014) 158-164.
[29] S. Elbasuney, Sustainable steric stabilization of colloidal titania nanoparticles, Applied Surface Science 409 (2017) 438-447.
[30] H. J. Krier, J. M.Peuker, N. Glumac, Aluminum Combustion in Aluminized Explosives: Aerobic and Anaerobic Reaction. In 49th AIAA Aerospace Sciences Meeting including the New Horizons Forum and Aerospace Exposition, Orlando, Florida, 2011
[31] K. L. McNesby, B. E.Homan, J. J. Ritter, Z. Quine, R.Z. Ehlers, B. A. McAndrew, Afterburn Ignition Delay and Shock Augmentation in Fuel Rich Solid Explosives, Propellants Explosives Pyrotechnics, 35 (1)2010 57-65.
[32] P. BROUSSEAU, C. J. ANDERSON, Nanometric Aluminum in Explosives, Propellants Explosives Pyrotechnics, 27 (4) 2002 300-306.
[33] G. D KOZAK, B.N. KONDRIKOV, A. I. Sumin, Dependence of detonation velocity on charge density for foamed alumotol (Al/TNT) and TNT mixtures, Combust Explos Shock Waves, 34 (4) 1998 448-452

Physico chemical studies for water at Rosetta branch of river Nile, Egypt

M S Masoud[1], A M Ismail[1], H M kamel[*2], A S El-Kholany[2] and A A M ElKasas[1]

[1] Chemistry Department, Faculty of Science, Alexandria University, Egypt.
[2] Medical laboratory department, faculty of allied medical sciences, Pharos University in Alexandria, Egypt

Abstract. This work was carried out to study the quality of water of Rosetta branch. Five sites were chosen for sampling during the period of January 2013 to December 2013. Physical parameters (Temperature, electrical conductivity (EC), turbidity and pH) and chemical parameters (concentration of Cl-, Ca^{+2}, Mg^{+2}, NH^{4+}, SO^{4-2}, NO^{2-}, and some of trace metals, alkalinity, total hardness, Dissolved Oxygen(DO), Chemical Oxygen Demand(COD) and Biological Oxygen Demand(BOD) were discussed. These results were compared with the results of another study that were carried out at Rosita Branch at 2018. Statistical analysis of data was computed to show the correlation between some parameters through the whole sampling sites during the period of the study, using simple correlation coefficient.

Keywords: Rosetta branch - Physical and chemical parameters – Water Quality Index - Correlation Coefficients

1. Introduction

Egypt faces severe water scarcity in recent years. Population increase leads to higher water demand and rapid economic growth leads to ecosystem degradation. Ineffective irrigation methods combined with poor water use degrade water quality. The Nile River is the main source of fresh water in Egypt [1], which suffers from an annual water shortage of about 8 billion cubic meters (BCM). The Nile River has two ending branches, Rosetta and Damietta. The Rosetta branch is about 210 km long and about 190 meters wide [2].
Identifying the source (s) of pollution and developing appropriate management strategies is essential to minimize potential public health risks [3]. Surface water quality in an area can be affected by point and unspecified point pollution [4].

Point source (PS) pollution occurs from a single specific source such as impacts from industries and wastewater treatment plants [5], while non-specific sources include runoff associated with a specific land use pattern such as urban areas (for example, water Rain, wastewater over flows), agriculture (such as fertilizers, pesticides, animal manure), or forest land uses [6]. The ingress of these sources into the water can represent improper drainage of toxic chemicals and pathogenic microorganisms; therefore, monitoring water quality and identifying health risks is essential to protect the population from water-borne diseases and to develop appropriate preventive measures. Since ecosystems such as rivers and lakes are affected by multiple sources, it is important to understand the spatial and temporal differences in the physical-chemical and microbiological parameters for assessment and management [7]. Nevertheless, the evaluation of several water quality parameters generates large and complex data sets and multivariate statistical techniques that are required to interpret the results [8-10]. Multivariate technologies such as cluster analysis (CA) and principle component analysis (PCA) have been successfully applied to better understand the water quality and environmental state of the studied systems [11, 12]. In addition to assessing water quality, multivariate technologies have many applications such as determining potential factors / sources affecting water systems [13, 14], reduction of data [15, 16], spatial and temporal changes [13], and aggregation sampling sites [17].

Most of the resource quality uses several indicators of water quality, the most important of which are the Water Quality Index (WQI), the Water Pollution Index (WPI), and the River Habitat Survey (RHS) [18, 19].

In the Rosetta branch, a mathematical model will be constructed that integrates data as input parameters. This model will create a degree that describes the spatial distribution. In addition, the timing of the water quality status will be determined. This template will be presented to the public as well as to decision and policy makers [3, 4]. There are several previous studies that used models and methods of water quality indicators to assess the state of water quality [5,20].

2. Materials and methods

2.1. Materials

All chemicals used were of appropriate purity grade. They were obtained from "Hach, Panreac, Merck, BDH, Chem-lab, VWR, Fluka, Fisher Scientific and Sigma-Aldrich. All reagents utilized for chemical analysis were prepared according to "Standard Methods for the examination of Water and Wastewater", 21st ED [21].

2.2. Analytical Methods

2.2.1. Sampling.

As a result of the health survey conducted on the Rosetta branch, during the period from January 2013 to December 2013, five sites along the branch were selected for regular sampling, including the intakes of the four treatment plants. Sample sites of Rosetta branch are shown in table 1.

Table 1. Sample sites along Rosetta branch

Site no.	Site name	latitude	longitude	Site position (Km)
1	Mahallet Diyay	31°1`47.13``E	30°43`21.13``E	159 after El-Kanater El-Khyria
2	Mahallet Abu Ali	31°5`55.93``N	30°41`8.66``E	168 after El-Kanater El-Khyria
3	Desouq	31°7`50.64``N	30°38`14.85``E	174 after El-Kanater El-Khyria
4	Fuwwah	31°12`4.32``N	30°34`1.82``E	185 after El-Kanater El-Khyria
5	Mutubas	31°16`54.97``N	30°31`17.75``E	202 after El-Kanater El-Khyria

IOP Conf. Series: Materials Science and Engineering **975** (2020) 012010 doi:10.1088/1757-899X/975/1/012010

2.2.2. Physical Parameters.

2.2.2.1. pH and Temperature.

The pH values were measured using an analytical pH meter fitted with a glass electrode and a temperature probe. By using standard buffer solutions of pH values 4, 7 and 10, the instruments were calibrated daily and to ensure a slope of more than 0.95 [21].

2.2.2.2. Turbidity.

It was measured by a nephelometric method (NTU) using turbidimeter (HACH, 2100N, IS, HACH Co., USA). The turbidity standards used to calibration of the instruments, it measured weekly [21].

2.2.2.3. Electrical Conductivity.

It was measured by a Laboratory method using Conductivity meter (WTW, Cond 720, WTW Co., USA). This instrument was calibrated daily using a standard of 1413 µs/cm at 25°C to ensure over 95% efficiency [21].

2.2.3. Chemical Parameters.

2.2.3.1. Chloride.

The chloride contents were determined by using the familiar Mohr`s method [21].

2.2.3.2. Alkalinity.

Total alkalinity (measured at pH 4) was determined by using bromocresol green and methyl red indicator [21].

2.2.3.3. Total hardness.

EDTA and its sodium salts form a chelated soluble complex when added to a solution of certain metal cations. If a small amount of a dye such as Eriochrome Black T is added to an aqueous solution containing Ca and Mg ions at pH of 10.0±0.1, the solution becomes wine red. If EDTA is added as a titrant, the Ca and Mg will be complexed and when all of the Ca and Mg have been complexed the solution turns from wine red to blue [21]. Total hardness was determined as recorded by titrating against EDTA using Eriochrome Black T as an indicator [21].

2.2.3.4. Calcium.

Calcium can be determined directly, with EDTA, when the pH is made sufficiently high that the magnesium is largely precipitated as the hydroxide and an indicator is used that combines with calcium only. Several indicators give a color change when all of the calcium has been complexed by the EDTA at a pH of 12 to 13 [20]. Calcium was determined by titration against EDTA standard using muroxide as an indicator [21].

2.2.3.5. Magnesium.

Magnesium may be estimated as the difference between hardness and calcium as $CaCO_3$,"Calculation method" [20]. Mg Hardness mg/l [total hardness (as mg $CaCO_3$/l) - calcium hardness (as mg $CaCO_3$/l)]

2.2.3.6. Ammonia.

It was determined by "Nesslerization method" according to SMWW. Color was measured spectro-photo-metrically using spectrophotometer (Cecil CE7400, USA) at 420 nm [21].

2.2.3.7. Nitrites.
These were determined through formation of a reddish purple azo dye produced at pH 2.0 to 2.5 by coupling diazotized sulfanilamide with N-(1-naphthyl-ethylenediamine dihydro-chloride (NED dihydro-chloride) [90].
These were determined by measuring the UV absorption using spectrophotometer (CecilCE7400, USA) at 543 nm [21].

2.2.3.8. Dissolved oxygen.
DO meter (WTW, Oxi 730, USA) with a membrane electrode was used for measuring DO [21].

2.2.3.9. Chemical Oxygen Demand (COD).
Most types of organic matter are oxidized by a boiling mixture of chromic and sulfuric acids. A suitable volume of sample and the reagents ($K_2Cr_2O_7$, H_2SO_4 and $HgSO_4$) was putted into digestion tubes [21]. The samples tubes were placed in COD digester (WTW CR3200, Germany) 150°C for 2 hours. After digestion the samples were slowly cooled to room temperature, the samples absorbance measured using (Thermo, Orion AQ2040, USA).

2.2.3.10. Biological Oxygen Demand (BOD).
BOD bottles were filled with the samples at 20°C and neutralized to pH 6.5 to 7.5. Bottles were closed with stopper and incubated in incubator (VELP FOC 225E., Italy) at 20°C for 5 days in a BOD Track (VELP BMS., Italy) that gave the reading directly as mg/l [21].

2.2.3.11. Sulfate.
Sulfate ions react with barium chloride solution to form insoluble barium sulfate, which is kept in suspension, by the buffer solution added [21]. The absorbance was recorded using spectrophotometer (Cecil CE7400, USA) at 420 nm.

2.2.3.12. Iron.
Iron is brought into solution, reduced to the ferrous state by boiling with acid and hydroxyl amine, using 1,10-phenanthroline indicator at pH 3.2 to 3.3 [21]. The absorbance was measured using spectrophotometer (Cecil, CE7400, USA) at 510 nm [17].

2.2.3.13. Manganese.
Persulate oxidation of soluble manganous compound to form permanganate is carried out in the presence of silver nitrate. The absorbance was recorded using spectrophotometer (Cecil, CE7400, USA) at 525 nm [21].

2.2.4. Water Quality Index.
Statistical Water Quality Index (SWQI)
The weighted arithmetic index method has been used for the calculation of WQI of water body. The quality rating or sub index (q_n) was calculated using the following expression:

$$q_n = 100\left[\frac{V_n - V_i}{}(S_n - V_i)\right] \quad (1)$$

Let there be n water quality parameters and quality rating or sub index (q_n) corresponding to n^{th} parameter is a number reflecting the relative value of this parameter in the polluted water with respect to its standard permissible value.
q_n: quality rating for the n water quality parameter
V_n: estimated value of the n parameter at a given sampling station.
S_n: standard permissible value of the n parameter.
V_i: ideal value of n parameter in pure water (i.e. 0 for all other parameters except the parameter pH and dissolved oxygen 7.0 and 14.6 mg/l respectively).

Assigned Unit Weight for various parameters is inversely proportional to the recommended standard (S_n) for the corresponding parameter.

$$W_n = \frac{K}{S_n} \tag{2}$$

W_n: unit weight for the n parameter.
S_n: standard value for n paramete
K: constant for proportionality.
The overall water quality index was calculated by aggregating the quality rating with unit weight linearity. [22-24]

$$SWQI = \Sigma q_n \frac{W_n}{\Sigma W_n} \tag{3}$$

2.2.5. Software.
2.2.5.1. MS Office 2007.
MS Excel 2007 software was used for calculation of standard deviation and average levels of water parameters in the present study, also used for generating graphs that summarize data. (Microsoft Co, USA)

2.2.5.2. Statistical Package for the Social Science (SPSS15) and (Minitab 14).
These two software statistical packages were employed for statistical treatment of data and predicting mathematical modeling, Minitab 14 was used to establish the model equation and to calculate the correlation between the water parameter and SPSS 15 software. It was used for conducting statistical analysis and generating tables and graphs that summarizes the data.

3. Results and Discussion
Rosetta branch is receiving several and different pollution types from different sources sewage and household effluents from El-Rahaway drains (more than $5X10^8$ m^3 daily effluents). Agricultural waste at Soble drain as well as liquid industrial waste generated by El-Malya and Soda Companies at Kafr El-Zyat City. All these pollutants have a prominent impact on the water environment of the Rosetta branch [25].

Water pollution is any chemical, physical or biological changes in water quality that have a harmful effect on any oragnism that drinks, uses, or lives in it. When humans drink contaminated water, it often has dangerous effects on their health. Water pollution can also make the water unsuitable for its intended use. In general, the use primary and secondary treatment systems are ineffective to remove pesticides residues and chlorinated organic pollutants as well as parasites, viruses, and other non- parasites microorganisms. As a result, these remnants of chemical and biological contamination may remain permanently in drinking water and still unable to obtain safe clean water, moreover, increased levels of chlorine usually added to filtered or raw water increase the concentration of chlorinated organic compounds that are known as carcinogenic and mutagenic [26-30].

3.1. Physical parameters

The mean values of seasonal and annual variation of the physical parameters in Rosetta branch during the period of the study are given in table 2, and represented example figure 1. the results show that:

1. The temperature values in different sites were ranged between 17.37-18.47, 25.37-25.87, 27.77-28.83 and 24.77-25.20 °C during winter, spring, summer and autumn, respectively. Water temperature varies with season, elevation, geographic location, climatic conditions, stream flow, streamside vegetation, ground water inputs, and water effluent from industrial activities. Water temperature rise when streamside vegetation is removed. This small variation of temperature would indicate a physically protected water supply, with closed underground reservoirs, connected with a source of underground origin or deep surface reservoir [31, 32]. The water temperature depends on latitude, altitude and means daily number of sunshine hours, and also affected by specific characteristic such as turbidity, wind force, plant cover, climatic condition and humidity [33].
2. The Turbidity values in different sites were ranged between 2.33-4.57, 1.68-5.04, 2.59-4.44 and 1.79-3.87 NTU (Nephlometric turbidity unit) during winter, spring, summer and autumn, respectively. The turbidity in water is the reduction of transparency due to the presence of particulate such as clay or silt, finely divided organic matter, plankton or other microscopic organisms. These cause light to be scattered and absorbed rather than transmitted in straight lines through the sample [33-36]. Erosion, waste discharge, and urban runoff can add suspended solids to a body of water. Agricultural runoff, in addition to directly increasing suspended solids, can also contribute to the growth of algae. After a storm or flooding, turbidity in surface water generally increases rapidly due to the increase in runoff. Bottom sediment may be stirred up by such actions as waves or currents, bottom-feeding fish, people swimming, wading, or storm runoff [29, 30].
3. The pH results show that, the values in different sites were ranged between 7.73-7.78. 7.65-7.71, 7.7-7.86 and 7.59-7.76 during winter, spring, summer and autumn, respectively. The pH values are within standard range set by the Egyptian Ministry of Water Resources and Irrigation (law 48/1982) [31]. The pH of water is a measure of the acid-base equilibrium and, in most natural water, is controlled by the carbon dioxide-bicarbonate-carbonate equilibrium system. An increased carbon dioxide concentration will therefore lower pH. The role of pH in water chemistry is also associated with corrosivity, alkalinity, acidity, chlorination, coagulation, CO stability [33-35].
4. The Electrical Conductivity results show that, the values in different sites were ranged between 610-631, 435-487, 405-414 and 546-553 μs/cm during winter, spring, summer and autumn, respectively. The increase in EC values were recorded during cold seasons is mainly attributed to the increase in cations and anions concentrations of the water, as result of low water level and discharge. But, the decrease in EC during hot seasons may be due to the sedimentation of suspended solid with organic salts causing decrease in chemical elements concentration [36].

IOP Conf. Series: Materials Science and Engineering **975** (2020) 012010 doi:10.1088/1757-899X/975/1/012010

Table 2. Seasonal and Annual variation of the water samples from Rosetta branch

TºC

Season / Site	Winter Average ±SD	Spring Average ±SD	Summer Average ±SD	Autumn Average ±SD	Annual Average ±SD
1	17.360±2.183	25.867±1.270	27.767±0.681	24.767±2.589	23.942±4.409
2	17.500±2.193	25.433±1.210	28.033±0.833	24.800±2.762	23.942±4.398
3	17.833±2.610	25.567±1.172	28.367±0.850	24.900±3.027	24.167±4.440
4	18.233±2.335	25.367±1.589	28.567±1.012	24.067±2.706	24.308±4.294
5	18.467±2.290	25.700±1.277	28.833±1.387	25.200±2.381	24.550±4.266
Regional average	17.880±2.322	25.587±1.303	28.313±0.952	24.947±2.693	24.182±4.362

pH

Season / Site	Winter Average ±SD	Spring Average ±SD	Summer Average ±SD	Autumn Average ±SD	Annual Average ±SD
1	7.730±0.072	7.690±0.104	7.697±0.029	7.763±0.058	7.720±0.068
2	7.757±0.058	7.710±0.090	7.700±0.026	7.743±0.067	7.728±0.060
3	7.770±0.036	7.693±0.085	7.707±0.072	7.627±0.112	7.699±0.087
4	7.783±0.067	7.693±0.074	7.717±0.086	7.587±0.118	7.695±0.106
5	7.780±0.070	7.653±0.064	7.863±0.292	7.663±0.040	7.740±0.160
Regional average	7.764±0.060	7.688±0.083	7.737±0.101	7.677±0.079	7.716±0.096

Turbidity

Season / Site	Winter Average ±SD	Spring Average ±SD	Summer Average ±SD	Autumn Average ±SD	Annual Average ±SD
1	4.567±0.482	5.043±0.387	4.437±0.592	3.873±0.878	4.480±0.680
2	4.480±0.151	4.350±1.421	4.183±0.535	3.767±0.725	4.195±0.773
3	4.290±0.451	4.247±1.556	3.363±0.911	3.083±0.132	3.746±0.970
4	3.393±0.300	4.307±1.741	3.000±0.646	3.033±0.696	3.433±1.017
5	2.327±0.241	1.677±0.137	2.590±0.792	1.793±0.369	2.097±0.554
Regional average	3.811±0.325	3.925±1.048	3.515±0.695	3.110±0.560	3.590±0.799

Electrical conductivity

Season / Site	Winter Average ±SD	Spring Average ±SD	Summer Average ±SD	Autumn Average ±SD	Annual Average ±SD
1	610.000±40.000	435.000±109.119	414.000±30.199	552.000±61.490	502.750±102.507
2	612.667±35.642	473.000±86.238	412.000±25.000	553.000±60.622	512.667±93.428
3	613.667±34.530	475.667±78.137	411.333±28.729	550.000±65.506	512.667±92.699
4	620.000±36.346	442.333±52.205	411.000±19.925	545.667±72.280	504.750±96.418
5	630.667±55.012	486.667±50.856	405.333±5.774	547.000±65.871	517.417±96.105
Regional average	617.400±40.306	462.533±75.311	410.733±21.925	549.533±65.154	510.050±96.231

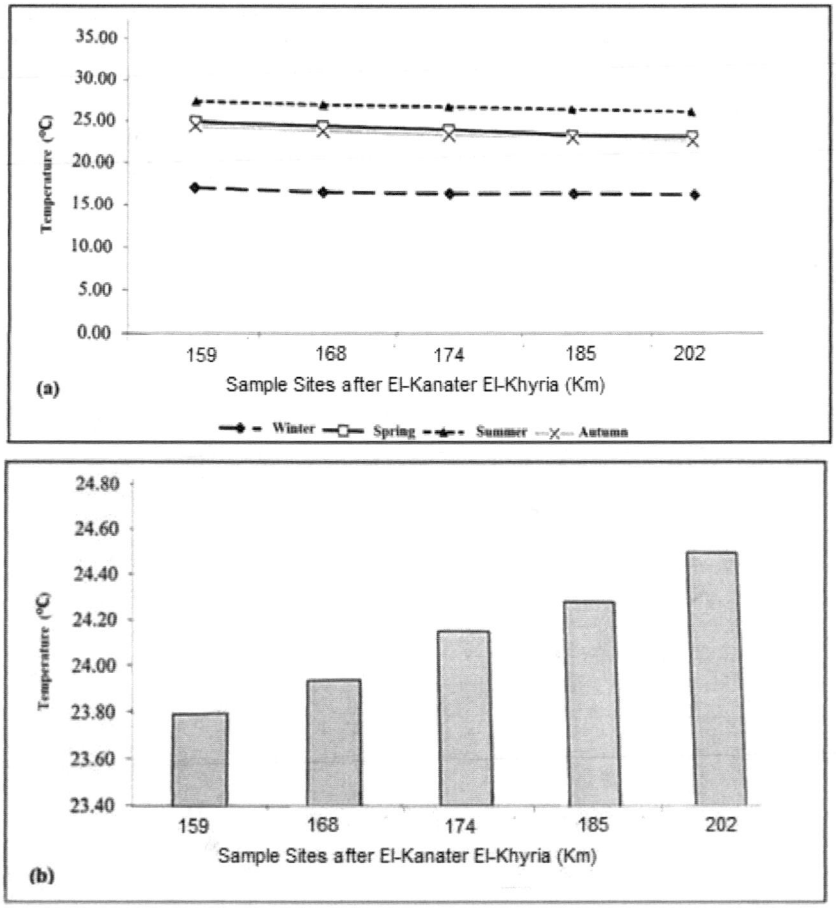

Figure 1. a Seasonal and (b) Annual variation of temperature (°C) from January 2013 to December 2013

3.2. Chemical parameters

tables 3-5, show the mean values of seasonal and annual variation of the Chemical parameters concentrations in Rosetta branch during the period of the study. The results show that:

1. The Chloride values in different sites were ranged between 68.00-78.67, 52.33-67.00, 41.33-44.33 and 64.67-69.00 mg/l during winter, spring, summer and autumn, respectively. The chloride values are within standard range set by Egyptian Ministry of Water Resources and Irrigation (law 48/1982) [32]. Chloride in surface and ground water from both natural and anthropogenic sources, such as run of containing road de-icing salts, the use of inorganic fertilizers, landfill leachates, septic tank effluents, animal feeds, industrial effluents, irrigation drainage, and seawater intrusion in coastal areas [38].
2. The alkalinity values (ranged between 208.00-238.00, 164.67-170.67, 164.00-164.67 and 180.00-184.00 mg/l during winter, spring, summer and autumn, respectively) are within stander range set by Egyptian Resources and Irrigation (law 48/1982) [32]. Alkalinity is buffering capacity of water body. Without this buffering capacity, any acid added to lake would immediately change its pH Alkalinity comes from rocks and soil, salts, certain plant activities, and certain industrial wastewater discharges (detergents and soup based products

are alkaline). If an area's geology contains large quantities of calcium carbonate ($CaCO_3$), limestone, water bodies tend to be more alkaline [37-39].

3. The total hardness values ranged between 208.00-220.67, 156.67-160.67, 144.00-148.00 and 169.33-172.67 mg/l during winter, spring, summer and autumn, respectively. The results declared that, the high values of total hardness content recorded during autumn and winter, this may be attributed to the low water level. Hardness may be considered as a physical or chemical parameter of water, it represents the total concentration of Ca and Mg ions, which reported as calcium carbonate. Surface raw water thought to be softener than ground water (more rain, less contact with soil minerals [40].
4. The calcium values ranged between 44.53-47.73, 36.00-39.20, 35.20-36.53 and 41.33-42.40 mg/l during winter, spring, summer and autumn, respectively. The results stated that, the high values of Ca^{+2} contents at different stations were recorded during winter, this may be attributed to the low water level through the brought period, or to the decaying and decomposing of phytoplankton and other microorganisms and liberated Ca^{+2} in different forms, or it is due to some re-dissolution of organisms and calcium compounds in the presence of CO_2 [41,42]. While the law Ca^{+2} contents recorded during summer attributed to the decrease of $CaCO_3$ solubility as a result of increase temperature and loss of CO_2, or due to its uptake by microorganisms and fishes living in the water branch, photosynthesis and carbohydrate metabolism [43].
5. The magnesium values ranged between 23.17-24.62, 16.04-16.69, 13.61-13.77 and 16.04-16.20 mg/l during winter, spring, summer and autumn, respectively. The high values of Mg^{+2} contents at different sites were recorded during winter, this may be attributed to the low water level through the brought period, or to the decaying and decomposing of phytoplankton and other microorganisms [41,42]. Magnesium is important element and needed by all biological cells for phosphate transfer involving adenosine tri-phosphate ATP and adenosine di-phosphate ADP [44].
6. The Ammonia values in different sites were ranged between 5.831-8.111, 0.961-1.812, 0.101-0.419 and 0.785-2.136 mg/l during winter, spring, summer and autumn, respectively. The highest concentration values of ammonia were recorded during cold season in winter this may be attributed to the low water levels during drought period and leached of fertilizers residues used on the agriculture into the aquatic environment of the branch via drains [46]. The lowest concentration of NH_3 were recorded during summer may be attributed to the dilution effect by the flood period, and to oxidation of NH_3 into NO_2^- and NO_3^- through nitrification process by the action of nitrosomonas bacteria genus according to the following equations [47,48]:

$$NH_4^+ + 1.5\,O_2 \longrightarrow NO_2^- + 2H^+ + H_2O$$
$$NO_2^- + 1/2\,O_2 \longrightarrow NO_3^-$$

Ammonia is an important parameter used to evaluate surface water quality. Small amount of ammonia occurs naturally, but a sudden increase in concentration may indicate sewage or industrial pollution. A rise in ammonia may be closely followed by an algae bloom with association taste and odor problems [49].

IOP Conf. Series: Materials Science and Engineering **975** (2020) 012010 doi:10.1088/1757-899X/975/1/012010

Table 3. Seasonal and Annual variation of the water samples from Rosetta branch

Cl^- (mg/l)

Season Site	Winter Average±SD	Spring Average±SD	Summer Average±SD	Autumn Average±SD	Annual Ave rage±SD
1	68.000±12.490	52.333±31.754	41.333±11.372	66.667±5.033	57.083±19.266
2	69.333±15.275	63.000±25.632	42.333±9.713	66.000±5.292	60.167±17.471
3	71.667±17.786	63.000±24.249	43.667±9.609	65.667±5.859	61.000±17.534
4	74.333±19.009	53.000±14.107	44.333±8.083	64.667±7.095	59.083±16.256
5	78.667±26.102	67.000±13.077	44.333±2.082	69.000±15.716	64.750±19.326
Regional average	72.400±18.132	59.667±21.764	43.200±8.172	66.400±7.799	60.417±17.971

NO_2^- (mg/l)

Season Site	Winter Average ±SD	Spring Average ±SD	Summer Average ±SD	Autumn Average ±SD	Annual Average ±SD
1	0.117±0.077	0.659±0.218	0.750±0.230	1.104±0.611	0.657±0.473
2	0.136±0.090	0.652±0.143	0.722±0.176	1.108±0.497	0.654±0.432
3	0.214±0.190	0.762±0.120	0.676±0.142	1.274±0.593	0.732±0.481
4	0.170±0.098	0.793±0.246	0.680±0.023	1.321±0.187	0.741±0.449
5	0.421±0.204	0.980±0.447	0.286±0.255	0.751±0.393	0.609±0.406
Regional average	0.211±0.132	0.769±0.235	0.623±0.165	1.112±0.456	0.679±0.448

SO_4^{-2} (mg/l)

Season Site	Winter Average ±SD	Spring Average ±SD	Summer Average ±SD	Autumn Average ±SD	Annual Average ±SD
1	68.453±1.872	45.827±6.108	37.733±3.946	49.173±0.873	50.297±12.212
2	69.123±2.595	46.487±6.031	36.000±4.045	50.730±0.435	50.585±12.930
3	69.770±6.226	45.203±7.005	34.733±3.668	51.523±0.676	50.308±13.981
4	68.440±4.525	43.340±7.216	32.867±3.121	53.580±0.505	49.557±14.254
5	67.257±3.805	44.130±10.895	31.000±3.464	51.910±5.396	48.574±14.817
Regional average	68.609±3.805	44.997±7.451	34.467±3.649	51.383±1.577	49.864±13.639

Alkalinity (mg/l)

Season Site	Winter Average ±SD	Spring Average ±SD	Summer Average ±SD	Autumn Average ±SD	Annual Average ±SD
1	208.000±9.165	166.667±25.403	164.000±19.079	184.000±11.136	180.667±23.608
2	215.333±14.189	167.333±21.939	164.000±16.371	182.667±11.372	182.333±25.425
3	222.667±15.535	166.667±24.194	164.667±13.614	180.000±12.000	183.500±28.389
4	232.000±20.785	164.667±25.716	164.667±11.547	183.333±16.773	186.167±33.188
5	238.000±27.785	170.667±18.903	164.000±12.166	181.333±13.317	188.500±34.600
Regional average	223.200±17.492	167.200±23.231	164.267±14.555	182.267±12.920	184.233±29.042

IOP Conf. Series: Materials Science and Engineering **975** (2020) 012010 doi:10.1088/1757-899X/975/1/012010

Table 4. Seasonal and Annual variation of the water samples from Rosetta branch

Ca^{+2}(mg/l)

Season Site	Winter Average ±SD	Spring Average ±SD	Summer Average ±SD	Autumn Average ±SD	Annual Average ±SD
1	44.533±3.946	36.000±4.157	35.200±2.117	42.400±2.117	39.533±5.019
2	44.533±3.946	37.333±3.946	35.467±2.013	42.133±1.848	39.867±4.625
3	45.067±2.444	37.333±3.946	35.733±2.444	41.333±1.222	39.867±4.432
4	45.600±2.117	36.267±3.695	36.000±2.117	41.600±2.400	39.867±4.749
5	47.733±5.787	39.200±2.884	36.533±2.444	41.333±2.013	41.200±5.301
Regional average	45.493±3.648	37.227±3.726	35.787±2.227	41.760±1.920	40.067±4.825

Mg^{+2}(mg/l)

Season Site	Winter Average ±SD	Spring Average ±SD	Summer Average ±SD	Autumn Average ±SD	Annual Average ±SD
1	23.490±1.403	16.686±1.964	13.608±2.430	16.200±1.964	17.496±4.170
2	23.490±1.403	16.362±2.297	13.608±2.430	16.200±1.964	17.415±4.221
3	23.490±1.012	16.362±2.297	13.770±2.191	16.038±1.752	17.415±4.133
4	23.166±1.223	16.038±2.227	13.770±2.023	16.200±1.840	17.294±4.010
5	24.624±3.958	16.524±1.752	13.770±1.562	16.038±1.752	17.739±4.778
Regional average	23.652±1.800	16.394±2.107	13.705±2.127	16.135±1.855	17.472±4.262

Mn^{+2} (mg/l)

Season Site	Winter Average ±SD	Spring Average ±SD	Summer Average ±SD	Autumn Average ±SD	Annual Average ±SD
1	0.097±0.005	0.070±0.014	0.053±0.007	0.073±0.007	0.073±0.018
2	0.089±0.006	0.063±0.010	0.041±0.007	0.069±0.009	0.066±0.019
3	0.076±0.006	0.056±0.004	0.036±0.006	0.063±0.005	0.058±0.016
4	0.075±0.015	0.054±0.012	0.032±0.010	0.059±0.009	0.055±0.019
5	0.066±0.015	0.048±0.019	0.026±0.009	0.053±0.006	0.048±0.019
Regional average	0.081±0.009	0.058±0.012	0.037±0.008	0.063±0.007	0.060±0.018

Iron (mg/l)

Season Site	Winter Average ±SD	Spring Average ±SD	Summer Average ±SD	Autumn Average ±SD	Annual Average ±SD
1	0.374±0.075	0.263±0.020	0.253±0.040	0.263±0.017	0.288±0.064
2	0.363±0.079	0.257±0.014	0.248±0.032	0.252±0.014	0.280±0.063
3	0.346±0.086	0.250±0.009	0.242±0.028	0.246±0.010	0.271±0.060
4	0.342±0.099	0.241±0.012	0.239±0.034	0.236±0.012	0.264±0.065
5	0.330±0.105	0.241±0.019	0.226±0.035	0.229±0.011	0.256±0.065
Regional average	0.351±0.089	0.250±0.015	0.241±0.034	0.245±0.013	0.272±0.063

NH_4^+ (mg/l)

Season Site	Winter Average ±SD	Spring Average ±SD	Summer Average ±SD	Autumn Average ±SD	Annual Average ±SD
1	8.111±1.178	1.812±1.548	0.419±0.276	2.132±1.671	3.119±3.278
2	7.952±1.547	1.775±1.376	0.398±0.230	2.097±1.566	3.056±3.225
3	7.601±1.529	1.400±0.744	0.301±0.128	2.136±1.375	2.860±3.084
4	7.429±2.951	1.595±1.277	0.299±0.113	1.697±1.414	2.755±3.244
5	5.831±2.404	0.961±0.698	0.101±0.062	0.785±0.193	1.920±2.612
Regional average	7.385±1.922	1.509±1.136	0.304±0.162	1.769±1.244	2.742±3.089

Table 5. Seasonal and Annual variation of the water samples from Rosetta branch

Total Hardness (mg/l)					
Season / Site	Winter Average ± SD	Spring Average ± SD	Summer Average ± SD	Autumn Average ± SD	Annual Average ± SD
1	208.000±15.620	158.667±18.475	144.000±10.392	172.667±4.163	170.833±27.269
2	208.000±15.620	160.667±16.166	144.667±8.083	172.000±5.292	171.333±26.469
3	209.333±10.263	160.667±16.166	146.000±7.211	169.333±4.619	171.333±26.095
4	209.333±10.066	156.667±14.189	146.667±5.033	170.667±9.018	170.833±26.319
5	220.667±30.746	166.000±12.490	148.000±3.464	169.333±7.024	176.000±31.760
Regional average	211.067±16.463	160.533±15.497	145.867±6.837	170.800±6.023	172.067±27.583

DO (mg/l)					
Season / Site	Winter Average ± SD	Spring Average ± SD	Summer Average ± SD	Autumn Average ± SD	Annual Average ± SD
1	4.410±0.698	4.393±0.525	3.240±0.867	4.943±1.823	4.247±1.141
2	4.467±0.666	4.847±0.389	3.170±1.100	4.660±1.693	4.286±1.150
3	4.650±0.685	5.107±0.463	2.587±0.905	3.303±1.381	3.912±1.318
4	4.807±0.704	4.993±0.696	2.747±1.416	2.920±1.287	3.867±1.420
5	4.803±0.208	5.917±0.772	2.767±1.663	2.643±1.380	4.033±1.749
Regional average	4.627±0.592	5.051±0.569	2.902±1.190	3.694±1.512	4.069±1.356

BOD (mg/l)					
Season / Site	Winter Average ± SD	Spring Average ± SD	Summer Average ± SD	Autumn Average ± SD	Annual Average ± SD
1	2.700±0.557	4.467±0.577	3.567±0.551	3.400±1.311	3.533±0.958
2	2.733±0.503	4.600±0.917	3.667±0.635	3.600±1.311	3.650±1.030
3	2.667±0.577	4.800±1.039	4.233±0.945	3.900±1.473	3.900±1.217
4	2.667±0.577	4.867±0.379	4.700±0.872	4.000±1.400	4.058±1.184
5	2.867±0.404	5.167±0.404	5.033±0.551	4.267±1.617	4.333±1.225
Regional average	2.727±0.524	4.780±0.663	4.240±0.711	3.833±1.423	3.895±1.123

COD (mg/l)					
Season / Site	Winter Average ± SD	Spring Average ± SD	Summer Average ± SD	Autumn Average ± SD	Annual Average ± SD
1	4.500±0.928	13.535±1.750	6.859±1.059	9.524±1.905	8.605±3.723
2	4.970±0.915	13.939±2.777	7.051±1.221	8.571±3.123	8.633±3.952
3	4.848±1.050	15.000±3.248	8.467±1.890	9.750±3.683	9.516±4.438
4	8.081±1.750	17.381±1.352	10.930±2.027	10.000±3.500	11.598±4.144
5	8.687±1.225	17.816±1.394	11.439±1.252	10.159±3.849	12.025±4.103
Regional average	6.217±1.173	15.534±2.104	8.949±1.490	9.601±3.212	10.075±4.072

IOP Conf. Series: Materials Science and Engineering **975** (2020) 012010 doi:10.1088/1757-899X/975/1/012010

7. The Nitrite values in different sites were ranged between 0.117-0.421, 0.652-0.980, 0.286-0.750 and 0.751-1.321 mg/l during winter, spring, summer and autumn, respectively. The results obtained of NO_2^- concentrations Show that, the nitrite were increased during autumn and decreased during winter. The sharp decrease of nitrite content during winter, might be due to principally the increase of oxidation rate of nitrite which converts nitrite into nitrate and as a result of denitrification biological reduction of nitrite which converts nitrite into cellular amino acids by the photosynthetic plankton algae and by transaminase enzyme which ultimately provide the fats, fatty acids, amino acids, nucleic acids, protein, organic acids and other organic compounds necessary for growth and reproduction of these organisms. This occurs during drought period [50].
8. The results of Dissolving Oxygen (DO) shows that, the values in different sites were ranged between 4.41-4.81, 4.39-5.92, 2.59-3.24 and 2.64-4.94 mg/l during winter, spring, summer and autumn, respectively. The highest values of DO were recorded during spring may be attributed to, the increase of photosynthesis activity, which liberate a significance amount of oxygen to surrounding water ecosystem, as well as related to the abundance of phytoplankton during this season in which the photosynthetic process regarded a main source of oxygen [51,52]. However, the lowest DO values were recorded during summer season, this may be due to the elevation in temperature during summer leading to the decrease the solubility of atmospheric oxygen surrounding to water [52, 53].
9. Biochemical Oxygen Demand (BOD) results shows that, the values in different sites were ranged between 2.67-2.87, 4.47-517 357.5.03 and 3.40-4.27 mg/during winter, spring, summer and autumn, respectively. The minimum values of BOD were recorded during winter this explained basically due to the drought period during this season. This agreed with that reported [54]. The higher BOD values recorded during spring may attribute to the photosynthetic activity and abundance of phytoplankton leading to an increase of DO there for BOD will increase [42]. High levels of BOD measurement refer to the wastewater contain organic materials that can be degradable by different organisms. This decomposition of organic materials can deplete dissolved oxygen supplies in receiving water body of aquatic life resulting in axoic or anaerobic conditions Methane, amines, and sulfide are produced in anaerobic water, causing the water to acquire an unpleasant odor, taste, appearance and less oxygen is available for higher animals such as fishes [54].
10. Chemical Oxygen Demand (COD) results shows that, the values in different sites were ranged between 4.50-8.69, 13.54-17.82, 6.86-11.44 and 8.57-10.16 mg/l during winter, spring, summer and autumn, respectively. This parameter is not requested for monitoring water supplies but used sometimes for evaluation of polluted raw water. Extremely useful in the determination of domestic waste and polluted waters, COD determination provides a measure of oxygen equivalent of the portion of the organic matter in a sample that is susceptible to oxidation by strong chemical oxidant. The high COD during the one year monitoring period may be due to loaded variety of samples with organic matter [55].
11. The Sulfate values in different sites were ranged between 67.257-69.770, 43.340-46.487, 31.00-37.733 and 49.173-53.580 mg/l during winter, spring, summer and autumn, respectively. The observed results of SO_4^{-2} concentration represented that, the higher values of SO_4^{-2} content during winter, may be due to, the death of microorganisms during drought period and to the decomposition of aquatic plants. This may also due to oxidation of S^{-2} into SO_4^{-2} which depends on the high concentration of dissolved oxygen during this season. This result agreed with that reported [55]. On the other hand, the lower SO_4^{-2} values were recorded during summer may be attributed to the elevation of temperature during this season leading to, two processes take place, Sulfate will be reduced directly by the action of sulfate reducing active bacteria into sulfide, and deposition of SO_4^{-2} to the sediment as sulfate salts. The above mentioned conclusion coincided with that reported [56].

12. The Iron values in different sites were ranged tween 0.330-0.374, 0.241-0.263, 0.226-0.253 and 0.229-0.263 mg/l during winter, spring, summer and autumn, respectively. The minimum values of iron were recorded during summer, may be attributed to that iron adsorbed by clay minerals, suspended matter, surface microorganisms and metals oxide as iron oxide under high temperature [57]. Generally, the distributions dynamics of iron in water ecosystem depends on the dissolved oxygen and present as insoluble $Fe(OH)_3$ or ferric oxide. The ferrous form can only exist in the absence of oxygen and the ferric forms are almost completely insoluble, in other words, in oxygenated water, iron is precipitated as ferric salts [58,59].
13. The Manganese values in different sites were ranged between 0.066-0.097, 0.048-0.07, 0.026-0.053 and 0.053-0.073 mg/l during winter, spring, summer and autumn, respectively, figure 2. The results of manganese concentration revealed that, the high values were recorded during winter and autumn seasons; this may be attributed to the effect of the drought period. However, the low water level and the slow motion of water current would facilitate in the excretion of manganese from this aquatic plants in addition to dissolution of sediment manganese and release to water during spring [41]. The low values of manganese during summer and spring may be attributed to the removal of manganese from aqueous phase to solid phase during precipitation of Mn^{+2} as MnO_2 or by adsorption of suspended particles during summer and spring. Also, the dilution effect of the flood period during autumn can add more explanation [60].

3.3. Correlation Coefficients

The most familiar measure of dependence between two quantities is the Pearson product-moment correlation coefficient, or "Pearson's correlation". If we have a series of n measurements of X and Y written as xi and yi where i =1, 2, n, The correlation coefficient is Written as:

$$r_{xy} = \frac{n\sum x_i y_i - \sum x_i \sum y_i}{\sqrt{n\sum x_i^2 - (\sum x_i)^2}\sqrt{n\sum y_i^2 - (\sum y_i)^2}} \tag{4}$$

The Pearson correlation coefficient ranges from -1 to +1. It is +1 in the case of a perfect positive (increasing) linear relationship (correlation), -1 in the case of a perfect decreasing (negative) linear relationship (anti-correlation), and some value between in all other cases, indicating the degree of linear dependence between the variables. As it approaches zero there is less of a relationship (closer to uncorrelated). The closer the coefficient is to either -1 or +1, the stronger the correlation between the variables [61].

Statistical analysis of data was computed to show the correlation between some parameters through the whole sampling sites during the period of the study, using simple correlation coefficient, table 6.

IOP Conf. Series: Materials Science and Engineering **975** (2020) 012010 doi:10.1088/1757-899X/975/1/012010

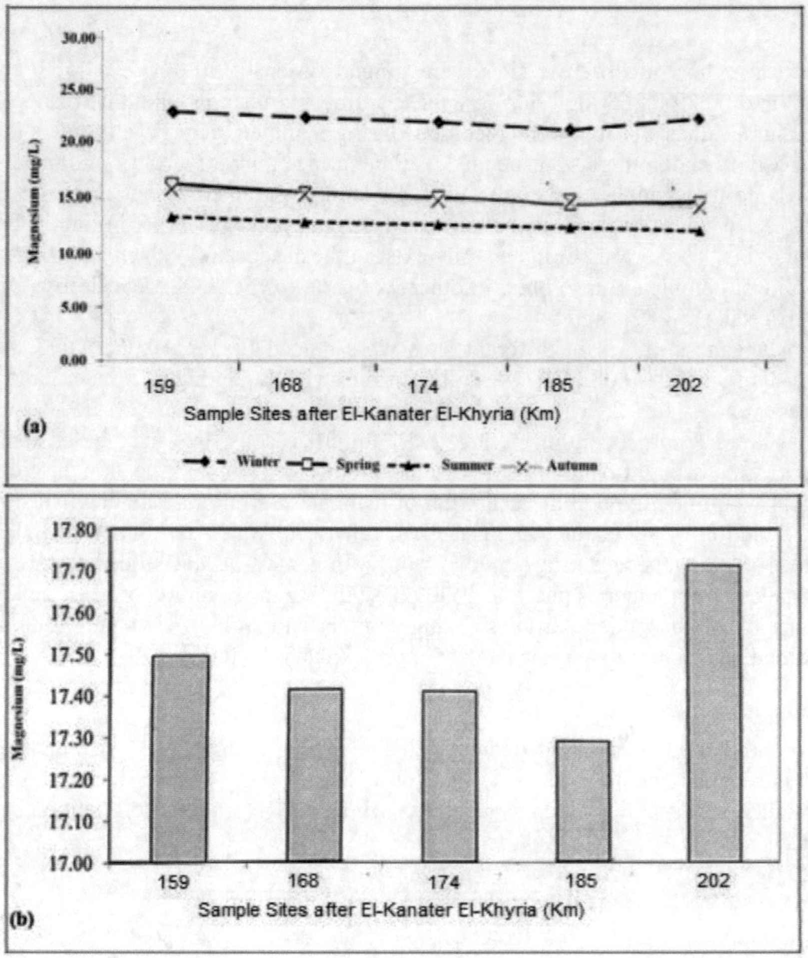

Figure 2. a) Seasonal and (b) Annual variation of magnesium (mg/l) from January 2013 to December 2013

Table 6. Person correlation coefficient for different parameters of Rosetta branch

parameters		r	parameters		r	parameters		r
Temperature	E.C.	-0.877	NH_4^+	NO_2^-	-0.529	total hardness	Ca^{+2}	0.899
	alkalinity	-0.864		BOD	-0.622		Mg^{+2}	0.955
	Cl^-	-0.667		COD	-0.427		SO_4^{-2}	0.890
	NH_4^+	-0.924	BOD	COD	0.788		Cl^-	0.743
	NO_2^-	0.515		NO_2^-	0.521	alkalinity	total hardness	0.923
	total hardness	-0.936	Ca^{+2}	Mg^{+2}	0.728		Ca^{+2}	0.864
	Ca^{+2}	-0.766	Fe^{+3}	Mn^{+2}	0.730		Mg^{+2}	0.858
	Mg^{+2}	-0.945	E.C.	Cl^-	0.780	Ca^{+2}	Cl^-	0.728
	SO_4^{-2}	-0.877		NH_4^+	0.782		SO_4^{-2}	0.770
	DO	-0.420		Ca^{+2}	0.827	Mg^{+2}	Cl^-	0.668
	Fe^{+3}	-0.821		Mg^{+2}	0.854		SO_4^{-2}	0.869
	Mn^{+2}	-0.769		SO_4^{-2}	0.861	Cl^-	SO_4^{-2}	0.670
				Fe^{+3}	0.635			
				Mn^{+2}	0.727			

Positive correlations have been obtained between:

- Electrical conductivity and each of anions and cations parameters. This indicates that the increasing in ions concentration leading to increase of EC.
- Calcium and each of chloride and sulfate, this indicates that calcium may be found as calcium chloride or a calcium sulfate.
- Magnesium and each of chloride and sulfate, this indicates that magnesium may be found as magnesium chloride or magnesium sulfate.
- Total alkalinity and each of calcium and magnesium may be due to their association to form calcium and magnesium carbonates or bicarbonates.
- Total hardness and total alkalinity indicating that total hardness is mostly temporary hardness (carbonate hardness).
- Total hardness and each of calcium, magnesium, chloride, and sulfate as expected.
- Both calcium and magnesium were taken the same behavior with different parameters in the correlation matrix analysis, also iron and manganese found to have the same behavior with different parameters in the correlation matrix.
- BOD was positively correlated with nitrogen compounds (nitrite) as expected because nitrogen compounds is essential to growth of organisms where the nutrient limits the primary productivity of the living bodies

Negative correlations have been obtained between:

- Dissolved oxygen was negatively correlated with temperature in Rosetta branch as expected.
- Ammonia and nitrite indicate that strong opposite relationship between the two parameters and the inter conversion between them.
- Ammonia and BOD and COD as expected.
- Temperature and ammonia attributed to during hot season the oxidation of ammonia into NO_2^- and NO_3^- through nitrification process by the action of nitrosomonas bacteria takes place.

3.4. Analysis of variance

To estimate the amount of variation due to certain causes (temporal and/or spatial reason); using analysis of variance (ANOVA) examination the different (usually called source of variations) operating simultaneously on a response to decide which effect are statistically significant and to estimate their contribution to the variability of response. For these reasons, a two way ANOVA of independent variable used in the present study showed the existence of seasonal and/or spatial difference of Rosetta branch.

ANOVA tests were calculated for all parameters of Rosetta branch. The results indicate that the effect of seasons were significant (there were significant difference between seasons) for all parameters except temperature and pH are not significant. The effect of sites was significant (there were significant difference between sites) for temperature, turbidity, ammonia, BOD, COD, iron and manganese. However, the effect was not significant (there were no significant difference between sites) for pH, electrical conductivity, chloride, alkalinity, total hardness, calcium, magnesium, nitrite, dissolved oxygen and sulfate.

3.5. Statistical Water Quality Index (SWQI)

The calculation of statistical water quality index, eleven important Parameters were chosen. The SWQI has been calculated by using the Law 48/1982 amended in 2013, Decree NO. 49, "Quality of fresh water". The weighted arithmetic index method has been used for the calculation of WQI of water body, table 7. Water quality index and status of water quality was presented in table 8 and figure 3.

IOP Conf. Series: Materials Science and Engineering **975** (2020) 012010 doi:10.1088/1757-899X/975/1/012010

WQI was calculated for five sites along Rosetta branch during the period of the study figure 3, it shows that:

- Quality of water in summer ranged from 26.268 at site (5) to 46.64 at site (1) (good water quality).
- Water in spring in site (5) show (poor water quality) and in site (3,4) show (very poor water quality)
- In autumn and winter the water is (unsuitable for drinking) for all sites except site (5) show (poor water quality) as indicates

Table 7. Calculated Assigned unit weight for each and status of parameter with their Egyptian standard limits "Law 48/1982 amended in 2013, Decree NO. 4"

Sr. No.	Parameters	Standard	Unit Weight
1	pH	6.5-8.5	0.118
2	TDS	<500	0.002
3	Ammonia	<0.5	2.000
4	DO	>5.0	0.167
5	COD	<6.0	0.167
6	BOD	<10.0	0.100
7	Sulfate	<200	0.005
8	Iron	<0.5	2.000
9	Manganese	<0.2	5.000

Table 8. Water quality index water quality

Index level	Quality
0-25	Excellent water quality
26-50	Good water quality
51-75	Poor water quality
76-100	Very poor water quality
>100	Unsuitable for drinking

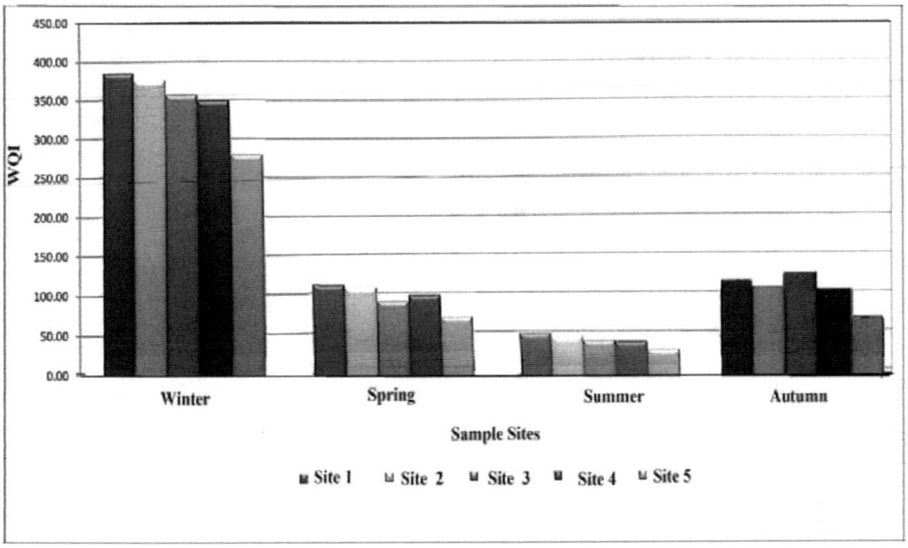

Figure 3. SWQI of Rosetta Branch at 5 sites

4. Conclusions and Recommendations:

The results show that:

- The pollution appears in the fresh water of the Rosetta branch with sewage and domestic wastes of El-Rahawy drain and agricultural wastes.
- Most of physical and chemical parameters are changeable and affected by seasonal variation.
- The degree of pollution of the branch is visible in the gradually increasing of the following parameters: electrical conductivity, chlorides, alkalinity, BOD, COD, hardness, sulfate, ammonia, iron, and manganese along the branch.
- Increasing of ammonia is the most indication of agricultural pollution.
- Increasing of electrical conductivity, chloride, COD and BOD indicates the sewage pollution.
- BOD and COD indicate the activation of microorganisms and presence of organic compounds where microorganisms use dissolved oxygen during oxidation process of it.
- WQI was calculated for the five sites along Rosetta branch during the period of the studies where:
  1. Good water quality in summer is in all sites.
  2. Poor water quality in spring in site (5) and very poor quality in sites (3,4).
  3. Unsuitable water for drinking in autumn and winter in all sites except site (5) show poor water quality.

5. References

[1] Ibrahim M, Al-Zyoud S and Elhaddad E, 2018 Open Journal of Geology 8 (2) 161-173.
[2] Badr MH, Elewa A, Shehata MB, Mohamed LF, and Abdelaziz GS, 2006 Bulletins of the Environmental Research 9 35-54.
[3] Carroll SP, Dawes L, Hargreaves M, and Goonetilleke A, 2006 Queensland University of Technology: Caboolture Shire Council, QLD, Australia 1–93.
[4] Nnane DE, Ebdon JE, and Taylor HD, 2011 Water Res. 45 2235–46.
[5] Hill MK, 2010 Understanding Environmental Pollution, 3rd ed.; Cambridge University Press: Cambridge, UK.

[6] Bu H, Meng W, and Zhang Y, 2014 Environmental Monitoring and Assessment 1863619–32.
[7] Razmkhah H, Abrishamchi A, and Torkian A, 2010 Journal of Environmental Managment 91 852–60.
[8] Li S, Li J, and Zhang Q, 2011 Journal of Hazardous Materials 195 306–17.
[9] Mei K, Liao L, Zhu Y, Lu P, Wang Z, Dahlgren RA, and Zhang M, 2014 Environmental Science and Pollution Research 21 8036–51.
[10] Wang X, Cai Q, Ye L, and Qu X, 2012 Quaternary International 282 137–44.
[11] Shrestha S, and Kazama F, 2007 Environmental Modelling & Software 22 464–75.
[12] Singh KP, Malik A, and Sinha S, 2005 Analytica Chimica Acta 538 355–74.
[13] Ogwueleka TC, 2015 Environmental Monitoring and Assessment 187 (3) 137.
[14] Sheikhy NT, Ramli MF, Aris AZ, Sulaiman WN and Fakharian K, 2014 Environmental Monitoring and Assessment 186 5797–15.
[15] Mustapha A, Aris AZ, Juahir H, Ramli MF and Kura NU, 2013 Environmental Science and Pollution Research 20 5630–44.
[16] Sharma A, Bora CR and Shukla V, 2013 Natural Resources Research 22 283–96.
[17] Gatica EA, Almeida CA, Mallea MA, Del Corigliano MC and Gonzalez P, 2012 Environmental Monitoring and Assessment 184 7257–74.
[18] Milanovi´c A, Urosev M and Milijasevic D, 2006 Bulletin of the Serbian Geographical Society 86 53–61.
[19] Milanovi´c A, Milijaˇsevi´c D and Brankov J, 2011 Carpathian Journal of Earth and Environmental Sciences 6 269–77.
[20] Panigrahy BP, Singh PK, Tiwari AK, Kumar B and Kumar A, 2015 Journal of Biodiversity and Environmental Sciences 6 33-9.
[21] Clesceri LS, Greenberg AE and Eaton AD, 2005 American Public Health Association, Washington DC, UAS.
[22] Dhirendra MJ, Alok K and Namita N, 2009 Rasayan journal of chemistry 2 (2) 195-203.
[23] Shilipi S and Saloni g, 2010 Archives of Applied Science Research 2 (4) 157-64.
[24] Yogendra K and Puttaiah ET, 2007 Shimoga Town, Karnataka, the 21th World Lake Conference 342-46.
[25] Abdo MH, 2013 Life Science Journal 10 (4) 493-501.
[26] Graun GF, Bull RJ, Clark RM, Doull J, Grabow W, Marsh GM, Qkun DA, Regli S, Sobsey MD and Symons JM, 1994 Journal of Water Supply: Research and technology-Aqua 43 (5) 207-18.
[27] Payment P, 1993 Washington International life Sciences Institute 99-113.
[28] Craun GF, Bull RJ, Clark RM, Doull J, Grabow W, Marsh GM, Okun DA, Regli S, Sobsey MD and Symons JM, 1994 Journal of Water Supply: Research and technology-Aqua 43 (4) 192-99.
[29] Nieuwenhuijsen MJ, Toledano MB, Eaton NE and Elliott P, 2000 a review Occupational Medicine 57 73-85.
[30] Golfinopulos SK and Arhonditsis GB, 2002 Chemosphere 47 1007-18.
[31] Jonh De Zuane PE, 1997 2nd John Wiley & sons Inc. New York. USA 120-21.
[32] Masoud MS, Elewa AA and El-Halim AM, 2002 Journal of Saudi Chemical Society 6 (3) 377-98.
[33] Viner AB and Smith IR, 1973 Proceedings of the Royal Society of London. Series B. Biological Sciences B184 235-70.
[34] Chen RH, Li FP, Zhang HP, Jiang Y, Mao LC, Wu LL and Chen L, 2016 Science Total Environmental 15 (553) 519-23.
[35] Hoda AE, Ahmed M, Laila M and Nermin N, 2007 Journal of High Institute of Public Health 37 (1) 140-54.

[36] Masoud MS, El-Aewa AA and Awad FK, 1998 Bulletin Faculty of Science, Asiut University 27 2B 23-36.
[37] Rahmani AR, 2008 Journal of Health Science 8 (1) 18-24.
[38] Jonh De Zuane PE, 1987 Environmental Protection Agency, Office of Drinking Water.
[39] Masoud MS, El-Samra MI and El-Sadawy MM, 2005 The Egyption Science Magazine 2 (3) 70-8.
[40] Masoud MS, El- Awi A and Abd El-Halim A, 2001 The 2nd Conference and exhibition for Life and Environment, Alexandria, Egypt 339.
[41] Mohamed AG, El Safty AM and Siha MS, 2013 Egyptian Journal of Occupational Medicine 37 (1) 95-119.
[42] Mohamed MM, Ahmed E, Ahmed MM, Hamdy A and Mohsen Y, 2018 Egyptian Journal of Aquatic Biology & Fisheries 22(5) 461-74.
[43] Alewa AA, Masoud MS and Awad FK, 1998 French-Lebanese Conference on the Water and health, Beirut, Lebanon.
[44] Stumm W and Morgan JJ, 1970 New Emphasizing Chemical Equilibria in Natural Water, Wiley Interscuence, New York.
[45] Magboul M, Jamal T, Mutwakil M, Mohammed S, Sulieman H, Hager M and Mushtaha E, 2017 Eurasian Journal of Soil Science 6 (3) 285-94.
[46] Abd El-Star AM and Morgan JJ, 2008 Egyptian Journal of Aquatic Reasearch 34 (2) 130-48.
[47] Kaiser GE and Wheaton FW, 1983 Journal of the World Mariculture Society 14 402-324.
[48] Masoud MS and Al-Saify MH, 2006 24th Annual Water Treatment Technology Conference, Abu-Qir Fertilizers.
[49] Kathleen F and Alan L, 1999 4th Ed. American water association USA 99 (177) 91-5.
[50] Munawar M, 1970 I. the Biotope. Hydrobiol 35 127-62.
[51] Saad MAH, 1980 Water Supply and Management 4 81-92.
[52] Cole GA, 1979 2nd C. V. Mosby Company, Toronto, London 401.
[53] Saad MAH, 1978 Estuarine and Coastal Marine Science 6 503-13.
[54] Raymond PA and Cole JJ, 2003 Science 301 88-90.
[55] John De Zuane PE, 1997 2nd Ed., John Wiley and Sons Inc. New York, USA.
[56] Vernberg FJ and Vernberg WB, 2001 University of South Carolina Press.
[57] Hassona AFA, 1989 M. Sc. Thesis, Faculty of Science al-Azhar University.
[58] Ruttner F, 1963 3rd Ed. University of Toronto Press 245.
[59] Masoud MS, Zaid MA and Haddad A, 2003 Bulletin of Macedonia of Chemistry and Technologists 22 (2) 143-54.
[60] Gohar MA, 1998 Faculty of Science Menofiya University, Egypt.
[61] Pathak H and Limaya SN, 2011 A Case Study, Ovidius University Annals of chemistry 22 (2) 87- 93.

Rain Transmission Losses Assessment in Arid Environment, Egypt: Numerical and Experimental Study

Adel Bakheet[1], Ahmed Sefelnasr[2] and Mohamed Saber[3]

[1]PhD. Environmental Geology, water and microbiology lab, Chemistry Administration, bakheet_adel@yahoo.com
[2]Associate professor of Hydrogeology, Geology Department, Assiut University
[3] Senior Researcher, Water Resources Research Center, DPRI, Kyoto University, japan

Abstract. In water-scarce regions, Transmission Losses Assessment from rainfall is required for the Numerical and Experimental Study of groundwater recharge. This Study is an attempt to assess Transmission Losses and the infiltration rate at Wadi El-Assiuti, Egypt, field infiltration experiments were conducted to measure the infiltration and porosity at the target area, such as using a double-ring infiltrometer. Determination and Quantify the infiltration rate of the soil is very vital as input parameters of the hydrological modeling, irrigation planning and other natural or man-made Processes .Steady state infiltration rates have been determined in selected sites of Wadi El-Assiuti. Accurate determination of infiltration rates is fundamental for trustworthy prediction of surface runoff and groundwater recharge. Double ring infiltrometer was used to measure infiltration rate at selected sites. An important part of the study was the accurate and consistent measurement of infiltration rates. .The infiltration rate was calculated according to the installation and operating instruction for the double ring in ASTM 2003. Final infiltration rates were 18 cm h^{-1}, for the first observation site, 24.2 cm $h^{-1}$for the second site and 59 cm $h^{-1}$for the third site. The analysis indicated that the soil for the three selected areas is sandy soil. Infiltration rates were taken at 0 to70 minutes of 10 minutes intervals. The infiltration rate at the third site is high compared to the other two selected sites. After performing the experiments, soil samples have been gathered and analyzed in the laboratory for porosity analysis. The results of this study can therefore be applied in the prediction of saturated hydraulic conductivity of the surface layers and groundwater recharge, and in developing or selecting the most efficient irrigation methods. The results from this work have been also involved in a numerical hydrological model to estimate the Transmission Losses form rainfall at Wadi El-Assiuti using Hydro-BEAM.

Key Words: Transmission Losses, Double ring infiltrometers, hydrological modeling, Wadi El-Assiuti, Hydro-BEAM.

1. Introduction

Not only Transmission losses reduce flood magnitudes but also play a vital role in groundwater recharge in arid areas. Assessment of the transmission losses as well as groundwater recharge for actual runoff events requires extensive data which are generally not available for non- experimental watersheds. The data shortage results in wide uncertainties in hydrological model of flash flood, groundwater recharge and transmission losses.

Recognition of hydrological processes of Wadi system is so important due to the importance of the water resources in the arid regions. In the arid areas, it is renowned that there are many serious problems, for instances, the shortage of water resources which lead to critical effects on the economic activity and human live. Another serious problem is increasing the losses which represent in the evaporation, initial and transmission losses. Despite of the high importance of water in arid and semi-arid regions, hydrological records and data have historically been sorely limited. Moreover,

IOP Conf. Series: Materials Science and Engineering **975** (2020) 012011 doi:10.1088/1757-899X/975/1/012011

countries of the arid areas are encountered the problem of overpopulation, So the demand of water for the agricultural and domestic purposes is necessary to overcome such problem. For aforementioned reason Egypt was selected as arid region to apply the Numerical and experimental study, especially in the Wadi system like Wadi EL-Assiuti in the eastern Desert. The hydrological modeling programs are very powerful tools to simulate the transmission losses process in the rainfall-runoff analysis.
The relationship between Wadi flow, transmission losses and groundwater recharge depend on the underlying geology. The alluvium implied the Wadi bed is functional in minimizing evaporation loss through capillary rise (the coarse structure of alluvial deposits minimizes capillary effects). According to reference [1] and [2] surface water and groundwater interactions depend strongly on the local characteristics of the underlying alluvium and the extent of their connection to, or isolation from, other aquifer systems.

Transmission losses in semiarid watersheds promote singularity about the spatial and temporal nature of surface water–groundwater interactions compared to humid basins. Because of transmission losses, the nature of surface water–groundwater interactions can be limited to brief periods during runoff events and to specific areas associated with the runoff production and downstream routing [3]. The rate of loss is linearly related to the volume of surface discharge as mentioned in reference [4] and [5] provided evidence that.

Reference [6] showed that losses are high when the alluvial aquifer is fully saturated, and small once the water table drops below the surface. Reference [7] provided an analysis of groundwater rise due to transmission loss for an experimental reach in Wadi Tabalah, S.W. Saudi Arabia and stated that about average 75% of bed infiltration reaches the water table.
High quality research is needed, specially to investigate spatial rainfall, infiltration and groundwater recharge from ephemeral flows. One reason of this paper focusing on the arid and semiarid regions is to estimate the transmission losses and its effect on both surface and subsurface water. Consequently, one part of this paper is an integrated numerical model based on sporadic precipitation and under conditions of limited data where the watershed modeling was developed by using GIS tool. Furthermore, surface runoff and stream routing modeling based on using the Kinematic wave approximation, the initial losses modeling estimated by applying SCS method and transmission losses modeling estimated with applying Walter's Equation [4], and groundwater modeling based on the linear storage model.

2- Experimental and Methodology

2.1 *Location and importance the study area*

The area under interest is Wadi El-Assiuti. Wadi El-Assiuti is located in the middle of the Egyptian Eastern Desert as shown in figure 1.Wadi El-Assiuti as a watershed basin is located directly East of the River Nile between longitudes31°12'&32°30′ E, and latitudes27°00'&27°48′ N. This basin is considered a sub-basin of the whole River Nile basin.

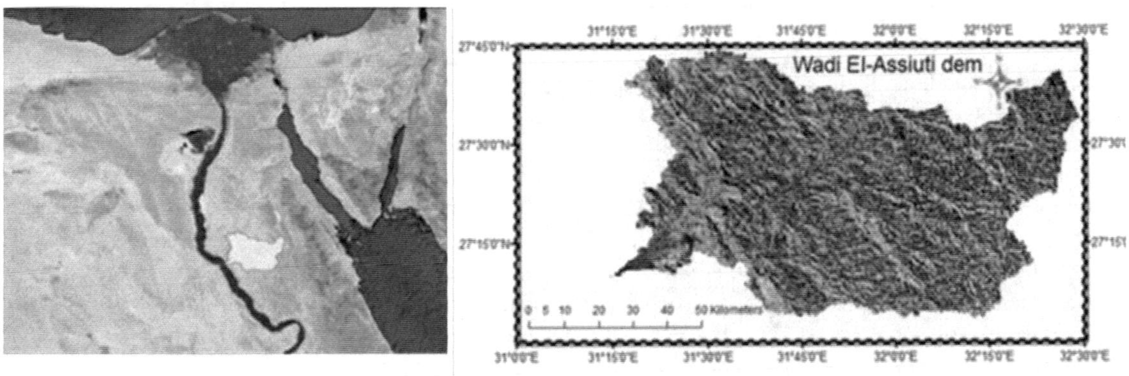

Figure 1. Location map of Wadi El-Assiuti watershed

IOP Conf. Series: Materials Science and Engineering **975** (2020) 012011 doi:10.1088/1757-899X/975/1/012011

The total area of Wadi El-Assiuti Catchment is 6051.40 km^2, the perimeter is 532.722 Km and the length of the main channel is 160.154 km. Most of its area is a desert except some part of urbanization, and very small areas of agricultures which are sealed to Assiut city along the River Nile. Therefore, studying this area is important due to the spread of populations and consequently the need of water resources for agricultural, domestic and manufactory purposes.
Wadi El-Assiuti area has been submitted for development improvements over the past centuries, where many of the former studies were applied and many of projects were established. Wadi El-Assiuti represents the arid conditions, its drainage system is ephemeral streams, also the rainfall is very rare in space and time. Currently, the establishing of new town, which will be in the near future crowded by populations, triggers the importance of hydrological models for water resources management and flood threat control.

2.2 The double ring infiltrometer measurement evaluation

Two different methods of measurements are commonly used with the ring-infiltrometers: constant-head and falling-head. Numerical modeling suggests that constant-head measurements give an overestimate compared to falling-head measurements for coarser textured soils and falling-head measurements tend to underestimate the actual infiltration rate [8].

Field measurements of infiltration rate using a double ring infiltrometer at three observation sites have been conducted at Wadi El-Assiuti as shown in figure 2.Two field trips were conducted for measuring infiltration rate using a double ring infiltrometer. The infiltration rate was calculated according to the installation and operating instruction for the double ring in [9].In general, the infiltration rate curve acted as theory predicted with high initial infiltration rates and slow approach to the steady state infiltration state.

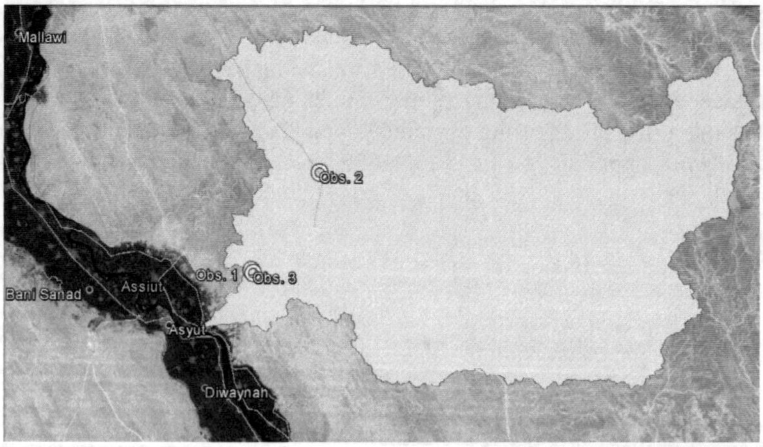

Figure 2. Double ring infiltrometer observation sites (1, 2 and 3) at Wadi El- Assiuti

2.3 Physical model (rainfall simulator)

The runoff was measured using a rainfall simulator at three observation sites at Wadi El-Assiuti as shown in figure 3. All experiments were performed with a portable rainfall simulator as depicted in figure 4. Portable rainfall simulators have been widely used for various purposes [10] because they make it possible to gather data under controlled conditions over relatively short periods of time [11].During our field observations at Wadi El-Assiuti, the soil infiltration rate is estimated from the rate of advance of the wetting front, separating wet and dry soil. The relationship between the soil infiltration rate and the progress of the advancing area wetted by water at a constant flow rate is described as follows: Initially, the soil infiltration rate is very high and a small area of the slope can absorb all of the supplied water. With soil infiltration rate decreasing in the previously wetted soil

IOP Conf. Series: Materials Science and Engineering **975** (2020) 012011 doi:10.1088/1757-899X/975/1/012011

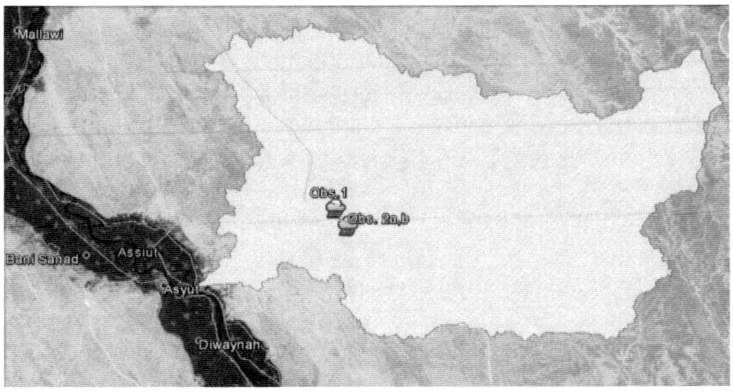

Figure 3. Rainfall simulator observation sites at Wadi El- Assiuti

A

B

C

D

F

G

Figure 4. (A, B, C, D, F and G) Field observations of infiltration experiments, runoff was measured using a rainfall simulator at Wadi El-Assiuti.

zone, the same water flow cannot be totally absorbed within the same soil surface area. Thus, the water flow advances over the surface and the wetted area increases gradually. The soil infiltration rate keeps decreasing with time until it reaches its steady value when the area wetted by the constant flow rate no longer expands but also remains constant. The advancing process of the in-flow water over the soil surface indicates that the water supply is sufficient to penetrate the soil in the wetted area [12].

2.4 The porosity analysis

The porosity together with the depth of the alluvium determines the maximum volume that can be absorbed by the alluvium as shown in figure 5. With regard to influence the infiltration rates porosity does not have as much weight as the other parameters. The final infiltration rate is not influenced by the porosity but by the initial infiltration rates. Low values for porosity will reduce the initial infiltration rates whereas large values will increase the initial infiltration rates. Together with the antecedent moisture index Antec, porosity is used to calculate the initial moisture content.

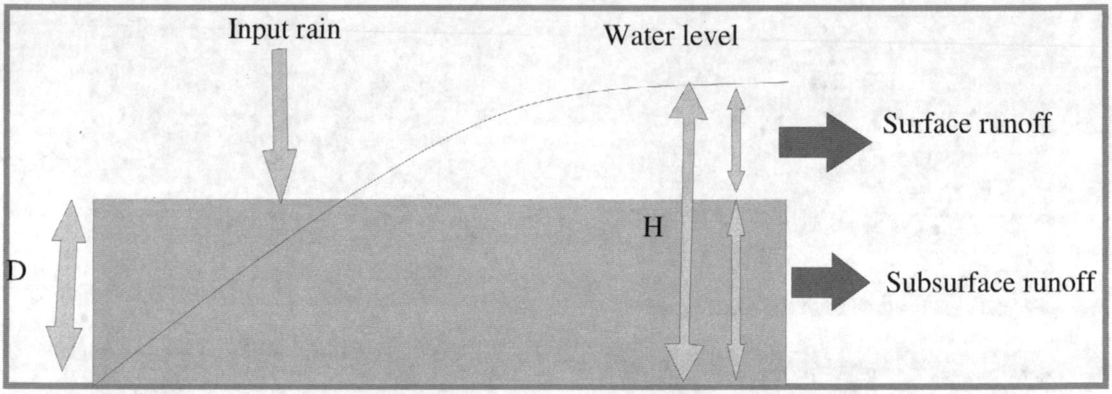

Figure 5. Conceptual model of effective soil depth *D* is layer depth, *H* water depth

Effective soil depth = porosity × soil depth

$$H = h_s + h_a \quad (1)$$

$$q_s = \alpha h_s^m \text{ Surface flow only Manning formula} \quad (2)$$

$$q_a = \alpha h_a \text{ Subsurface flow only Darcy's formula} \quad (3)$$

Where: h_s surface water level, h_a water depth in soil q : discharge per unit length of flow

2.5 Hydro-BEAM model

The Hydrological River Basin Environment Assessment Model (Hydro-BEAM) was originally developed by Kojiri et al. [13] and has been used in many studies as a river basin environment model for assessing water quantity, sediment, reservoir operations, and the impacts of climate change and anthropogenic activity (e.g., [14,15]). However, all of these applications were limited to environments with humid conditions until the model was later modified for flash flood simulations in the arid valleys [17, 18] and semi-arid basins [16].

The most beneficial worthiness of distributed hydrological models such as Hydro-BEAM is their representation of the spatial variations in watershed characteristics and hydrological processes. The watershed under investigation is divided into a number of unit mesh cells (in this study, approximately 5457 mesh cells with a resolution of 1 km). The unit mesh CELL is divided into two pairs of rectangular hill slopes and one river channel. The river channel geometry (depth and width) is related to the upstream area and is tested using Google Earth and DEM data. Vertically, each mesh is represented by a combination of one surface layer and three subsurface layers. The integrated kinematic wave model is used to describe the flow over the surface layer A and throughout the channel by considering the diverse surface characteristics of the different land use types. The subsurface layers (layers B, C and D) are modeled using the linear storage model. The Natural Resources Conservation Service (NRCS) method is used to calculate the initial losses in the target catchments [19], and a regression formula developed for the arid regions and valleys by Walters [4] is adopted to estimate the instantaneous transmission losses with runoff.

2.6 Hydro-BEAM model calibration

For several reasons, the calibration of rainfall-runoff models in wadi systems is hampered by a lack of data [20, 21]. To our knowledge, no flow rate data have ever been recorded in Wadi El-Assiuti. In this study, model parameterization is performed based on the calibrated model parameters conducted by [18] in Wadi Al-Khoudh in Oman. Applying a calibrated hydrological model from another wadi to a wadi under similar conditions has been done by some researchers, such as [22], who transferred the SWAT model calibration in Wadi Girafi in Palestine to all major wadis in the Saini Peninsula and the Eastern Desert in Egypt.

Both Al-Khoudh and El-Assiuti wadis had arid conditions with high average potential evapotranspiration values. However, the mean annual rainfall is higher in Wadi Al-Khoudh (150mm). Both Al-Khoudh and El-Assiuti wadis had very similar land use settings; both of them had areas mainly consist of desert and bare mountain land (comprising 90-95%) of the total area with very few urban and agricultural activities mainly in the downstream regions (5-10%). Therefore, these similarty in land use and climatic conditions could support the application of the Wadi Al-Khoudh parameter settings to Wadi El-Assiuti.

3. Results and Discussion

3.1 Double ring infiltrometer measurement

The measured steady state infiltration rates for the selected experiments are presented in table 1

Table 1. Field measurements of double ring infiltrometer at Wadi El-Assiuti, Egypt.

	LOCATION		elevation m	Initial Infiltration rate cm/h	Final Infiltration rate cm/h	Wetting front depth cm
	North	East				
Site1	27°16' 52.1"	31° 22'19.3"	101	24	18	75
Site2	27°27'53.87"	31°31'31.76"	222	60	24.2	98
Site3	27°16'33.34"	31°22'35.15"	88	62.4	59	108

The field measurements of infiltration rate were plotted against time as shown in figures 6. A, B and C.Figure 6 shows temporal changes in infiltration rates measured at different sites at Wadi El-Assiuti. The infiltration rate decreased rapidly at the beginning of the measurement, as well as, it showed marked fluctuations. All infiltration curves showed initial rapid declines, then; the infiltration fluctuated and decreased toward the end of the experiment. The infiltration rates in the last of the experiment were averaged for each plot. The results were as follows: final infiltration rates (basic infiltration) were 18 cm h^{-1}, for the first observation site, 24.2 cm$h^{-1}$for the second site and 59 cm $h^{-1}$for the third site. An explanation for the third site increase in infiltration rates involves macro pores. Macro pores in the surface soil, formed by soil fauna and the decay of plant roots; it is a common feature of vegetation cover lands. Water can enter macro pores when the surface soil layer is saturated or partially saturated [23]; consequently, the infiltration rate will increase. The increased infiltration rates in the experiment may have reflected a change in water flow paths under such conditions. A higher infiltration rate (59cm h^{-1}) was observed on the third observation site, where there is

IOP Conf. Series: Materials Science and Engineering **975** (2020) 012011 doi:10.1088/1757-899X/975/1/012011

agricultural land with vegetation cover. For observation site with lower vegetation cover, the infiltration rate was 18cm h^{-1} at the first site and 24.2cm h^{-1} at the second site. These results suggest that dense surface vegetation cover increases the infiltration rate significantly. Agricultural soils usually show wide spatial and temporal changes in pore characteristics due to changes in soil texture, structure, horizonation, root growth [24].

Infiltration data can be analysed according to a number of infiltration models. One such model is that developed by [25] and can be stated as:

$$I = K\,t + S\,t^{\frac{1}{2}} \tag{4}$$

Where I, is the cumulative infiltration (cm), S is the soil water sorptivity (cm/ h),K is the saturated hydraulic conductivity, and t is the time (h).
By regressing the cumulative infiltration data collected in the field based on equation 5. One can estimate the values of the parameters $K$and S. The infiltration rate i (cm h^{-1}) can be computed from equation 5 as follows:

$$i = \frac{dI}{dt} = K + \frac{1}{2} S\, t^{1/2} \tag{5}$$

The infiltration rate (i), can be approximated by K as time increases [26].

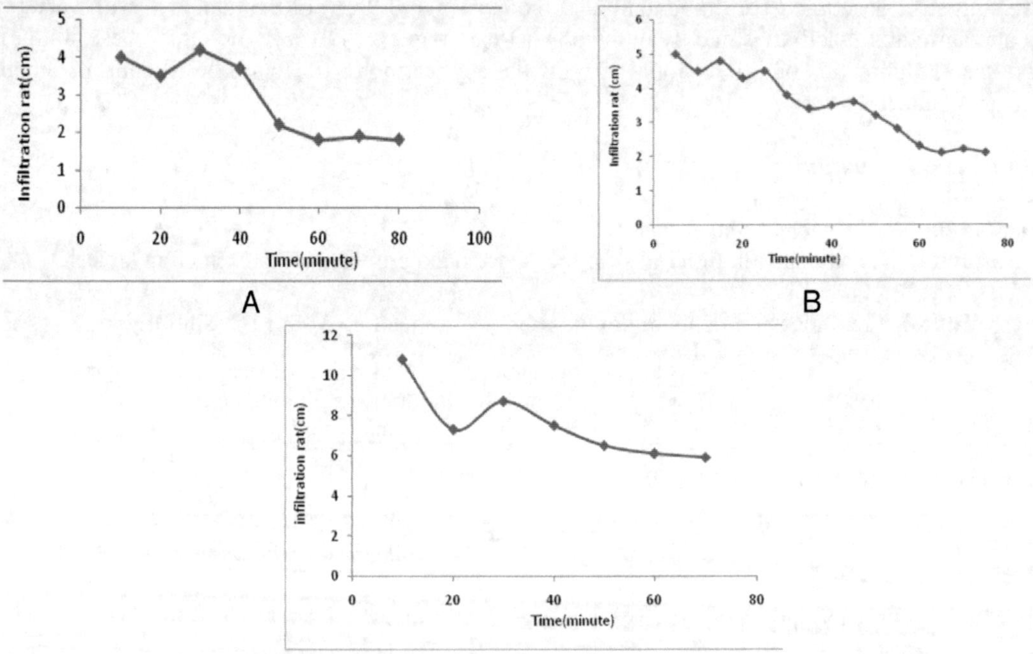

C

Figure 6. (A, B and C) Infiltration rate curve at observations sits (1, 2 and 3) at Wadi El- Assiuti

Cumulative infiltration and time were recorded for each test. figure 7. (a, b, c) illustrates a linear relationship between the cumulative infiltration and time. In addition it illustrates that the cumulative infiltration increased with time. This occurred because more water accumulated in the soil as the time increase.

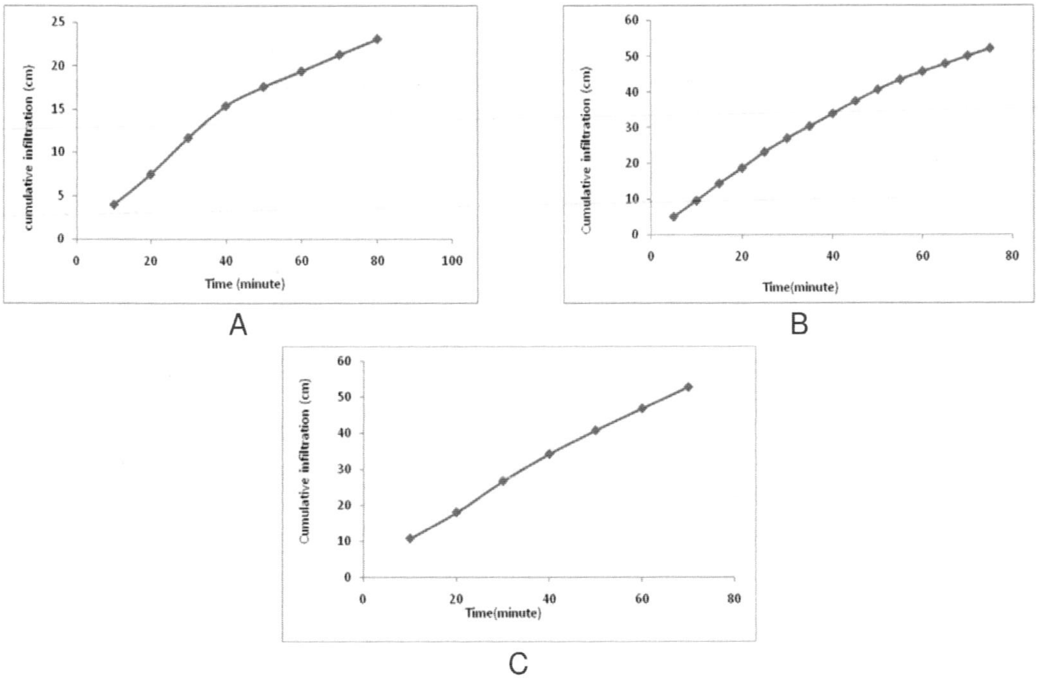

Figure 7. (A, B and C) Cumulative infiltration at observation sites (1, 2 and 3) at Wadi El-Assiuti

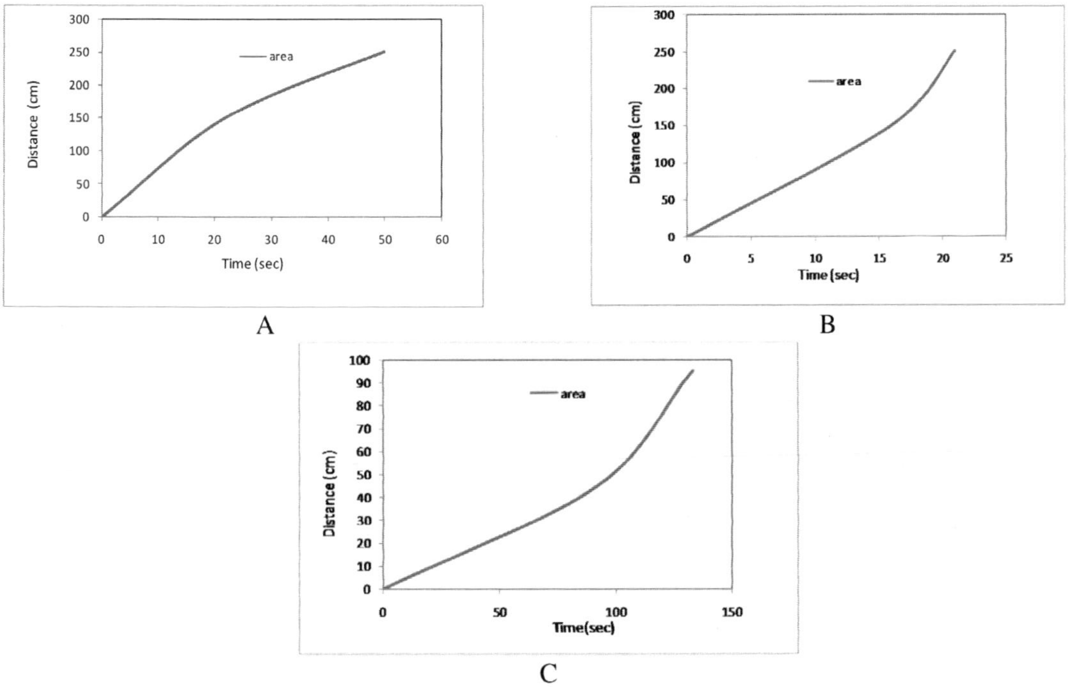

Figure 8. The distance traveled by runoff over time during the experiment of runoff simulatorat Wadi El-Assiuti.

IOP Conf. Series: Materials Science and Engineering **975** (2020) 012011 doi:10.1088/1757-899X/975/1/012011

3.2 Physical model (rainfall simulator) results

During the first experiment water flows 150 cm distance from the upstream to the downstream point after 22 sec then 250 cm after 50 sec. in the second experiment runoff flows about 150 cm distance in 16 sec, then 250 cm in 21 sec. In addition the third experiment water travels 45 cm in 92 sec then 95 cm in 133 sec as shown in figure 8.

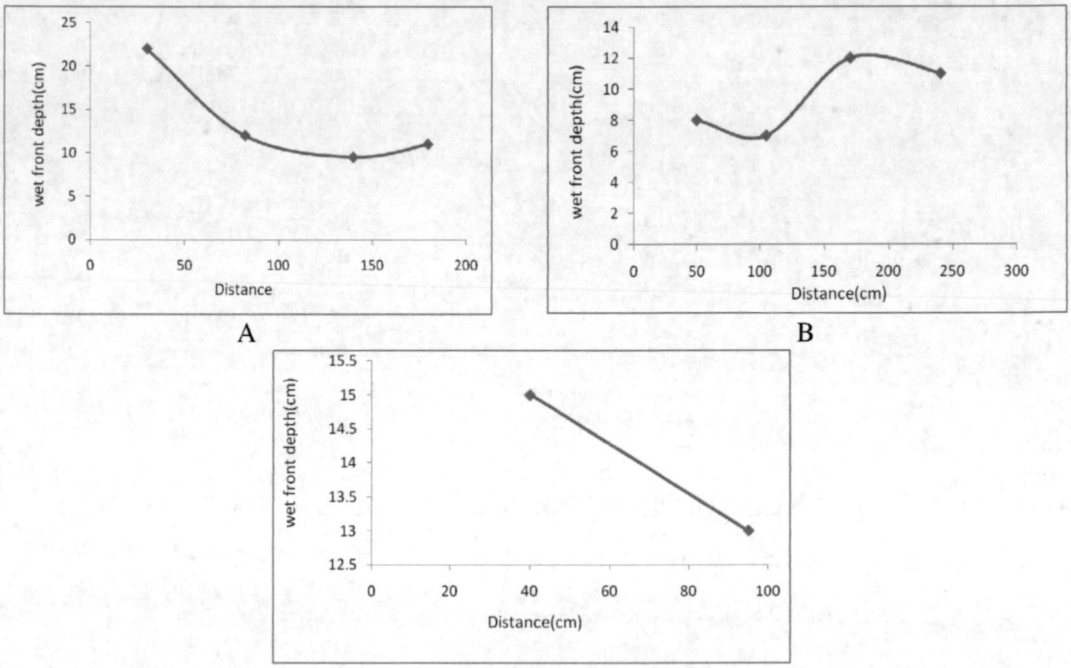

A B

C

Figure 9. Wet front depth over distance during the experiment of runoff simulator at Wadi El-Assiuti.

Additionally, wetting depth after162 sec of rainfall in the first experiment reached a depth of more than 22 cm, in the second experiment the wetting front was 12 cm after 145 sec. In the third experiment the wetting depth was 15 cm after 133 sec. so the distance against the wet front was plot as shown in figure 9. Additionally transmission losses are calculated and listed in

Table 2. Rainfall simulator results

	Volume of water in (rain) cm^3	Volume of water out (runoff) cm^3	Transmission losses cm^3	Transmission losses %	slope
site 1	38278.2	0	38278.2	100	8.53
site 2	20496	10744.5	9751.5	47.58	3.43
site 3	31894.2	0	31894.2	100	

3.3 The porosity results

Geomorphology of the region controls the alluvium material of the channel, and transmission losses depend primarily on the bed and over-bank materials. Actually, the coarse materials have a high porosity which enables significant transmission losses. This high porosity leads to a high hydraulic conductivity which plays one of the main roles in the transmission losses phenomenon. Reference [27] demonstrated that transmission losses are influenced by the quantity and texture of the channel alluvium. The increase in porosity of the soil increased percolation of water hence water runoff was reduced. The measured porosity at different observation sites at Wadi El- Assiuti was shown in figure 10, and then listed in table 3.

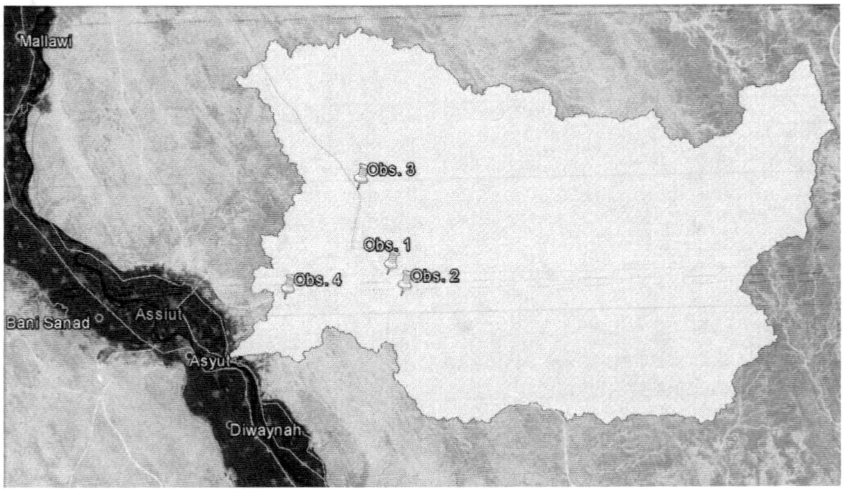

Figure 10. Porosity observation sites at Wadi El- Assiuti

Table 3. Porosity

observation site	site		porosity
	North	East	
site 1	27°19' 08.00"	31°35' 02.90"	0.3867
site 2	27°16' 52.90"	31°36' 43.50"	0.4533
site 3	27°27' 53.87"	31°31' 31.76"	0.4267
site 4	27°16' 33.34"	31°22' 35,15"	0.4267

These experimental results of transmission losses and porosity could be very useful for developing the physical models for valley system especially for the parameters calibrations. The measured porosity at different points along the channel bed varied between 0.3867 and 0.4533. Higher porosity increases infiltration and percolation rates and the water-holding capacity of the soil.

3.4 Hydro-BEAM model results

The flash flood events occurred in January, 2004, February, 2006, January 2008 and January 2010 had severe impacts on many valleys throughout Egypt. The simulation results (Figure 11) described below Highlights the assessment of transmission losses in order to determine its implications for groundwater recharge potential.

The simulated hydrographs of the selected events show that transmission losses take only a few hours to reach its peak value, before it gradually decrease to return to a zero status (Figure 11). Variations can be noticed in the values of the hydrograph and in the shape of the hydrograph from one event to another due to differences in the rainfall spatial patterns and intensities . The behavior of the hydrograph is directly linked to the pattern of the hyetograph. Even within the same storm, the transmission losses rate can vary between locations due to the spatiotemporal variability of rainfall and the geomorphologic parameters of the upstream catchment.

IOP Conf. Series: Materials Science and Engineering **975** (2020) 012011 doi:10.1088/1757-899X/975/1/012011

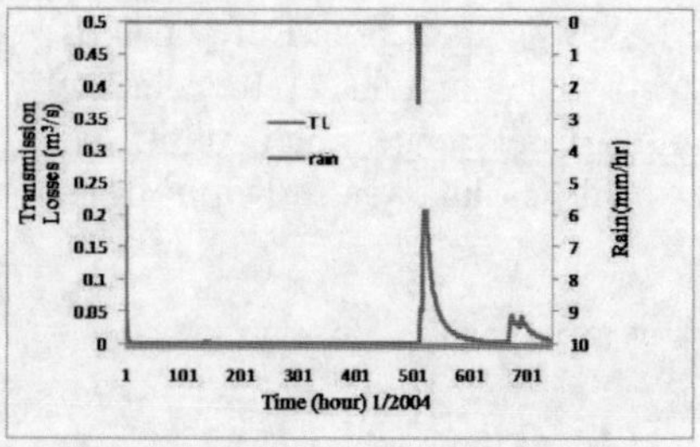

Hourly transmission losses (January, 2004).

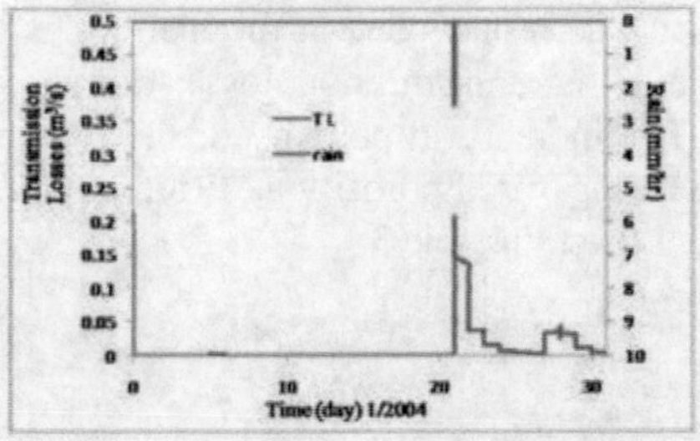

Daily transmission losses (January, 2004).

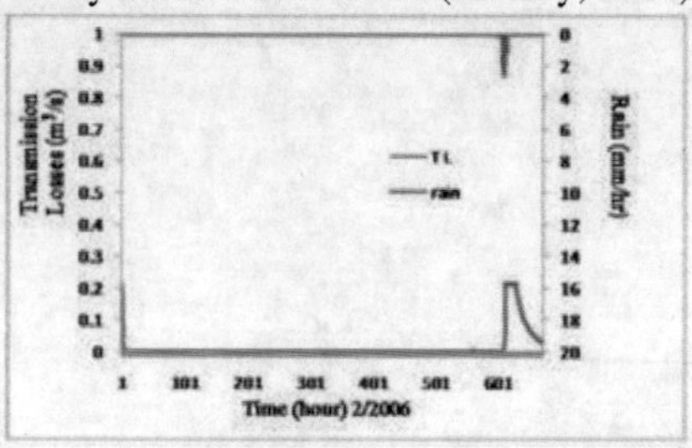

Hourly transmission losses (February, 2006).

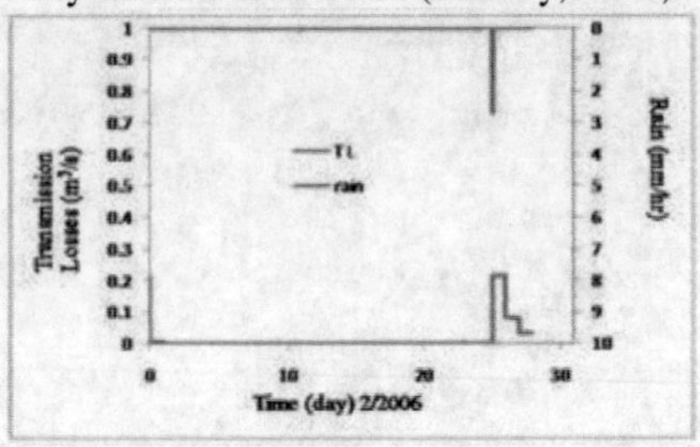

Daily transmission losses (February, 2006).

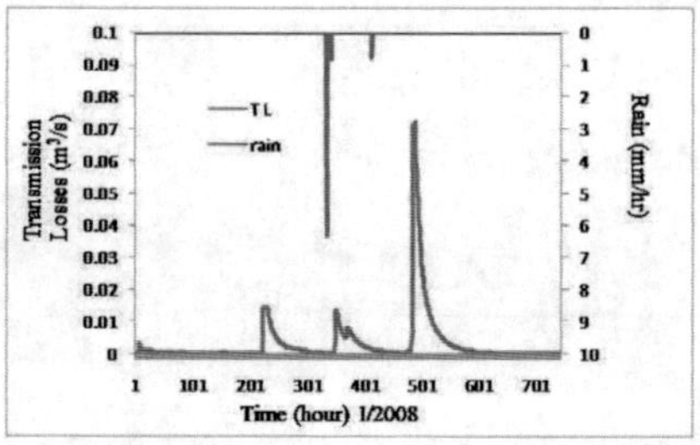

Hourly transmission losses (January, 2008).

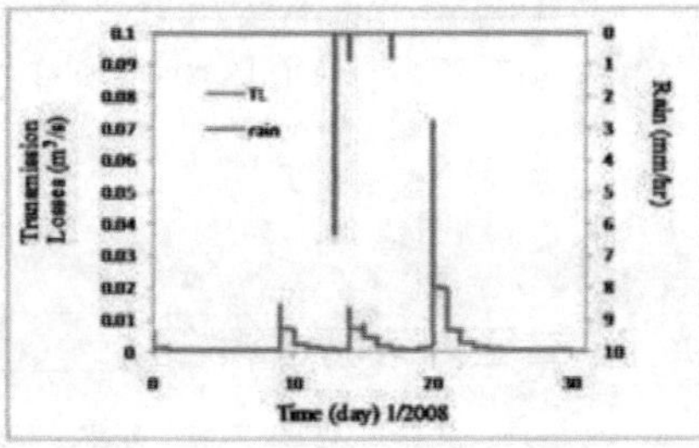

Daily transmission (January, 2008).

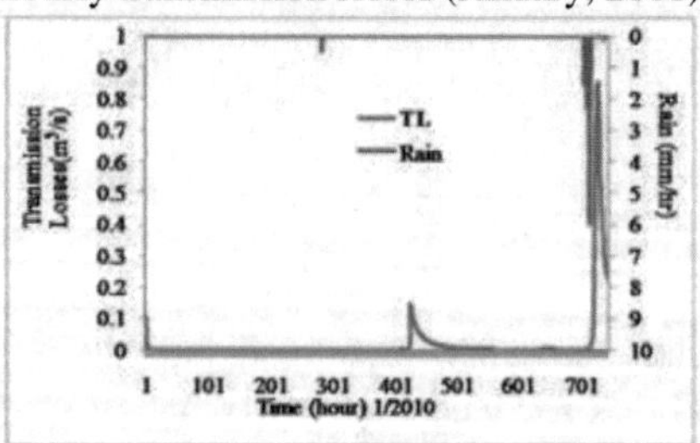

Hourly transmission losses (January, 2010).

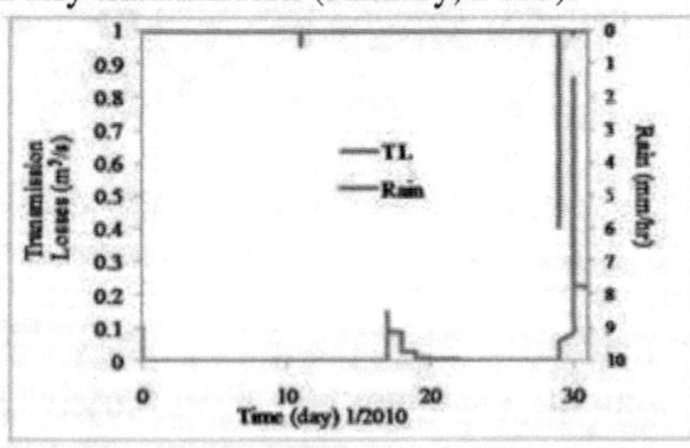

Daily transmission losses (January, 2010).

Figure 11. Hourly and Daily transmission losses simulation hydrograph and rainfall hyetograph in Wadi El-Assiuti

The geographic distribution maps of rainfall input and the simulated transmission losses values of the target events are shown in Figure 12 and Figure 13. Despite being affected by significant rainfall, the Wadi El-Assiuti sub-catchments do not record any surface runoff due to their high losses. The transmission losses distribution maps and simulated hydrographs record different patterns for each event, especially in their upstream regions, due to the spatiotemporal variability of rainfall. In the 2010 event, the upstream area recorded higher transmission losses than the downstream area due to the greater rainfall intensity in the upstream region, in contrast to the 2004, 2006 and 2008 events.

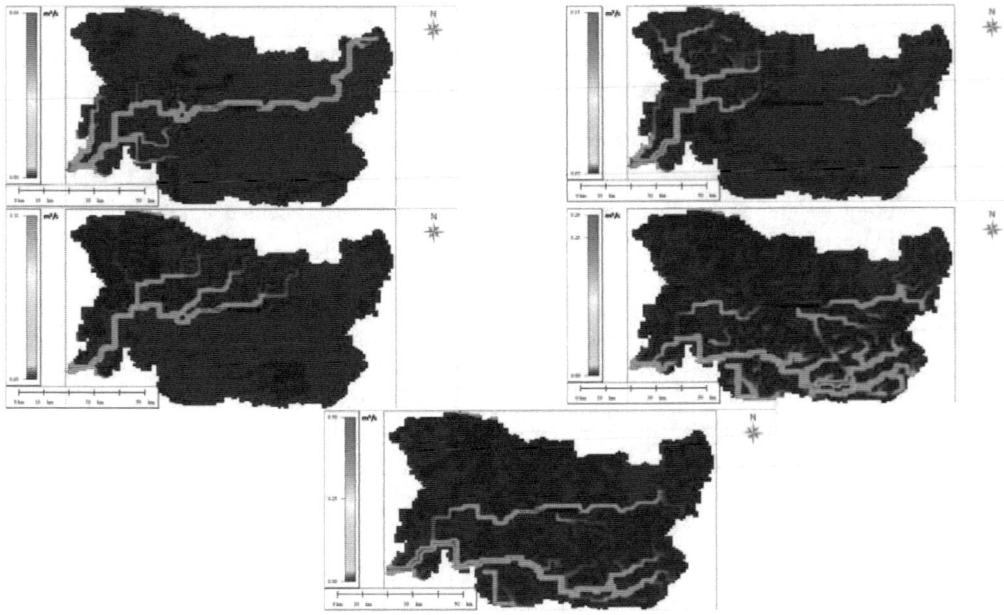

Figure 12. Distribution maps of transmission losses at Wadi El-Assiuti, Egypt (8-24 Jan. 2008).

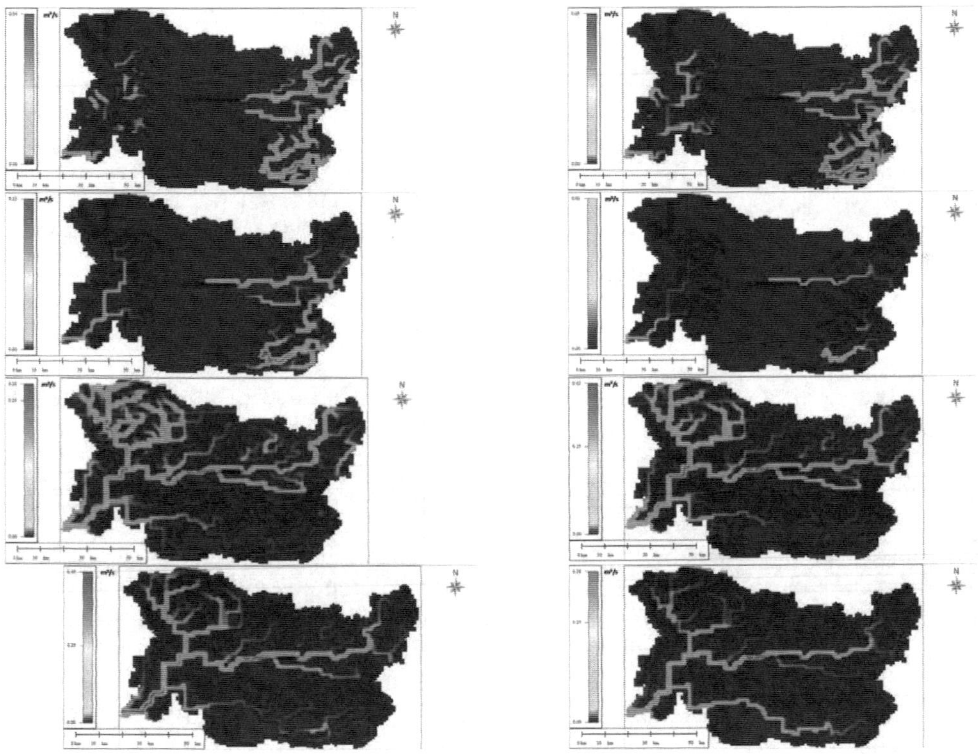

Figure 13. Distribution maps of transmission losses at Wadi El-Assiuti, Egypt (17-18 Jan. 2010).

The estimating of transmission losses is important due to their effect on channel flow reduction as well as their influence on recharging the groundwater of alluvial and quaternary aquifers in wadi El-

Assiuti. The results indicate that estimated transmission losses are affected by the volume of surface runoff as a translation of Walter's equation. In real flood events, the volume and percentages of total transmission losses are usually bigger than the total volume of surface runoff which pass the outlet; which indicates how significant transmission losses are as processes in wadi systems. Synthetic storms which have much higher rainfall than real one, the volume and percentages of transmission losses are smaller than volume and percentages of surface runoff which pass the outlet; transmission losses percentage increasing rate is smaller than the surface runoff which pass outlet increasing rate as the rainfall intensity increases. This is because the channel bed capacity for surface water infiltration is limited and more surface runoff will pass without losses.

4. Summary and conclusions

The soil infiltration rate is an important factor in analysis of the hydrological system that determines the fraction of rain water entering the soil or irrigation and the amount of runoff generated that is responsible for subsequent soil erosion. Two methods to measure the soil infiltration rate are introduced together with corresponding models. These can be considered as good methods in soil infiltration rate measurement. The initially high soil infiltration rate can be measured using double ring infiltrometer. The rainfall simulator method can be used to analyze the effects of raindrops, slope, surface seal formation, etc., on soil infiltration rate. This method is capable of measuring the soil infiltration rate that can be used to quantify the effects of many factors on soil, as well as measuring transmission losses.

The transmission losses and runoff were measured using a rainfall simulator at three observation sites at Wadi El-Assiuti. All experiments were performed with a portable rainfall simulator. Portable rainfall simulators have been widely used for variety purposes .The rainfall simulator method was thus able to provide a new way to obtain a good estimate of the initial losses and transmission losses under artificial rainfall. It provides a useful tool to investigate losses and associated infiltration rate under rainfall conditions

A variety of field tests exists for determining the infiltration rate of a soil. Infiltration rate was calculated according to the installation and operating instruction for the double ring in ASTM 2003. Final infiltration rates were 18 cm h^{-1}, for the first observation site, 24.2 cm $h^{-1}$for the second site and 59 cm $h^{-1}$for the third site. The analysis indicated that the soil for the three selected areas is sandy soil. Infiltration rates were taken at 0 to70 minutes of 10 minutes intervals. The infiltration rate at the third site is high compared to the other two selected sites.

The Hydro- BEAM model estimates transmission losses and groundwater recharge from runoff events in arid Environment wherever hydrological and meteorological records are scanty for direct estimation of transmission losses and groundwater recharge from runoff events. The transmission loss can be evaluated using Walter's equation and the result is reasonable due to its agreement. It is concluded that transmission losses participate as the main if not the only source of recharge to the subsurface. It is noticed that the volume of transmission losses is affected by the volume of surface runoff as evidence that the rate of transmission losses is linearly related to the volume of surface discharge.

Application of the model to another arid reach whose transmission losses and groundwater recharge were directly estimated, results in good agreement between the transmission losses for this reach and that for a reach of Wadi El- Assiuti of similar size and flow volumes. This agreement indicates that the transmission losses along the Wadi El- Assiuti reaches are reasonable.

The transmission losses were subdivided into channel moistening, which subsequently evaporates, and groundwater recharge. Estimation of actual evaporation from channel alluvial fill indicated that large event evaporation is substantially smaller than the transmission losses. It can be concluded, therefore, that most of the lost volume percolates towards aquifers. The mean annual groundwater recharge from runoff events in the Wadi El-Assiuti watershed is estimate as 0.12 MCM. Considering the litho-stratigraphy of the Wadi El- Assiuti watershed, the groundwater recharge could be distributed between the underlying lithologic groups.

The incorporation of geological and lithological information into a quantitative hydrological model helps surmount a shortage in hydrological and meteorological records. This new approach was successfully applied for additional watercourses in Wadi El-Assiuti and appears suitable for application in other similar arid areas throughout the world.

Hydro-BEAM is a multilayer hydrological model, four layers (A-D ; A-Layer is composed of the surface and soil surface layer. kinematic wave model and Manning equation are used to estimate the surface runoff and roughness coefficient in each mesh of the watershed basin. B-D-Layers are subsurface layers, which are evaluated using linear storage model, with the assumption of that the flow in each of B and C layers toward the river, but D-layer is considered as groundwater storage. It makes the ground-water zone which does not exert influence in river flow. When storage water content reaches to thickness and becomes saturated state, water content flows into the upper layer of model as returns style.

Hydro-BEAM consists mainly of three main modeling parts; climatic modeling, watershed modeling and the main program modeling. Numerous numerical codes for transmission losses and surface runoff analysis have been developed. Specific 50 source codes were created and saved using Parallel programming, next all source file must be compile. Implement of all codes can be done by using Intel Parallel Studio

As an extension of this study, we plan to use the assessed transmission losses and surface runoff in some Arabian valleys

References

[1] Wheater H.S., Woods Ballard B. and Jolley T.J., (1997) An integrated model of arid zone water resources: evaluation of rainfall-runoff simulation performance. In: Sustainability of Water Resources under Increasing Uncertainty, IAHS Pubn. No.24

[2] Telvari A., Cordery I. and Pilgrim D.H. (1998) Relations between transmission losses and bed alluvium in an Australian arid zone stream. In Hydrology in a Changing Environment. Eds Howard Wheater and Celia Kirby, Wiley, Vol II, pp 361-36

[3] Boughton W. C. and Stone J. J. (1985) Variation of runoff with watershed area in a semi-arid location, J. of Arid Environment, 9, pp. 13–25.

[4] Walters M.O. (1990) Transmission losses in arid region. J. of Hydraulic Engineering, 116, 1, pp. 127-138.

[5] Jordan P.R. (1977) Streamflow transmission losses in Western Kansas. Jul of Hydraulics Division, ASCE, 108, HY8, pp. 905-919.

[6] Andersen N. J., Wheater H. S., Timmis A. J. H. and Gaongalelwe D. (1998) Sustainable development of alluvial groundwater in sand rivers of Botswana. In Sustainability of Water Resources under Increasing Uncertainty, IAHS Pubn. No. 240, pp. 367-376.

[7] Sorman A. U. and Abdulrazzak M. J. (1993) Infiltration - recharge through Wadi beds in arid regions. Hydr. Sci. Jnl., 38, 3, 173-186.

[8] Gregory J. H., Dukes M. D., Miller G .L. and Jones P. H. (2005) Analysis of double-ring infiltration techniques and development of a simple automatic water delivery system, Applied Turfgrass Science, doi: 10.1094/ATS-2005-0531-01-MG

[9] A.S.T.M. D3385-03 (2003) Standard Test Method of Infiltration Rate of Soils in Field Using Double-Ring Infiltrometer. ASTM International, Ed.

[10] Walsh R.P., Coelho C.O., Elmes A., Ferreira A.J., Goncalves A.J., Shakesby R.A., Ternan, J.L., and Williams A.G. (1998) Rainfall simulation plot experiments as a tool in overland flow and soil erosion assessment, north-central Portugal. Geoökodynamik 19 (3–4), 139–152.

[11] Navas A., Alberto F., Machín J., and Galán A. (1990) Design and operation of a rainfall simulator for field studies of runoff and soil erosion. Soil Technology 3 (4), 385–397.

[12] Mao L. L., Lei T. W., Li X., Liu H., Huang X. F., Zhang Y. N. (2008) A linear source method for soil infiltrability measurement and model representations. J. Hydrol. 353, 49–58.

[13] Kojiri T.; Tokai A.; and Kinai, Y. (1998)Assessment of river basin environment through simulation with water quality and quantity. Annu. Disaster Prev. Res. Inst. Kyoto Univ., 41, 119-134.

IOP Conf. Series: Materials Science and Engineering **975** (2020) 012011 doi:10.1088/1757-899X/975/1/012011

[14] Kojiri T.; Hamaguchi T.; and Ode M. (2008) Assessment of global warming impacts on water resources and ecology of a river basin in Japan. /. *Hydro-Environ. Res.,* 1, 164-175.

[15] Sato Y.; Kojiri T.; Michihiro Y.; Suzuki Y.; and Nakakita E.(2013) Assessment of climate change impacts on river discharge in Japan using the super-high-resolution MRI-AGCM. Hydroi. Process.**,** 27, 3264-3279.

[16] Saber M.; and Yilmaz K. (2016)Bias correction of satellite-based rainfall estimates for modeling flash floods in semi- arid regions: Application to Karpuz River, Turkey. Nat. Hazards Earth Syst. Sci. Discuss.**,** 2016, 1-35.

[17] Abdel-Fattah M.; Kantoush S.A.; and Sumi T. (2015) Integrated management of flash flood in wadi system of Egypt: Disaster prevention and water harvesting. *Annu. Disaster Prev. Res. Inst. Kyoto Univ.***,** 58, 485-496.

[18] Saber M.; Hamaguchi T.; Kojiri T.; Tanaka K.; and Sumi T.(2015) A physically based distributed hydrological model of wadi system to simulate flash floods in arid regions. Arab. /. Geosci. 2015**,** 8, 143-160.

[19] SCS. (1997) Soil conservation service, chapter 5, stream flow data. In Hydroiogy*Nationai Engineering Handbook;* USDA: Washington, DC, USA,.

[20] McIntyre N.; and Al-Qurashi A. (**2009**) Performance of ten rainfall-runoff models applied to an arid catchment in Oman. *Environ. Model. Softw.***,** *24, 726-738.*

[21] Sen Z. (***2008)***Modified hydrograph method for arid regions. *Hydroi. Process.* ***2008,*** *22, 356-365.*

[22] Milewski A.; Sultan M.; Yan E.; Becker R.; Abdeldayem A.; Soliman F.; and Gelil K.A. (**2009**) A remote sensing solution for estimating runoff and recharge in arid environments. /. Hydrol.**,** 373, 1-14.

[23] Weiler M. and Naef F. (2003) Simulating surface and subsurface initiation of macropore flow. Journal of Hydrology 273, 139–154.

[24] Reynolds W.D., Elrick D.E., and Youngs E.G. (2002) Single-ring and double- or concentric-ring infiltrometers. In: Dane, J., Topp, G. (Eds.), Methods of Soil Analysis, Part 4, Physical Methods. SSSA Book Series No. 5, Madison, WI, USA, pp. 821–826

[25] Philip J.R. (1957a) The theory of infiltration: 1.The infiltration equation and its solution. Soil science 83, 345–357.

[26] Chow V. T., Maidment D. R. and Mays L. W. (1988) Applied Hydrology, McGraw-Hill, New York, U.S.A.

[27] Keppel R. V. and Renard. K. G. (1962) Transmission losses in ephemeral sand-gravel bed streams of the southwest. Proc. ASCE, Journal of the Hydraulics Division, Vol. 88, Hy., 3, pp. 59-68.

Synthesis and characterization of zinc-imidazolate coordination polymer and its use for photocatalytic degradation of methylene blue dye under visible light irradiation

Mostafa Mohsen[1, 2], **Ibrahim Naeem**[1], **Mohamed I. Awaad**[1] **and Ahmad Baraka**[1]

[1] Chemical Engineering Department, Military Technical College, Kobry Elkobbah, Cairo, Egypt
[2] the corresponding author Moustafa.klm@gmail.com

Abstract. Zinc-Imidazole coordination polymer (Zn-Im CP) was facilely synthesized by hydrothermal method from zinc acetate and imidazole as ligand. The structure of the synthesized Zn-Im CP was analyzed by using scanning electron microscopy (SEM), EDX, FTIR, and powder X-ray diffraction (PXRD). Surface area of Zn-Im CP was determined by BET analysis of N_2 adsorption. The photocatalytic performance of Zn-Im CP using visible light irradiation source was investigated using UV–Vis spectrophotometer for the degradation of methylene blue. The pseudo first order reaction kinetic model was applied on the degradation data which gave the rate constant of 0.006 min^{-1}. The results indicate that Zn-Im CP exhibits good photocatalytic performance for the photodegradation of methylene blue following indirect mechanism pathway.
Keywords: Photocatalytic degradation, Visible light, Coordination polymer, Dye, imidazole.

1. Introduction

Textile industry and alike produce dyes effluents with a serious environmental impact due to their toxicity and carcinogenic effect and they show persistency because of difficulty of natural degradation. Various methods have been suggested to handle the dye removal from waters including biodegradation, coagulation, adsorption, advanced oxidation process and membrane–based separation processes. The short comings inherited in the preceding processes drive researchers in the last decade to explore extensively the field of photocatalysis and to continually develop new photocatalysts for the degradation of organic pollutants, especially dyes, in wastewater and effluents[1, 2]. Photocatalysis process is a common and clean method for water decontamination from different pollutants such as organic dyes. It has several advantages over other decontamination methods such as the use of the natural oxidant (O_2), its action at normal conditions (temperature and pressure),ease separation of catalysts and its availability for recycling[3].

Classical semiconductors such as TiO_2, ZnO, and CuO have proved their good photocatalytic properties in the degradation of organic dyes such as methylene blue (MB), methyl orange (MO), rhodamine B (RHB), and other hazardous materials. The band gaps (BG) of TiO_2 and ZnO are 3.2 and 3.17 eV respectively. Therefore, such catalysts are only active under irradiation with ultraviolet light ($\lambda < 387$)[3, 4].The development of new visible light photocatalytic materials, characterized by narrow band gaps, is a pressing demand in the field of photocatalysis. In this context, coordination polymers (CPs) and metal-organic frameworks (MOFs) have taken more attention as candidates as photocatalytic materials by the virtue of their unique properties of large surface area and/or small band gap [5].

Many researchers have developed CPs and MOFs for photocatalytic degradation of organic dyes. To name a few, Liming Fan *et al.* synthesized five different coordination polymers based on rigid tetra carboxylic acids and imidazole linkers using Cu, Ni, Co, and Zn as nodes. The produced CPs achieved degradation efficiency of up to 97.3%for MB after 2 hours[6]. Lei-Lei *et al.* constructed five different Zn(II)/Cd(II) coordination polymers derived from a new linear carboxylate/pyridyl ligand; (pyridin-3-yl)methyl 4-(2-(4-((pyridin-3-yl) methoxy) phenyl) diazenyl) benzoate. The photocatalytic activities of the produced CPs were evaluated by degradation of MB (initial concentration: 12.5 ppm) under UV-Vis light irradiation and the results showed that a degradation efficiency of 95% with MB in 120 min could be reached [7].

Imidazole and substituted imidazoles have been considered as building blocks and linkers for the construction of many coordination networks. Typically, they are widely used to synthesize MOFs with zeolite-like structures. These MOFs are called ZIFs and are intended for specific functions[8, 9].The N-imine atom of the imidazole ring bears lone pair of electrons and can serve as an electron donor for Lewis acidic metal ions, whereas the N-pyrrole atom can be converted to Lewis base by deprotonation under basic conditions forming imidazolate anion that has the ability to act as a bidentate bridging ligand [10]. In this context, ammonia solution (33%wt., pK_a= 37) which is more basic than imidazole (pK_a= 14.5) has been used in the synthesis process for the deprotonation of imidazole before coordination.

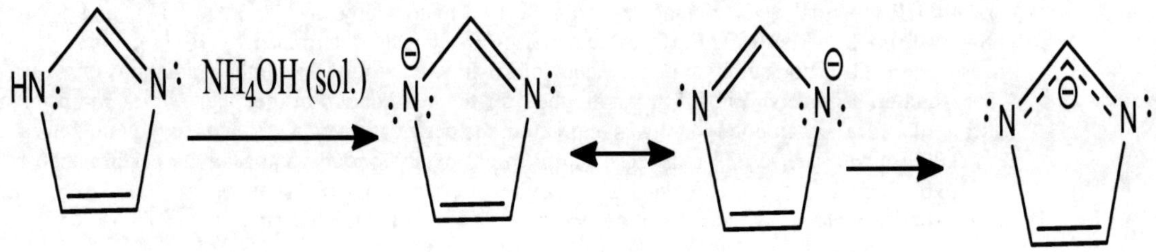

In this work, we synthesized Zn-Im CP and then evaluated its photodegradation activity towards organic dyes under visible light irradiation. MB, which is difficult to be degraded naturally, was utilized as model dye contaminant, and the kinetics of the catalytic photodegradation reactions was investigated as well.

2. Experimental

2.1. Synthesis of Zn-Im

Imidazole (extra pure, Oxford), zinc acetate dehydrate (extra pure, Sigma), ammonia solution (33%wt., Piochem), and distilled water were used in this work. Imidazole (2 mmol, 136 mg/10 ml distilled water)and ammonia solution (3 drops, 33%wt.) were added in a 100 ml Pyrex glass reactor. The mixture was left for 10 minutes and then zinc acetate (1 mmol, 219 mg/10 ml distilled water) was added resulting in a slight turbidity. The reactor was tightly closed and placed into a pre-heated oven at 70°C. After about two hours, the product (Zn-Im CP) started to precipitate as white hollow cubic crystals. The sample was collected and washed four times with distilled water. Finally, the Zn-Im CP sample were dried at 80°C, and then stored for later use; characterizations and photocatalytic degradation experiments.

2.2. Characterization of Zn-Im

IOP Conf. Series: Materials Science and Engineering **975** (2020) 012012 doi:10.1088/1757-899X/975/1/012012

The synthesized Zn-Im CP was characterized using several techniques: (i) crystalline nature was examined by powder X-ray diffraction (PXRD, Shimadzu XD-l) and structure determination was accomplished by phase match search of the experimental PXRD spectrum using QualX software against Cambridge Crystallography Open Database (COD), (ii) chemical structure was investigated by FTIR (using standard KBr pellet method, recording range 500 to 3500 cm^{-1}, Jasco FT/IR 4100. The surface morphology of Zn-Im CP was visualized using SEM (Zeiss EVO-10microscopy) and the coupled (EDX) was applied to analyze the chemical composition. Surface area of Zn-Im CP was determined by BET analysis (NOVA Station A). UV-VIS/DRs spectra were recorded using (Jasco V 530 spectrometer, Japan).

2.3. Assessment of photocatalytic versus photolytic decomposition

To obtain relevant information about the photocatalytic degradation, it was important to carry out some preliminary photolysis experiments to examine its efficiency without catalyst. The decomposition of MB by visible light was studied in two cases. The first case was thc decomposition in the absence of catalyst which is called the photolysis process and the second case was the decomposition in the presence of catalyst which is called the photocatalytic process. The photolytic decomposition of MB (25 ml of aqueous solution, concentration 5 ppm) was conducted in a 100 ml beaker top-exposed to the led lamp radiation (light spectrum in the range 400-800 nm, 35 W). To perform the photocatalytic decomposition experiments, the same amount and concentration of MB with specific amount of Zn-Im CP (50 mg) were loaded into the 100ml beaker and irradiated by the led lamp radiation. In both experiments, aliquots of 3 ml were sampled using a micropipette at specified time intervals. The aliquots samples were analyzed immediately using carry Agilent 60 UV–Vis spectrophotometer.

The photodecomposition efficiency of the contaminant (η) was calculated from the following equation:

$$\eta = \left(\frac{C_o - C_t}{C_o} \right) \times 100$$

Where C_o is the initial concentration of the contaminant (mg/l), and C_t is the concentration at time (t).

The effect of pH on the photodegradation process was studied as solution pH markedly influences the overall efficiency of the photocatalytic processes. To study the influence of solution pH on the initial photocatalytic decomposition rate of contaminants, the photocatalytic decomposition experiments of MB were carried out at three different pH values to cover the whole pH range of aqueous solutions; 4 (acidic), 7.5 (nearly neutral) and 10 (basic).

3. Result and discussion

3.1. Characterization of Zn-Im CP

Figure.1 shows the SEM image of Zn-Im CP where particles have a shape of being regular-ribbed interestingly with longitudinal holes.

The particles edges are about 20μm. According to EDX spectrum, figure 2. the synthesized Zn-Im CP is composed of the elements (% atom) of Zn (5.5), C (55), N (35) and O (4.18) and no other elements were detected. The elemental ratios can be given as: Zn(1): C(10): N(7): O(0.76) Accordingly, carbon to nitrogen ratio in Zn-Im CP (about 1.5:1) is similar to that of imidazole molecule which suggests the intact presence of imidazole in Zn-Im CP structure.

Figure 1. SEM image of the as-synthesized Zn-Im CP

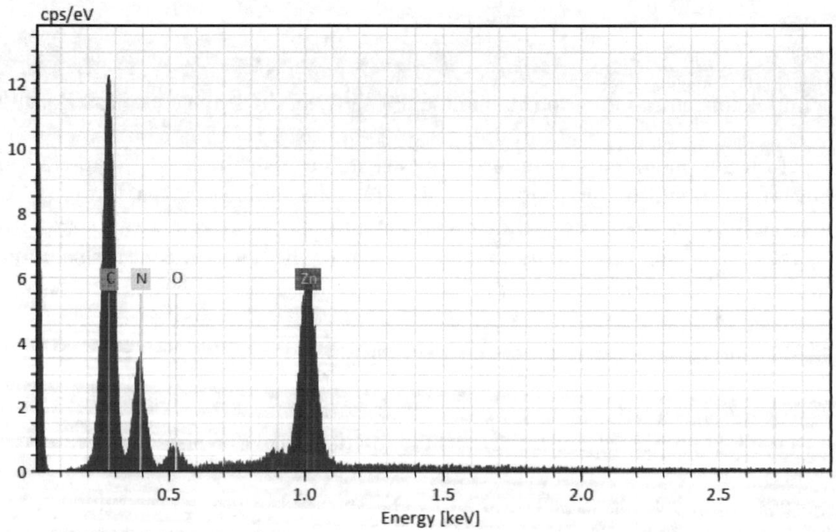

Figure 2. EDX spectrum of Zn-Im CP

The FTIR spectra of imidazole and Zn-Im CP are shown in figure.3. The spectrum of imidazole illustrates its known characteristic peaks: C–H stretching vibrations in the regions 3090-3140 cm^{-1} and 2900-2950 cm^{-1}.The two peaks appearing at 3095 and 2900 cm^{-1} obviously disappeared in the new Zn-Im CP spectrum giving an indication of developing new structure of imidazole with increased rigidity due to coordination with the metal nodes[11, 12].

The N–H stretching vibration over the broad region from 2500-3000 cm^{-1} with two characteristic peaks at 2901 and 2995cm^{-1} [13-15].When imidazole reacts completely with metallic ions, these peaks have been reported to disappear from the CPs spectra confirming the modification of imidazole structure due to deprotonation process and formation of coordination bonds with zinc.

The peaks at 1457and 1365cm^{-1} in imidazole should be designated to C–N stretching mode of vibration [12, 14].These peaks have been shifted to 1491 and 1367cm^{-1}in the Zn-Im CP spectrum which means that imidazole moiety is still found in the new product. The stretching modes of C=N and C=C are observed in the broad region from1500 to 1600 cm^{-1} with two characteristic peaks at 1540 and 1569 cm^{-1}. These two peaks are observed at 1585 and 1610 cm^{-1} in the case of CP. Such observed shift is relevant to higher energy band implying an increase of imidazole ring rigidity due to the formation of new coordination chains that affect the ring coherence [11, 14, 15].

The C–H out-of-plane-bending strong peaks appearing at 1048 and 1095 cm^{-1} in imidazole are shifted to 1086 cm^{-1} in the Zn-Im CP which manifest the existence of intact imidazole in the structure[15, 16].The presence of the broad band at the region from 3200 to 3550 cm^{-1} is an indication for the existence of OH group [13, 17].

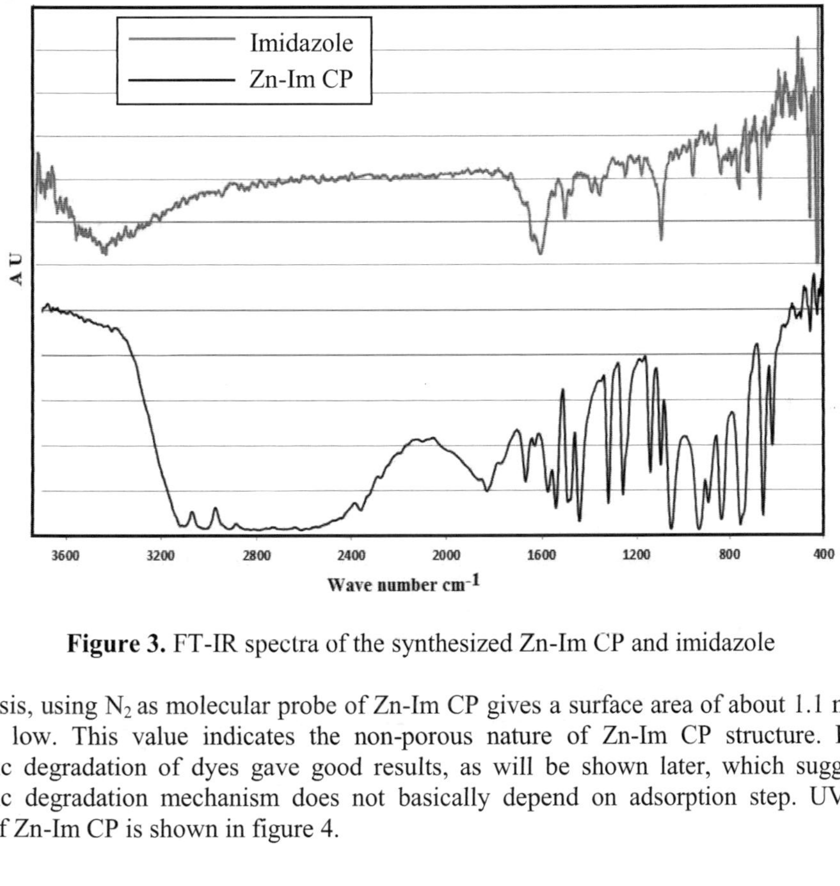

Figure 3. FT-IR spectra of the synthesized Zn-Im CP and imidazole

BET analysis, using N_2 as molecular probe of Zn-Im CP gives a surface area of about 1.1 m^2/g which is considerably low. This value indicates the non-porous nature of Zn-Im CP structure. However the photocatalytic degradation of dyes gave good results, as will be shown later, which suggests that the photocatalytic degradation mechanism does not basically depend on adsorption step. UV-Vis diffuse reflectance of Zn-Im CP is shown in figure 4.

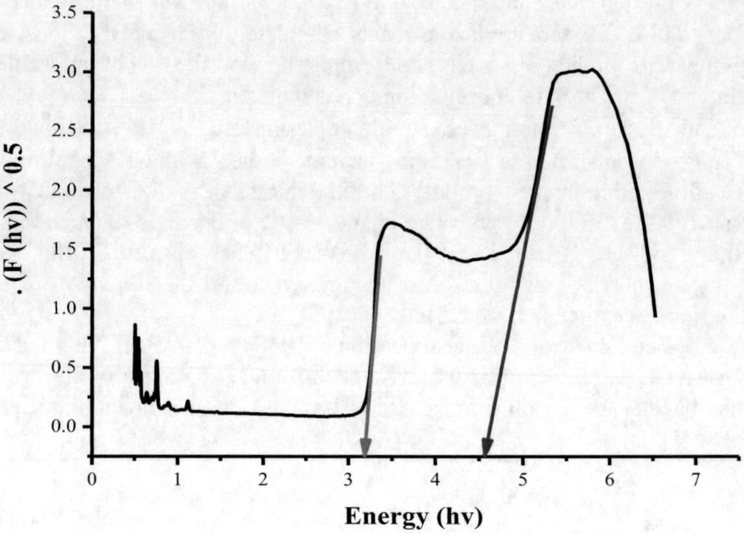

Figure 4. Diffuse reflectance for Zn-Im CP

The band gap value (E_g) was calculated from the plot of $(F(hv))^{0.5}$ versus hv where $F=(1-R)/(2R)$, R is reflectance , h is Plank's constant, and v is the frequency [18, 19]. The calculated E_g values are 3.15 and 4.6 eV. The smaller band gap, 3.15 eV, nominates the Zn-Im CP to be active under the irradiation with visible light.

Figure 5 shows XRD pattern of Zn-Im CP. It is clear that the synthesized Zn-Im CP is highly crystalline due to the sharp peaks of its spectrum. This high crystallinity provides the privilege of high charge mobility upon excitation. In addition, it shows the suitable choice of ammonia as deprotonating agent in the synthesis.

To determine the structure of the synthesized Zn-Im CP, its powder XRD pattern was exported to the phase identification software (QualX, version 2.24) which uses the crystallography open database (COD) for search match. The search match resulted in zinc imidazolate structure, chemical formula ($C_{12}H_{12}N_8Zn_2$), with figure of merit value of 0.82. The cif file of the compound was imported from Cambridge Crystallography Data Centre (CCDD) and further refined against the experimental PXRD spectrum via Materials Studio software (Biova, USA) to extract the experimental parameters using Pawley refinement. table 1 gives the structure parameters of Zn-Im CP where its cell is tetragonal with volume of 6788.81A^3 and it crystallizes in I 41 C D space group. The formula unit of the compound is $[Zn(Im)_2]_n$ and there are 16 formula unit in the cell. table 2 lists the experimental parameters of PXRD measurement.

Figure.6 compares the experimental PXRD pattern of the proposed CP to the experimental one. It is obvious that the two patterns highly coincide and the difference between them is very small. This is reflected in the lower value of R_{wp} (10.30%), which is a measure of the degree of matching between the simulated pattern and experimental one. Figure. 7 visualizes the chemical structure of the synthesized Zn-Im CP where Zn atoms (in red) are tetrahedrally coordinated to four imidazole rings.

IOP Conf. Series: Materials Science and Engineering **975** (2020) 012012 doi:10.1088/1757-899X/975/1/012012

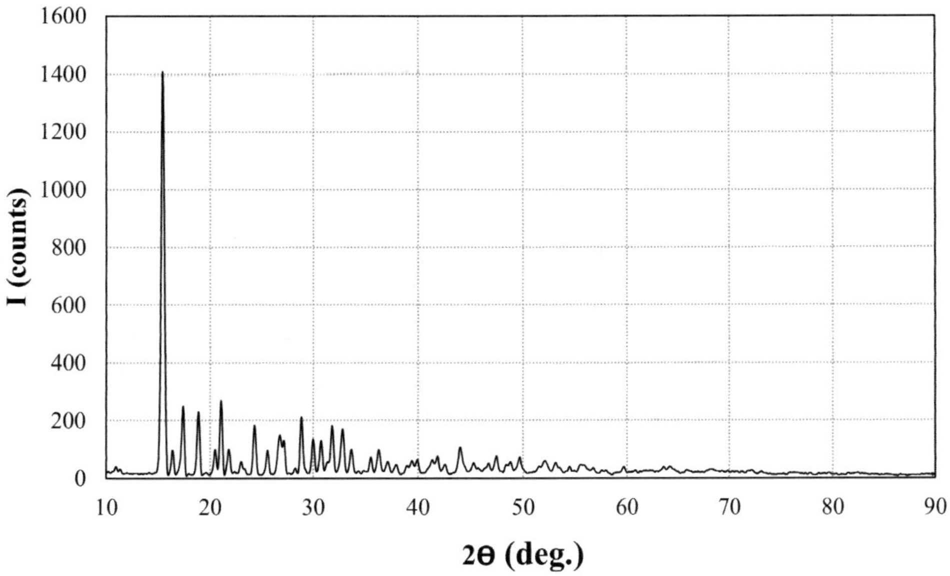

Figure 5. XRD powder patterns of Zn-Im CP

Table 1: structural parameters of Zn-Im CP

Parameter	Value	Parameter	Value
Lattice Type	Tetragonal	Space Group	I 41 C D
a	23.36	Cell Volume (A^3)	6788.81
b	23.36	Density (g/cm^3)	1.56
c	12.4	2θ Range (degrees)	5.02-89.98
α	90.0	Step Size (degrees)	0.04
β	90.0	Number of Reflections	734
γ	90.0	Final R_p	7.09%
Z	16	Final R_{wp}	10.30%

Table 2: PXRD experimental parameters

FWHM	
U	1.06144 ± 0.04530
V	-0.43825 ± 0.02825
W	0.14136 ± 0.00421
Profile Function: Pseudo-Voigt	
NA	0.00000 ± 0.01133
NB	-0.00000 ± 0.00040
Line Shift Instrument Geometry: Bragg-Brentano	
Zero Point	0.17007 ± 0.00172
Asymmetry Correction: Berar-Baldinozzi 2θ Limit: 30 (degrees)	
P1	-0.01770 ± 0.03911
P2	0.17957 ± 0.00815
P3	0.11348 ± 0.07320
P4	-0.38424 ± 0.01747

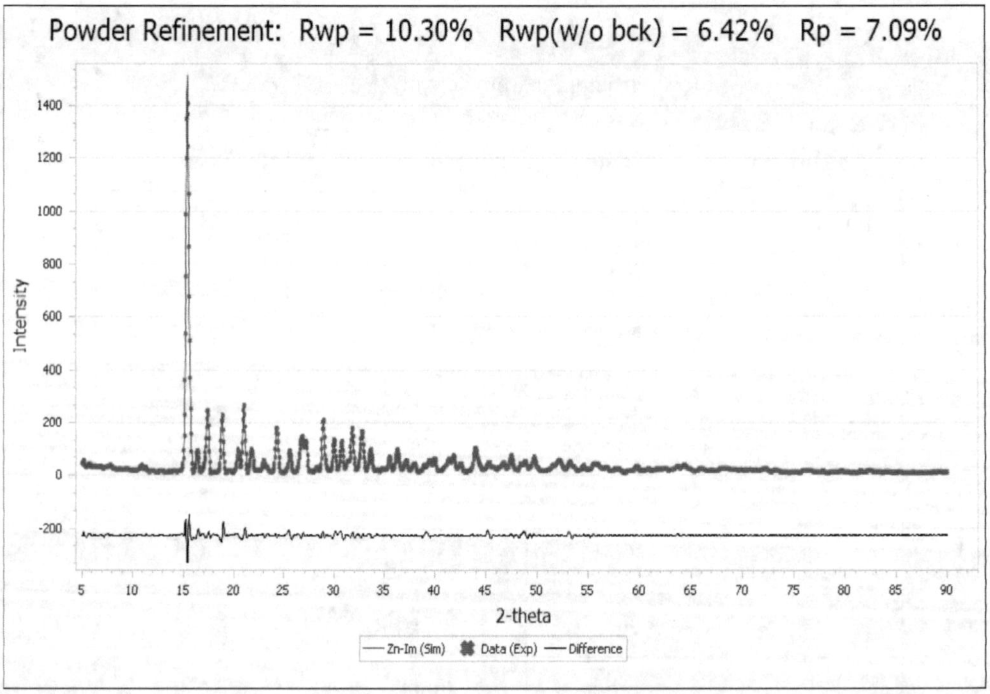

Figure 6.PXRD spectrum: a) experimental; b) simulated; and c) difference

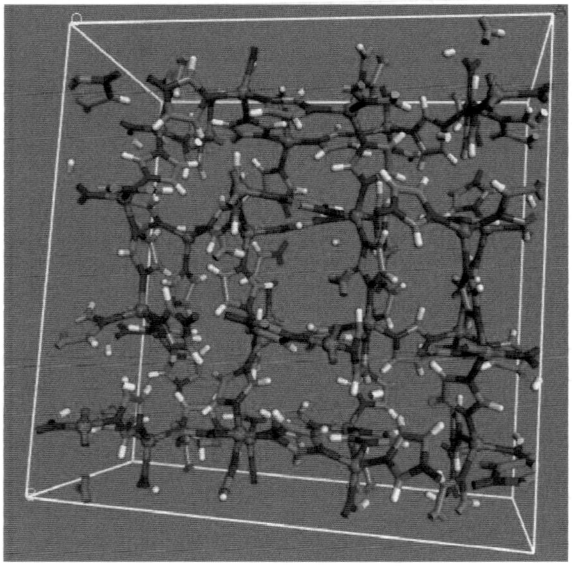

Figure 7: Chemical structure of Zn-Im CP

3.2. Photocatalytic degradation of MB by Zn-Im CP

3.2.1. Photolytic versus photocatalytic degradation of the MB.

Figure.8 shows the change in decomposition efficiency with time for MB under visible light in the absence/presence of Zn-Im CP. In absence of Zn-Im CP experiment indicated a decomposition efficiency of 66 % was recorded after 300 min. as a result of photolysis process. In presence of Zn-Im CP experiment derived decomposition through photocatalytic process which raises the efficiency to 90% within the same period of illumination.

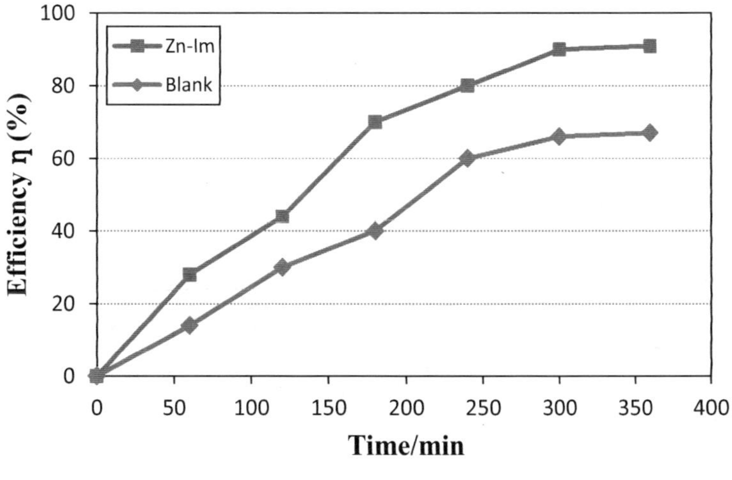

Figure 8. Photolytic and photocatalytic degradation of the MB solution (pH= 7.5).

IOP Conf. Series: Materials Science and Engineering **975** (2020) 012012 doi:10.1088/1757-899X/975/1/012012

3.2.2. Effect of pH on the photocatalytic process

The pH dependence of photocatalysis can be associated with changes of the surface charge of the photocatalyst, net charge of pollutant, changes in its adsorption modes, and amount of produced hydroxyl radicals. These changes can lead to a modification of the overall rate of decomposition [20]. Additionally, the change in pH may cause intermediates formation that could poison the photocatalyst [20, 21].

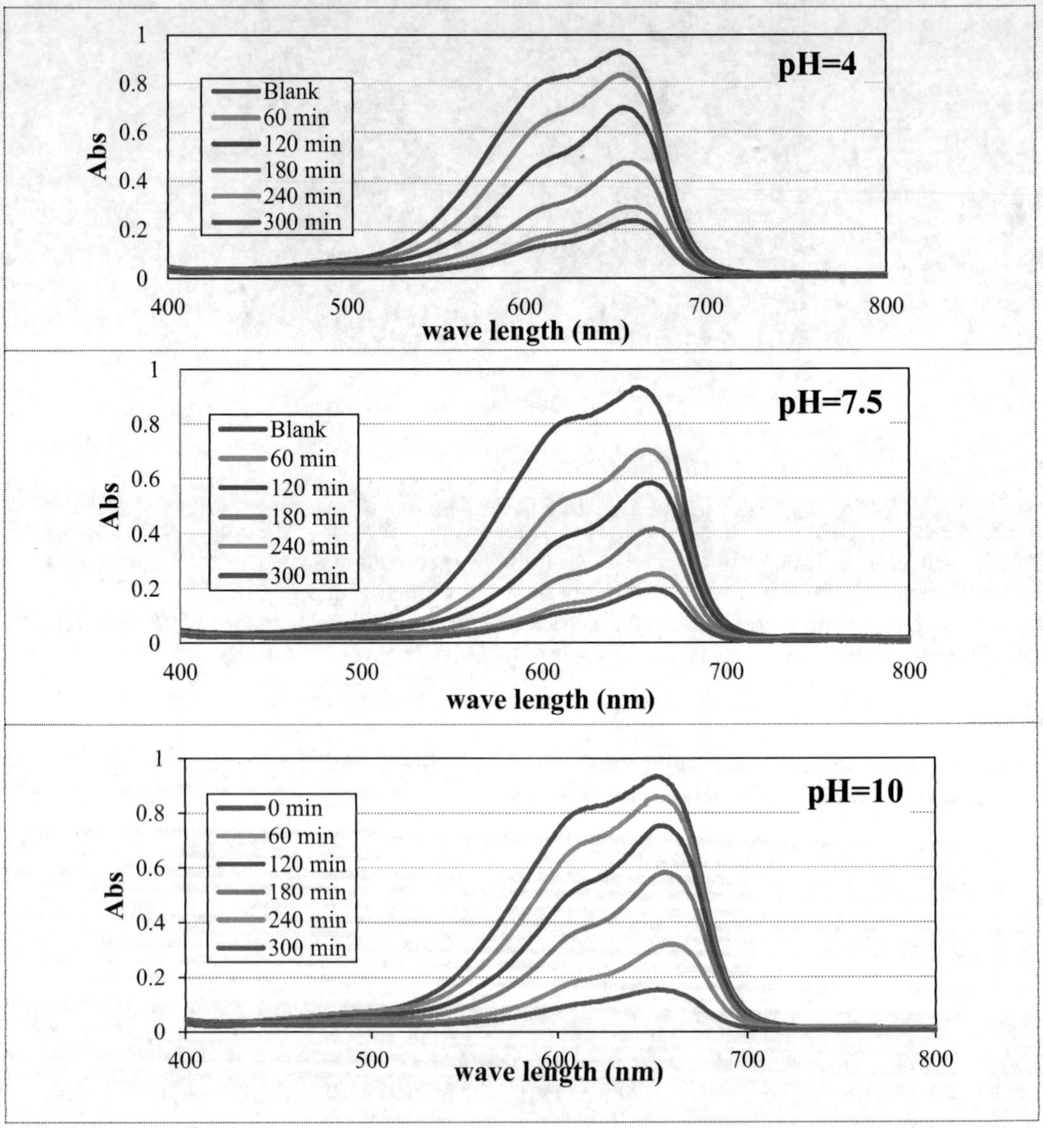

Figure 9.UV–vis absorption spectra of MB solution during photo catalytic degradation using (Zn-Im CP) catalyst under visible light at different pH.

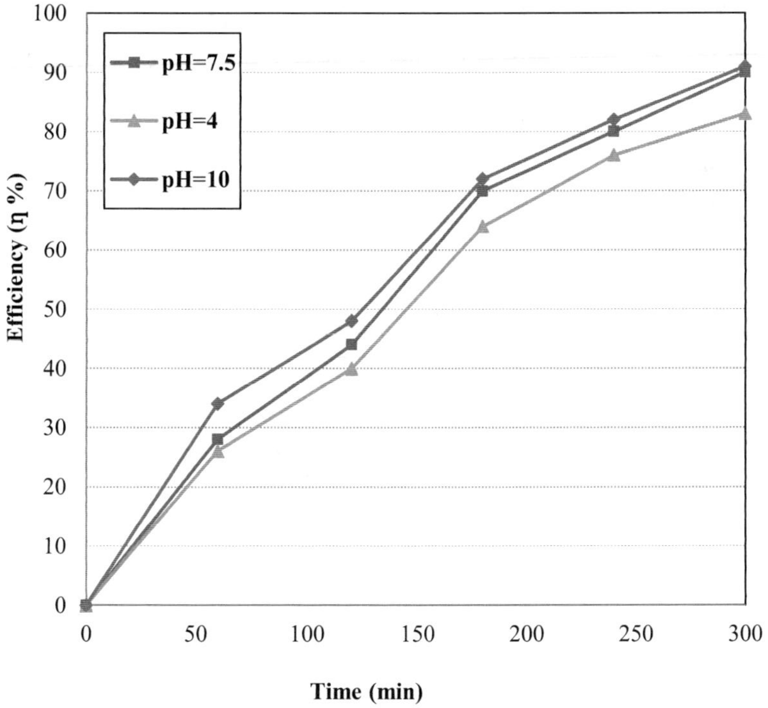

Figure 10. Photo catalytic degradation of the MB solution at different pH

To study the effect of pH on the degradation efficiency, three sets of experiments were performed fixing all experimental conditions except for the pH. For each experiment, specific amount of the solution was withdrawn at definite time intervals and the absorbance of the solution was measured using UV-VIS spectrometer over the range 400-800 nm. The results of these experiments are shown in figure 9.The results illustrate that Zn-Im CP photo-degraded MB effectively under visible light irradiation as indicated by the significant reduction in MB peak at 664 nm after irradiation period of 300 minute.

For each pH value, the maximum absorbance, occurring at 664 nm, was extracted for all curves and converted into concentration unit which was finally plotted against its corresponding time. The output is displayed in figure 10 which shows the change in MB concentration with time during visible light irradiation in the presence of Zn-Im CP at different pH values. It is observed that the degradation rate at all pH is very close to each other which indicate the non-effectiveness of changing medium pH on our CP.

3.2.3. Assessment of degradation with respect to initial pH

The pH of the solution plays an important role in the degradation rate of dyes. At high pH (basic medium), the degradation rate at first 120 min was higher than others, that's might be due to the presence of plenty of hydroxyl groups (^{-}OH) which is the main source of ($^{\bullet}OH$) that is responsible for dyes degradation according to degradation mechanism scheme given below [4, 22].In contrast, in acidic medium, the presence of plenty of H^{+} will cause repulsion forces with the cationic dye (MB^{+}) limiting its approach to the Zn-Im CP surface, hence yielding low degradation rate.

3.2.4. Photocatalytic degradation mechanism for MB

A simplified model of photocatalytic degradation based on indirect mechanism pathway is presented in Figure. 11. When the Zn-Im CP was irradiated with the visible light, the electrons (e^-) in the valence band (VB) were excited to the conduction band (CB), at the same time the same number of holes that have been produced (h^+) were left in VB. At VB the plenty of h^+ interact with water molecules providing ultimately $^{\bullet}OH$ which oxidizes of MB. Simultaneously, at CB, the excited electrons inject into adsorbed oxygen molecules on the surface of the Zn-Im CP to form $^{\bullet}O_2^-$ which converts to $^{\bullet}OH$ causing MB oxidation.

In order to investigate the kinetics of the photocatalytic degradation in neutral medium, the pseudo first order model was applied, $\ln(C_o/C)=kt$, where C_o and C are the initial and the remaining concentrations of MB at regular time intervals respectively and k is the rate constant. Plots of ln (C_o/C) against t were drawn for pH 7.5. From the plots, k was determined at pH=7.5 According to fitted equation, k= 0.006 min^{-1}.Table 3, gives the rate constants of photocatalytic degradation of MB (initial pH=7.5) by Zn-Im CP compared to some other photocatalysts, indicates a moderate performance of Zn-Im CP.

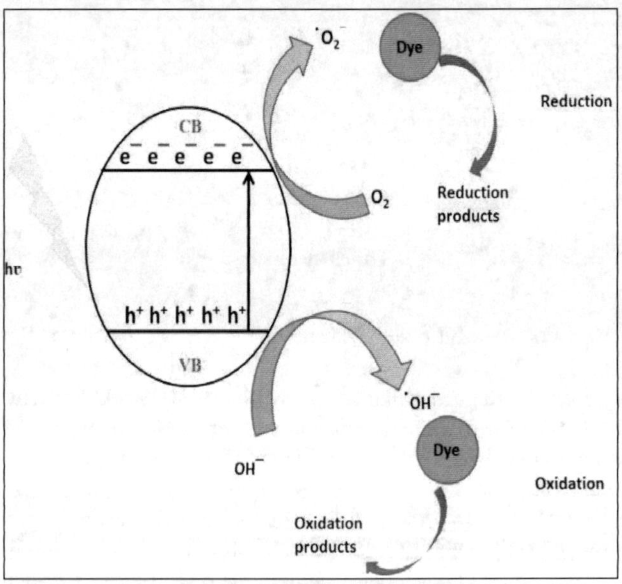

Figure 11.Simplified photocatalytic reaction mechanism of the MB solution on CPs.

Table 3: Rate constants of photocatalytic degradation of MB (initial pH=7.5)

Dye	k (min^{-1})*			
	This work	[18]	[23]	[24]
MB	0.006 (R^2=0.979)	0.0110	0.00435	0.00194

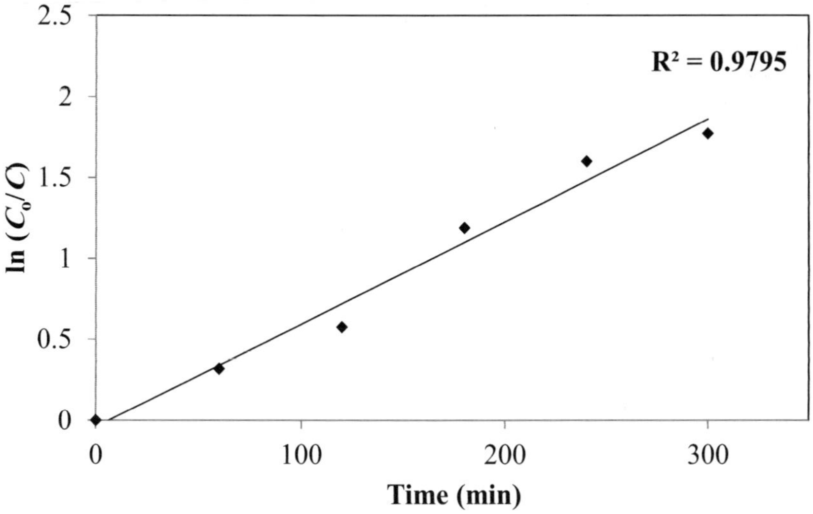

Figure 12.First-order kinetic plot for the photocatalytic degradation of MB using Zn-Im CP as catalyst under visible light irradiation.

4. Conclusion

The coordination polymer Zn-Im CP based on zinc acetate and imidazole was simply synthesized using ammonia as deprotonating agent at comparatively low temperature, 70°C. Zn-Im CP is highly crystalline which provides the privilege of high charge mobility upon excitation, hence suitable photocatalysis. Zn-Im CP shows two band gap values where E_g = 3.15 eV should be responsible for excitation under visible irradiation. For the three initial pH, 4, 7.5, and 10, Zn-Im CP show stable photodegradation of MB from 83%-91%. The rate constant is 0.006 min^{-1} which is comparable to the rates of many other photocatalysts.

References

[1] Yang, Y. and J. Luan, *Synthesis, property characterization and photocatalytic activity of the novel composite polymer polyaniline/Bi2SnTiO7.* Molecules, 2012. **17**(3): p. 2752-2772.

[2] Rauf, M. and S.S. Ashraf, *Fundamental principles and application of heterogeneous photocatalytic degradation of dyes in solution.* Chemical engineering journal, 2009. **151**(1-3): p. 10-18.

[3] Muhammad, A.S., et al., *Effects of Operating Parameters on Photocatalytic Degradation of Methylene Blue or Phenol on Supported Titanium-Based Catalysts.* Asian Journal of Chemistry, 2015. **27**(1): p. 313-320.

[4] Sakthivel, S., et al., *Solar photocatalytic degradation of azo dye: comparison of photocatalytic efficiency of ZnO and TiO2.* Solar energy materials and solar cells, 2003. **77**(1): p. 65-82.

[5] Wen, L.-L., et al., *Structures, photoluminescence, and photocatalytic properties of six new metal–organic frameworks based on aromatic polycarboxylate acids and rigid imidazole-based synthons.* Crystal Growth and Design, 2009. **9**(8): p. 3581-3589.

[6] Fan, L., et al., *Coligand syntheses, crystal structures, luminescence and photocatalytic properties of five coordination polymers based on rigid tetracarboxylic acids and imidazole linkers.* CrystEngComm, 2015. **17**(48): p. 9413-9422.
[7] Liu, L.-L., et al., *Construction of five Zn (II)/Cd (II) coordination polymers derived from a new linear carboxylate/pyridyl ligand: design, synthesis, and photocatalytic properties.* Dalton Transactions, 2016. **45**(31): p. 12352-12361.
[8] He, Y., et al., *Multifunctional metal–organic frameworks constructed from meta-benzenedicarboxylate units.* Chemical Society Reviews, 2014. **43**(16): p. 5618-5656.
[9] Eddaoudi, M., et al., *Zeolite-like metal–organic frameworks (ZMOFs): design, synthesis, and properties.* Chemical Society Reviews, 2015. **44**(1): p. 228-249.
[10] Chen, S.-S., *The roles of imidazole ligands in coordination supramolecular systems.* CrystEngComm, 2016. **18**(35): p. 6543-6565.
[11] Hu, Y., et al., *In situ high pressure study of ZIF-8 by FTIR spectroscopy.* Chemical Communications, 2011. **47**(47): p. 12694-12696.
[12] Ramasamy, R., *Vibrational spectroscopic studies of imidazole.* Armenian Journal of Physics, 2015. **8**(1): p. 51-55.
[13] Omrani, A., et al., *Study on curing mechanism of DGEBA/nickel–imidazole system.* Thermochimica Acta, 2008. **468**(1-2): p. 39-48.
[14] Drolet, D.P., et al., *FT-IR and XPS study of copper (II) complexes of imidazole and benzimidazole.* Inorganica Chimica Acta, 1988. **146**(2): p. 173-180.
[15] Morzyk-Ociepa, B., E. Różycka-Sokołowska, and D. Michalska, *Revised crystal and molecular structure, FT-IR spectra and DFT studies of chlorotetrakis (imidazole) copper (II) chloride.* Journal of Molecular Structure, 2012. **1028**: p. 49-56.
[16] Fernández-Bertrán, J., et al., *Proton transfer in solid state: mechanochemical reactions of imidazole with metallic oxides.* Journal of Solid State Chemistry, 1999. **147**(2): p. 561-564.
[17] Suh, M.P., B.Y. Shim, and T.-S. Yoon, *Template syntheses and crystal structures of nickel (II) complexes of hexaaza macrocyclic ligands with pendant functional groups: formation of a coordination polymer.* Inorganic Chemistry, 1994. **33**(24): p. 5509-5514.
[18] Yang, Y.-J., et al., *Synthesis and characterization of two 3D supramolecular cadmium (II) coordination polymers based on aromatic polycarboxylate and semi-rigidity bis (imidazole) ligands.* Polyhedron, 2018. **156**: p. 72-79.
[19] Nada, A.A., et al., *Elaboration of nano titania-magnetic reduced graphene oxide for degradation of tartrazine dye in aqueous solution.* Solid State Sciences, 2018. **78**: p. 116-125.
[20] Carp, O., C.L. Huisman, and A. Reller, *Photoinduced reactivity of titanium dioxide.* Progress in solid state chemistry, 2004. **32**(1-2): p. 33-177.
[21] Evgenidou, E., K. Fytianos, and I. Poulios, *Photocatalytic oxidation of dimethoate in aqueous solutions.* Journal of Photochemistry and Photobiology A: Chemistry, 2005. **175**(1): p. 29-38.
[22] Doong, R.-A., et al., *The influence of pH and cadmium sulfide on the photocatalytic degradation of 2-chlorophenol in titanium dioxide suspensions.* Water research, 2001. **35**(12): p. 2873-2880.
[23] Luan, J. and Y. Zhuang, *Synthesis, Structural Property, Photophysical Property, Photocatalytic Property of Novel ZnBiErO4 under Visible Light Irradiation.* Materials, 2018. **11**(2): p. 303.
[24] Pu, Y., et al., *Visible-Light-Induced Degradation of Methylene Blue by SrBi3VO8 Nanoparticles.* Journal of the American Ceramic Society, 2015. **98**(8): p. 2528-2533.

Synthesis, characterization and application of analcime to control nitrate ions from the ground water samples from Wadi El-Assiuti – Egypt as a low-cost and locally available adsorbent

Ahmed A. Abdelmoneim[1], **Mohamed Abdul-Moneim**[2], **Ahmed A Geies**[3], **and Seham O. Farghaly**[1]

[1]Geology Departments, Faculty of Science, Sohag University, Sohag, Egypt.
[2]Geology Department, [3]ChemistryDepartment, Faculty of Science, Assiut University, Assiut, Egypt.

Abstract. In this study, the sorption behavior of synthetic (Analcime) zeolites with respect to nitrate ions has been studied in order to consider its application to purity ground water. Analcime or also can be called analcite ("weak") is a kind of grey, white or colorless tectosilicates minerals. It is hydrated sodium aluminum silicates which exist in cubic form in crystalline. Analcime was successively synthesized from kaolinite as a raw material using the fusion with NaOH method. The conditions of hydrothermal crystallization (zeolitization) were found to be at temperature of 170 C°, and time span between 36 h and 72 h for kaoline with the molar composition of $6Na_2O$: $0.75Al_2O_3$:$30SiO_2$. $780H_2O$.The synthetic materials have been characterized by X-ray diffraction (XRD), scanning electron microscope (SEM), Fourier transform infrared spectroscopy (FT-IR), thermo gravimetric (DTA/TGA) analysis and Surface area and porosity of synthesized analcime .The results indicate that the crystallization of analcime not affected by the hydrothermal reaction time. Also analcime was tested as adsorbents for the removal of nitrate ions from the ground water samples from Wadi El-Assiuti – Egypt .The adsorption capacities of nitrate ions by analcime, as a function of its concentration, were determined at room temperature by varying analcime concentration for each water samples. During the process, all the other parameters (pH and contact time) were kept constant with respect to the initial concentration of nitrate ions in the water samples. It was found analcime has good removal efficiency obtained at pH 6-7.6, adsorbent dosage 10-12 g/L, and contact time 60 min. The Langmuir constants model for NO_3^- ion sorption on the adsorption isotherms is fitted well. The R_L value in the present investigation was equal or less than one, indicating that the adsorption of NO_3^-ion by analcime is favorable.
Keywords. Analcime, Ground water, Characterization, Element contents, Water treatment.

1. Introduction

Groundwater begins with precipitation that seep into the ground .It refers to the water that is stored in the pervious, porous and permeable rocks referred to as aquifers, everywhere in the world. Depending on the rock type and formation, groundwater is found in the ground within the depth of 100meters and in some places up to 1000 meters deep. A great percentage of people worldwide use this source of water for their agricultural, domestic and industrial purposes [1].Variation in groundwater chemistry is mainly a function of the interaction between the groundwater and the mineral composition of the aquifer materials through which it moves. Hydro-chemical processes, including dissolution, precipitation, ion exchange, sorption, and desorption, together with the residence time occurring along the flow path, control the variation in chemical composition of groundwater [2]. Nitrate, due to its high water solubility, is possibly the most widespread groundwater contaminant in the world, seriously threatening supplies of drinking water and promoting eutrophication [3]. A high concentration of nitrate in drinking water leads to production of nitrosamine, which is related to cancer and increases the risk of diseases such as methemoglobinemia in newborn infants [4-6]. Hence, reduction of nitrates in drinking water to permissible levels is mandatory For these reasons, the US Environmental Protection Agency (EPA) and the World Health Organization (WHO) have defined a maximum contamination level (MCL) of 10 mg N-NO3-/L in water. [7-8]. Various technologies have been used to remove nitrate from groundwater, including reverse osmosis, ultrafiltration, ion exchange, ion-exchange-membrane bioreactors, catalytic reduction, electro-dialysis, activated carbon, land disposal, chemical de-nitrification, and microbiological treatment [3], [9-10]. Zeolites are crystalline hydrated aluminum silicates with a framework structure containing pores that are occupied by water and by alkali and alkaline earth cations. Due to their high cation-exchange ability as well as molecular sieve properties, natural zeolites (cheap materials, easily available in large quantities in many parts of the world) show special importance in water and gas purification, adsorption and catalysis [11]. Zeolites have also an advantage over filter materials with significant permeability and are widely used for water purification by filtration [12-13]. Low-silica zeolites (LSZs) with Al/Si ratio of 1:1 are relatively rare [14]. These zeolites exhibit the highest ion exchange capacity and have important industrial applications. They are used as adsorbents or substitute for conventional phosphatic builder. Analcime is a hydrated sodium aluminum silicate which exists in cubic crystals. The chemical formula of analcime is $NaAlSi_2O_6{\cdot}H_2O$ [15]. with small amount of potassium and calcium. The structure and chemical properties are more similar to the feldspathoids, even though they are classified to the zeolite mineral. Natural analcime can be found in the analcime basalt and other alkaline igneous rocks. Analcime can be used as an ion-sieve [16]. One of the analcime characteristics is exchange ions at room temperature, and increasing temperature make the ion-exchange more easily. This is due to the presence of smaller pores in analcime structure [17]. Generally, zeolite minerals are rare in the entire world, therefore over 200 synthetic zeolites have been synthesized either using chemicals or natural materials. In our study, it is aimed to putificate nitrate ions from ground water by adsorption on synthesized analcime and to investigate the kinetics and equilibrium parameters involved during this adsorption. The parameters in this study included initial concentration, contact time, temperature, adsorbent dosage and influence of pH . Moreover, our work is to investigate the features of nitrates ion sorption from ground water by using synthetic zeolite analcime under static and dynamic conditions.

2. Hydrogeological setting of Wadi Assiuti

Wadi EL-Assiuti area represents a segment of the Nile valley in Upper Egypt. It is location on the fringes of the flood plane east of Assiut city. The area is bounded by latitudes 27° 5` N and 27°20` N and longitudes 31° 10`E and 31° 25`E. The area is located in an arid region with almost no rainfall. Temperature varies from 5 C° in winter to 45°c in summer. El fatah district is a part of Wadi EL-Assiuti area in Assiut governorate from its eastern side figure 1a, b. Several studies were previously carried out on this study area for dealing with its evaluation of its ground water potentiality. The study area is a rectangular flat area of about 400 Km^2. It is bounded from the west by the Nile River and from the other sides by the limestone plateau that is dissected by a great number of wadis. Groundwater in this aquifer is characterized by fresh water. The salinity from 800 to 1000 mg/L. The salt assemblages are: Ca $(HCO_3)_2$, Mg (HCO_3), $NaHCO_3$, Na_2SO_4 and NaCl which indicate a clear resemblance to the salt assemblages of surface water. The study of hydro-chemical characteristics in Wadi EL-Assiuti revealed on the high concentration of nitrate in some studied wells relative to [7-8].

3. Materials and Methods

Thirteen groundwater samples were collected from Wadi EL-Assiuti, for physical and chemical analyses. The samples were collected in clean polyester bottles which were thoroughly rinsed with sample water and tightly sealed and labeled after collection. These samples were kept in refrigerated and transferred to the laboratory for analysis.

All analyses were carried out in the geochemical laboratory of geology department, Faculty of Sciences, Sohag and Assiut University. The average of the analyses results are given in table 1a,b. Analcime was prepared by two method using kaolinite raw materials as a source of Al and Si and

NaOH (Merck, Darmstadt, Germany) were used for zeolite synthesis. The various characteristics of the synthesized analcime were also identified include The X-ray diffracto-grams of analcime was obtained with a Philips X-ray unit (Phillips generator PW-1710) and Ni-Filtered CuKα radiations. The morphology of analcime was examined by scanning electronic microscopy (SEM) using a Jeol JSM-5400 LV-ESM.

The FTIR spectra of the synthesized analcime that were obtained on a Shimadzu 2110 PC Scanning meter within the limits of 4,000–500 cm-1 with a resolution of 0.5 cm-1 (KBr pellets, CAN/KBr = 1/3 by mass, 6 atm.). (TGA) and differential thermal gravimetric (DTG) were carried out in air with Shimadzu DTG-60 at heating rate of 10°C/min. The chemical composition of the raw materials kaolinite are given in table 2.

Table 1.a. Results of the physical and chemical analyses of groundwater samples from Wadi EL-Assiuti

Well No	Units	pH	TDS ppm	Error %	EC µs/cm	Heavy metals			
						Fe^{+2}	Mn^{+2}	Cd^{+2}	Pb^{+2}
1	ppm					0.2	0.07	0.1	Nil
	epm	8.1	6003	2.7	9380				
2	ppm	8.8	1536	-3.2	2400	Nil	0.05	Nil	Nil
	epm								
3	ppm	8.8	198.4	2.5	310	Nil	Nil	Nil	Nil
	epm								
4	ppm	8.3	3168	2.8	4950	0.14	Nil	Nil	Nil
	epm								
5	ppm	9.1	198.4	1.6	310	Nil	Nil	Nil	Nil
	epm								
6	ppm	8.1	4480	2.5	7000	0.16	Nil	Nil	Nil
	epm								
7	ppm	8.2	1633	-2.4	2530	Nil	Nil	Nil	Nil
	epm								
8	ppm	8.5	198.4	-1	3100	0.09	Nil	Nil	Nil
	epm								
9	ppm	9.2	217.6	-4	340	Nil	Nil	Nil	Nil
	epm								
10	ppm	8	3955	-0.06	6180	0.3	Nil	Nil	Nil
	epm								
11	ppm	8.8	1740	0.15	2720	0.1	Nil	Nil	Nil
	epm								
12	ppm	8.3	2112	-0.6	3300	0.2	Nil	Nil	0.2
	epm								
13	ppm	8.4	198.4	1.4	3100	0.1	Nil	Nil	0.08
	epm								

IOP Conf. Series: Materials Science and Engineering **975** (2020) 012013 doi:10.1088/1757-899X/975/1/012013

Table 1.b. Results of the physical and chemical analyses of groundwater samples from Wadi EL-Assiuti

Well No	Units	Major Cations				TH	Major Anions			
		Ca^{2+}	Mg^{2+}	Na^{1+}	K^{1+}		HCO^{1-}_3	Cl^1	SO_4^{2-}	NO_3^{1-}
1	ppm	72	180	1700	17.6	921	490	2680	165	42
	epm	3.6	15	74	0.5		4.8	76	3.4	0.7
2	ppm	40	5.7	475	8.8	123	159	680	165	17
	epm	2	0.47	21	0.2		2.6	19	3.4	0.3
3	ppm	23	4.6	28	5.6	76	51	45	14	28
	epm	1.2	0.4	1.2	0.1		0.8	1.3	0.3	0.5
4	ppm	70	61	990	11	426	145	1480	165	80
	epm	3.5	5	43	0.3		2.4	42	3.4	1.3
5	ppm	22	4.5	33	6.5	74	25	50	25	41
	epm	1.1	0.4	1.4	0.2		0.4	1.4	0.5	0.7
6	ppm	82	61	1520	2.2	456	148	2275	165	94
	epm	4.1	5	66	0.1		2.4	64	3.4	1.5
7	ppm	75	5.4	535	21	209	84	800	165	74
	epm	3.7	0.4	23	0.5		1.4	23	3.4	1.2
8	ppm	87	5.4	685	13	240	95	1025	165	100
	epm	4.3	0.4	30	0.3		1.3	29	3.4	1.6
9	ppm	32	4.12	32.4	7.3	97	34	46	18	100
	epm	1.6	0.3	1.4	0.2		0.5	1.3	0.4	1.6
10	ppm	72	5.8	1590	13	204	84	2370	165	92
	epm	3.6	0.4				1.4	67	3.4	1.5
11	ppm	64	5.2	69	0.3		72	1011	19	79
	epm	3.2	0.4	675	11.	181	1.2	29	0.4	1.3
12	ppm	64	5.2	28	0.3		72	1011	165	72
	epm	3.2	0.4	746	13	181	1.2	29	3.4	1.2
13	ppm	82	5.6	33	0.3	228	91.5	1075	165	21
	epm	4.1	0.5	721	12.6		1.5	30	3.4	0.4

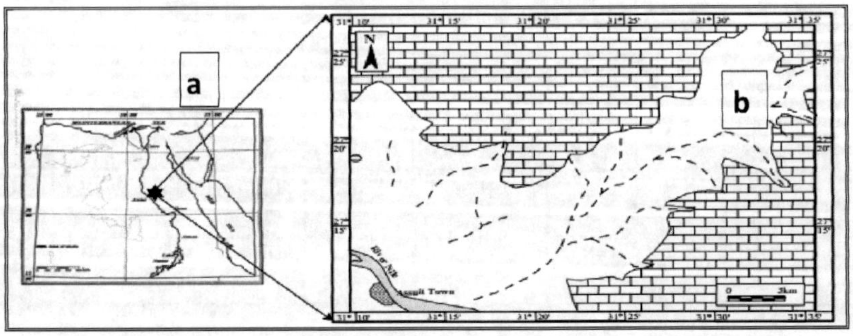

Figure 1a, b. Simplified geological and Location map of Wadi EL-Assiuti

Table (2). Chemical composition of kaolinite

Composition	SiO_2	TiO_2	Al_2O_3	Fe_2O_3	MnO	MgO	CaO	Na2O	K2O	P_2O_5	SO_3
Content (wt. %)	49.2	2.82	32.97	6.82	0.24	2.2	9.43	0.8	0.05	0.13	0.26

3.1. Synthesis of analcime

By using alkaline fusion (after) [18] to synthetic analcime using kaolinite NaOH pellets was mixed with the calcinated Kaoline by ratio 1.2: 1 and the mixture was fused at 600 C° for 1 h . The alkaline reagent added to the starting material acts as an activator agent during fusion. The alkali-fused products were then dissolved in water (H2O/alkali fused starting material ratio)=(1:4.9), generally under stirring conditions until the reaction gels were homogenized aged for 24 h under static conditions. The mixtures were transferred into poly tetra fluoro-ethylene (PTFE) bottles of 200 ml and a stainless steel autoclave, then crystallized under static conditions at 170 C° for 5, 3, 1.5, days .After removal from the oven they were quenched in cold water and the product recovered by vacuum filtration, washed well with distilled water until the pH reach up to 11 and dried at 80 C° overnight.

Analcime was synthesized through a hydrothermal method using the following reaction condition: $6Na_2O$: $0.75Al_2O_3$:$30SiO_2$. $780H_2O$. At first, 4.75g of sodium hydroxide were dissolved in 10 g water. To this solution about 3.575 of kaolinite were added. The mixture was stirred (200-300 rmp), thereafter 161.25 g of water were added and stirred.

The gel was then transferred to poly tetra fluoro ethylene (PTFE) bottles (200 ml) and a stainless steel autoclave, and then put in oven at 170 C° for 36h.After removal from the oven, quenched in cold water, and the product recovered by vacuum filtration, washed with distilled water and dried at 80 C° overnight.The synthetic zeolite was identified and characterize by the next measurements .From the previous methods its noted that the crystallization of analcime not affected by the hydrothermal reaction time as shown in table 3.Also the synthetic zeolite (analcime) had the same X-ray diffraction this indicate the previous note as shown in figure 2.

Table (3). Synthesis conditions and crystallinity data of synthesized analcime.

Sample name	Type of clay	Natural materials	Synthesis time (days)	Synthesis temperature (K)
Analcime -1	kaoline	-	5	443
Analcime -2	Kaoline	-	3	443
Analcime -3	kaoline	Sand	1.5	443

3.2. Characterization 0f synthetic analcime

The various characteristics of the resulting analcime include: X-ray diffraction, scanning electron microscopy, Fourier Transform Infrared (FT-IR) spectroscopy and thermal gravimetric analysis.

3.2.1. X-Ray Diffraction.

The identification and characterizations of the raw material and synthetic zeolite shown in fig.(2). a 2θ range of 5-55°.Kaoline is the predominant mineral phase in the raw material which can be identified by its characteristic XRD peaks at 12.34° and 24.64° 2θ Fig.(2a). However, minor mineral impurities, such as quartz, illite, muscovite and halloysite, also occur. The X-ray diffracto-grams of the synthesized give reflection peaks at 15.76°, 25.93° and 30.48°As shown in the XRD patterns figure 2b, which is consistent with the analcime [19-20].

IOP Conf. Series: Materials Science and Engineering **975** (2020) 012013 doi:10.1088/1757-899X/975/1/012013

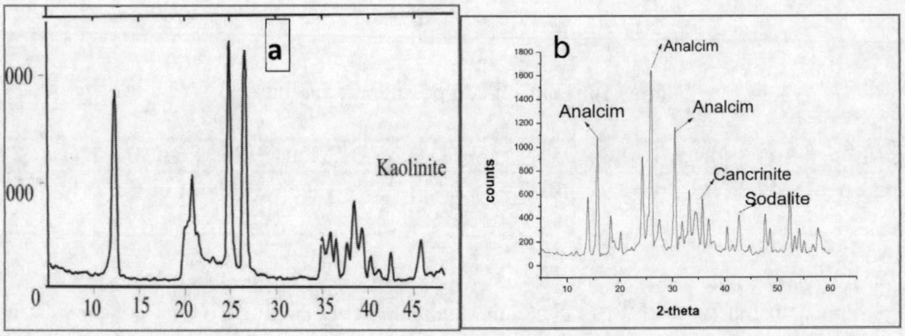

Figure 2. XRD patterns of a) Raw material kaolinite b) Synthetic analcime

3.2.2. Scanning Electron Microscope (SEM).
The morphology of the raw material and synthetic zeolite were examined as shown in figure 3a, b. The main morphological features observed in the synthesized analcime from met kaolinite are, spherical and cubical.

Figure 3. SEM micrographs of (a) Raw material kaolinite ;(b) Synthetic analcime

32.3. Fourier Transform Infrared (FT-IR) Spectroscopy.
The characterization of raw material (kaolinite) and synthetic analcime with transmission Fourier transform infrared spectroscopy (FT-IR) is described in table 4 and figure 4 a, b. The characteristic OH-stretching vibrations of kaolinite band at 3460.71 cm^{-1} .The symmetric stretch (753-71 cm^{-1}), double ring vibration (63-44 cm^{-1}), T-O bending modes (44.04 cm^{-1}),or the internal linkage vibrations of TO4 (T=Si or Al) tetrahedral and to symmetrical stretching respectively. The bands at (1677 .26 cm^{-1}) is assigned to the water in the channels of kaoline. FT-IR spectroscopy is used to probe the structure of the synthesized Analcime and monitor reactions in zeolite pores. Specifically, structural information can be obtained from the vibrational frequencies of the zeolite lattice observed in the range between 200 and 1500 cm^{-1}, (21)..In general, each zeolite has characteristic infrared pattern. However, some common features are observed which, include the asymmetric (950 - 1250 cm^{-1}) and symmetric stretch (660 - 770 cm^{-1}), double ring vibration (500 - 650 cm^{-1}), T-O bending modes (420 –

500 cm^{-1}), and possibly opening modes (400 - 420 cm^{-1}). The FT-IR-spectral data of the synthesized analcime is presented in figure 4b.The double rings (D4R and D6R) in the framework structures of the zeolitic (500 - 650 cm^{-1}), is near to 653 (s) cm^{-1}, which is characteristics Analcime, [18]..The bands at 462 (s) cm^{-1} of analcime, is near the absorption bands within the range 420–500 cm^{-1} which are related to the T–O–T bending of vibration mode (T = Al, Si) respectively. These absorption bands characterizing T-O bending vibrations may shifted to lower frequencies due to decreasing Si/Al ratio in the internal linkage due to the different length of the Al-O and Si-O bands [21-22]. The band 724 (s) cm^{-1} of analcime is near from the bands in the range 720–790 cm^{-1} is associated with symmetric stretching vibration of 4-membered rings. Fig. (4b). this band should be assigned to the 4-membered ring vibrations. Because these rings contain the lowest number of members of all rings occurring in the zeolite structure, therefore the bands due to these rings occur at relatively high wave numbers in the pseudo lattice band range [23].The band1650 (w), of analcime, is near the bands at 1647 and 1648 cm (Lewis sites) region is assigned to the zeolitic water in the channels of zeolite [24].The bands at spectra 3446, 3460 and 3482 are attributed to the asymmetric stretching mode of molecular water coordinated to the edges of the zeolite channels [25-26], and the band3602 cm^{-1} of analcime is near the bands that are attributed to the asymmetric stretching mode of molecular water coordinated to the edges of the zeolite channels [25] m [26]. The bands that located at the lowest wave numbers466 and 377 cm^{-1}, are corresponding to the characteristic bending vibrations carried out in the 4-membered rings. table (4), [27].

Table 4. Fourier Transform Infrared (FT-IR) Spectroscopy of Synthetic analcime

	Characteristic band (cm^{-1})	**Analysis of bands**
Analcime	3602 (b)	O—H stretching
	1650 (w)	Bending H—O—H
	1467 (m)	Na-T stretching
	939 (m)	the asymmetric Al –O stretch of sodalite
	724 (s)	4- or 6-membred double rings
	653 (s)	the symmetric Al –O stretch of sodalite
	462 (s)	Sending vibrations of Si-O and Al– O of the tetrahedral

3.2.4. Thermal Gravimetric Analysis (TGA) of the synthesized zeolites

The thermal behavior of synthesized analcime (TGA) and (DTG) are shown in table 5, Figure 5. The synthesized product shows up to four dehydration steps. The position of these DTG peaks and the number of dehydration steps can be attributed to the different compensating cation-water binding energies. As well as to the different energy associated with the diffusion of the desorbed water through the porous structure of the synthesized product. Their weight loss percentages reflect the water loss from the zeolite structure, and the amount of desorbed water is related with the number of compensation cations in the framework of the zeolite [28].

IOP Conf. Series: Materials Science and Engineering **975** (2020) 012013 doi:10.1088/1757-899X/975/1/012013

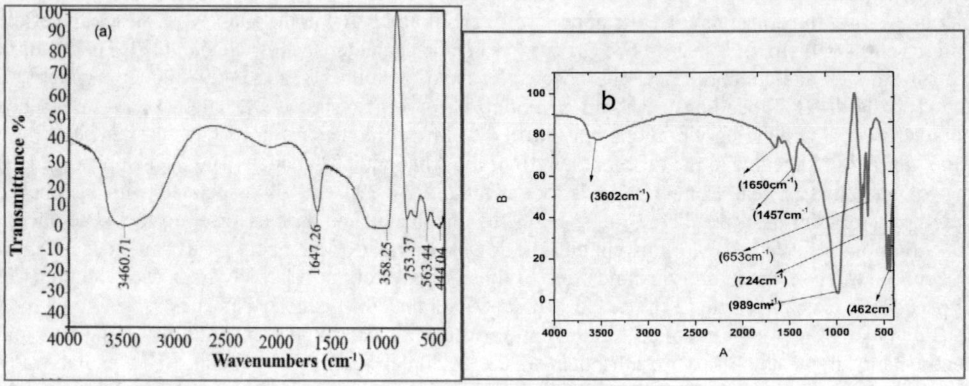

Figure 4. Fourier transforms infrared spectroscopy (FT-IR) of
(a) Raw material kaolinite, (b) Synthetic analcime

The peaks observed between 39-52 °C correspond to surface water in zeolitic materials; the peaks observed between 100 and162 °C indicate zeolitic water, although in some cases in this temperature range up to two peaks occur, which can be explained by the heterogeneous nature of the synthesized product. TGA curves also, showed a small weight loss in the range 0-2 % starting at 90°C until 150°C which may be attributed to loss of observed moisture and entrapped solvents. The thermographs of analcime are given in figure 5.

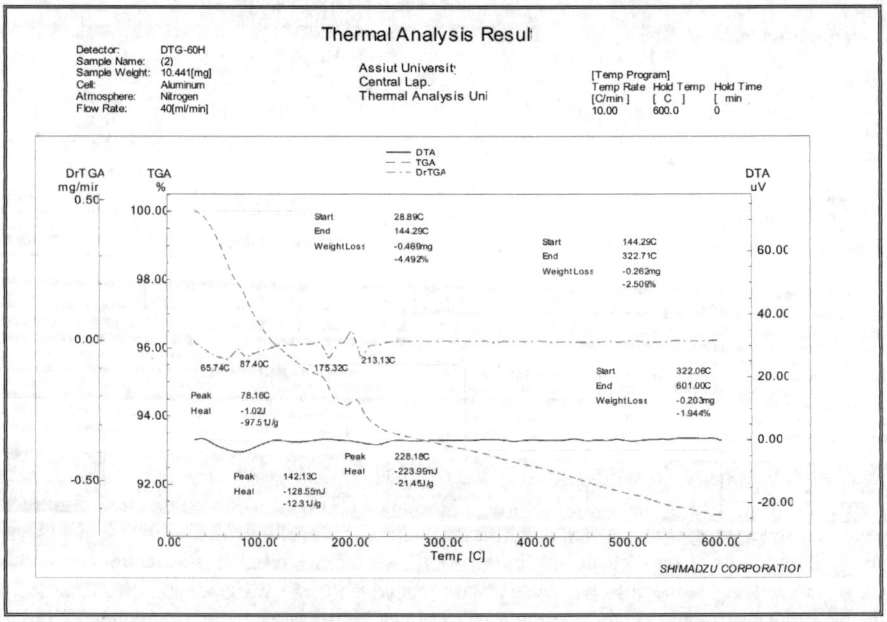

Figure 5. Thermal analysis of the synthesized Analcime

Table (5): Thermal analysis of the synthesized zeolites

Type of zeolites	Temperature (°C) for various percentage decompositions*			
	2%	4%	6%	8%
Analcime	85	140	235	470

4. Surface area and porosity of the synthesized Zeolites

4.1. Adsorption Related Properties (Surface Area, Pore Volume, Microspore Surface Area and Pore Size Distribution).

Physical adsorption measurements are widely used to determine the adsorption characteristics such as; surface area, pore volume and pore size distribution. The first step in the presentation of the adsorption isotherm is the identification of the isotherm type and hence the natures of the adsorption process. In this way it is possible to obtain information about the type or pore structure of the adsorbent and this is the most commonly used computational procedure for quantitative evaluation of the isotherm. The physical-sorption isotherms are generally in six forms and presented in figure 6.

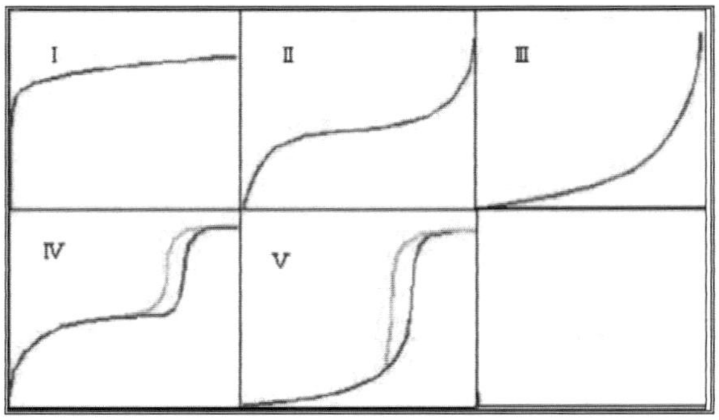

Figure 6. Types of isotherms

Nitrogen gas adsorption-desorption isotherms were measured at -196°C using Nova 3200 instrument (Quantachrom Instrument Corporation, USA). Test samples were thoroughly outgassed for 2h at 250°C to a residual pressure of 10^{-4} torr, and the weight of the outgassed sample was that used in calculations. The specific surface area, S_{BET} was calculated by applying the BET equation. The porosity of the samples was determined from the desorption curves using Nova enhanced data reduction software, Ver. 2.13. A highly pure nitrogen gas was used as adsorb ate. The surface area and porosity of the samples were determined on the basis of nitrogen adsorption and desorption measurement. The measured isotherms exhibit types III isotherms with closed hysteresis loops at intermediate relative pressure. As shown in figure 7. The pore size distribution for these samples could be determined from the desorption data. The specific surface area is derived by applying the BET method in the conventional range of relative pressure with cross-sectional area of N_2 = 16.2 Å^2. The values of S_{BET}, total pore volume, V_P (calculated as liquid volume at $P/P^O = 0.95$), BET-C constant and the average pore diameter, $r_H = V_p/S_{BET}$ (parallel plate chap) for each Synthesized Zeolite are given in table 6

Table 6. Textural properties of the Synthesized Zeolites

Sample	SBET m^2/g	C-BET	Total Pore Volume cc/g	Average Pore diameter Å	Exter. Sur. Area m^2/g	Microp. Sur. Area m^2/g
Analcime	13.35	-55.79	0.04915	147.235	22.177	0.0000

For mesopores, in view of character of the hysteresis loop, the analysis of the samples showed a single peak distribution that is effectively located at the average pore radius as seen in table 7 indicating a mesoporous structure. BET-C is a constant, which is related exponentially to the heat of adsorption of the first layer. In practice, the value of C can be used to define the measure of heat of adsorption. A high value of C (= 100) is associated with a sharp knee in the isotherm. If C value is low «20 the sharp knee cannot be identified as a single point in the isotherm [29]. The value of C is neither too low nor too high, i.e. 20 < C < 200, and that the BET plot is linear for the P/Po region taken for calculations table 9. The value of C may be taken as an indication that basically monolayer-multilayer formation is operative and is not accompanied by any meaningful micro pore filling which is usually associated with an increase in the value C above 200.The pore size analysis is achieved using the corrected model-less method developed for mesopores. In view of character of the hysteresis loop, the analysis of the samples showed a single peak distribution that is effectively located at the average pore radius as seen in table 8, indicating a mesoporous structure. The value of C is neither too low nor too high, i.e. 20 < C < 200, and that the BET plot is linear for the P/Po region taken for calculations figure 8. The value of C may be taken as an indication that basically monolayer-multilayer formation is operative and is not accompanied by any meaningful micro pore filling which is usually associated with an increase in the value C above 200 figures 9 and 10.

Table 7. Textural properties of the synthesized zeolites

Type of Zeolite	slope	intercept	Correlation coefficient	C-BET	Total surface area in cell m^2	Specific surface m^2/g
Analcime	265.47	-4.674426	0.997330	-55.792	1.1083m²	13.

Table 8. BJHC cumulative pore size distribution (desorption) of the synthesized analcime

Type of Zeolite	Total Pore Volume (cc/g)	Average pore diameters (Å).
Analcime	0.04915	147.238.

IOP Conf. Series: Materials Science and Engineering **975** (2020) 012013 doi:10.1088/1757-899X/975/1/012013

Table 9. t-method—deBoer (adsorption) of the synthesized analcime

Type of Zeolite	External Surface Area m^2/g	Microspore Surface Area m^2/g	Microspore Volume cc/g	Correlation Coefficient
Analcime	22.1765	0.0000	0.00000	0.905983

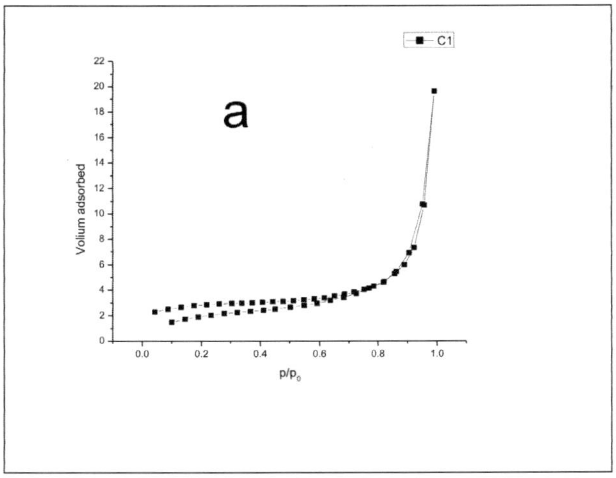

Figure 7. Isotherm (Adsorption/Desorption) of analcime

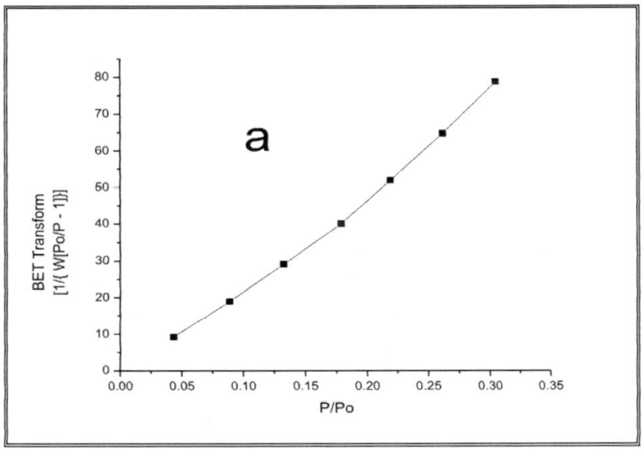

Figure 8. Multipoint BET (Adsorption) of Analcime

IOP Conf. Series: Materials Science and Engineering **975** (2020) 012013 doi:10.1088/1757-899X/975/1/012013

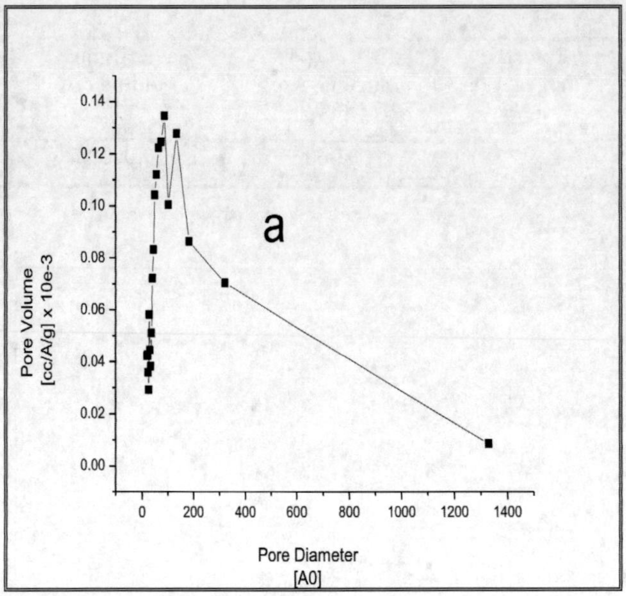

Figure 9. BJHdV/Dd (Desorption) of Analcime

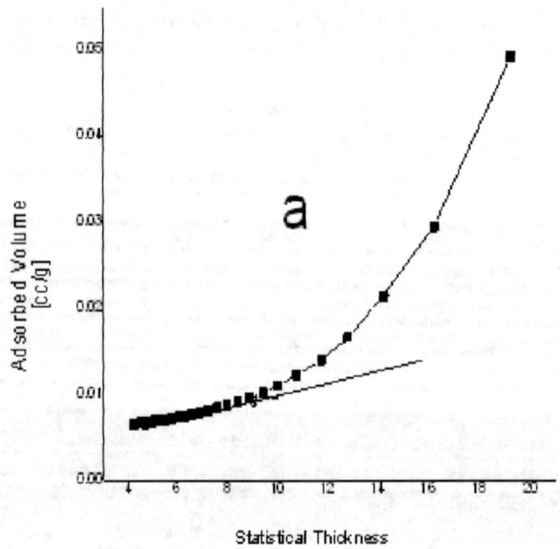

Figure 10. tMethod deBoer (Adsorption) of analcime

5. Ground water treatment

5.1 Nitrate adsorption experiments

The nitrate adsorption experiments were conducted using contaminated ground water samples from Wadi El Assiuti of known initial concentration. Batch experimental mode was adopted due to its simplicity. In the 1st set of experiments,

1-50 mℓ of ground water was taken in 100 ml glass beaker.

2- A dose of 0.5 g of synthetic analcime was added

3- The beaker with contents was placed in a thermostatic shaker operated at 200 r/min

4- After 1 h the samples were filtered through a Whatman Filter Paper No. 41.

In a second set of experiments, the effects of adsorbent mass, pH, temperature and time on nitrate removal from water were explored by varying the initial solution pH from (6±0.2) to (10±0.5) and time from 1h to 3h. A final set of experiments involved the use of groundwater samples were analyzed for nitrate levels before contacting them with different doses of sorbents. The equilibrium nitrate concentration was then determined.

6. Results and Discussion

6.1 Effect of adsorbent mass

The number of active sites relates to adsorbent mass and hence sorption performance. An increase in the solid concentration increases the surface area of the adsorbent, which in turn increases the number of binding sites for the same liquid volume and thus the total amount of metal that is removed increases (30).The effect of adsorbent mass on the nitrate percentage removal under room temperature for 1h and at solution pH is shown in figure 11. It is clear that an increase in adsorbent mass resulted in an increase in the removal of nitrates from water.

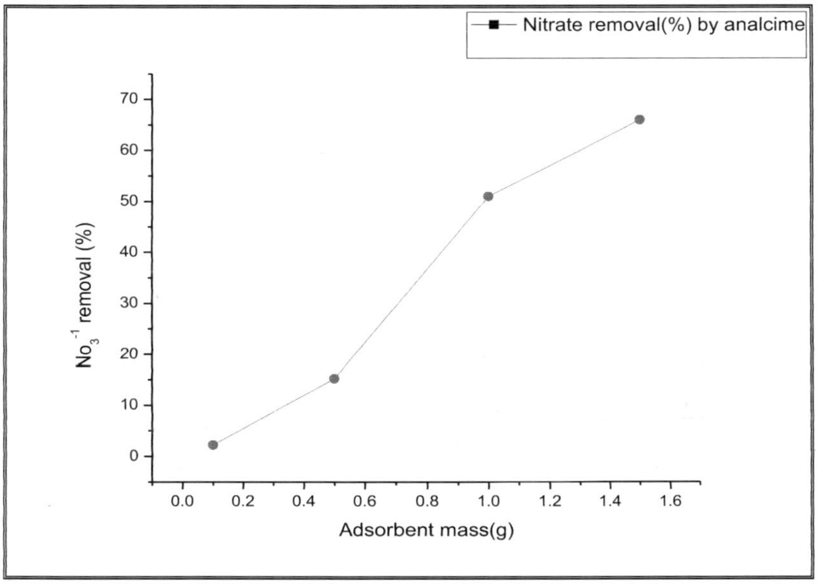

Figure 11. Effect of adsorbent mass on nitrate removal from ground water sample by analcime

6.2. Effect of pH

Solution pH is one of the factors that have been found to significantly affect the sorption process [31]. A study was therefore undertaken to ascertain to what extent pH affected the sorption of nitrate table 10 is showing initial solution pH (6-8) before and after sorbent addition pH adjustment. The pH was adjusted at equilibrium for system (solution–solid adsorbent). The effect of pH changes on the removal of nitrate have been was examined at room temperature, fixed time (1h) and absorbent (analcime) dosage 0.5g. Results reveal that nitrate removal from ground water is marginally affected by water pH. With no electro kinetic data at this stage, it can only be hypothesized that the surface charge on synthetic (analcime), was significantly affected by pH changes. A confirmation is required to ascertain this. The increasing in nitrate adsorption with decreasing in pH from initial solution pH (6-8) and the highest removal efficiency occurred at pH from (6-6.5). It could be due to interaction during passive transport in the pores and competition between OH^{1-} and NO_3^{1-} anions for active sites. One disadvantage of using adsorption processes in removal of anionic contaminants from water inheres in most sorption media's inability to perform adequately at higher pH .Indeed several researchers have reported a significant reduction in media performance in nitrate removal from water at near neutral and alkaline pH (6-7.6), the pH range typical of natural systems such as groundwater (32, 33, 34). A media, such as synthetic (analcime) used in this study, whose performance does not deteriorate with pH changes is desirable for water treatment. Meanwhile the pH of the system (solution-solid) after sorption was found to increase when the initial pH was acidic and to decrease when the initial pH was alkaline. This is due to the buffering effect of the zeolite. Sorbents normally shift solution pHs or their point-of-zero charge (pH pzc) [31]. As shown in table 10.

Table 10. Effect of pH on Nitrate removal from ground water sample by analcime

Initial pH	Final pH (sol+sorbent)	NO_3^{-1}Removal (%) by
8	10.4	15
8	9.75	54
6	7.6	78
6	6.5	79
6	6-5.8	58

6.3. Effect of temperature

Usually, at elevated temperatures the ions uptake is higher due to higher affinity of the mineral for the metal and/or an increase in the active sites of the solid. At higher temperature the energy of the system facilitates the attachment of ions on the mineral's surface [35]. Analcime can exchange ions at room temperature. When the temperature increases, the ion-exchange occurs more easily. The reason is that there are smaller pores in analcime. The effect of temperature was investigated in a range of 25-45 ºC at 1h, dosage 0.5g and solution pH is shown in figure 12 for nitrate adsorption by analcime, it was observed that the decreasing of nitrate concentrations in the water with increasing temperature. Based on that, the nitrate adsorption process is chemical in nature [33], [36]. Consequently, with increasing temperature the retarding specific or electrostatic interactions becomes weaker and the ions become smaller because solvation is reduced. As shown in figure 12.

6.4. Effect of time

The effect of contact time on nitrate adsorption form ground water samples using synthetic analcime was given in table 11 and figure 13. Experiments were performed with varying time between 0.25h to 1.5h at normal solution pH (8-8.5), under room temperature and the sorbent mass is 0.5 g per 50 mℓ water.

IOP Conf. Series: Materials Science and Engineering **975** (2020) 012013 doi:10.1088/1757-899X/975/1/012013

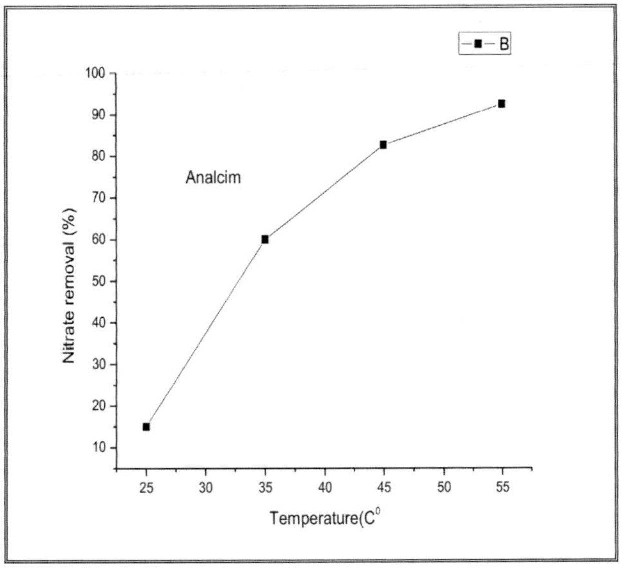

Figure 12. Effect of temperature on nitrate removal from ground water sample by analcime

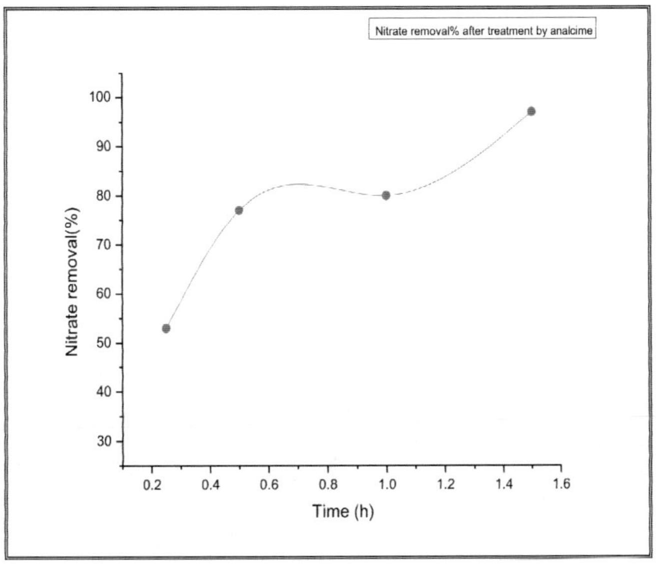

Figure 13. Effect of time on nitrate removal from ground water sample by analcime

IOP Conf. Series: Materials Science and Engineering **975** (2020) 012013 doi:10.1088/1757-899X/975/1/012013

Table (11): Effect of Time on the nitrate removal

No_3^{-1} conc. (ppm)before treatment	Time (h)	No_3^{-1}Removal (%) by Analcime
92	1/4	53
	1/2	77
	1	80
	11/2	97

From the results it is clear that with analcime adsorbent the adsorption was rapid in the first 40 min followed by gradual increase with time until equilibrium was attained figure 13. The fast adsorption at the initial stage was probably due to the initial concentration gradient between the adsorbate in the solution and the number of vacant sites available on the adsorbent surface. The attainment of equilibrium adsorption might have been due to reduction in the available active adsorption sites on the adsorbent with time resulting to limited mass transfer of the adsorbate molecules from the bulk liquid to the external surface of adsorbent [37]. So that the ability of both analcime and cancrinite for nitrate removal are affected by time as shown in table 11 and figure 13.

6.5. Mechanism of nitrate removal by analcime

Many studies have been carried out for the removal of nitrate from groundwater for making it potable. However, some of the methods fall short of being reliable and cost effective. Such as Reverse osmosis, Distillation, biological de-nitrification. And adsorption, ion exchange. Adsorption is a phenomenon of adhesion of ions or molecules from one phase into another. Adsorption depends on the surface charge of the media used to adhere the ions called as adsorbent and the charge of ions to be adhered called as adsorbate .Adsorption is most widely researched and still evolving process in water and wastewater treatment. Zeolites have the characteristic of a high adsorption capacity at the lower partial pressure and higher temperature because of the electric field in the cavities and polar actions of zeolites. Also, the zeolite itself is also a kind of polar substance. Also, the water molecules can be bonded to the ions and framework ions by aqueous bridges in the voids of natural zeolites. The water forms bridges between the outer cations. Nitrate is a polyatomic compound of nitrogen with three molecules of oxygen and one molecule of nitrogen bonded by a ligand. Most of the nitrate is in the form of ammonium nitrates or magnesium nitrates which are found in commonly used agricultural fertilizers. The cations in the salts are adsorbed by the clay particles having negative surface charge. This negative charge repulses the anions like nitrate. Thus, the nitrate readily mixes with pore water. For removing nitrate from water ,adsorbents with positive surface charge are employed which gives more removal efficiency. Materials with Al and Fe ions are more efficient to remove anion due to their amphoteric nature. They can remove cations and anions depending on the environment like pH, Zeolites are another novel material for removal of impurities and pollutants from water [33]. Analcime (ANA) [$NaAlSi_2O_6{\cdot}H_2O$] has the structure and chemical properties similar to the feldspathoids, even though they are classified to the zeolite mineral., classified as such due to its low silicon content. However, cancrinite is classified also as a zeolite, due to its open pore structure, which confers molecular sieve properties [18]. Analcime has a Cubic; tetragonal, orthorhombic, or monoclinic, pseudo cubic, with degree of ordering which can incorporate cations such as Na^+, K^+ and Ca^{2+}, as well as inorganic anions such as CO_3^{2-}, OH^- and NO_3^- and water molecules [18]. Analcime can exchange ions at room temperature. When the temperature increases, the ion-exchange occurs more easily. The reason is that there are smaller pores in analcime

6.5.1 Adsorption Isotherm of NO31- ion by synthetic analcime.

Langmuir and Freundlich are the best models to explain the trend of adsorption based on the essence of absorbents saturated with adsorbate after enough contact time [38-41]. The study were carried out by contacting 0.5g of the previous synthetic zeolites with different concentrations (17,42,74,94 and 100) ppm from solutions under room temperature, for 1 h .and at solution pH on the shaker. The final concentration of nitrate in solution was determined by measuring the absorbance by spectrophotometer (using kits). The data was fitted into the following isotherms: Langmuir, Freundlich, which are shown in tables 12,13,14 and figures 14 a, b.

Langmuir adsorption isotherm is given as equation 1

$$q_e = q_{\max} \times C_e \times K/1 + C_e \times K \quad (1)$$

Where qe is the amount of the solute adsorbed per mass unit of adsorbent (mg/g); Ce is the equilibrium concentration of the aqueous phase (mg/L); and $q\max$ is the maximum. Amount adsorbed. The parameters, $q\max$ and$K$, are Langmuir constants. And qe can be calculated buy equation 2

$$qe = (C_i - C_e)/X \quad (2)$$

Where qe is the amount of the solute adsorbed per mass unit of adsorbent (mg/g), Ci is the initial concentration in the aqueous phase (mg/L), Ce is the concentration in the aqueous phase at equilibrium (mg/L), V is the volume of the aqueous phase (L), and X is the weight of the absorbent (g).

Freundlich isotherm is an empirical equation. The nonlinear form of Freundlich's equation is written as equation 3 [42].

$$qe = K_f C_e^{1/n} \quad (3)$$

Where qe is the amount of the solute adsorbed per mass unit of adsorbent (mg/g), Ce is the equilibrium aqueous pollutant concentration (mg/L), Kf is Freundlich constant, and n is Freundlich exponent, Kf is the indicator of adsorption capacity, and $1/n$ is a measure of intensity of adsorption. Freundlich is an exponential equation which can be used to determine the values of parameters. The logarithm conversion is one way to determine Freundlich Parameters. Therefore converts to log

$$qe = \log K_f + 1n \log C_e$$

By plotting log q_e against $\log C_e$, the values of Freundlich parameters can be determined. The slope of the line is the exponent and the logarithm of intercept is the constant of Freundlich parameters.

Based on the data given in table 12,13, the plots of log qe versus log C_e, and $1/q_e$ versus $1/C_e$ were depicted to show the Langmuir and Freundlich isotherms, respectively figures 14 a, b.

The results showed that Langmuir adsorption isotherm had better matching on data with $R^2 = 0.8$ than Freundlich adsorption isotherm_and this confirmed that the sorption data fitted into Langmuir ,Freundlich isotherm for sorption of nitrate ion onto the synthetic zeolites (analcime) under the conditions used in the above experiment. Also the characterization of the Langmuir equation can be explained in terms of the equilibrium parameters (R_L) which is a dimensionless constant referred to as separation factor or equilibrium parameter equation 4 [43].

$$R_L = 1/1 + (K_L + C_0) \quad (4)$$

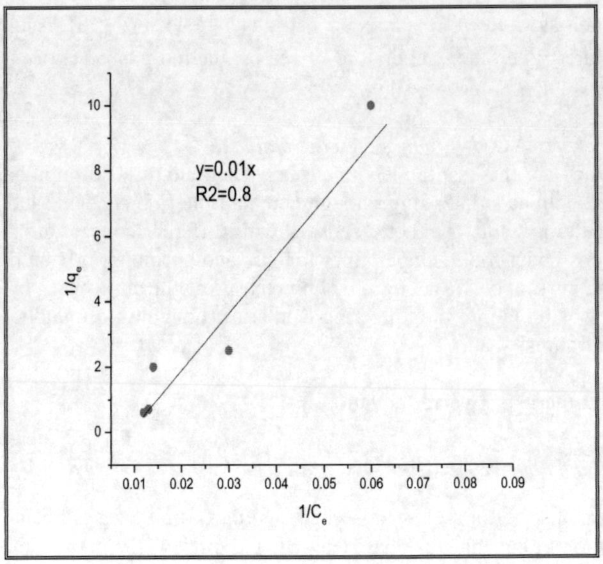

Figure 14, a. Langmuir adsorption

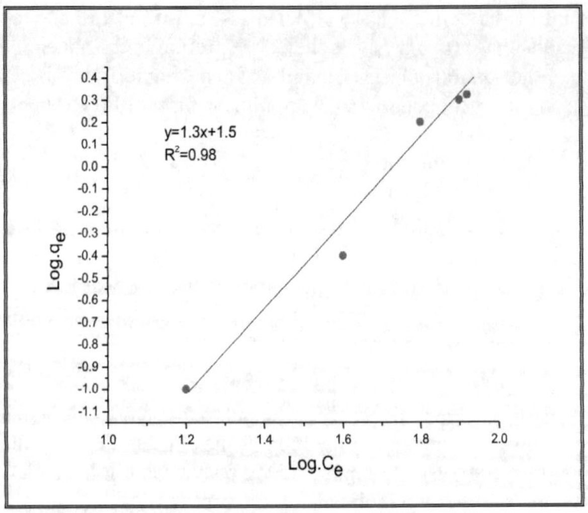

Figure 14, b. Freundlich adsorption isotherm

The above parameters describe the nature of the adsorption: irreversible ($R_L = 0$); favorable ($0 < R_L < 1$); linear ($R_L = 1$); unfavorable ($R_L > 1$) [44]. From the data calculated in table (14),in the case of nitrate ion are favorable ($0 < R_L < 1$); And From this research work, the maximum monolayer coverage capacity (q_o) from Langmuir Isotherm model was determined to be 0.01for nitrate ion and K_L (Langmuir isotherm constant) is 0 for the same ion. This again confirmed that the Langmuir isotherm was favorable for sorption of nitrate ion onto the synthetic zeolite (analcime) under the conditions used in this study [45-46].

Table 12. Parameters for plotting Langmuir, Freundlich, and Adsorption Isotherms of nitrate ion by analcime.

C_0 ($mgNO_3^-/L$)	C_e ($mgNO_3^-/L$)	q_e (mg/g)	$1/C_e$	$1/q_e$	Log. C_e	Log. q_e
17	16	0.1	0.06	10	1.2	-1
42	38	0.4	0.03	2.5	1.6	-0.4
74	69	0.5	0.014	2	1.8	0.2
94	76	1.4	0.013	0.7	1.9	0.3
100	80	1.6	0.012	0.6	1.92	0.322

Table 13. Linear regress ion equation's Langmuir and Freundlich isotherm and constants for adsorption of nitrate ion by analcime.

<table>
<tr><th>Metal ion</th><th rowspan="2">Type of zeolites</th><th colspan="4">Langmuir adsorption</th><th rowspan="2">Langmuir equations</th></tr>
<tr><td rowspan="6">No_3^{-1}</td><td>q_m(mg/g)</td><td colspan="2">K_L(L/mg)</td><td>R^2</td></tr>
<tr><td rowspan="5">Analcime</td><td>0.01</td><td colspan="2">-</td><td>0. 8</td><td>Y=0.01x</td></tr>
<tr><td colspan="4">Freundlich adsorption</td><td rowspan="2">Freundlich equations</td></tr>
<tr><td>1/n</td><td>n</td><td>K_F</td><td>R^2</td></tr>
<tr><td>1.1</td><td>0.9</td><td>1.5</td><td>0.9</td><td>Y=1.1x+1.5</td></tr>
</table>

Table 14. R_L values for nitrate ion concentrations by analcime

No_3^-conc. (ppm) before treatment	17	42	74	94	100
R_L value	0.05	0.023	0.013	0.0099	0.0089

7. Conclusions

In this research, analcime was successfully synthesized from cheaper raw materials, such as clay minerals (kaolinite) by using the fusion with NaOH method. The conditions of hydrothermal crystallization (zeolitization) were found to be at temperature of 170°C, and time span between 36 h and 72 h for kaoline. The synthetic materials have been characterized by X-ray diffraction (XRD), scanning electron microscope (SEM), Fourier transform infrared spectroscopy (FT-IR) and thermo gravimetric (DTA/TGA) analysis. Also we used analcime for nitrate removal from groundwater. The first stage of experiments was carried to Batch experimental tests for the ability of analcime for

IOP Conf. Series: Materials Science and Engineering **975** (2020) 012013 doi:10.1088/1757-899X/975/1/012013

nitrate removal from for the removal of nitrate ions from the ground water samples from Wadi El-Assiuti – Egypt at initial nitrate concentrations 92 ppm, at solution pH initial pH values (6–8), at time 1h and under normal temperature. It is exhibited the highest nitrate removal efficiency about 79-80% at best pH values of the ground water samples of the studded area (6–6. 5), Adsorbent mass0.5g , at temperature 55 ºC for time (1.5h), and initial nitrate concentration of 92mg/L NO^{-3}. The Langmuir, Freundlich constants model for NO^{-3} ion sorption on the adsorption isotherms are fitted well. The RL value in the present investigation was (0 < RL< 1); indicating that the adsorption of NO^{-3} ion by analcime is favorable.

8. References

[1] Magdy H. E., Moustafa M. A. and Hosam A. S. (2012): Impact of hydrochemical Processes on Groundwater Quality, Wadi Feiran, South Sinai, Egypt ,Australian Journal of Basic and Applied Sciences, vol.6(3),pp 638-654.

[2] ApodacaL. E, Jeffrey B. B, Michelle C. S. (2002). Water quality in shallow alluvial aquifers, Upper Colorado River Basin, Colorado, 1997. J. Am. Water Res. Assoc.vol. 38 (1) ,pp.133-143.

[3] Bhatnagar A., Sillanpää M. . (2011): A review of emerging adsorbents for nitrate removal from water, Chemical Engineering Journal, Volume 168, pp. 493-504.

[4] Mazeikiene A., Valentukeviciene M., Rimeika M., Matuzevicius AB and Dauknys R.(2008): Removal of nitrates and ammonium ions from water using sorbent zeolite (clinoptilolite). J. Environ. Eng. Landscape Manage.vol. 16 (1), pp. 38-44.

[5] Bouwer.H (1989): Agricultural contaminants: Problems and solutions. Water Environ. Technol.vol.1,pp. 292-297.

[6] Samatya S., Kabay N., Yuksel Ü., Arda M . and Yuüksel. (2006): Removal of nitrate from aqueous solution by nitrate selective ion exchange resins. Reactive Functional Polym. 66 (11) 1206-1214.

[7] US EPA; (2014): Water: Basic Information about Regulated Drinking Water Contaminants: Basic Information about Nitrate in Drinking Water. Available on: http://water.epa.gov/drink/contaminants/basicinformation/nitrate.cfm, Accessed 17 June

[8] World Health Organization (WHO) (2007): Nitrate and nitrite in drinking water, back ground document for development of WHO Guidelines for Drinking water Quality, WHO/SDE/WSH/07.01/16 /Rev/1.

[9] Tofighy M. A.; Mohammadi T. (2012): Nitrate removal from water using functionalized carbon nanotube sheets, Journal of Hazardous Materials, Vol. 90, pp. 1815-1822.

[10] Santafé-Moros A.; Gozálvez-Zafrilla J. M.; Lora-García J. (2005): Performance of commercial nanofiltration membranes in the removal of nitrate ions, Desalination, Vol. 185, pp. 281-287 .

[11] Inglezakis V. J., (2005): The concept of ''capacity'' in zeolite ion–exchange systems, Journal of Colloid and Interface Science vol. 281, pp. 68–79.

[12] Sakaalauskas, Valentukeviciene.(2003):investigation into the influence of natural powdered zeolite into drinking water treatment at Druskininkai water workIII ,Journal of Environemental Engineering and Ladscepe mangmeneII vol.4,pp.169-178.

[13] Brannvall E.; Mažeikienė, A.; Valentukevičienė, M. (2006): Experimental research on sorption of petroleum products from water by natural clinoptilolite and vermiculite, Geologija 56: 5–12. ISSN 1392–110X.

[14] Healey A. M., Johnson G.M., Weller M.T. (2000): The synthesis and characterization of JBW-type zeolites. Part A: Sodium/potassium aluminosilicate, Na2K [Al3Si3O12].0.5H2O. Microporous and Mesoporous Mater. Vol.37, pp.153-163.

[15] Gatta G .Nestola, Fabrizio B., Tiziana B. (2004) : Elastic behavior, phase transition, and pressure induced structural evolution of analcime,journalAmerican Mineralogist,vol.91(4),pp. 568-578.

[16] Fabbrizio, Alessandro S., Max W G., Detlef E., Jost.(2008): Experimental determination of radium partitioning between leucite and phonolite melt and 226 Ra-disequilibrium crystallization ages of leucite,journal Chemical Geology ,vol. 255(3),pp. 377-387

[17] Azizi, Seyed N. T., Salma E. (2013): Cu-modified analcime as a catalyst for oxidation of benzyl alcohol: experimental and theoretical. Journal Micro porous and Mesoporous Materials, vol. 167.pp. 89-93

[18] Ríos C. A., Williams C. D. and Fullen M. A. (2009): Nucleation and Growth History of Zeolite LTA Synthesized from Kaolinite by Two Different Methods. Applied Clay Science, vol.42, pp.446-454. http://dx.doi.org/10.1016/j.clay.2008.05.006.

[19] Treacy MMJ, Higgins JB. (2001): Collection of simulated XRD powder patterns for zeolites. Published on behalf of the Structure Commission of the 'International Zeolite Association', journal Powder Patterns, vol.203, pp. 204

[20] Zhao H, Deng Y, Harsh J B., Flury M., and Boyle J. S. (2004) :Alteration of kaolinite to Cancrinite and Sodalite by Simulated Hanford Tank Waste and its impact on Cesium retention.claysandclayminerals,vol.52,pp.1-13.

[21] Li G. H. (2005): FT-IR Studies of Zeolite Materials: Characterization and Environmental Applications. Ph.D. Thesis, Graduate College, the University of Iowa, Iowa City, 162.

[22] Mohammed A. A, Shakir I., K. and Esgair K. K. (2013): The Use of Prepared Zeolite Y from Iraqi Kaolin for Fluid Catalytic Cracking of Vacuum Gas Oil. Journal of Engineering, vol.19, pp.1256-1270.

[23] Wang, Haiyan T., Elizabeth A H., Yining. (2006): Investigations of the adsorption of n-pentane in several representative zeolites journal of Physical Chemistry B, vol.110 (16), pp. 8240-8249

[24] Blanco C., Gonzalez, Pesquera F., Benito C., Mendioroz I. S. and Pajares J.A. (1989) : Differences between One Aluminic Palygorskite and Another Magnesic by Infrared Spectroscopy. Spectroscopy Letters, vol. 22, pp.659-673. http://dx.doi.org/10.1080/00387018908053926.

[25] Zhu X.; Choo K.-H.; Park J.-M. (2006): Nitrate removal from contaminated water using polyelectrolyte enhanced ultrafiltration, Desalination, Vol. 193, pp. 350-360.

[26] Faghihian H., and Godazandeha N., (2009): Synthesis of Nano Crystalline Zeolite Y from Bentonite. Journal of Porous Materials, vol.16, pp.331-335. http://dx.doi.org/10.1007/s10934-008-9204-0

[27] Mozgawaet W., Jastrzębski W., Handke M. (2005): Vibrational spectra of D4R and D6R structural units. Journal of Molecular Structure, vol.744–747, pp.663–670.

[28] Reyes, Carlos A. R. W., Craig A., Oscar M. C. (2009): Nucleation and growth process of sodalite and cancrinite from kaolinite-rich clay under low-temperature hydrothermal conditions. Materials Research,vol.16(2),pp. 424-438.

[29] Sing, Kenneth SW. (2013): Assessment of surface area by gas adsorption.J. Adsorption by Powders and Porous Solids: Principles, Methodology and Applications,pp. 237-263

[30] Ören, Hakan K.A., Abidin.(2006): Factors affecting adsorption characteristics of Zn^{+2} on two natural zeolites, j. Journal of Hazardous Materials, vol.131(1-3), pp. 59-65

[31] Onyango Ms. and Matsuda H. (2006): Fluoride removal from water using adsorption technique. In: Tressaud A (ed.) Fluoride and Environment: Agrochemical, Archaeology, and Green Chem. And Water (Advances in Fluorine Science, Vol. 2 Elsevier B.V., the Netherlands. Pp.1-48.

[32] Chatterjee S., Lee Mw, Lee Ds and Woo Sh. (2009): Nitrate removal from aqueous solutions by cross-linked chitosan beads conditioned with sodium bisulfate. J. Hazardous Mater. Vol.166 (1) 508-513.a.

[33] Chatterjee S. and Woo Sh (2009): The removal of nitrate from aqueous solutions by chitosan hydrogel beads. J. Hazardous Mater. Vol.164 (2-3) pp.1012 -1018.

[34] Cengeloglu, Yunus A., Gulsin T., Ali K., Izzet D., Nesim. (2008): Removal of boron from water by using reverse osmosis. Separation and Purification Technology, vol.64 (2), pp. 141-146

IOP Conf. Series: Materials Science and Engineering **975** (2020) 012013 doi:10.1088/1757-899X/975/1/012013

[35] Arief V. O., Trilestari K., Sunarso J., N.Indraswati, Ismadji S. (2008): Recent progress on biosorption of heavy metals from liquids using low cost biosorbents: characterization , biosorption parameters and mechanism studies, Clean Soil Air Water, vol.36 ,pp.937–962

[36] Inglezakis V. J.; Zorpas A. A; Loizidou M. D.; Grigoropoulou, H. P. (2003): Simultaneous removal of metals Cu^{+2}, $Fe^{+3}$and Cr^{+3} with anions $SO4^{-2}$ and $HPO4^{-2}$ using clinoptilolite, Microporous and Mesoporous Materials vol.61,pp. 167– 171.

[37] Sofia M., Sodeh S. , Seyed J.A., and Mohammad O. (2014): Removal of heavy metals from aqueous solution using platinum nanoparticles/Zeolite-4A, J.Environ Health Sci Eng,vol.12(7).

[38] Hamdi W. F, Gamaoun D. E. (2013): Pelster, and M. Seffen, "Nitrate sorption in an agricultural soil profile," Applied and Environmental Soil Science, vol.2013, Article ID 597824, p.7

[39] Qafoku N. P, Sumner M. E. and Radcliffe D. E. (2000): "Anion transport in columns of variable charge subsoils: nitrate and chloride," Journal of Environmental Quality, vol. 29(2), pp. 484–493,

[40] Bhatnagar, A. Kumar, E. and Sillanp¨a¨a, M. (2010): "Nitrate removal From water by nano-alumina: characterization and sorption studies," Chemical Engineering Journal, vol. 163(3), pp. 317– 323

[41] Hizal J. and Apak R. (2006): "Modeling of cadmium (II) adsorption on kaolinite-based clays in the absence and presence of humic acid," Applied Clay Science, vol. 32, (3-4) pp. 232–244.

[42] Zheng C. and Bennett G.D. (2002): Applied Contaminant Transport Modeling, JohnWiley & Sons, New York, NY, USA, 2nd edition,

[43] Pandey P. K., Sharma S. K., Sambi S. S. (2010): Kinetics and equilibrium study of chromium adsorption on zeoliteNaX, International Journal of Environmental Science & Technology, vol.7 (2), pp. 395-404

[44] Rozada F.; Otero M.; García A. I.; Morán. (2007)"A. Application in fixed-bed systems of adsorbents obtained from sewage sludge and discarded tyres, Journal Dyes Pigment, vol. 72,(1), pp.47–56.

[45] Annul, Nuriersoz, Mustafa, (2006): Adsorption characteristics of heavy metal ions onto a low cost biopolymeric sorbent from aqueous solutions, Journal of Hazard Mater, vol. 136 , (2) , pp. 272-280

[46] Zhao D., Zhou J. and Liu N. (2006): Preparation and Characterization of Mingguang Palygorskite Supported with Silver and Copper for Antibacterial Behavior. Applied Clay Science, vol.33, pp.161-170. http://dx.doi.org/10.1016/j.

Ionic Liquids based on Different Chain Fatty Acids as Green Corrosion Inhibitors for C-steel in Produced Oilfield Water

Y Atef[1] and A Ghanem[2]
[1] Research Laboratory, Egyptian Chemistry Administration, Cairo, Egypt.
[2] PVT Laboratory, Production Department, Egyptian Petroleum Research Institute (EPRI), Cairo, Egypt.
e-mail: jasy9@yahoo.com

Abstract. The work presents the investigation of fatty acids based ionic liquid as green corrosion inhibitors for C-steel in produced oilfield water. Different-chain length fatty acids C6, C12 and C18-1 are mixed with di-ethanol amine (DEA) to synthesize some ionic liquids and evaluated the corrosion rates on C-steel surface using weight loss method and potential time measurements. The prepared ionic solutions were shown to be effective in reducing the corrosion rates as sustainable corrosion inhibitors. The polar unit is indicated as the reaction center for the adsorption process. Generally, the inhibition efficiency increases with increase the inhibitor concentration.
Keywords: Carbon steel, ionic liquids, fatty acids, corrosion inhibitors, adsorption.

1. Introduction

Corrosion is an irreversible and spontaneous destruction of metal or alloy through chemical or electrochemical reaction with the environment [1]. Therefore, corrosion has drawn considerable academic and industrial attention [1-3]. The corrosion of carbon steel (C-steel) caused by the presence of produced oilfield water is considered a common problem across petroleum industries [4]. Produced oilfield water (also called formation water) exists in gas and oil reservoirs and it's rich in huge amounts of dissolved salts like chlorides and sulfates besides corrosive dissolved gases like CO_2 and H_2S.

The research within the sector of green corrosion inhibitors has been addressed toward the goal of using cheap effective molecules at low or zero environmental impact. Green approach towards corrosion inhibition of carbon steel in produced oilfield water using ionic liquids which replaced the inhibitors supported toxic materials produced from petrochemical resources to reduce the pollution of the environment [5,6]. Moreover, the using of ionic liquids has attracted considerable attention because of their interesting properties like low volatility, non-inflammability, non-toxic nature, high thermal/ chemical stability and their ability to adsorb on the metallic surface. The massive number of possible cation/ anion combinations allows an excellent variety of tunable interactions; also their unique properties and characteristics make it a promising and a favorable to work within several fields [7], their physicochemical properties, mainly, good electrical conductivity, solvent transport, and a comparatively wide electrochemical window [8]. Introducing different functional groups into cations, which provides an excellent deal of ILs with new structures, can markedly change its physicochemical properties, and it also affords more choices for applications of ILs in electrochemical devices.

The present work is to produce green corrosion inhibitors supported fatty acids ionic liquids such as hexanoic or caproic acid, lauric acid and oleic acid which are prepared by the reaction with diethanol amine. Moreover, the ability of the prepared materials to inhibit the corrosion of steel in oilfield water by applying different concentrations from the prepared inhibitors and tested it with different chemical techniques.

2. Experimental

2.1. Materials

2.1.1. Chemical composition of carbon steel.

The experiments were performed using carbon steel with chemical composition that given in table 1.

Table 1. Chemical composition of carbon steel by wt. %.

Chemical Composition	C	Mn	Si	S	P	Fe
wt. %	0.12	0.5	0.17	0.6	0.046	Rest

2.1.2. Ionic liquids synthesis

Diethanol amine, DEA (0.1 mole) was loaded into 100 ml two-necked round bottom flask on a magnetic stirrer. The system was equipped with a reflux condenser with vigorous stirring at constant rate under nitrogen. The same molar ratio of various fatty acids C_6, C_{12} and $C_{18\text{-}1}$ (0.1 mole) was added individually and drop wise to the system for about 15 minutes. The reaction mixture was allowed to stir for twenty-four hours. The product was treated with acetone, removed by evaporation under vacuum and dried for 24 hours at 45 °C. The molecular structures of the prepared ionic liquids as inhibitors are shown in table 2.

Table 2. Molecular structure of the prepared ionic liquids.

Chemical Name	**Molecular Structure**	**M.wt**
IL-1: DEA-C_6 bis(2-hydroxyethyl)ammonium hexanoate	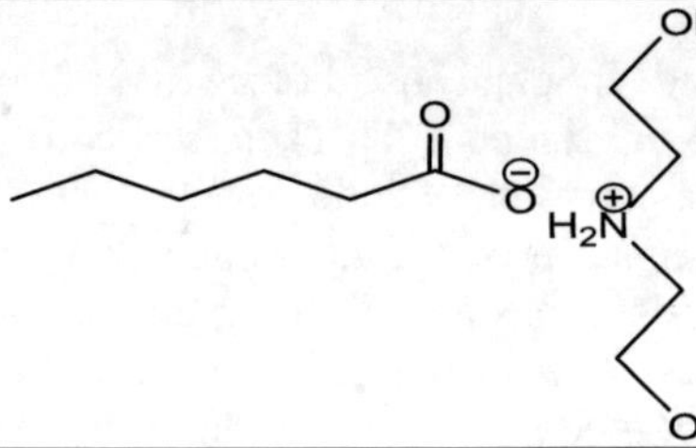	221.2
IL-2: DEA-C_{12} bis(2-hydroxyethyl)ammonium dodecanoate	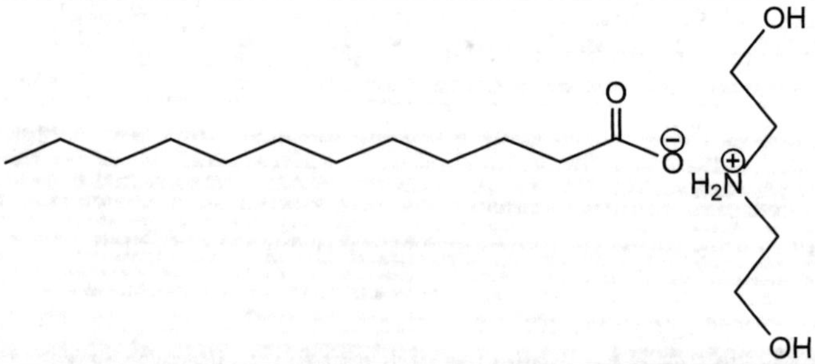	305.46
IL-3: DEA-$C_{18\text{-}1}$ bis(2-hydroxyethyl)ammonium oleate	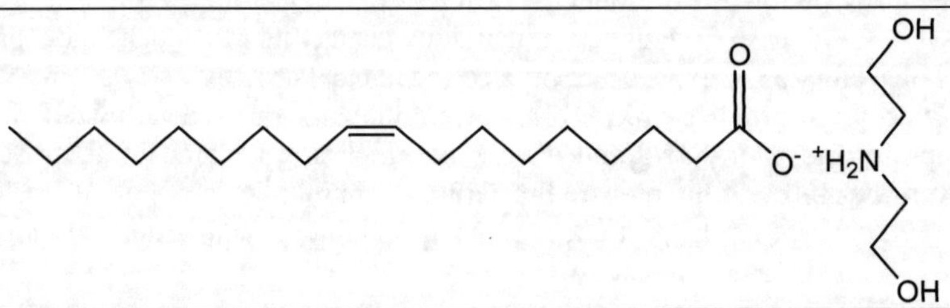	387.6

IOP Conf. Series: Materials Science and Engineering **975** (2020) 012014 doi:10.1088/1757-899X/975/1/012014

2.1.3. FTIR analysis.

Infrared spectra of the three prepared ILs as represented in figure 1, show that each one of the three ILs exhibits a broad band at the range of 3100–3400 cm^{-1}, assigned to –NH stretching vibration by a secondary amine. The strong vibrational bands in the range of 2800–2950 cm^{-1} are assigned to C-H symmetric and asymmetric stretching vibrational motion of methylene and methyl groups within the alkyl chains of various ionic liquids. The presence of a sharp peak at 3006 cm^{-1} in IL-3 of monounsaturated fatty acid (oleic acid) only is attributed to =C-H stretching [9]. The characteristic peaks of fatty acid-ionic liquids are ascertained by the presence of the symmetric and asymmetric stretching vibrations of the carboxylate group (C=O & COO) for the three ILs at the range of 1552–1558 cm^{-1}, 1624–1653 cm^{-1} & 1450–1457 cm^{-1}, respectively. Thermo Scientific Nicolet iS50 FTIR Spectrometer used in this work and performed at Chemistry Administration, Cairo, Egypt.

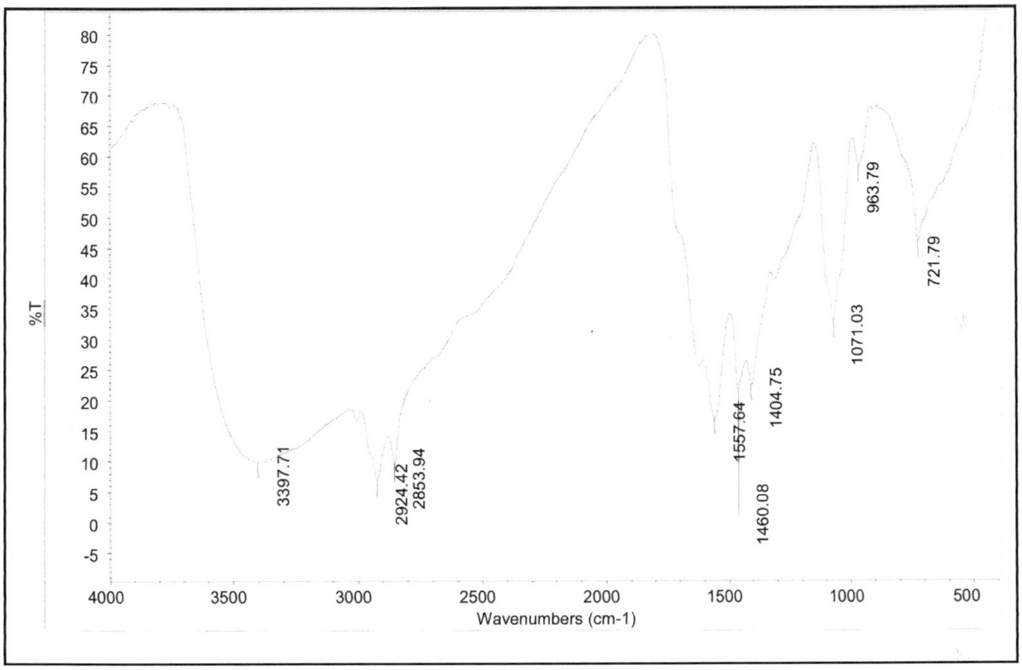

Figure 1. FTIR spectra of DEA-C_{18-1} IL.

2.1.4. ^{1}H-NMR analysis.

The three ionic liquids nearly exhibit an equivalent chemical shifts as shown in figure 2(a,b,c), which, are often summarized as following: the peaks appeared at chemical shifts (δ, ppm) of 0.79 – 0.81 are assigned to the terminal methyl group (t, 3H, CH_3 –); (δ, ppm) of 1.18–1.99 are assigned to (m, 2H, CH_2–); (δ, ppm) of 2.75 – 2.79 are assigned to (P, 2H, CH_2 –CH_2 –NH_2^+); (δ, ppm) of 3.5 – 3.53 are assigned to (t, 1H, CH_2 - OH) [10]. The downfield shift from δ, ppm adequate to 5.24 just in case of IL3 of oleic acid to 5.3 in IL2 of lauric acid and eventually to 6.0 in the case of IL1of hexanoic acid are due to the difference of alkyl chain length in each ionic liquid. The data of Nuclear Magnetic Resonance (^{1}H NMR spectra) were performed using Joel 500 MHz, at National Research Center (NRC), Cairo, Egypt.

IOP Conf. Series: Materials Science and Engineering **975** (2020) 012014 doi:10.1088/1757-899X/975/1/012014

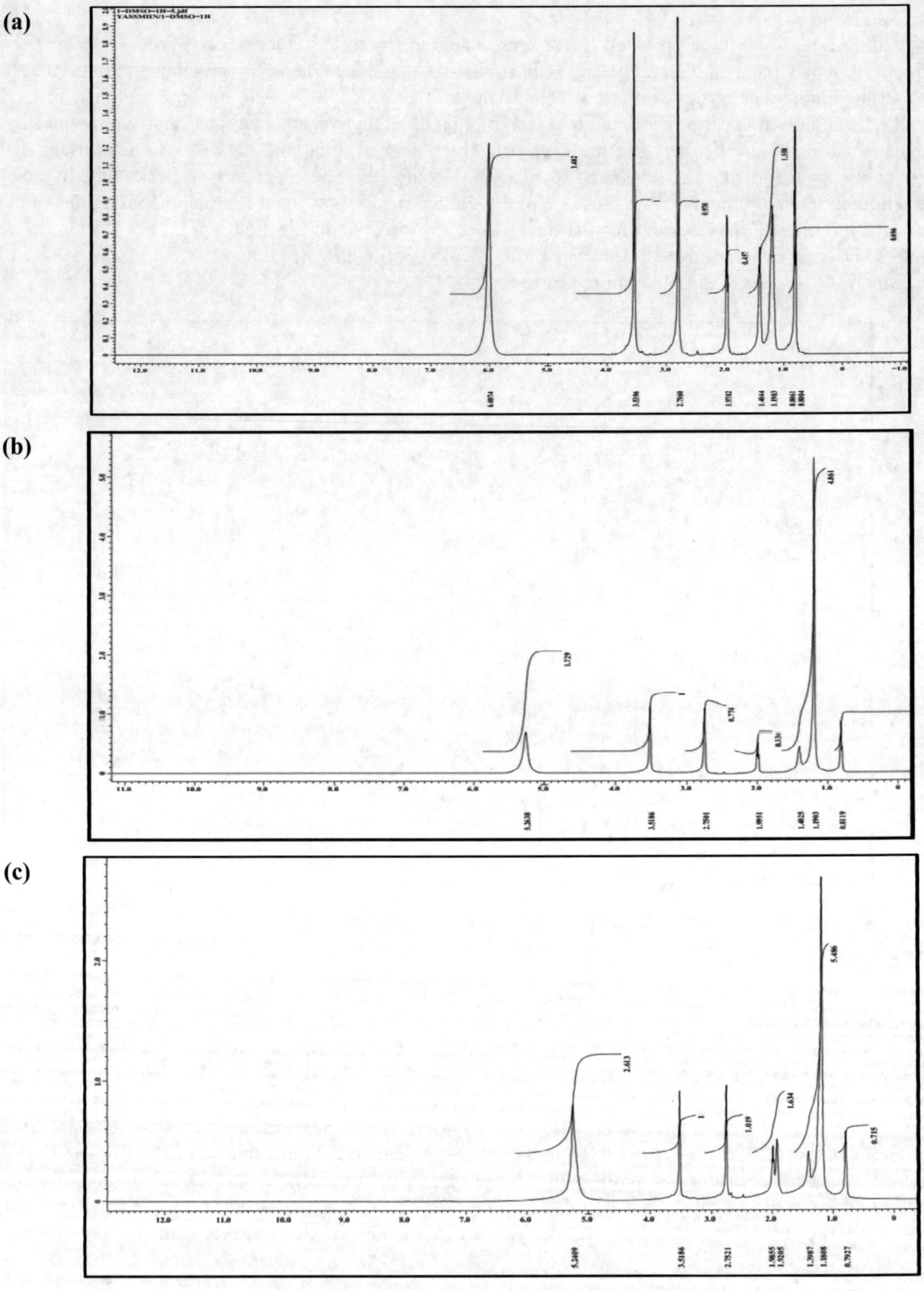

Figure 2. ^{1}H-NMR spectra of (a) DEA-C_6, (b) DEA-C_{12} and (c) DEA-C_{18-1} ILs, respectively.

IOP Conf. Series: Materials Science and Engineering **975** (2020) 012014 doi:10.1088/1757-899X/975/1/012014

2.1.5. Preparation of solutions.
The oilfield formation water used in this work has the chemical analysis listed in table 3. It diluted by 50% with distilled water and used as an aggressive solution for all experiments. The prepared ionic liquids were prepared in molarity, M, by using the same concentration of formation water. All experiments were performed at room temperature, 25 °C, in equilibrium with the atmosphere.

Table 3. Chemical composition of oilfield formation water.

Test	Result
Chloride Salinity	19148 ppm
Ca++	561 ppm
Mg++	146 ppm
pH	7.5

2.2. Methods
2.2.1. Weight loss measurements.
The dimensions of carbon steel coupons 7.5 x 1.2 x 0.1 cm included two holes of radius 0.4 cm, with total surface area of 17.74 cm^2 of two sides used for weight loss measurements and obtained from the oil field region (the study area). Before all measurements, the exposed area was mechanically abraded with different grades of emery papers (280–1200). The specimens were completely immersed in 100 ml of the test solution from of 1 to 24 hours. The weight loss for each specimen is taken as the difference in weight before and after immersion using digital electronic balance with sensitivity of ±0.1 mg.

2.2.2. Potential time measurements.
The potential time curves were performed by open circuit and record the potential shifting with and without different concentrations of ionic liquids for 3 hrs. The carbon steel electrode of geometric area 0.12 cm^2 is employed for the measurements and the saturated silver/silver chloride electrode (Ag/AgCl) used as a reference electrode for referring to all potentials.

3. Results and Discussion

3.1. Weight loss measurements
The weight loss of carbon steel in diluted formation water by 50% with and without the addition of various concentrations of inhibitors was compiled in figures 3, 4 & 5. Inspection of those figures reveal that, the linear variation of weight loss with time in uninhibited and inhibited 50% formation water indicates the absence of insoluble surface films during corrosion i.e. the inhibitors are first adsorbed on the metal surface and thus impede corrosion either by merely blocking the reaction sites anodic and cathodic or by altering the mechanism of the anodic and cathodic processes [11].

The carbon steel surface is corroded by the aggressive Cl^- and S^{2-} anions which are existed in oilfield formation water. The corrosion rates calculated from the slope of the fit linear for wt. loss-time curves as shown in equation (1). It's clear that the corrosion rate decreased by adding increased concentrations from DEA-C_6, DEA-C_{12} and DEA-C_{18-1} ILs to the aggressive solution, as shown in figure 6. The inhibition efficiencies (IE%) and surface coverage (θ) obtained from weight loss calculated from equations (2), (3), respectively. It is obvious that the IE increases with increasing the inhibitor concentration [12] in the following order: DEA-C_{18-1} ILs > DEA-C_{12} ILs> DEA-C_6 ILs. The inhibition efficiencies values are listed in table 4.

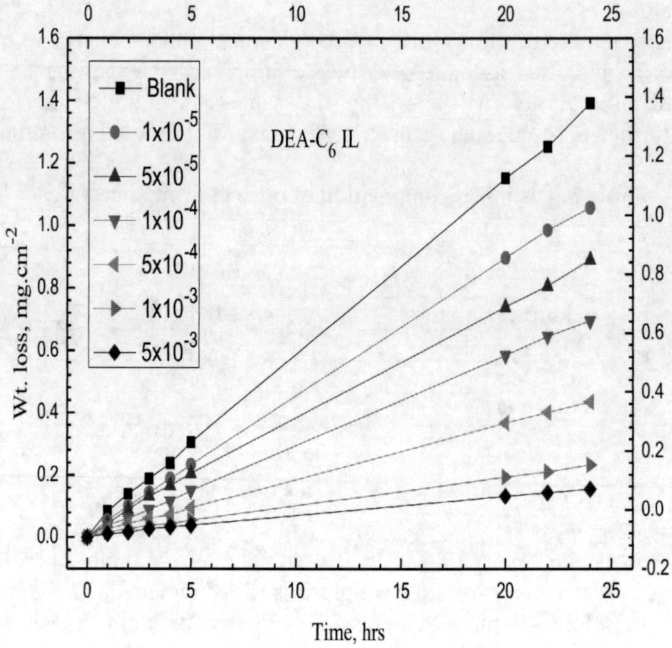

Figure 3. Plot of Wt. loss-time curves for carbon steel in different concentrations of DEA-C_6 IL.

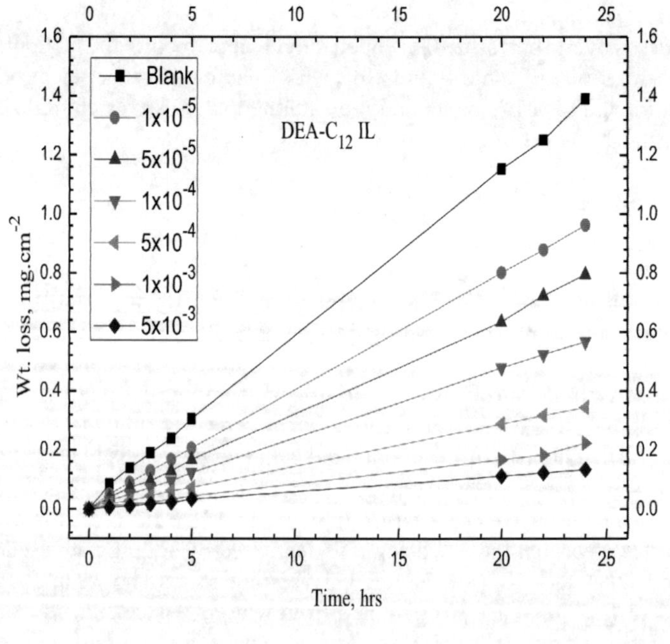

Figure 4. Plot of Wt. loss-time curves for carbon steel in different concentrations of DEA-C_{12} IL.

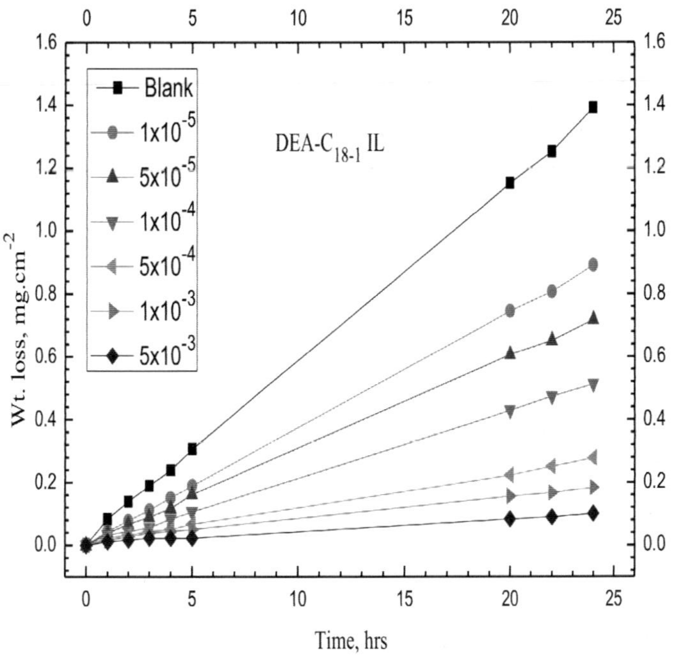

Figure 5. Plot of Wt. loss-time curves for carbon steel in different concentrations of DEA-C_{18-1} IL.

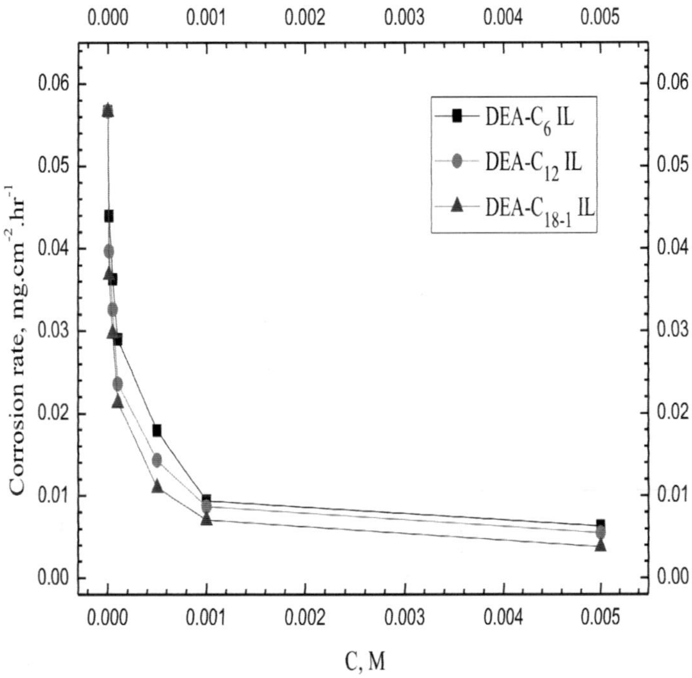

Figure 6. Corrosion rate (C_R) for corrosion for carbon steel as a function of concentration of DEA-C_6, DEA-C_{12} and DEA-C_{18-1} ILs with 50% oilfield formation water.

$$C_R = \frac{W_o - W_i}{W_o} \tag{1}$$

where W_o and W_i are the wt. losses in absence and presence of the inhibitor, respectively.

$$\%IE = \frac{CR_o - CR_i}{CR_o} \times 100 \tag{2}$$

$$\theta = \frac{CR_o - CR_i}{CR_o} \tag{3}$$

where CR_o and CR_i are the corrosion rates in the absence and presence of the inhibitor, respectively.

Table 4. Rate of corrosion (C_R) and inhibition efficiency (IE) for corrosion of carbon steel in different ionic liquids.

C, M	DEA-C_6 IL		DEA-C_{12} IL		DEA-C_{18-1} IL	
	C_R, mg.cm^{-2}.hr^{-1}	IE, %	C_R, mg.cm^{-2}.hr^{-1}	IE, %	C_R, mg.cm^{-2}.hr^{-1}	IE, %
Blank	56.7x10^{-3}	-	56.7x10^{-3}	-	56.7x10^{-3}	-
1x10^{-5}	44x10^{-3}	22.39	39.7x10^{-3}	29.98	36.8x10^{-3}	35.097
5x10^{-5}	36.3x10^{-3}		32.6x10^{-3}	42.5	29.7x10^{-3}	47.62
1x10^{-4}	29x10^{-3}	35.97	23.6x10^{-3}	58.38	21.3x10^{-3}	62.43
5x10^{-4}	17.9x10^{-3}	48.85	14.3x10^{-3}	74.78	11x10^{-3}	80.6
1x10^{-3}	9.4x10^{-3}	68.43	8.7x10^{-3}	84.65	7.1x10^{-3}	87.48
5x10^{-3}	6.3x10^{-3}	83.42	5.5x10^{-3}	90.3	3.8x10^{-3}	93.3

3.1.1. Adsorption isotherm.

Since corrosion is a spontaneous process, relative rate of corrosion among a given series of metals is related to the change in standard Gibb's free energy ΔG^o. A more negative value of ΔG^o related with higher spontaneity of reaction [13-16]. It corroded in order to form stable corrosion products [17-19]. The values of the degree surface coverage θ were evaluated at different concentrations of the DEA-C_6, DEA-C_{12} and DEA-C_{18-1} ILs according to equation 4.

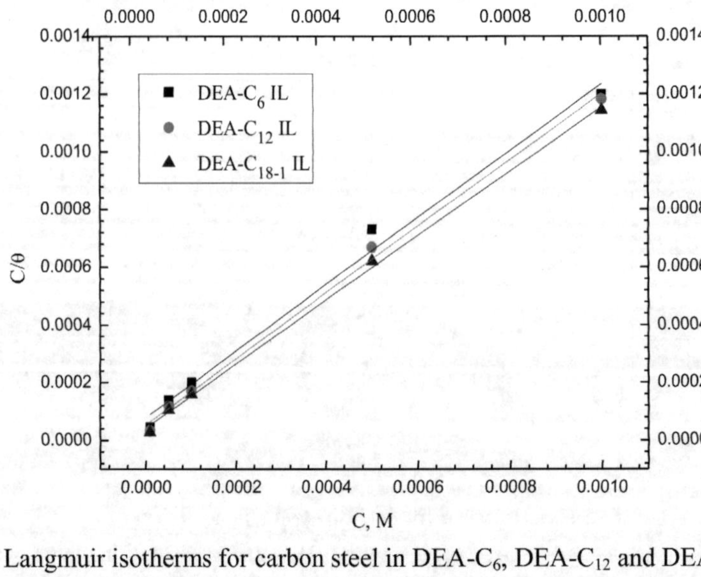

Figure 7. Langmuir isotherms for carbon steel in DEA-C_6, DEA-C_{12} and DEA-C_{18-1} ILs.

$$C_{inh} /\theta = 1/K_{ads} + C_{inh} \tag{4}$$

where K_{ads} is the equilibrium constant of the adsorption/desorption process.

Table 5. Langmuir adsorption isotherm of DEA-fatty acids ILs on carbon steel in 50% oilfield formation water.

ILs inhibitors	Langmuir adsorption isotherm			
	K_{ads}, M	$\Delta G°_{ads}$, kJ mol^{-1}	Slope	R^2
DEA-C_6	$1.27x10^4$	−30.58	1.15	0.987
DEA-C_{12}	$1.89x10^4$	−31.47	1.14	0.995
DEA-C_{18-1}	$2.38x10^4$	−32	1.11	0.998

Figure 7, represents the relationship between C/θ and C. Straight lines were obtained with intercept of K_{ads}^{-1} with the slopes of ~ 1, suggesting that the adsorption of all ILs on the carbon steel surface follow Langmuir isotherm. From table 5, it can be seen the values of K_{ads} and the related free energy of adsorption $\Delta G°_{ads}$ that evaluated by the following equation:

$$\Delta G°_{ads} = -2.303\ RT \log (55.5\ K) \tag{5}$$

where, R is the universal gas constant, T is that the temperature (273 K) and K is that the equilibrium constant of adsorption. The value of 55.5 is that the concentration of water within the solution in mol/L. The negative values refer to the spontaneity of the adsorption process and therefore the film stability on the C-steel surface. In addition to, the high K_{ads} values indicate a fast adsorption process on the metal active sites. This phenomenon is due to the fast orientation of the high electronic density zone of the molecule located over the metal ion. The calculated values of $\Delta G°_{ads}$ ranged from −30.58 to −32 kJ mol^{-1}, listed in table 5, suggesting competitive physisorption and chemisorption processes of the active constituents present in DEA-C_6, DEA-C_{12} and DEA-C_{18-1} ILs on carbon steel surface [20].

3.2. Potential-time Curves
Corrosion studies by open circuit potential in 50% oilfield formation water with and without using DEA-C_6, DEA-C_{12} and DEA-C_{18-1} ILs as corrosion inhibition of carbon steel shows the simulation curves has the same trend, figures 8, 9 & 10, respectively. At the beginning the potential decreases rapidly in case of aggressive solution up to 10 minutes of exposure and slow gradually by increasing concentrations of inhibitors then the potential remains constant and it results in a marked decrease in potential development. These data have been elucidated that two counter-acting processes happened, the first one being a breaking of the electrode surface adsorbed layer, It can be seen that carbon steel behavior either with and without inhibitor addition, and as a result delayed-action corrosion takes place moving the E_{OC} to more negative data till constancy. The steady-state potential proceeds towards less negative values with raising ILs concentrations. These potential data clarify that ILs having high efficiency as corrosion inhibitors. The variation of the steady state potential E_{st} with the logarithm of the molar concentration of ILs is shown in figure 11, whereby Est. varies with log C according to the following equation:

$$E_{st} = a_1 + b_1 \log C_{inh} \tag{6}$$

Where, a_1 and b_1 are constants depending on the nature of the metal and type of anions used.

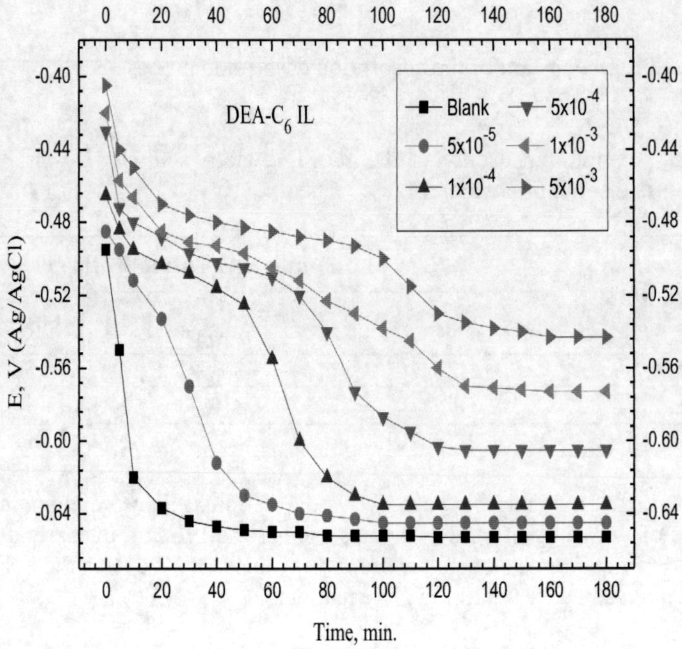

Figure 8. Plot of potential-time curves for carbon steel in different concentrations of DEA-C_6 IL.

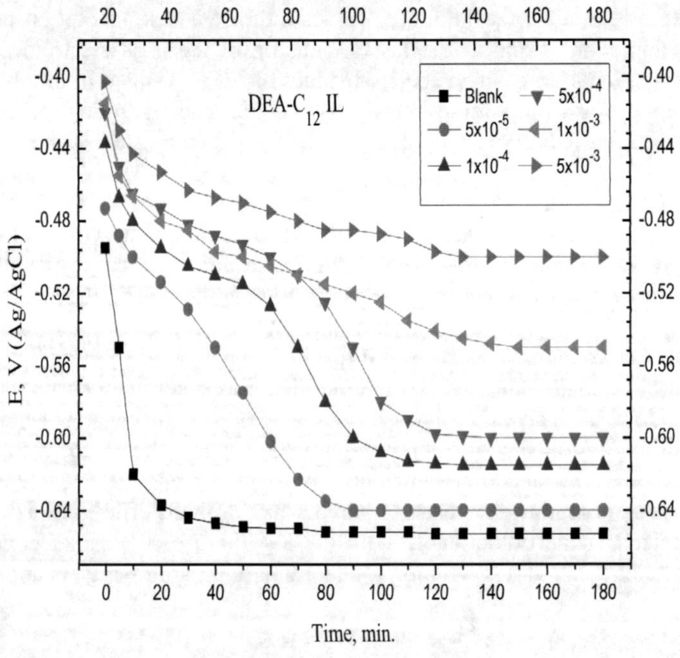

Figure 9. Plot of potential-time curves for carbon steel in different concentrations of DEA-C_{12} IL.

IOP Conf. Series: Materials Science and Engineering **975** (2020) 012014 doi:10.1088/1757-899X/975/1/012014

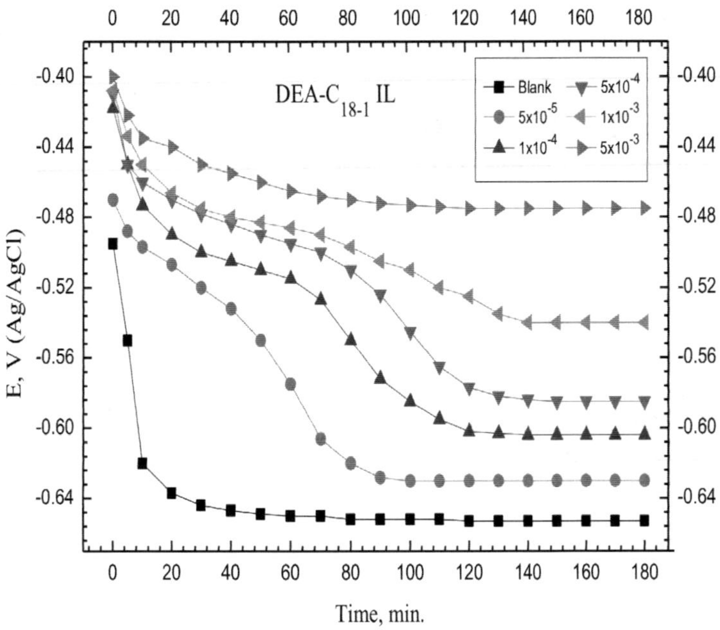

Figure 10. Plot of potential-time curves for carbon steel in different concentrations of DEA-C_{18-1} IL.

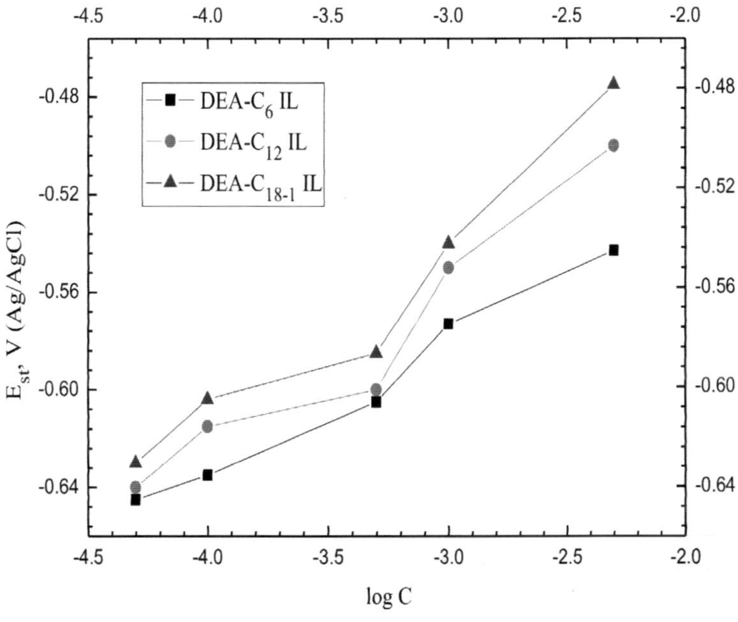

Figure 11. The steady-state potential, E_{st}, as a function of log C for carbon steel in DEA-C_6, DEA-C_{12} and DEA-C_{18-1} ILs.

IOP Conf. Series: Materials Science and Engineering **975** (2020) 012014 doi:10.1088/1757-899X/975/1/012014

From the equation (6), it can be described a general phenomenon, regardless of whether the anodic reaction results in film destruction or film repair which was derived theoretically [21] on the basis of the specific adsorption of the anions on the surface of the metal or the metal oxide. The rate of oxide film destruction for carbon steel in absence and presence of DEA-C_6, DEA-C_{12} and DEA-$C_{18\text{-}1}$ ILs was concluded from the linear relationship between the open circuit potential of the C-steel and the logarithm of immersion time, t, for DEA-C_{12} IL as in figure 12, and the relation:

$$E = a_2 - b_2 \log t \tag{7}$$

Where a_2, and b_2 are constants for oxide film destruction. It has been suggested that the driving force for surface oxide film destruction or formation on the metal is the free energy change of the reaction between the metal and the test solution. This process is assumed to proceed via solid state mechanism by the influence of a sufficiently high electric field to cause ionic migration and hence sustain self-diffusion of metal ions through semiconducting oxide films [22]. Based on this notion a relationship between E and log t has been theorized [23] as given by the following equation:

$$E = \text{const.} + 2.303\ \delta/\ \beta \log t \tag{8}$$

Where δ represents the rate of oxide film thickening or destruction per unit decade of time, and β is a constant which is identified as [23, 24]:

$$\beta = (nF/RT)\ \alpha\ \delta \tag{9}$$

Where α is the transfer coefficient similar to that encountered in normal electrochemical reactions [23, 24], $(0 < \alpha < 1)$ and δ is the width of the energy barrier surmounted by the ion during transfer.

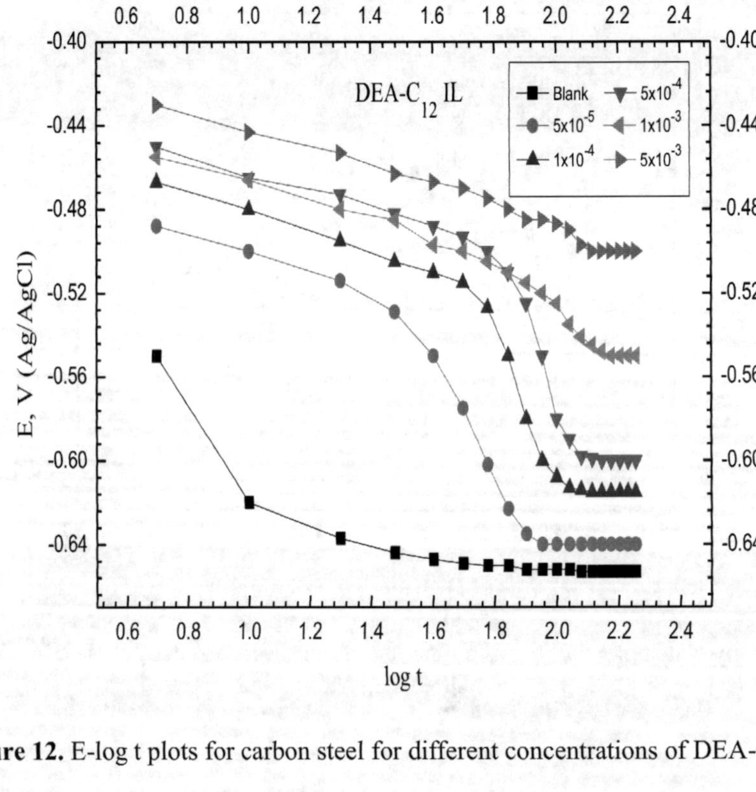

Figure 12. E-log t plots for carbon steel for different concentrations of DEA-C_{12} IL.

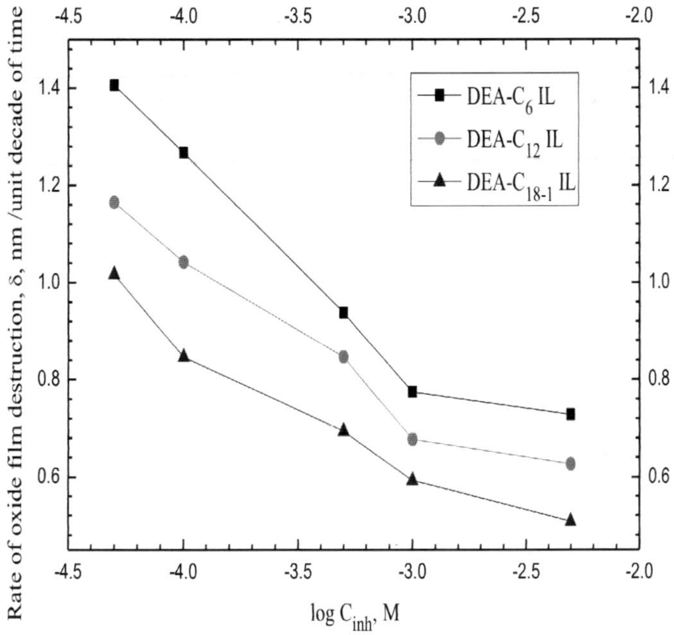

Figure 13. Rate of oxide film destruction, δ, as a function of log C for carbon steel in DEA-C_6, DEA-C_{12} and DEA-C_{18-1} ILs with 50% oilfield formation water.

Table 6. Rate of O-destruction, δ (nm/unit decade of time) and the inhibition efficiency, IE (%) for potential time curves.

C, M	DEA-C_6 IL		DEA-C_{12} IL		DEA-C_{18-1} IL	
	δ	IE	δ	IE	δ	IE
Blank	2.44	-	2.44	-	2.44	-
5x10⁻⁵	1.41	42.52	1.16	52.38	1.01	58.45
1x10⁻⁴	1.26	48.2	1.04	57.41	0.84	65.37
5x10⁻⁴	0.94	61.63	0.84	65.37	0.69	71.61
1x10⁻³	0.77	68.35	0.67	72.3	0.59	75.76
5x10⁻³	0.73	70.22	0.62	74.38	0.5	79.22

From the values of the slopes of the straight lines of figure 12, as in DEA-C12 IL, relating the variation of E with log t, one readily calculate the values of the rate of oxide destruction, δ in nm/unit decade of time to different concentrations of DEA-C6, DEA-C12 and DEA-C18-1 ILs as shown in figure 13. It is found that the rate of O-destruction decreases with increasing the concentration of all ionic liquids inhibitors, as demonstrated in table 6. Also, the inhibition efficiency is calculated from the equation:

$$IE = [1 - (\delta inh/\delta o)] \times 100 \quad (10)$$

where δo and δinh are the rates of oxide film destruction of the passive oxide film on the carbon steel surface in absence and presence of inhibitor, respectively. The values of the percentage inhibition efficiency, IE, are computed as a function of ionic liquids concentrations and shown in figure 14. It increases of the order DEA-C18-1 ILs > DEA-C12 ILs> DEA-C6 ILs.

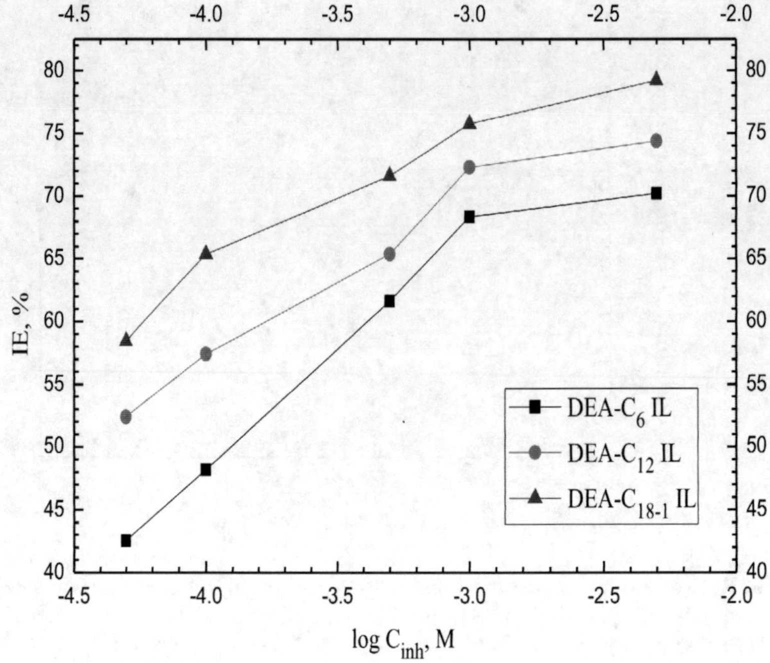

Figure 14. Inhibition efficiency, IE, for carbon steel in DEA-C_6, DEA-C_{12} and DEA-C_{18-1} ILs with 50% oilfield formation water.

3.3. Mechanism of Corrosion Inhibition

Literature survey reveals that several synthetic ionic liquids have been used as effective corrosion inhibitors for mild and carbon steel in various electrolytic media [20,25]. Two generic species DEA molecules in combination with fatty acids have been tested according to this methodology as ionic liquids. The ionic liquid under taken in the study inhibits metallic corrosion by adsorbing on the surface as resulted previously which the mechanism obeyed Langmuir adsorption isotherm [20]. This suggests that the metal surface exposed to the aggressive formation water is positively charged. The positively charged surface can attract Cl^- and S^{2-} ions, which in turn will attract DEA molecules species through electrostatic interactions. This causes the adsorption of a DEA species on the metal surface, which prevents carbon steel to ionized from entering the aggressive formation water. The fatty acid ionic liquids as lubricants interact with steel surface under the stress and prepare the thin film of low shear-strength, which reduces the friction and protect the surface against the undesirable wear [9]. The syntheses of DEA-C_6, DEA-C_{12} and DEA-C_{18-1} ionic liquids are confirmed by FTIR and NMR analyses. The viscosity of these ionic liquids increases with increasing the chain length of fatty acid anions and was attributed to increasing van der Waals interaction between the associated methylene units. The introduction of unsaturation sites in fatty acid anions significantly diminished the van der Waals interaction owing to steric constraint and reduced the viscosity. The inherent polar nature of fatty acid anions facilitates their interaction promptly with steel surface under the boundary lubrication and forms the stable chemical thin film of ionic liquids which reduces both the friction and the wear. Further, significantly low cost, environmentally-friendly nature, renewable and sustainable resources of fatty acid precursors promise immense potential of these ionic liquids for various applications.

4. Conclusions

The investigated ionic liquids of DEA-C_6, DEA-C_{12} and DEA-C_{18-1} showed excellent alternative corrosion inhibition with eco-friendly properties for the protection of carbon steel in 50% oilfield formation water, which was confirmed by wt. loss and potential time curves.

The inhibition mechanism was attributed to the strong adsorption ability of this ionic liquid to form a protective layer on the metal surface, reducing the active sites and the steel dissolution at the anodic sites by the aggressive environment.

Prepared ionic liquids gave a highly inhibition efficiency at rather low concentrations (i.e. $5x10^{-5}$).

This was further confirmed by several parameters which calculated in this study as K_{ads}, ΔG_{ads} and δ values.

The performance of corrosion inhibition of two techniques is in agreement with each other ranged of the order: DEA-C_{18-1} ILs > DEA-C_{12} ILs> DEA-C_6 ILs.

5. References

[1] Revie R W and Uhling H H, Corrosion and corrosion control: Wiley 4th Eds., (2007) 1-3.
[2] Masadeh S and Miner J, Mater. Character. Eng., 3 (2015) 271-276.
[3] Panchenko Y M and Marshakov A I, Corros. Sci., 109 (2016) 217-229.
[4] Deyab M A, Inhibition activity of Seaweed extract for mild carbon steel corrosion in saline Produced oilfield water, Desalination, 384 (2016) 60-67.
[5] Yuan J, Giordano C and Antonietti M, Chem. Mater., 22 (2010) 5003-5012.
[6] Carneiro J, Tedim J, Fernandes S C M, Freire C S R, Gandini A, Ferreira M G S and Zheludkevich M L, Surf. Coat. Tech., 226 (2013) 51-59.
[7] Ferraz R, Branco L C, Prudencio C, Noronha J P and Petrovski Z, Ionic Liquids as Active Pharmaceutical Ingredients, Chem. Med. Chem., 6 (2011) 975-985.
[8] Tsuda T and Hussey C L, Electrochemical applications of room-temperature ionic liquids, Electrochemical Society Interface, 16 (1), (2007) 42-49.
[9] Rashi Gusain and Khatri Om P, Royal Society of Chemistry Advances, 6 (2016) 3462-3469.
[10] Ayman M Atta, Gamal A El-Mahdy, Hamad A Al-Lohedan and Abdel Rahman O Ezzat, Molecules, 20 (2015) 11131–11153.
[11] Abdallah M, Corros. Sci., 46 (2004) 1981.
[12] Li X, Deng S, Fu H and Mu G, Inhibition effect of 6-benzylaminopurine on the corrosion of cold rolled steel in H_2SO_4 solution, Corros. Sci., 51 (2009) 620-634.
[13] Bakkar A and Ataya S, Corros. Sci., 85 (2014) 343-351.
[14] Pavithra M K, Venkatesha T V, Punith Kumar M K and Tondan H C, Corros. Sci., 60 (2012) 104-111.
[15] Obot I B, Macdonalda D D and Gasem Z M, Corros. Sci., 99 (2015) 1-30.
[16] Popova A, Christov M and Zwetanova A, Corros. Sci., 49 (2007) 2131-2143.
[17] Alaoui Mouayd A, Orazem M E, Sutter E M M, Tribollet B and Koltsov A, Corros. Sci., 82 (2014) 362-368.
[18] Popova A, Christov M and Vasilev A, Corros. Sci., 94 (2015) 70-78.
[19] Finšgar M and Jackson J, Corros. Sci., 86 (2014) 17-41.
[20] Yesudass S, Lukman O Olasunkanmi, Indra Bahadur, Mwadham M Kabanda , Obot I B and Eno E Ebenso, Journal of the Taiwan Institute of Chemical Engineers, 64 (2016) 252-268.
[21] Abd El Kader J M and Shams El Din A M, Br. Corros. J., 14 (1979) 40-45.
[22] Young L, Anodic Oxide Films, Academic Press, London, (1961).
[23] Abd El-Kader J M, Abd El-Wahab F M, El-Shayeb H A and Khedr M G A, Br. Corros. J., 16 (1981) 111-114.
[24] El-Taib Heakal F, Ghoneim A. A, Mogoda A S and Awad Kh M, Corros. Sci., 53 (2011) 2728-2737.
[25] Sami Ben Aoun, Der Pharma Chemica, 5 (3), (2013) 294-304.

Thermal and potentiometric studies for complexes of Zr and Hf with some barbituric acid derivatives

M S Masoud [1], A S Hassan[1], N M Desouky[1], H M kamel [2], A S El-Kholany*[2]

[1] Chemistry Department, Faculty of Science, Alexandria University, Egypt.
[2] Medical Laboratory Department, Faculty of Allied Medical Sciences, Pharos University in Alexandria, Egypt. E-mail address: elkholany73@yahoo.com

Abstract. The complexes of Barbital (Bal), Thiobarbituric acid thione (TBA), Thiobarbituric acid thiol (2-TBA), 2-Thiouracil (TU) with Zr (IV) and Hf (IV) were prepared. The structures and mode of bonding were characterized by elemental analysis, IR and electronic spectroscopy. Thermal analyses of the prepared complexes were done by DTA, TG, DSC techniques. The thermodynamic parameters were evaluated. The change of entropy values, $\Delta S^{\#}$, for all the complexes lie within the range -0.267,-0.299 KjK^{-1} mol^{-1}. The thermal processes proceed in complicated mechanisms with ordered transition states. In general, the complexes dissociated after losing small molecules such as H_2O, Cl_2 and CO. The decomposition is usually ended by the metal moiety. The DSC technique explained the effect of temperature on physical properties of some complexes. The glass transition phase doesn`t appear in all the complexes, while the crystallization temperature (Tc) was 78^0C-322.12^0C. Potentiometric titration studies calculated the dissociation constants of the ligands, the stability constants of the complexes and the distribution of species at different pH values.
Keywords. IR, DTA, TGA, DSC, Potentiometric titration, Complexes, Zr, Hf, Barbituric, Acid Derivatives.

1. Introduction

Pyrimidine ring has an important role in several biological systems, where they exist in nucleic acids, many vitamins, coenzymes, and antibiotics [1-2]. The nucleic acid is related to antimetabolites used in anticarcinogenic chemotherapy [3]. Metal complexes of pyrimidine are extensively studied for their great verity of biological activity ranging from antitumoral, antiviral, antibacterial, antimalarial, chemotherapeutic activity, etc. [4-5]. Furthermore, pyrimidines are reported as corrosion inhibitors for steel [6-7].

Barbiturates and thiobarbiturates represent important classes of medicinal compounds. Barbituric acid or 2,4,6(1H,3H,5H)-pyrimidinetrione is the lead compound of all barbiturates[1]. The synthesis of simple barbituric or thiobarbituric acids involves the condensation of malonic acid esters with thiourea or urea were used in pharmaceutical, these compounds continue to attract attention due to the wide range of biological activities. The use of compounds includes sedatives, hypnotics and antiepileptic drugs [8]. Furthermore, barbiturates and thiobarbiturates are reported to have antibacterial, antiurease and antioxidant activities[9]. In addition, some derivatives bearing the pyrimidine-2, 4, 6-trione unit have a potential therapeutic effect against diet-induced non-alcoholic fatty liver disease [10]. Some compounds include a thiobarbiturate unit that is used as inhibitors [11].The substitution on the pyrimidine-2,4,6-trione moiety is the effective pharmacophore in several different compounds

IOP Conf. Series: Materials Science and Engineering **975** (2020) 012015 doi:10.1088/1757-899X/975/1/012015

showing antitumor, anti-invasive and antiangiogenic effects [12], and also as promising compounds for amyotrophic lateral sclerosis [13]. Pyrimidine-2,4,6-triones attached to aryl hydrazones have been reported to be Ribosomal S6 Kinase 2 (RSK-2) inhibitors [14].

The thiobarbituric acid consists of a pyrimidine cyclic structure. These compounds have been described as privileged structures, as they provide various points of attachment for a diverse array of structural elements that can be used to target receptor agonists or antagonists owing to the versatile these compounds are more often used for the mankind Most of the thiobarbiturate derivatives are bilogicaly active and have many applications in agrochemicals further as pharmaceutical like herbicides, sedative, fungicidal, antiviral agents antioxidant, anti-inflammatory, antidepressant, antibacterial and antitumor, ,etc. [14-15]. The pyrimidine nucleus itself is unreactive toward electrophilic substitution reactions because of the -I and -M effects of the two nitrogen atoms while 2-thiouracil nucleus (1) is active because of the electron releasing effect of -OH and -SH groups[16-17]. Thus, 2-thiouracil could be chloro-sulphonated in a good yield [18-19].

Numerous inorganic and organometallic complexes of Zr have been described with zircon ($ZrSiO_4$), being its most widely recognized inorganic form [20-23]. Zirconium can exist in several oxidation states including (II), (III) and (IV), which is its preferred oxidation state [24]. Zirconium (II) complexes are known, but they typically require p-donor ligands to enhance stability even under inert atmosphere conditions, and even fewer reports describing the Zr (III) oxidation state exist [24].

The synthesis and characterization of a series of five heteroleptic hafnium and zirconium complexes bearing chloro ligands as well as differently substituted phenoxy benzoxazole ligands were reported [25]. All complexes had deen analyzed with regard to their structural chemistry. Moreover, their emission properties were studied, including emission spectra and excited-state lifetimes.

The original intention for this study was to introduce heavy atoms, which are in principle capable of influencing the excited state transitions as has been demonstrated earlier for so-called triplet-harvesting complexes including examples such as tris (pyridylphenyl) iridium (II) $Ir(ppy)_3$ [26], osmium(II) polypyridine complexes or a range of platinum complexes with extended heterocyclic ligands [27-29]. In our laboratory, Masoud et al reported a detailed structural chemistry, thermal and potentiometric studies of some nucleic acid complexes [30-40].

2. Experimental

2.1. General method for the preparation of the metal complexes

Ligands used are barbituric acid, barbiatal, thiobarbituric acid (thiol and thione) and 2-thiouracil, to prepare zircoium and hafnium complexes. Ligand was dissolved in hot distilled water, metal chloride was dissolved in ethanol added 1 with shaking for 5-7 min and a complex started to precipitate. Complexes were left 2 days to complete precipitation. The precipitated product was filtered off, washed with water and dried in vacuum at 40 °C. The analytical data are collected in table 1.

2.1.1Analysis.

a) Chloride analysis: is determined by Mohr and Volhard methods
b) N and S analysis: These are determined at the micro analytical data center at Cairo University.
c) Metal ion analysis: The metal ion contents were determined successfully gravimetrically.

2.1.2 Preparation of solutions:

a) An accurate CO_2-free 0.01 M solution of NaOH was prepared by diluting standard 1.0 M solution. The exact concentration of NaOH was determined by titrating against standard potassium hydrogen phthalate. Standard NaOH solution was analyzed from time to time. The solution was preserved in a waxed bottle fitted with a $CaCl_2$ tube.
b) 0.1 M solution of KCI was prepared to adjust the ionic strength of solutions.

2.2. Instruments and procedures:

2.2.1. Infrared spectra.

The KBr disc infrared spectra of the ligands and their metal complexes are measured over the frequency range 500-4000 cm^{-1} using BRUKER TENSOR 37 FT-IR spectrophotometer. Calibration of frequency readings is made with polystyrene film (1602±1 cm^{-1}).

Table 1. Analytical data of the prepared complexes

Complex	Reaction	N		S		Metal		Cl	
		Calc.	Found	Calc.	Found	Calc.	Found	Calc.	Found
HfCl.2Bal	$HfCl_4$ + Bal(1:1)	9.6	10.1	---	---	30.6	30.6	6.1	6.9
ZrCl.Bal	$ZrCl_4$ + Bal(1:1)	9.0	9.6	---	---	29.3	30.1	11.4	11.5
[Hf.TU.H_2O]Cl_2	$HfCl_4$ + TU(1:1)	7.1	7. 5	8.1	8.5	---	---	17.9	17.7
[Zr.TU.H_2O]Cl_3	$ZrCl_4$ + TU(1:1)	8.1	8.3	9.3	9.7	26.5	26.6	31.0	31.1
[Hf.BA.H_2O]Cl_2	$HfCl_4$ + BA(1:1)	7.1	7.3	---	---	45.1	45.2	17.9	17.7
ZrCl.2BA	$ZrCl_4$ + BA(1:1)	14.6	15.1	---	---	23.8	24.1	9.2	9.4
$HfCl_2$.2BACl_2	$HfCl_4$ + BA(1:2)	14.7	15.1	---	---	23.4	24.2	9.3	9.5
[ZrCl.2BA.H_2O]Cl	$ZrCl_4$ + BA(1:2)	14.9	15.2	---	---	16.1	16.5	12.6	12.6
[Hf.2TU.2H_2O]Cl_2	$HfCl_4$ + TU(1:2)	10.2	10.8	11.4	11.7	31.5	31.7	12.7	12.6
[$ZrCl_4$.2TU]3H_2O	$ZrCl_4$ + TU(1:2)	6.7	7.1	7.7	7.8	21.9	22.5	34.2	34.9
[Hf.2TBA]Cl	$HfCl_4$ + TBA(1:2)	11. 2	11.3	12.8	12.9	35.5	36.6	7.0	6.2
$ZrCl_4$.2TBA	$ZrCl_4$ + TBA(1:2)	7.4	7.6	8.5	9.1	24.1	24.4	37.7	38.4
HfCl.2(2-TBA)	$HfCl_4$ + 2-TBA(1:2)	11.2	11.7	12.8	13.1	35.5	35.7	7.0	6.7
[Zr.2(2-TBA)]Cl	$ZrCl_4$ + 2-TBA(1:2)	13.5	13. 6	15.4	15.7	21.9	22.1	8.6	8.9

2.2.2 Thermal analysis.

Differential thermal analysis (DTA), the thermo-gravimetric analysis (TGA) and differential scanning calorimetry (DSC) are carried out for some complexes using SEIS STA PT 1000 at the micro analytical data center at Cairo University, Egypt. The rate of heating was 10°C/min.

2.2.3. Potentiometric measurements.

These were carried out using a Jenway-3505 pH-meter. The electrode system was calibrated before and after each serious of pH measurements under the same conditions using standard buffers of pH`s 4.0, 7.0 and 10. The titration cell consisted of water jacketed 150 ml vessel fitted with a polyethylene stopper in which appropriately located holes are present, one of them allowed the insertion of a micro-burette containing CO_2 free NaOH.

Another hole was used to insert the electrode. Also there are holes for the inlet and outlet of the nitrogen gas that was purified as follows: it was passed through a train of bubbles containing successively silver nitrate to get rid of any traces of oxygen, followed by copper sulphate solution and alkaline pyrogallol to get rid of the oxygen too. The gas was passed through a tube containing $CaCl_2$ to get rid of moisture and CO_2.

The potentiometric titration procedure to evaluate the dissociation constants of the organic compounds was carried out by introducing the appropriate volume of the organic compound into the titration cell followed by 5 ml KCl solution of an ionic strength 0.1 M, and completed to 50 ml with doubly distilled water. Before each titration the solution in the titration vessel was left for about 15 min to attain the desired constant temperature (25°C) which is controlled by using a thermostatic device. During the whole titration, purified nitrogen gas was slowly bubbled in the solution.

The same potentiometric titration experiments were applied for studying the complexes equilibria in 75%, 50% and 25% (V/V) ethanol-water. This was constructed as follows, the complex solution {2 ml 10^{-3} M metal chloride + 5 ml 10^{-3} M ligand + 5 ml 0.1 M KCl + (29, 19 and 9 ml) ethanol and completed to 50 ml by distilled water} was titrated against standard NaOH. The experiments were done using the same conditions of titrating the organic compounds.

Correction of the pH-readings with a correction term δ for mixed solvents (ethanol-water) calculated by the following equation: ($\delta = pH_x - pH_S$) Where pH_x and pH_s are the values of unknown and standard buffer solutions, δ = 0.28 for 25% (ethanol - water), δ= 0.35 for 50% (ethanol - water) and δ=0.45 for 75% (ethanol- water) [41].

3. Results and discussion

3.1. Infrared spectra

3.1.1. Infrared spectra of the Iigands.

The IR spectra of solid BA, TBA, 2-TBA, TU and Bal are aimed to be investigated to determine the assignment of the vibrational frequencies especially of the fundamental groups (C-O, N-H, C=S and OH), table 2. This facilitates the tracing of keto-enol tautomerism phenomena.

The ring in BA is significantly distorted from planarity; the methylene part of the ring has a boat configuration [42]. Thus, the symmetry of the isolated molecule is C_5 of C_{2V}. The latter is not possible in the solid because of the manner in which the molecules are linked by hydrogen bonds.

The solid BA contains pairs of molecules linked through two hydrogen bond bands, between the N-H groups in the 3 positions and carbonyls in the 2 position are in a ribbon structure through hydrogen bonds between the N-H groups in 1 position and carbonyls in the 4 position. The carbonyl in the 6 position is not involved in the hydrogen bonding [43-46]

TBA is a heterocyclic compound containing thioamide group -NH-C=S) can exist in either the thione (TBA) or the thiol (2-TBA) form [47]. In general, the protropic tautomers of hetero aromatic compounds comprise a mobile hydrogen atom to be moved from one site to another.

For TBA, the absence of ν S-H stretching at 2500 cm^{-1} and the presence of an absorption at 1546 cm^{-1} due to ν C-N suggests the thione (A) tautomer is more predominated in the solid rather than the thiol (B):

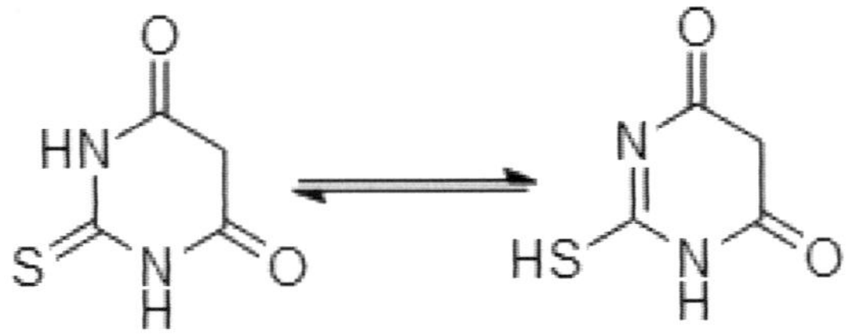

The strong band at 1297.9 cm^{-1} is due to δ O-H, the ν C=S and ν C-N are identified by the presence of strong bands at 1242.8 and 1163.7 cm^{-1} which refer to the existence of the following tautomers:

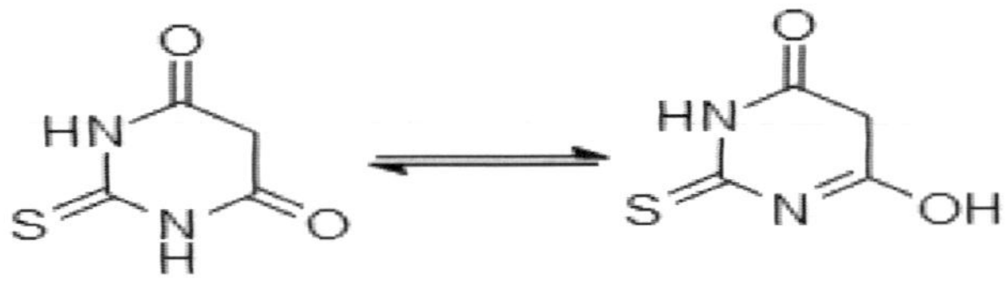

For (TBA), the disappearance of a well-defined bands at the range 3565-3450 cm^{-1} of a hydrogen bond of the type N-H.....O and appearance of a broad band at 3564.50 cm^{-1} of O-H type is an evidence of the existence of 2-thiobarbituric acid. The appearance of νS-H stretching at 2596.27 cm^{-1} and the band at 1567.53 cm^{-1} due to νC=N suggests the thiol tautomer.

IOP Conf. Series: Materials Science and Engineering **975** (2020) 012015 doi:10.1088/1757-899X/975/1/012015

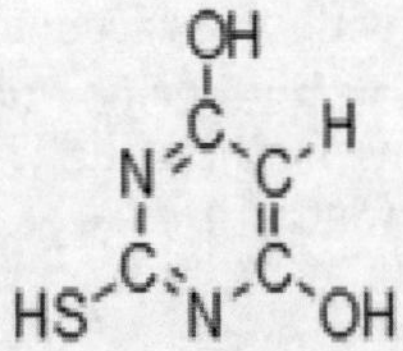

The vibrational spectrum for (TU) has been interpreted assuming planar geometry and C_s point group symmetry. The oxygen atom attached to C_4 takes part in the hydrogen-bonding of the type N-H....O, the S-atom does not participate in the hydrogen- bonding.

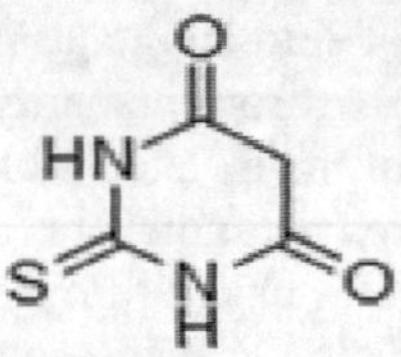

The presence of well-defined splitting three bands of stretching vibrations for barbital of the NH groups are present in the frequency range 3080.91- 3210 cm^{-1} and three CH stretching modes associated with the ethyl groups is located at the range 2984.43-2866.91 cm^{-1}

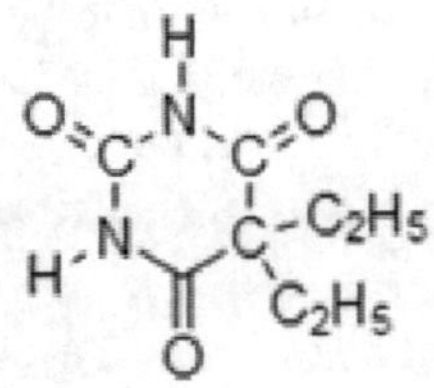

3.1.2. Infrared spectra of the prepared metal complexes.

It is necessary to investigate the infrared spectra of the complexes compared to that of the free ligand, Table (2), Figures (1-5) In general, the metal complexation may increase or decrease the vibrational frequency of the coordinated functional group depending on the strength of π-interaction occurring between the metal ion and the π-electron of the functional groups.

IR spectra suggest the participation of nitrogen and oxygen atoms on complexation. For [Hf.BA.H_2O]Cl_2 complex, the bands at 1413 and 1233 cm^{-1} are red shifted which represent C-N stretches and the bands at 3187-3099 cm^{-1} which represent NH groups symmetric and asymmetric vibrations are disappeared and the band at 1717 cm^{-1} is red shifted which suggest the contribution of the oxygen atoms on complexation, a very broad band at 3408 cm^{-1} has appeared which may represent the contribution of H_2O on complexation and/or hydrogen bonding. The 2Cl atoms are in the outer coordination sphere.

For $HfCl_2$.4BA and [ZrCl.3BA.H_2O)CI complexes, the atom band on at 3099 cm^{-1} is disappeared indicating the contribution of one nitrogen atom on complexation. The bands at 1619, 1413 and 1233 cm^{-1} are blue shifted slightly for ZrCl.2BA and $HfCl_2$.4BA complexes confirming the nitrogen atom involvement on complexation. For [ZrCL.3BA.H_2O]CI complex, the band at 3562 cm^{-1} is red shifted by 30 cm^{-1} which may indicate H_2O is involved on complexation and/ or existence of hydrogen bonds. The strong band at 1717 cm^{-1} in the free ligand assigned as C=O becomes of different shapes on complexation, i.e. still exists at 1717 cm^{-1} in a strong feature in [ZrCl.3BA.H_2O]Cl complex and in ZrCl.2BA and $HfCl_2$.4BA complexes in a broad feature. Cl atoms of [Zr.3BA.H_2O]Cl_2 complex are outer and inner the coordination sphere, but for ZrCl.2BA and $HfCl_2$.4BA complexes are in the inner sphere.

IOP Conf. Series: Materials Science and Engineering **975** (2020) 012015 doi:10.1088/1757-899X/975/1/012015

The [Hf.TU.H_2O]Cl_2, [Zr.TU.H_2O]Cl_3 and [Hf.2TU.3H_2O]Cl_2 complexes, gave IR bands at 1422 and 1564 cm^{-1} are blue and red shifted, respectively, which imply that nitrogen atom is contributed on complexation, the band at 1701 cm^{-1} is blue and red shifted which indicated the involvement of the oxygen atom on complexation. The band at 1213 cm^{-1} which represents C=S is red shifted to suggest the contribution of the sulphur atom on complexation. All Cl atoms are in the outer coordination sphere.

For [$ZrCl_4$.TU].3H_2O complex showed a very broad band at 3421.45 cm^{-1} which may represent the contribution of H_2O on complexation and/ or hydrogen bonding. The band at 1422 cm^{-1} is red shifted which indicates that the nitrogen atom is contributed on complexation, the band at 1701 cm^{-1} is blue shifted and may refer to the involvement of the oxygen atom on complexation.

The band at 3450 cm^{-1} is strongly red shifted by ~40 cm^{-1} for [Hf.2TBA]CI and $ZrCl_4$.TBA complexes, the band at 1163 cm^{-1} is blue shifted indicating the nitrogen atom contribution on complexation. The band at 1646 cm^{-1} (the average band at 1693, 1643 and 1603 cm^{-1}) is red shifted which suggests the involvement of the oxygen atom in complexes, the band at 1443 cm^{-1} is disappeared. Furthermore, for [Hf.2TBA]Cl complex the band at 1242 cm^{-1} is disappeared and the band at 3104 cm^{-1} is blue shifted, for $ZrCl_4$.TBA complex the band at 1242 cm^{-1} is red shifted and the band at 3186 cm^{-1} is disappeared.

The band at 3099 cm^{-1} is blue shifted for HfCl.2(2-TBA) and [Zr.2(2-TBA)]Cl complexes. The band at 1162 cm^{-1} is blue shifted indicating the nitrogen atom contribution on complexation. The band at 3565 cm^{-1} is strongly red shifted by ~170 cm^{-1} for the complexes and the band at 1663 cm^{-1} (the average band at 1720, 1650 and 1618 cm^{-1}) is strongly red shifted for HfCl.2(2-TBA) complex and blue shifted for [Zr.2(2-TBA)]CI complex to suggest the involvement of the oxygen atom in complexes. The band at 1437 cm^{-1} is red shifted. Furthermore, for HfCl.2(2-TBA) complex the band at1239 cm^{-1} is disappeared.

IR spectra suggest the participation of nitrogen and oxygen atoms on complexation. For HfCl.2Bal complex, the band at 1461 cm^{-1} is blue shifted representing C-N stretches and the bands at 3210- 3081 cm^{-1} representing NH groups symmetric and asymmetric vibrations are changed in their shape and red shifted by ~ 40 cm^{-1}. The bands at 1766 and 1719 cm^{-1} are blue shifted and the band at 1679 cm^{-1} is red shifted which suggest the contribution of the oxygen atoms on complexation. The Cl atom is in the inner coordination sphere.

For ZrCl.Bal complex, the band at 1461 cm^{-1} in Bal is red shifted representing C-N stretches, the bands at 1766 and 1719 cm^{-1} are blue shifted and the band at 1679 cm^{-1} is red shifted which refers to the contribution of the oxygen atoms on complexation. The Cl atom is in the inner coordination sphere. This complex gives very weak changes in IR spectra.

Table 2. Some fundamental Infrared bands

compound	νO-H	νN-H	νS-H	νC-H	νC=O	νC=N	νC-N	νC=S
BA	3562, 3477	3187, 3099	-----	2878	1749, 1717	1619	1413, 1233 1192	------
[Hf.BA.H_2O]Cl_2	3409	-------	-------	-------	1622	-------	1360, 1292	-------
ZrCl.2BA	3562, 3478	3188, 3099	------	2877	1750, 1717	1622	1416, 1236 1193	-------
$HfCl_2$. 4BA	3562, 3478	3189	------	2877	1717	1622	1416, 1236 1193	-------
[ZrCl.3BA.H_2O]Cl	3533, 3477	3188	-------	2880	1749, 1718	1620	1414, 1234 1193	------
Bal	-------	3210, 3166 3081	-----	2984, 2943 2867	1766, 1719 1679	-----	1460, 1324 1241	------
HfCl.2Bal	-------	3249, 3087	-------	2973, 2946 2878	1769, 1722 1676	-------	1463, 1324 1238	-------
ZrCl.Bal	-------	3210, 3167 3080	-------	2985, 2945 2865	1767, 1720 1678	-------	1460, 1324 1240	-------
TBA	3565, 3450	3187, 3105	-----	2920, 2877	1693, 1643 1604	1546	1164	1443 1242
[Hf.2TBA]Cl	3408	3189, 3110	-------	2887	1630	-------	1186	------
$ZrCl_4$.TBA	3601, 3411	3104	-------	2880	1626	-------	1190	1226
2-TBA	3565	3099	2596	2872	1720, 1650 1618	1568	1162	1437 1239
HfCl.2(2-TBA)	3383	3103	-------	2874	1618	-------	1165	1406
[Zr.2(2-TBA)]Cl	3396	3101	2598	2873	1721, 1651 1619	1568	1163	1431 1239
TU	3490	3084, 3047	------	2926	1701	1565	1423	1213
[Hf.TU.H_2O]Cl_2	3463	3084, 3047	-------	2928	1705	1563	1421	1212
[Zr.TU.H_2O]Cl_3	3484	3082, 3046	-------	2926	1699	1562	1421	1210
[Hf.2TU.3H_2O]Cl_2	3485	3081, 3044	-------	2924	1696	1562	1419	1210
[$ZrCl_4$.TU]3H_2O	3421	3086	-------	2930	1705	1564	1421	1213

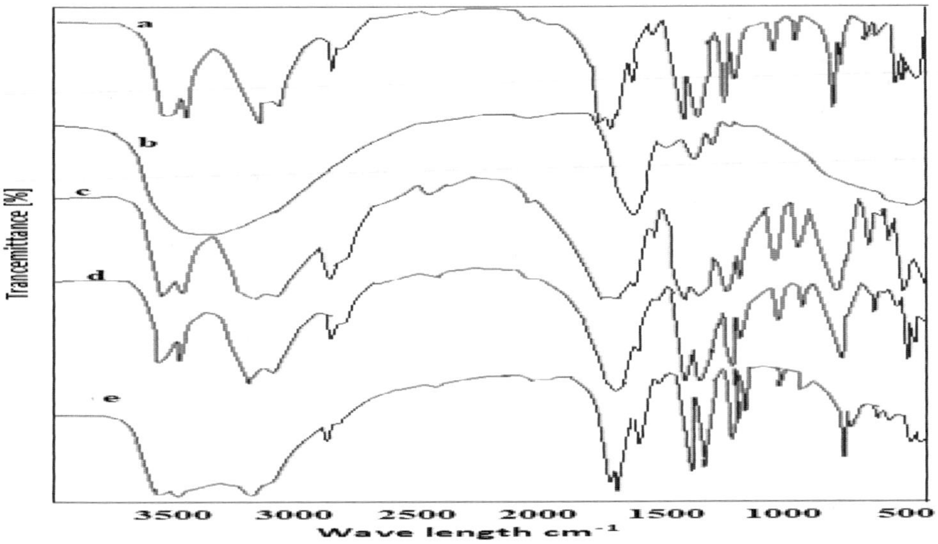

Figure 1. Infra-Red spectra of Barbituric acid (BA) and its complexes
a)BA b) [Hf.BA.H_2O]Cl_2 c) HfCl2. 4BA
d) ZrCl.2BA e) [ZrCl.3BA.H_2O]Cl

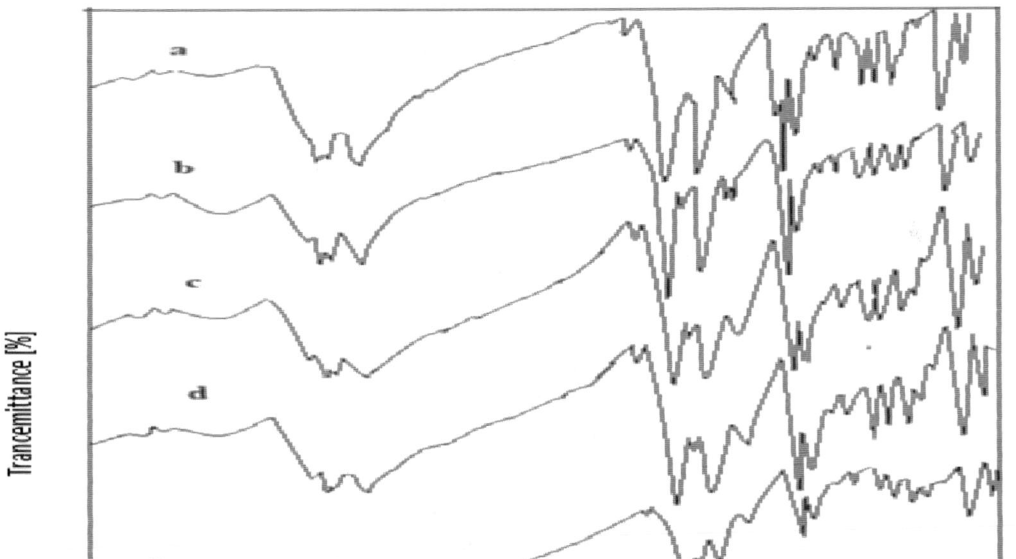

Figure 2. Infra-Red spectra of Thiouracil (TU) and its complexes:
a)TU b) [Hf.TU.H_2O]Cl_2 c) [Hf.2TU.3H_2O]Cl_2
d) [Zr.TU.H_2O]Cl_3 e) [$ZrCl_4$.TU]3H_2O

IOP Conf. Series: Materials Science and Engineering **975** (2020) 012015 doi:10.1088/1757-899X/975/1/012015

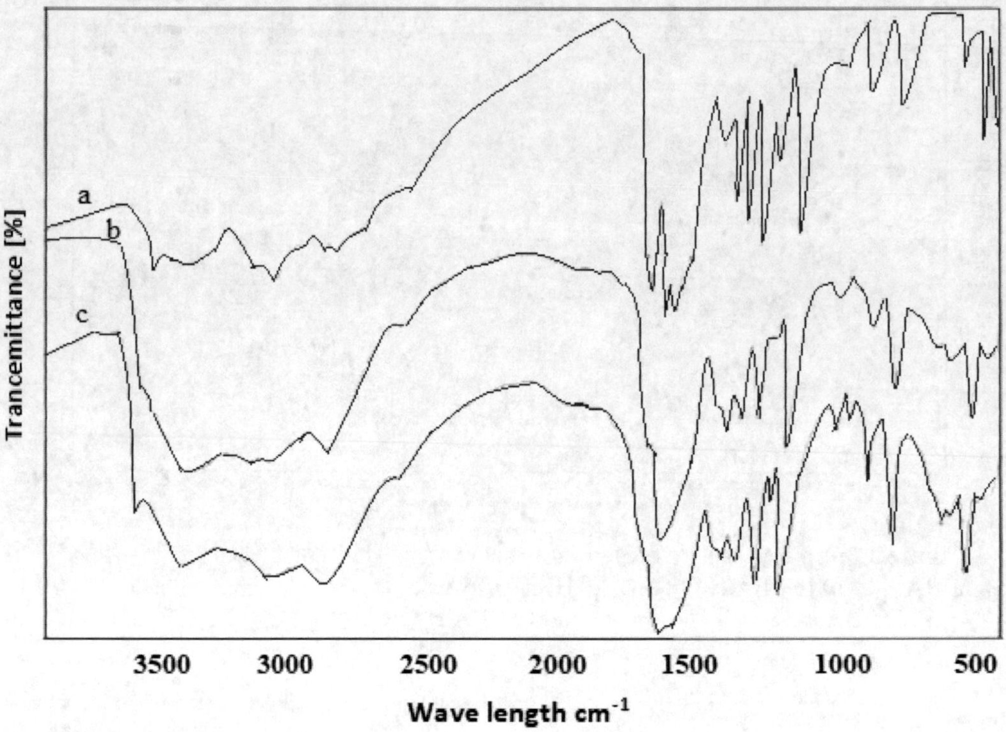

Figure 3. Infra-Red spectra of Thio barbituric acid(TBA) and its complexes:
a)TBA b) [Hf.2TBA]Cl c) $ZrCl_4$.TBA

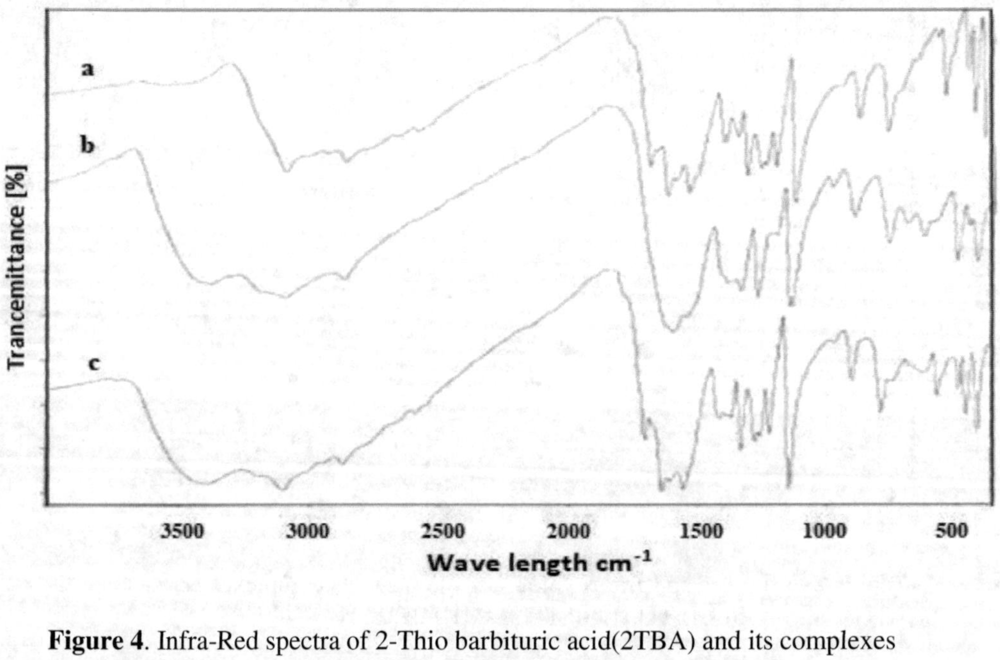

Figure 4. Infra-Red spectra of 2-Thio barbituric acid(2TBA) and its complexes
a) 2-TBA b) HfCl.2(2-TBA) c) [Zr.2(2-TBA)]Cl

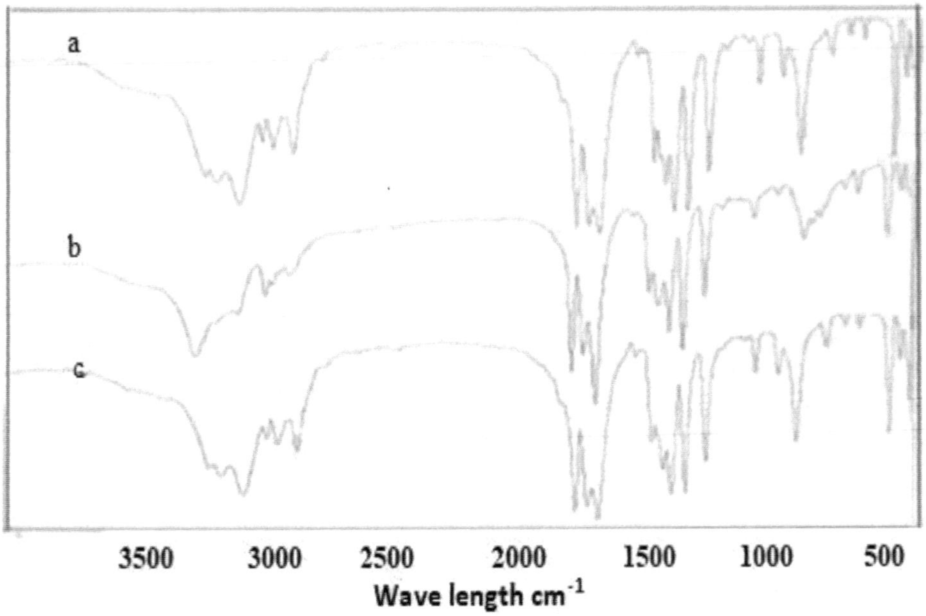

Figure 5. Infra-Red spectra of Barbital (Bal) and its complexes:
a)Bal b) HfCl.2Bal c) ZrCl.Bal

3.2. Thermal analysis

3.2.1 DTA and TGA.

The order of the reaction process (first-order, second-order, .. etc.) will be determined from the symmetry of the DTA curve. The asymmetry of the peak is simply a/b and also the reaction order is calculated as follows:

$$S = 0.63\ n^2 \quad \text{and} \quad n = 1.26\ (a/b)^{1/2} \tag{1}$$

The values of collision factor, Z, is obtained in case of Horowitz Metzger from the relation

$$z = \frac{\Delta E_a}{RT_m}\beta \exp\left(\frac{\Delta E_a}{RT_m^2}\right) \tag{2}$$

Also, the entropies of activation ($\Delta S^{\#}$) can be obtained from the relation:

$$z = \frac{kT_m}{h}\exp\left(\frac{\Delta S^{*}}{R}\right) \tag{3}$$

Where, R represents molar gas constant, β rate of heating (KS^{-1}), k is the Boltzmann constant and h is the Planck's constant.

From the DTA curves the heat of transformation, $\Delta H^{\#}$, could be calculated. For any phase transformation taking place at any peak temperature, T_m, the change in enthalpy ($\Delta H^{\#}$) is obtained from the following equation:

$$\Delta S^{\#} = \Delta H^{\#} / T_m \tag{4}$$

IOP Conf. Series: Materials Science and Engineering **975** (2020) 012015 doi:10.1088/1757-899X/975/1/012015

Thermal studies (DTA, TGA and DSC) were carried out for some of the prepared complexes (Zr and Hf), table 3, representative example for [ZrCl.3BA.H_2O]CI complex is given in figures 6 and 7.

The decomposition of the complex occurred in three steps. The first step started at 51.30 °C - 141.88 °C where the % loss was found to be 6.063% (calculated 6.317%) attributed to the loss of Cl atom in the outer sphere. In the second step between 141.88 °C - 304.16 °C, % loss of 76.852% was obtained due to fragmentation of H_2O and three barbituric moieties. The last step at 367.50°C - 513.07 °C is due to the loss of the second Cl atom leaving a residue of Zr moiety. The following scheme for the decomposition is suggested:

The DTA show two endothermic peaks with T_m = 74.45-118.37 °C. Two other endothermic peaks were obtained with T_m = 250.85 °C and 274.74°C attributed to the decomposition of the complex leaving Zr moiety as residue.

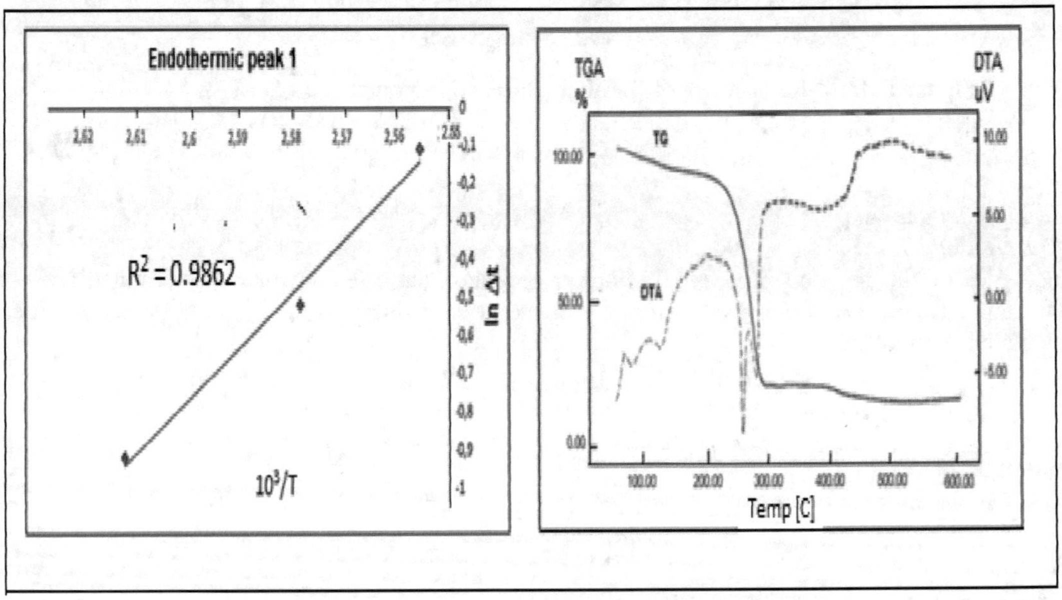

Figure 6. TG & DTA curves of [ZrCl.3BA.H_2O]CI complex

The TG and DTA curve of [Hf.2TBA]Cl complex, the decomposition occurred in three steps. The first step started at 42.15 °C-104.31 °C where the % loss was found to be 8.593% (calculated 7.121%) attributed to the loss of Cl atom in the outer sphere. In the second step between 104.31°C- 274.07°C, % loss of 31.254% (calculated 31.896%) was obtained due to fragmentation of one TBA moiety and O atom from the second ligand. The last step at 274.07°C-585.93°C is due to the loss of 2-TU leaving a residue of Hf moiety, (found 26.545%, calculated 25.276%).

The TG and DTA curves of HfCl.2(2-TBA) complex, it is clear that the complex showed a multistage mechanism. A step between 32.78 °C- 137.52 °C is due to the loss of CO molecule with % loss of (found 5.817%, calculated 5.6179%). The second step between 137.52°C-248.12°C is attributed to the loss of H_2O molecule and Cl atom (found I1.729%, calculated 10.732%). The third step between 248.12°C 404.81°C with a weight loss of found 19.855%, (calculated 20.060%) is due to the fragmentation of $C_3H_3N_2S$. The last step between 404.81°C-601.44°C, % weight loss of found 23.237%, (calculated 22.111%) is due to the loss of the rest of the complex ending with the formation of HfO_2.

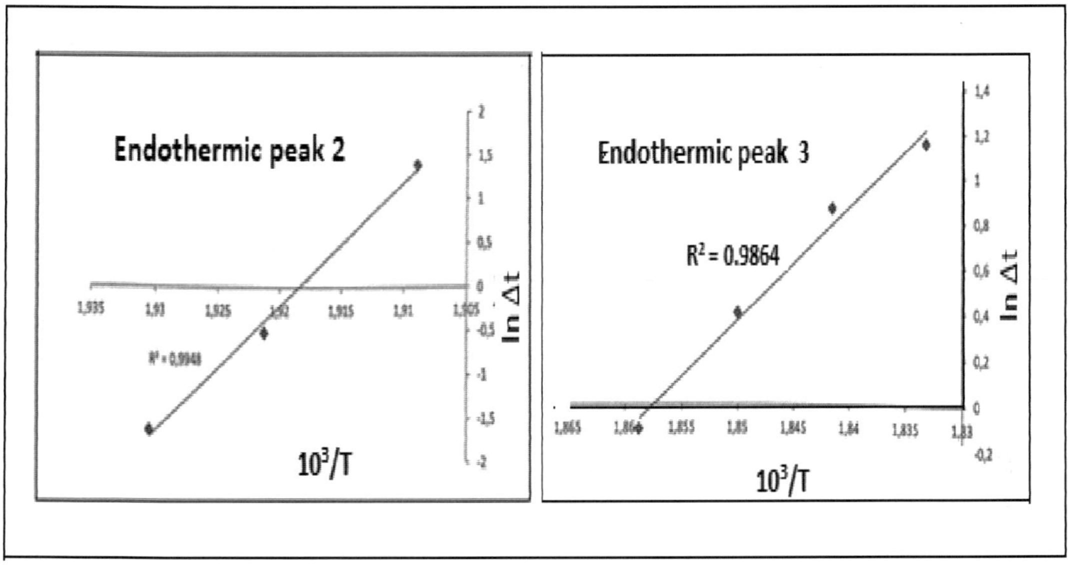

Figure 7. Plot constructed from DTA curves of [ZrCl.3BA.H_2O]CI complex

The TG and DTA curves of [Zr.2(2-TBA)]Cl complex is decomposed in a multistep form. The first step between 23.93°C - 134.80°C with % loss of (found 12.204%, calculated 12.758%) due to the loss of Cl atom in the outer sphere and O atom decomposed from one of the two ligands. The second step between 134.80°C - 273.39°C is assigned to the loss of one 2-TBA (found 34.324%, calculated 34.508%). The third step between 273.39 °C - 568.94°C is due to the loss of 2-TU leaving a residue of Zr atom moiety, (found 33.483%, calculated 30.620%).

Table 3. The DTA analysis data of the complexes

Complex	Type	a	b	Slope	T_m °K	ΔE_a (KJ mol^{-1})	n	α_m	$\Delta S^{\#}$ KJ K^{-1} mol^{-1}	$\Delta H^{\#}$ (KJ mol^{-1})	10^3Z (Sec^{-1})
[ZrCl.3BA.H_2O]Cl	Endo	1.4	0.9	-14.32	118.37	119.07	1.57	0.55	-0.27	-32.43	0.121
	Endo	0.7	0.6	-139.77	250.85	1162.08	1.36	0.57	-0.27	-67.11	0.555
	Endo	1.4	2.1	-50.22	274.74	417.49	1.03	0.63	-0.28	-76.25	0.183
[Hf.2TBA]Cl	Endo	0.7	1.1	-20.98	242.36	174.45	1.01	0.63	-0.28	-68.51	0.087
	Endo	1.2	0.7	-53.53	418.77	445.01	1.65	0.54	-0.28	-118.93	0.128
	Exo	2.4	2.1	-13.50	533.13	112.23	1.35	0.58	-0.30	-159.65	0.025
		2.4	2.1	-56.02	533.13	465.73	1.35	0.58	-0.29	-153.34	0.105
[$ZrCl_4$.TU]3H_2O	Endo	1.0	2.5	-17.21	310.95	143.10	0.80	0.67	-0.29	-89.70	0.055
HfCl.2(2-TBA)	Endo	1.2	0.6	-16.65	310.78	138.41	1.78	0.52	-0.29	-89.74	0.054
		1.2	0.6	-72.80	310.78	605.24	1.78	0.52	-0.28	-85.93	0.234
	Exo	2.1	2.0	-16.27	536.58	135.24	1.29	0.58	-0.30	-159.91	0.030
		2.1	2.0	-56.78	536.58	472.05	1.29	0.58	-0.29	-154.33	0.106
[Zr.2(2-TBA)]Cl	Endo	1.0	1.0	-42.40	251.40	352.51	1.26	0.59	-0.28	-69.75	0.168
	Exo	0.3	1.9	-93.70	482.58	779.05	0.50	0.75	-0.28	-135.94	0.194
		0.3	1.9	44.93	482.58	-373.51	0.50	0.75	-------	---------	-0.093
	Exo	2.6	1.3	-41.97	516.65	348.92	1.78	0.52	-0.29	-149.57	0.081

IOP Conf. Series: Materials Science and Engineering **975** (2020) 012015 doi:10.1088/1757-899X/975/1/012015

For the DTA of this complex, an endothermic peak with T_m= 251.40°C was obtained due to the loss of one of the two ligands and two exothermic peaks were obtained with T_m=482.58°C and T_m=516.65°C respectively. The first exothermic peak occurred with phase transition and the second exothermic peak is broad, verifying the decomposition of the complex in a multistep form.

The TG and DTA curves of [$ZrCl_4$.TU].3H_2O complex, the decomposition occurred in two steps. The first step started at 37.88°C-179.75°C where the % loss was found to be 6.064% attributed to the loss of 3H_2O molecules from the outer sphere. In the second step between 179.75°C-373.23°C, % loss of 82.88% is due to the loss of the rest of the complex leaving a residue of Zr moiety.

3.2.2 Differential scanning calorimetry (DSC)

DSC curves and physical properties obtained for the complexes are presented in figures 8-10 and table 4 and 5. The glass transition phase does not appear in all of the complexes studied. Above the glass transition, the samples have a lot of mobility. When they reach the correct temperature, they gain enough energy to convert to very ordered arrangements, i.e. crystals. When samples fall into these crystalline arrangements, they give off heat. However, this drop in the heat flows as a big dip in the plot of heat flow versus temperature. The temperature at the lowest point of the dip represents the samples crystallization temperature. Measuring the area of the dip, gives the latent energy of crystallization for the samples. Since samples give off heat when it crystallizes, crystallization is an exothermic transition.

The crystallization temperatures (T_c) of the complexes studied were around 78°C-322.12°C. All metal complexes measured have two crystallization temperatures except of [$ZrCl_4$.TU].3H_2O complex has one crystallization temperature. The heat flow is shown in units of heat (g) supplied per unit time (0). The heating rate is temperature increase (T) per unit time (t).

Heat capacity (C_p) can be determined by dividing the heat flow (q/t) by the heating rate (T/t). That is to say, heat supplied divided by the temperature increase. It is the heat absorbed by a closed system of constant composition on raising the temperature 1K.

The variation of C_p versus T can be represented using Debye model as the following relations: Where "a" is the slope of the line and "b" is the intersection of the line with y-axis (C_p axis). C_p is the specific heat at constant volume (γ) a constant equals 10-4 (cal/gram.mole.K^2).

By plotting C_p/T as y-axis and T^2 as x-axis, figure 9, a straight line with a slope equals α and the intersection with y-axis gives the coefficient (γ) is obtained, hence the Debye temperature can be determined.

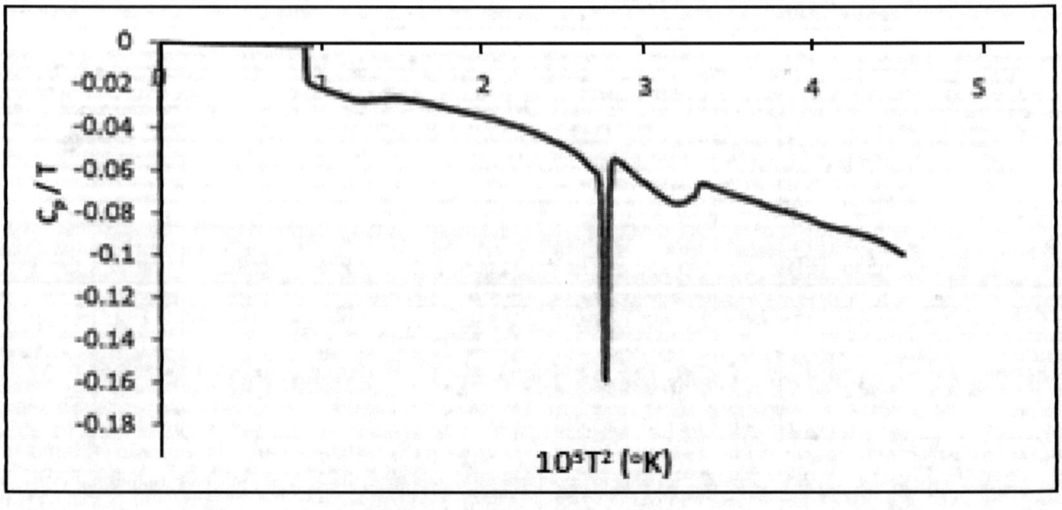

Figure 8. DSC curve for [ZrCl.3BA.H_2O]CI complex

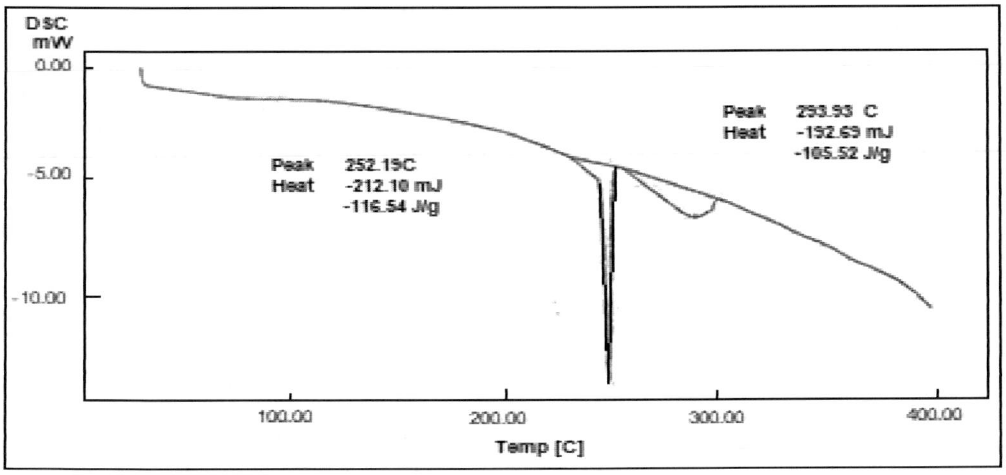

Figure 9. DSC curve C_p/T vs. T^2 for [ZrCl.3BA.H_2O]CI complex

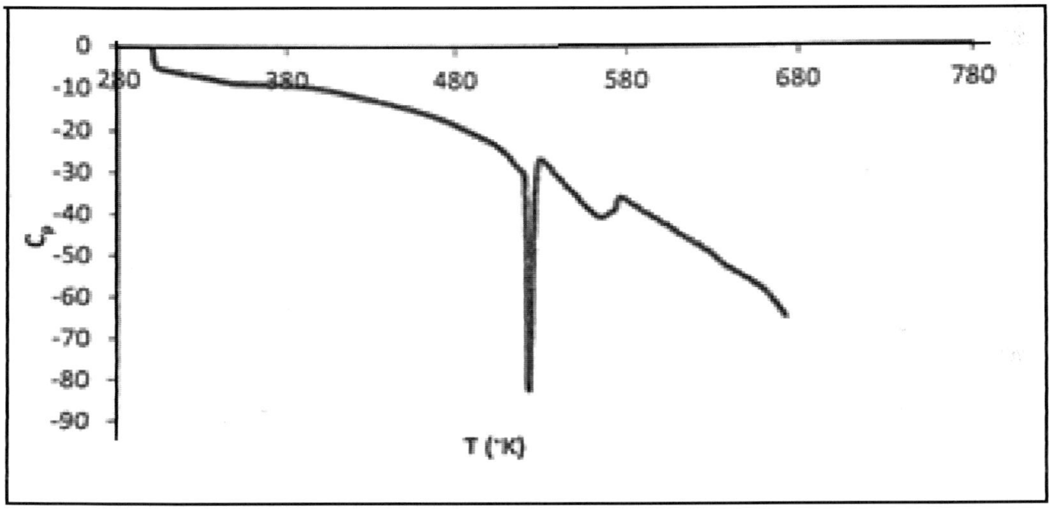

Figure 10. Dependence of C_p on T for [ZrCl.3BA.H_2O]CI complex

Table 4. Crystallization temperature of the complexes

Complex	Crystallization temperature (T_c) (°C)
[ZrCl.3BA.H_2O]Cl	252.19, 293.93
[$ZrCl_4$.TU].3H_2O	322.12
[Hf.2TBA]Cl	78.21, 242.69
HfCl.2(2-TBA)	80.82, 315.83
[Zr.2(2-TBA)]Cl	210.79, 241.84

Table 5. Variation of a, b, α and γ for the complexes according to Debye model

Complex	$C_p = aT + b$		$C_p/T = \alpha T^2 + \gamma$	
	a	b	$\alpha \times 10^{-7}$	γ
[ZrCl.3BA.H_2O]Cl	-0.0760	17.806	-2	0.0049
	-0.0207	-1.271	-6	0.1216
	-0.1166	37.922	-2	0.0205
	-0.4288	201.01		
	-0.2558	112.32		
[$ZrCl_4$.TU]$3H_2O$	-0.1065	26.866	-2	0.0096
	-0.1369	48.097	20	-0.7650
	1.2223	-792.60		
[Hf.2TBA]Cl	-0.2949	85.897	-10	0.1001
	-0.0608	13.103	-1	-0.0094
	-0.3063	130.64	-3	0.0274
	-0.1912	80.906	-2	0.0191
HfCl.2(2-TBA)	-0.1655	4.5050	-7	0.5170
	-0.0618	14.981	-1	-0.0082
	-0.1555	57.914	-3	0.0239
	-0.1549	59.584	-2	0.0115
	-0.4247	210.90	-6	0.1290
	0.1229	-109.14	2	-0.1454
	-0.2979	159.07	-3	0.0716
[Zr.2(2-TBA)]Cl	-0.1400	35.611	-5	0.0208
	0.0130	-19.180	2	-0.0626
	-0.0629	11.497	-0.7	-0.0239
	-0.6771	317.13	-10	0.2135
	-0.2169	91.859	-2	0.0154
	-0.3347	172.43	-3	0.0369

3.3. Potentiometric titration studies

The potentiometric studies in this work are divided into two parts as follows:

(1) Experiments for the free ligands in presence of 25%, 50% and 75% (v/v) of ethanol-water media at 25°C.

(2) Experiments of the complex solutions in presence of 25%, 50% and 75% (v/v) in ethanol-water media at 25°C.

3.3.1. Potentiometric studies of the free ligands in presence of different percentages of ethanol-water media.

Acid-base titration techniques were used to determine the acidity and stability constants. The Potentiometric measurements depends on the determination of the average number of the protons related to the reagent, $\overline{n}_A$. This was resolved at different pH's using the simplified following equation [48]

$$\overline{n}_A = Y - \frac{V_i N^\circ}{V_\circ C_A} \quad (5)$$

Where, V_i refers to the volume of alkali needed to achieve a given pH on the titration curve, Vo is that the initial volume of the ligand, N° is that the alkali concentration, C_A is that the total concentration of the ligand and Y is that the number of replaceable hydrogen atoms in the ligand. The dissociation constants have been obtained by plotting $\overline{n}_A$ versus pH for the free ligands. One pK's value has been obtained for ligands with Y=1 by recording the pH values at $\overline{n}_A = 0.5$ [49]. Where similar results have been obtained; the data were collected in table 6.

Table 6. pK values for the ligands in Ethanol-water media at 25°C (KCl =0.1M).

Ligand	$\overline{n}_A$ - pH			Point-wise method			Algeberic method		
	75%	50%	25%	75%	50%	25%	75%	50%	25%
Bal	7.98	7.89	7.85	8.00	7.89	7.88	8.01	7.92	7.89
TBA-Thione	8.17	7.11	6.05	8.02	6.95	6.12	8.06	6.98	6.09
TBA-Thiol	8.22	8.12	7.45	8.22	8.12	7.35	8.25	8.15	7.38
TU	8.85	8.44	9.35	8.86	8.40	9.32	8.89	8.44	9.36

$$pH + log\frac{\overline{n}}{1-\overline{n}} = \log K_1\,,\ \overline{n} < 1 \tag{6}$$

Algebraic method has been applied to calculate the dissociation constants of ligands where the equilibrium involved as follows [50].

$$H_2A \leftrightarrow H^+ + HA^- \qquad k_1 = [H^+][HA^-]/[H_2A] \tag{7}$$

$$HA^- \leftrightarrow H^+ + A^{-2} \qquad k_2 = [H^+][A^{-2}]/[HA^-] \tag{8}$$

H_2A and parentheses represent the ligands and the molar concentrations. Since K1>K2, every dissociation stage was considered separately.
Where: C_A represents the total concentration of the ligand species and "a" represents the number of moles of base that added per mole of the ligand present. It follows of the organic co-solvent in the low pH buffer region:

$$C_A = [H_2A] + [HA^-] \tag{9}$$

$$aC_A + [H^+] = [HA^-] + [A^{2-}] \tag{10}$$

$$K_1 = \frac{[H^+](aC_A + [H^+])}{C_A - (aC_A + [H^+])} \tag{11}$$

In the high pH buffer region, the concentration of the acid from (H_2A) of the ligand is neglected and so:

$$K_2 = \frac{[H^+]((a-1)C_A - [OH^-])}{(2-a)C_A - [OH^-]} \tag{12}$$

All ligands in different percentages of ethanol-water media gave one pK's value based on potentiometric measurements indicating that different mechanisms are produced:

(1) Tautomerism produced of CH-CO skeleton.
(2) Tautomerism produced of NH-CO skeleton.
(3) Tautomerism produced of NH-CS skeleton.

IOP Conf. Series: Materials Science and Engineering **975** (2020) 012015 doi:10.1088/1757-899X/975/1/012015

The distinction within the pK value of the barbital to that of thiobarbituric skeleton because of that sulphur is less electronegative than oxygen. This makes thiobarbituric acid to be less acidic than barbital. In turn, gives higher pK value compared to barbital.

By increasing the percentage of ethanol from 25% to 50% and then to 75% gave the same trend however the curves shifted to a higher pH. These results in the pK values increase with increasing ethanol percentage due to the dielectric properties of the ethanol that decrease the dissociation of the ligand species, the higher concentration of the solvent results in ligand-ethanol interaction through hydrogen bond formation.

The potentiometric information shows that, the pK values depend on both the nature and also the proportion of the organic cosolvent. by increasing the organic cosolvent content within the medium the pK value increase. However, the acidity constants in a pure aqueous medium (Ka_1) associated with that in water-organic solvent mixtures (Ka_2) by the equation [51].

$$K_{a1} = K_{a2}\left(\frac{\gamma_{H^+}\gamma_{A^-}}{\gamma_{HA}}\right) \tag{13}$$

Where: y is that the activity coefficient of the subscripted species in a partially aqueous medium to that in a pure aqueous one. Since it is illustrious that the electrostatic effects of solvents operate only on the activity coefficients of charged species [52], one will expect that the rise within the quantity of the organic co-solvent within the medium can rise the activity coefficients of both H^+ and A^- ions. This can lead to a decrease in the acid dissociation constant (high pK value), that is consistent in keeping with the results obtained for the investigated compounds.
The effects of ethanol solvent on the behavior of those ligands are discussed using another scope. It was assumed that the J factor represents a solvent-transfer number characteristic of the tested chemical reaction that can be attributed to the transfer of the solvent. The subsequent equation is tested [53-54].

$$J \log [S] - \log K = -\Delta G/(2.303\ RT) - W \log [H_2O]/[S];$$
$$\log [H_2O/S] = X; \quad J \log [S] - \log K = Y$$

Where: [S] and ΔG represent the solvent concentration and the free energy respectively. The data are collected in table 7, Trail values of J = 1, 2, 3, 4 are used to find values of W for the gradients of Y vs. X (the slopes give the values of water molecules (W)). The data obtained may throw light on the role of equation and solvation during the dissociation, where the solute molecules are surrounded by water and solvent molecules in different manner. This in fact can`t, however, tell much about the nature of association between solute species and solvent molecules, whether the value of J is due to only the molecules in the primary solvation zone or includes molecules external to but closely associated with these.

For all ligands at different percentages of ethanol- Water mixture, the (W) values for J=I, 2, 3, 4 at 25°C are lower than that of J values to pinpoint more solvation. At TU the pK_1 at J=1 the equation and solvation are coexists due to W ≈ J.[54]

3.4 Complex formation studies
By scrutiny the pH titration curves of the free ligands thereto of the complex solutions, drop of the pH values occurs suggesting that the mechanism of complexation relies on proton liberation. An advantage was taken of the features of this curves ion (i) the Preliminary assessment of the complex stoichiometric, (ii) the final choice of the "best" set of formation constants for the system, based on graphical comparisons as developed above. In the calculation of the stability constants, it was assumed that factors like hydrolysis of metal ions and therefore the presence of nuclear hydrogen and hydroxyl bearing complexes was absent and could be neglected. This might be achieved by taking low concentration of the metal ion.

No side reactions between the metal and also the anionic part of the ligand take place. The metal ion complexing will not have an effect on the relative stability values and this factor becomes common throughout. The formation of ion-pair between the anionic species and the positive part of the strong electrolyte could be neglected by controlling the ionic strength of the solution, to be of low values. The activity coefficients of all species were constant. So, the concentration term may well be used rather than activities.

Table 7. X-Y data in different percentages of ethanol-water media at 25°C for variable values of (J)

	Y at 25°C						
Ligand	[s], %	log[s]	-X	J=1	J=2	J=3	J=4
Bal	25	1.398	-0.477	9.378	10.776	12.174	13.572
	50	1.699	0	9.589	11.288	12.987	14.686
	75	1.875	0.477	9.725	11.600	13.475	15.350
TBA-Thione	25	1.398	-0.477	9.568	10.966	12.364	13.762
	50	1.699	0	8.809	10.508	12.207	13.906
	75	1.875	0.477	7.925	9.800	11.675	13.550
TBA-Thiol	25	1.398	-0.477	9.618	11.016	12.414	13.812
	50	1.699	0	9.814	11.513	13.212	14.911
	75	1.875	0.477	9.325	11.200	13.075	14.950
TU	25	1.398	-0.477	10.248	11.646	13.044	14.442
	50	1.699	0	10.139	11.838	13.537	15.236
	75	1.875	0.477	11.225	13.100	14.975	16.850

The acid-base properties of the free ligands facilitate the investigation of the coordinating behavior of those ligands towards Zr(IV). It is apparent that the ligands under investigation have strong coordinating ability. The pK values of the free ligands have been affected by complexation. The free ligand exponent pL, the degree of formation of the system n and hence the stability constants of the metal ligand complexes could be calculated by the measurements during titration of the solution of chelating agent in the presence and in absence of metal ions with alkali could, Where C_M and C_L represent the analytical concentrations of the metal and ligand, respectively. Plotting the ¯n values against pL, values at ¯n equals to 0.5 [48], table 8, give the corresponding pK values.

Concordant results have been obtained on applying the point-wise calculation method [49]. On plotting log ¯n function against pL, straight lines are obtained from that log K values are obtained. table 8 shows the log K and stoichiometry of the complexes at 25°C.

Table 8. Stability constants of Zr(IV) complexes with ligands at 25°C

Ligand	$\overline{n}_A$ - pL			Point-wise method		
	75%	50%	25%	75%	50%	25%
Bal	7.59	7.04	6.41	7.58	7.00	6.42
TBA-Thione	7.94	6.65	6.25	7.93	6.66	6.31
TBA-Thiol	9.30	8.90	9.45	9.28	8.91	9.48
TU	7.83	7.54	7.08	7.86	7.55	7.09

Different values of the stability constants of the complexes are obtained in different concentrations of the complex species. The concentration distribution of various (Bal, TBA-thione, TBA-thiol and

IOP Conf. Series: Materials Science and Engineering **975** (2020) 012015 doi:10.1088/1757-899X/975/1/012015

TU) and their complexes species in solution as a function of pH were calculated and plotted by applying SPE and SPEPLOT computer programs [55], figures 11and12.

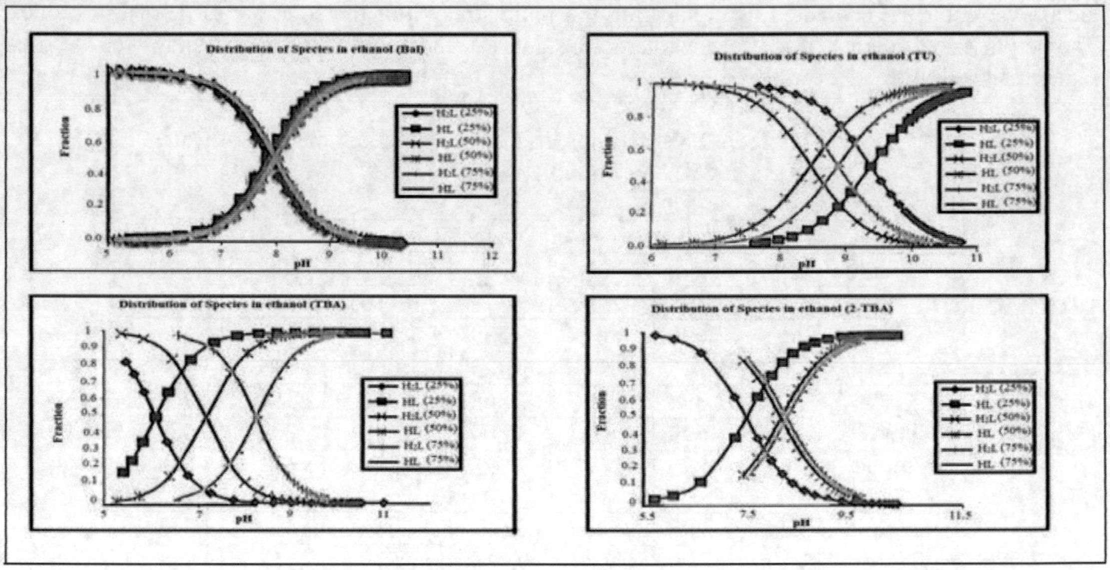

Figure 11. Distribution of species diagram of the ligands.

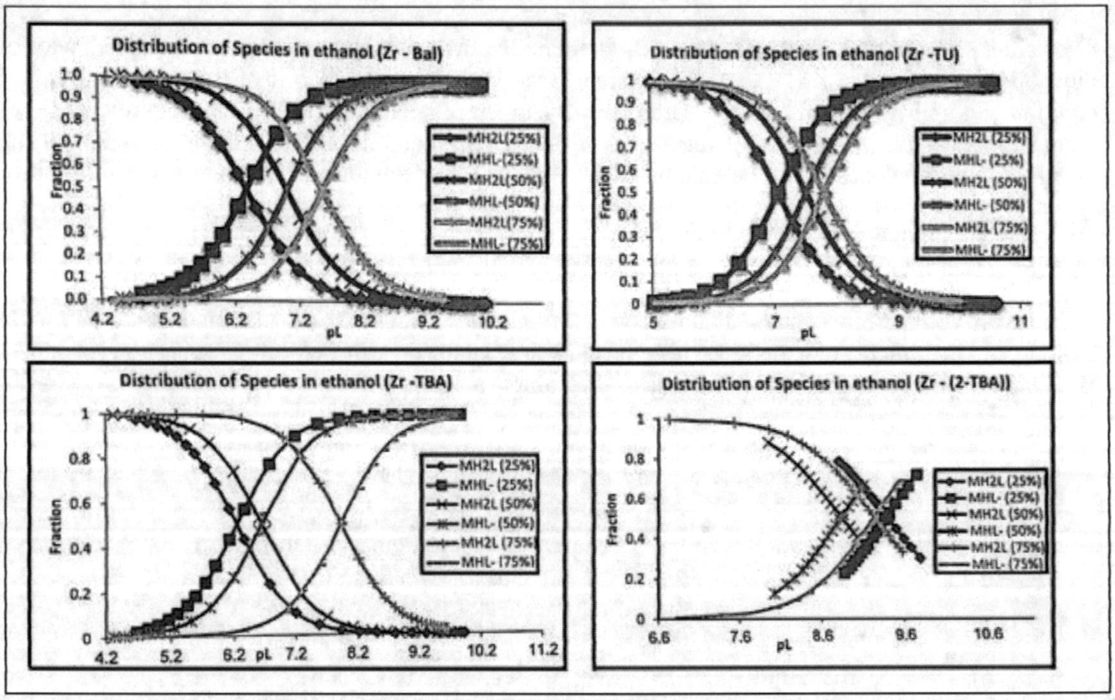

Figure 12. Distribution of species diagram of the complexes.

From the plotting of the ligand species fraction and the pH in different % ethanol, depending on the pK values, one can give the following points:

1. For Bal, the curves start with high % of [HL] species which decreased with increasing pH with appearance of [HL] species at pH around 5.09. The concentration of both [H_2L] and HL seems to be the same i.e. each with 0.5 fractions. With increasing pH the HL^- predominates till reach a maximum fraction at pH= 10.
2. For TU, the curves start with high % of [H_2L] species which decreased with increasing pH with appearance of [HL] species at pH around 6.19. The concentration of both [H_2L] and [HL^-] seems to be the same i.e. each with o.5 fraction, with increasing pH the [HL^-] predominates till reach a maximum fraction at pH=10.5.
3. For TBA- thione, the curves start with high % of [H_2L] species which decreased with increasing pH with appearance of HL species at pH around 5.63. The concentration of both [H_2L] and [HL^-] seems to be the Same i.e. each with 0.5 fraction, with increasing pH the [HL^-] predominates till reach a maximum fraction at pH 10.5.
4. For TBA-thiol, the curves start with high % of [H_2L] species which decreased with increasing pH with appearance of [HL^-] species at pH around 6.5. The concentration of both [H_2L] and [HL^-] seems to be the same i.e. each with 0.5 fraction, with increasing pH the [HL^-] predominates till reach a maximum fraction at pH=10.
5. By increasing the % of ethanol from 25% to 50% and then to 75% gave the same trend but the curves shifted to a higher pH. This leads to the fact that the pK values increase with increasing ethanol %, and this may be due to the dielectric properties of the ethanol that retard the dissociation of the ligand species.
6. On comparing the TBA with 2-TBA in (75% and 50% ethanol - water), it showed that the species distribution curves for TBA takes apart more than 2-TBA in case of HL -(2-TBA) at lower fraction and higher apparent pH vales. Meanwhile, the distribution of species in (75%, 50% and 25% ethanol -water) TBA and 2-TBA complexes show the same trend.
7. The study of the Zr-Bal complexes gave the same trend for the free ligand, where on increasing pH the existence of the [MH_2L] species decreased to a minimum at pH 9 while the [MHL^-] species Starts to increase in the pH range 4.29 -9 then predominates.
8. The study of the Zr- TU complexes gave the same trend for the free ligand, where on increasing pH the existence of the [MH_2L] species decreased to a minimum at pH =9 while the [MHL-] species starts to increases in the pH range 5.12-9 then predominate.
9. The study of the Zr-TBA complexes gave the same trend for the free ligand, where on increasing pH the existence of the [MH_2L] species decreased to a minimum at pH= 9.5 while the [MHL^-] species starts to increase in the pH range 4.2-9.5 then predominate.
10. The study of the Zr-2TBA complexes gave the same trend for the free ligand, where on increasing pH the existence of the [MH_2L] species decreased to a minimum at pH=9.5 while the [MHL^-] species starts to increase in the pH range 8-9.5.

4. Conclusion

- The infra-red spectra assigned the different modes of vibrations of the fundamental functional groups. The measurements referred that barbituric acid, thiouracil and barbital exist in the keto form, thiobarbituric acid is in keto and enol form.
- The change in entropy values, $\Delta S^{\#}$, for all complexes lie within the range -0.267, -0.299 $KJK^{-1} mol^{-1}$.
- The thermal processes proceed in complicated mechanisms with more ordered transition states.
- The DSC studies shows that the crystallization temperatures (T_c) of the complexes studied were around 78◦C – 322.12◦C.
- The stability constant for Zr(IV) with barbital was from (6.41-7.59), with thiobarbituric acid-thione (6.25-7.94), with thiobarbituric acid-thiol (8.90-9.45) and for Thiouracil (7.08-7.83) at different percentages of ethanol.

5. References

[1] Hueso-Urena F, Illan-Cabeza N A, Moereno-Carretero M N, Martinez-Matose J M and Ramirez-exposito M J, 2003 *J. Inorg. Biochem 94*, 326-334.
[2] Refat M S, El-Korashy S A and Ahmed A S, 2008 *Spectochim. Acta A*, *71*, 1084-1094.
[3] Tsunoda A, Shibisawa M, Yasuda Y and Nakao N, 1994 *Anticancer Res. 14*, 2637-2642.
[4] Casas J S, Castellans E E, Louce M D, Ellena J, Sanchez A, Sordo J and Taboada C,*2006 J. Inorg. Biochem* , *100*, 11, 1858-1860.
[5] Campbell M J, 1975 *Chem. Rev. 15*, 279-319.
[6] Masoud M S, Khalil E A, Ramadan A M, Gohar Y M and Sweyllam A, *2007 Spectrochim Acta A*, *67*, 669-677.
[7] Masoud M S, Awad M K, Shaker M A and El-Tahawy M M T, 2010 *Corros. Sci. 52*, 2387-2396.
[8] Uhlmann C, Fröscher W and Neurosci CNS. 2009 *Ther. 15*, 24-31.
[9] Sokmen B, Ugras S, Sarikaya H, Ugras H and Yanardag R, 2013 *App Bio. Biotech*, *171*, 2030-2039
[10] Ma L, Li S, Zheng H, Chen J, Lin L, Ye X, Chen Z, Xu Q, Chen T, Yang J, Qiu N, Wang G, Peng A, Ding Y, Wei Y and Chen L, 2011 *Eur. J. Med. Chem. 46*, 2003-2010.
[11] Uciechowska U, Schemies J, Neugebauer R C, Huda E M, Schmitt M L, Meier R, Verdin E, Jung M and Sippl W, 2008 *Chem. Med. Chem. 3*, 1965-1976.
[12] Maquoi E, Sounni N E, Devy L, Olivier F, Frankenne F, Krell H W, Grams F and Foidart J M, 2004 *Clin. Cancer Res. 10*, 12, 4038-4047.
[13] Xia G, Benmohamed R, Kim J, Arvanites A C, Morimoto R I, Ferrante R J, Kirsch D R and Silverman R B,2011 *J. Med. Chem. 54*, 2409-2421.
[14] McCluskey A J, Robinson P, Tim H, Janet L, Scott C and Edwards A K, 2002 *Tetrahedron Lett. 43*, 3117-3120.
[15] Kadoma Y and Fujisawa S, 2012 *Polymers*, *4*,2, 1025-1036.
[16] Langer M, Kratz F, Rothen-Rutishauser B, Wunderli-Allenspach H and Beck-Sickinger A G, 2001 *J. Med. Chem.44*, 1341-1348.
[17] Alyar S and Karacan N, 2009 J. Enzyme Inhib. *Med. Chem.* 24, 986-992.
[18] Fathalla A, Zaghary W A, Radwan H H, Awad S M and Mohamed M S, 2002 *Arch. Pharm. Res.* 25, 3, 258-269.
[19] Awad S M, Zohny Y M, Ali S A, S. Mahgoub and Said A M, 2018 *Molecules*, *23*, 11, 2913-2926.
[20] Cotton F A, Wilkinson G and Gaus P L, 1995 Basic Inorganic Chemistry, 3th ed. John Wiley & Sons, Inc.: Singapore.
[21] Blumenthal W B, 1962 *J. Chem. Educ. 39*, 604–610.
[22] Blumenthal W B, 1968 *Talanta*, *15*, 877–882.
[23] Blumenthal W B, 1973 *Am. Ind. Hyg. Assoc. J. 34*, 128–133.
[24] Nikunj B B, Darpan N P and Thaddeus J W, 2018 *Molecules*, *23*,3, 638-641.
[25] Kuhlmann L, Methling R, Simon J, Neumann B, Stammler H G, Strassert C A and Mitzel N, 2018 *Dalton Transactions*, 47, 11245-11252.
[26] a) Baldo M A, O'Brien D F, You Y, Shoustikov A, Sibley S, Thompson M E and Forrest S R, 1998 *Nature*, *395*, 151-154.; b) Sajoto T, Djurovich P I, Tamayo A B, Oxgaard J, Goddard W A and Thompson M E, 2009 *J. Am. Chem. Soc. 131*, 28, 9813-9822.
[27] Kober E M, Caspar J V, Lumpkin R S and Meyer T J, 1986 *J. Phys. Chem. 90*, 16, 3722-3734.
[28] Sanning J, Ewen P R, Stegemann L, Schmidt J, Daniliuc C G, Koch T, Doltsinis N L, Wegner D and Strassert C A, 2015 *Angew. Chem. Int. Ed. 543*, 786-791.
[29] Lubczak J and Lubczak R, 2018 *Int. J. Chem. Kinetics*, *50*,2, 122-132
[30] Masoud M S, Soayed A A and Ali A E, 2004 *Spectrochim. Acta A*, *60*, 1907-1915.

[31] Masoud M S, Haggag S S and Khalil E A, 2006 *Nucleosides, Nucleotides and Nucleic Acids*, *25*, 1, 73-87.
[32] Masoud M S, Hindawy A M, Soayed A A and Abd El-Kaway M Y, 2011 *J. Fluid Phase Equilibria*, *312*, 37-59.
[33] Masoud M S, Abd El-Kaway M Y, Hindawy A M and Soayed A A, 2012 *Spectrochimica Acta Part A*, *92*, 256-282.
[34] Masoud M S, Ramadan A M and El-Ashry Gh M, 2013 *Thermo Chimica Act*, *551*, 164-174.
[35] Masoud M S and Abd El-Kaway M Y, 2013 *Spectro Chim. Acta A*, *102*, 175-185.
[36] Masoud M S, Awad M K, Shaker M A, Ali A E and El-Tahawy M, 2013 *Research on Chemical Intermediates*, 6, 2741-2761.
[37] Masoud M S, El-Marghany A, Orabi A, Ali A E and Sayed R, 2013 *Spectrochimica Acta A*, *107*, 179-187
[38] Masoud M S, Ali A E and Abd El-Kaway M Y, 2014 *J. Thermal Analysis and Calorimetry*, *116*, 183-194.
[39] Masoud M S, Kamel H M and Ali A E, 2015 *Spectrochim. Acta A*, *137*, 1417-1425.
[40] Masoud M S, Elsamra R M I and Hemdan S S,2017 *J. Serb. Chem. Soc. 82*,7-8, 851-864.
[41] Gelsema W J, Deltgny C L , Remijnse A G and Bluleven H A, 1966 *Recueil* **,** *85*, 647-659.
[42] Foye W O, Lemke T L and Williams D A, 1995 *Principles of Med. Chem.*, 4th ed. Williams & Wilkins, Philadelphia, *88*, 89, 154-180.
[43] Bolton W, 1963 *Acta cryst. 16*, 166-173.
[44] Price W C, Bradley J E S, Fraser R D B and Quilliam J P, 1954 *J. pharm. and Pharmacol. 6*, 522-528.
[45] Wills Jr. J N, Cook R B and Jankow R,1972 *Anal. Chem. 44*, 1228-1234.
[46] Mesley R J,1970 *Spectrochim. Acta A* , *26*, 1427-1448.
[47] Bult A and Klasen H B,*1976 Lett.* 9, *2*, 81-94.
[48] Irving H and Rosstti H, 1953 *J. Chem. Soc.* 3397-3405.
[49] Jahagirdar D V and Khanolkar D D, 1973 *J. Inorg. Nucl. Chem. 35*, 3, 921-930.
[50] Charberk S and Martell A E,1952 *J. Am. Chem. Soc. 74*, 20, 5052-5056.
[51] Coetzee J F and Ritchie C D, 1969*Solute- Solvent Interactions*; Marcel Dekker: New York, 221.
[52] Denison J Tand Ramsey J B, 1955 *J. Am. Chem. Soc. 77*,9, 2615-2621.
[53] Mui K, McBryde W A E and Neiboer E,1974 *Can. J. Chem. 52*, 10, 1821-1833.
[54] Masoud M S and Abdullah A A, 1982 *J. Chem. Eng. Data, 27*,1, 60-62.
[55] Martell A E and Mutekaities R J,1992 *Determination and Use of Stability Constant*, VCH, New York.

IOP Conf. Series: Materials Science and Engineering **975** (2020) 012016 doi:10.1088/1757-899X/975/1/012016

Thermodynamic Analysis for Tracking the Effect of the Egyptian Zirconium Silicate on the slag properties of low hydrogen Welding Electrodes

A AbdElkarim[1], H M Abd El Aziz[2], Waleed A Mohrez[1] and H Mashaal[1]

[1]Nuclear materials Authority, Cairo, Egypt
[2]Mining and petroleum Dep., faculty of Engineering, Al-Azhar University, Cairo, Egypt

Abstract One of the most common types of welding electrodes, used in welding processes; are the low hydrogen electrode classification; E 7018. This electrode is applied for fabrication most carbon steel alloys, where it used for welding plate works of ships, equipment, tanks, boilers and pressure vessels, steel structures of buildings, bridges as well. One of the most common drawbacks resulting of using of this welding electrode is the difficulty of removing the slag between the deposited passes, which consumes time and money and hence decreasing the quality of welding process also one of the most drawbacks of this type of welding electrode is the rabid burning of coating flux at elevated amperes increase.

This study aimed to suggest rooted solution for these two problems by adding zirconium silicate to the flux recipe. Different percentages of Egyptian zirconium silicate were added to the coating flux recipes of type E7018 where the changes at both weldability and slag removing between deposited passes were noticed at different amperes. By using HSc simulation program of the thermodynamics data and expected phases in slag were studied. Microstructure samples of weld metal were imaged and analyzed, basicity index was calculated, XRF chemical analysis for all slag different samples, were determined. The study concluded that adding 5% of the Egyptian zirconium silicate to the flux clearly improved; the welding properties as well as the ease of slag removal.

Keywords. Zircon, E7018, thermodynamic, black sand and welding electrode

1. Introduction

The continues improvement of flux coated electrode of the type low hydrogen electrodes is indispensable to enhance the performance of welding and its slag removal, where such flux coated type is rich of calcium carbonate or calcium fluoride [1-3] ,the main criteria of success the recipe of low hydrogen electrode ;is to decrease the amount of hydrogen in weld metal ,where this target can be achieved by decreasing the absorption of humidity on the surface of flux coat [4] on the other hand all recipes of low hydrogen type shall keep basicity index not less than "2" to keep the slag mobility and weld deposition smooth as much as applicable [5] ,all applicators and manufacturers seeking to select electrodes, which possesses such potentiality of slag removal between passes to decrease the time of welding deposition especially in hot passes filling at large thicknesses of base metal .

Weld bead contour or profile of weld deposition can be affected by the slag chemistry and its physical properties one of the main parameters which affecting on weld profile is the high temperature slag , which called at some texts as "refractory slag ", which provides an advantage of rabid solidification before the weld metal to be solidified, forming a mold rounded wall that holds the molten metal in confined place

IOP Conf. Series: Materials Science and Engineering **975** (2020) 012016 doi:10.1088/1757-899X/975/1/012016

[6-10] Zirconium and hafnium compounds are widely used in today's technology to achieve such refractory slag and hence achieving such rabid solidification of the slag[11-14].

In this research we used a different recipes using different zirconium oxide percentage at the flux coating, and hence the basicity index of each recipe was changed as well ,where we used percentage of zirconium oxide 1,3,57,9 % by weight of flux coating ,all phases of flux dissociated phases were calculated by thermodynamic calculation method of "HSC chemistry 5 "to find the optimum condition and the most appropriate zirconium oxide percentage, which shall be added to the recipe to achieve most viable slag removal .

We also investigated the mechanical properties of the weld metal to track any variation of the designed mechanical properties of such electrode with such variation in the slag chemistry.

All investigations have been achieved to track and analysis the phases which formed in the slag; by XRF and chemical analysis, also soundness tests by non-destructive methods were done by radiographic.

2. Experimental work

2.1. Thermodynamic Calculations

Before proceeding adding zirconium silicate ore in the form of zirconium silicate to welding electrode flux coat, a comprehensive calculation was carried out in order to have a pre-estimation of the materials balance conditions.

Computer software for simulation all reactants and materials balance called" HSC chemistry-5" was used to calculate the thermodynamic parameters at different reaction temperatures. The calculations showed that it is possible to dissociate $ZrSiO_4$ to ZrO_2 and SiO_2 in accordance to the following equation:

$$ZrSiO_4 \longrightarrow ZrO_2 + SiO_2$$

As shown in table (1) thermodynamic data for previous equation it is noticed that the negativity of free energy value (Delta G) started at temperature 1200 C, which means that the dissociation temperature of $ZrSiO_4$ will dissociate spontaneously to direction of formation both ZrO_2 and SiO_2.

To ensure that All of dissociated ZrO_2quantity, which came from added zirconia to the flux coat, will be directly moved to the slag phase and was neither reduced to Zr metal and nor interfered inside the weld metal.

The thermodynamic calculations for the three expected reduction reactions of ZrO_2 with metals; Fe (existed in form of iron powder), Mn (in form of ferromanganese) and Si (in form of ferrosilicon) were done according to equations and the results were shown in tables (2,3 and 4)

$$ZrO_2 + Si = Zr + SiO_2$$

$$ZrO_2 + 2Fe = Zr + 2FeO$$

$$ZrO_2 + 2Mn = Zr + 2MnO$$

From tabulated data in table (2, 3 and 4) can deduce that Δ G at temperature range from 0 to 2000 ^{o}C is positively value ,which means that the reduction of ZrO_2 by previous three reducing metals (Si,Fe and Mn) will not be done and all quantity of ZrO_2 will directly oriented into slag and weld metal will not be interfered byZrO_2 .

By Using HSC chemistry-5 program the prediction phases of slag with changed adding quantity of $ZrSiO_4$ to flux recipe as 2,4 and 6% from weight, were done and shown in figures 1:a, B and C.

IOP Conf. Series: Materials Science and Engineering **975** (2020) 012016 doi:10.1088/1757-899X/975/1/012016

Table 1. Thermodynamic data of the dissociation reactions

T (C°)	ΔH (kJ)	ΔS (J/K)	ΔG (kJ)	K	Log(K)
0.000	12.590	7.637	10.504	9.797E-003	-2.009
200.000	13.033	8.835	8.853	1.053E-001	-0.977
400.000	13.210	9.148	7.052	2.836E-001	-0.547
600.000	14.519	10.765	5.119	4.940E-001	-0.306
800.000	14.062	10.292	3.017	7.131E-001	-0.147
1000.000	16.022	12.106	0.609	9.441E-001	-0.025
1200.000	24.754	18.161	-2.000	1.177E+000	0.071
1400.000	24.875	18.237	-5.638	1.500E+000	0.176
1600.000	25.116	18.373	-9.298	1.817E+000	0.259
1800.000	35.905	23.774	-13.383	2.174E+000	0.337
2000.000	38.556	24.995	-18.261	2.628E+000	0.420

Table 2. Thermodynamic calculation for possibility of ZrO_2 reduction by Si.

T (C°)	ΔH (kJ)	ΔS (J/K)	ΔG (kJ)	K	Log(K)
0.000	189.576	11.901	186.326	2.322E-036	-35.634
200.000	188.597	9.172	184.257	4.537E-021	-20.343
400.000	188.332	8.651	182.508	6.865E-015	-14.163
600.000	189.841	10.520	180.655	1.555E-011	-10.808
800.000	189.713	10.386	178.567	2.031E-009	-8.692
1000.000	195.036	15.196	175.690	6.183E-008	-7.209
1200.000	185.382	8.465	172.911	7.386E-007	-6.132
1400.000	184.710	8.036	171.265	4.495E-006	-5.347
1600.000	134.585	-21.719	175.268	1.294E-005	-4.888
1800.000	145.660	-16.175	179.192	3.053E-005	-4.515
2000.000	170.831	-4.390	180.811	6.995E-005	-4.155

IOP Conf. Series: Materials Science and Engineering **975** (2020) 012016 doi:10.1088/1757-899X/975/1/012016

Table 3. Thermodynamic calculation for possibility of ZrO2 reduction by Fe

T (C°)	ΔH (kJ)	ΔS (J/K)	ΔG (kJ)	K	Log(K)
0.000	565.228	47.546	552.241	2.431E-106	-105.614
200.000	568.244	56.144	541.679	1.566E-060	-59.805
400.000	569.447	58.350	530.169	7.192E-042	-41.143
600.000	568.497	57.195	518.558	9.454E-032	-31.024
800.000	563.382	51.992	507.586	1.957E-025	-24.708
1000.000	564.266	52.799	497.044	4.033E-021	-20.394
1200.000	556.519	47.461	486.601	5.556E-018	-17.255
1400.000	604.154	76.408	476.313	1.345E-015	-14.871
1600.000	577.646	61.798	461.890	1.314E-013	-12.881
1800.000	577.746	61.847	449.529	4.708E-012	-11.327
2000.000	600.083	72.326	435.676	9.723E-011	-10.012

Table 4.Thermodynamic calculation for possibility of ZrO_2 reduction by Mn

T (C°)	ΔH (kJ)	ΔS (J/K)	ΔG (kJ)	K	Log(K)
0.000	329.690	43.179	317.895	1.598E-061	-60.796
200.000	330.035	44.372	309.041	7.582E-035	-34.120
400.000	329.260	43.040	300.288	4.972E-024	-23.303
600.000	327.926	41.318	291.849	3.461E-018	-17.461
800.000	321.701	34.950	284.194	1.465E-014	-13.834
1000.000	323.249	36.388	276.922	4.340E-012	-11.362
1200.000	302.811	21.877	270.583	2.541E-010	-9.595
1400.000	272.359	1.928	269.134	3.955E-009	-8.403
1600.000	268.301	-0.367	268.988	3.151E-008	-7.502
1800.000	264.768	-2.161	269.248	1.643E-007	-6.784
2000.000	371.818	48.409	261.777	9.641E-007	-6.016

From figure 1 we can observed that all ZrSiO4 approximately dissociated to ZrO_2 and a part of it was reacted with CaO, which existed in the molten slag phase and formed $CaZrO_3$ compound, the melting point of phase $CaZrO_3$ (2340 C) is lower than Melting point of ZrO_2 (2710 C), which means that the formation of ZrO_2 compound in molten slag phase is better than $CaZrO_3$,where it leads to preventing an early burning of the electrode . From the figs we can concluded that adding 4% $ZrSiO_4$ to slag is the best conditions, where the presence of ZrO_2 with more quantity than $CaZrO_3$. The increasing of $ZrSiO_4$precent to 6% will lead to an increase percent of $CaZrO_3$ phase and decreasing ZrO_2 and hence leads to early burning to stick electrode and deteriorate the welding performance and increasing the consumption of welding electrodes.
From thermodynamic calculation we can conclude that: -

IOP Conf. Series: Materials Science and Engineering **975** (2020) 012016 doi:10.1088/1757-899X/975/1/012016

1- All $ZrSiO_4$ will be dissociating at 1000 to 1200C.
2- All $ZrSiO_4$ will not reduce to Zr metal and all quantity will be slag former.
3- The increasing of $ZrSiO_4$ higher than 5% is not useful.

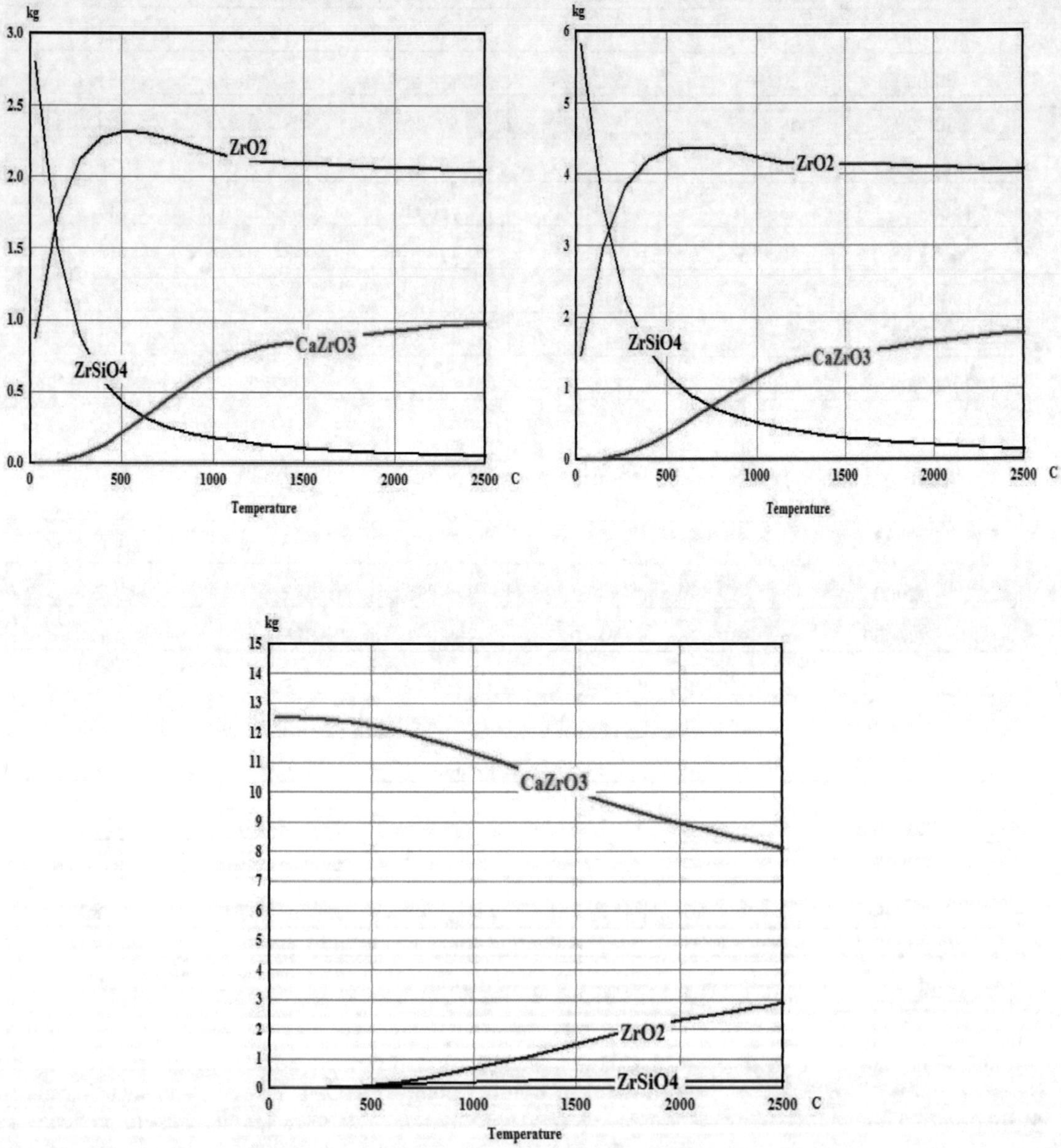

Figure 1. Prediction phases of slag with adding (a: 2% , b: 4% and C: 6%) $ZrSiO_4$

3. Experimental and Methodlogy Procedure

3.1 Manufacturing of welding electrode

The experimental work and methodology represents on the materials processing and balance to produce different recipes such recipes as investigated from thermodynamic data with different percentage of zirconium silicate with percentage of 1,3,5,7 % by weight.

The raw materials of zirconium silicate is a commercial grade with a purity of 99%with using screen analysis for controlling mesh size up to -500 micron from Rosetta black sand , Egypt .

we used two routes of mixing one by dry process for one hour by dry mixer rotating sparely with 100 RPM , then followed by dry mixing inside wet dryer for 30 minutes with adding bindery material as potassium silicate to make adhesion of the flux coat, then followed by high pressure extrusion process to coat the flux on the surface of core wire. then the extruded green electrodes were aerated at open air yard for 48 hours to solidify the flux coat locally by increasing its cohesion and adhesion ,then the electrodes packed inside electric furnace to temperature 350 °C for six hours to elaborate the flux coat from the absorbed humidity.

3.1. Wedling process

After drying and packing of the welding electrode we selected our samples of test with three diameters 3.25,4 and5 mm with flux coat diameter up to then we conducted deposition the weld metal with weaving technique

3.2 Testing and analysis

Investigation by XRF analysis for the sample were done , a co operation with fusion company for welding electrode a five recipe of flux with their contents were done the percent of adding $ZrSiO_4$ were changed as following in table 5.below

Table 5:- shows the number of samples and % of Zircon added

No of recipe	**% $ZrSiO_4$**
1	without
2	1
3	3
4	5
5	7

After the production of electrodes the performance test with a professional welder were done for 5 samples and XRF of slag phase for each sample was done to investigate the percentage of zirconia in the slag, mechanical and chemical test of the weld metal for each sample also was done to investigate if there any negative or positive impacts in the mechanical properties of the weld metal soundness test for the best sample were done.

4. Results and Discussions

4.1. Zircon ore Analysis

Run of mine sample of the zircon ore was analyzed to investigate and assure from the purity and percentage of zirconia in the ore by XRF device ,the analysis is as resulted at the tabulated data in table 6, where "zirconia "ZrO_2 and HfO_2 % showed that it is up to 67% and silicate SiO_2% up to 32.6% .

IOP Conf. Series: Materials Science and Engineering **975** (2020) 012016 doi:10.1088/1757-899X/975/1/012016

Table 6. XRF analysis of Zircon concentrate

ZrO_2	HfO_2	SiO_2	TiO_2	Al_2O_3	Fe_2O_3	CaO	MgO	P_2O_3	Ignation loss at 1025
65.47	1.32	32.58	0.05	0.06	0.05	0.03	0.05	0.13	0.14

4.2. Analysis of Slag and Metal Phases and welding performance

The results of analyse of this stage is divided into mechanical test of weld metal for each test sample, chemical analyses for both metal and slag phase and the monitoring of welding performance, welding amperage required for each test sample and slag removal as well on the other hand.

4.3. Welding performance and amperage required.

It was noticed that the test samples, showed various welding performance and variant amperage values with changing the percentage of added zirconium ore, when no addition of zirconium for sample No 1 it show immoderate welding performance and amperage ranging from 130,170,200 for diameters 3.25, 4, 5 mm respectively and showed severe difficulty of slag removal after each pass due to physical adhesion between slag and metal interface required mechanical gauging and brushing to clean the passes from the slag formed, whereas ,when an increase of the percentage of zircon ore from 1 ,3,5,7 % to samples 2,3,4,5 respectively it was noticed that the amperage required for burning increased from (130,170,200 amperes) with diameters 3.25, 4, 5 mm respectively to (140,180,210amperes) for sample 2 and (150,190,230 amperes) for sample 3 and (160,210,250 amperes) for sample 4,and (160,210,250 amperes) for sample 5 as shown in table 7 and was noticed also that there was clear enhancement of slag removal after each pass with increasing the percentage of zirconium directly proportional .

Table 7. Maximum Ampere for electrode burning

Sample No.	$ZrSiO_4$ %	Max. ampere, A		
		3.25 mm Electrode diameter	**4.00 mm Electrode diameter**	**5.00 mm Electrode diameter**
1	0	130	170	200
2	1	140	180	210
3	3	150	190	230
4	5	160	210	250
5	7	160	210	250

Where the most easy slag removal has been investigated in the sample number 5 with percentage of zirconium 5% as shown in figure 2 where an increase of zirconium as high temperature refractor mineral worked as mold wall contributed positively to prevent any physical and chemical bonding between slag and weld metal phase at the reaction interface ,this result is not only impacted to the welding performance positively but also impacted economically during the productivity of welding ,where such result will decrease the time of finishing the deposition especially at the large thicknesses.

4.3.1. Chemical Analysis of Metal and Slag Phases.

The chemical analysis of metal and slag phase was conducted where the analysis of slag phase for each test sample has been investigated to track the phase of zirconia, which dissociated from zirconium silicate.

The results in table 8 showed that the sample No 1 without any addition of zirconium silicate gives a zirconia up to 0.025 which may be come from other gangue minerals, but when increasing the zirconium silicate at samples 2,3,4,5 showed appearance of zirconia with values; (0.67,2.08,3.45,4.88%) respectively, whereas the balance of silicate impacted at the silicon oxide which appears with percentage (22.08,22.99,23.91,25.8%) for such samples 2,3,4,5 respectively as shown in tables 9 . The dissociation of zirconium silicate to zirconia and the percentage of each phase is agreed with the thermodynamic data which conducted by HSC chemistry 5.

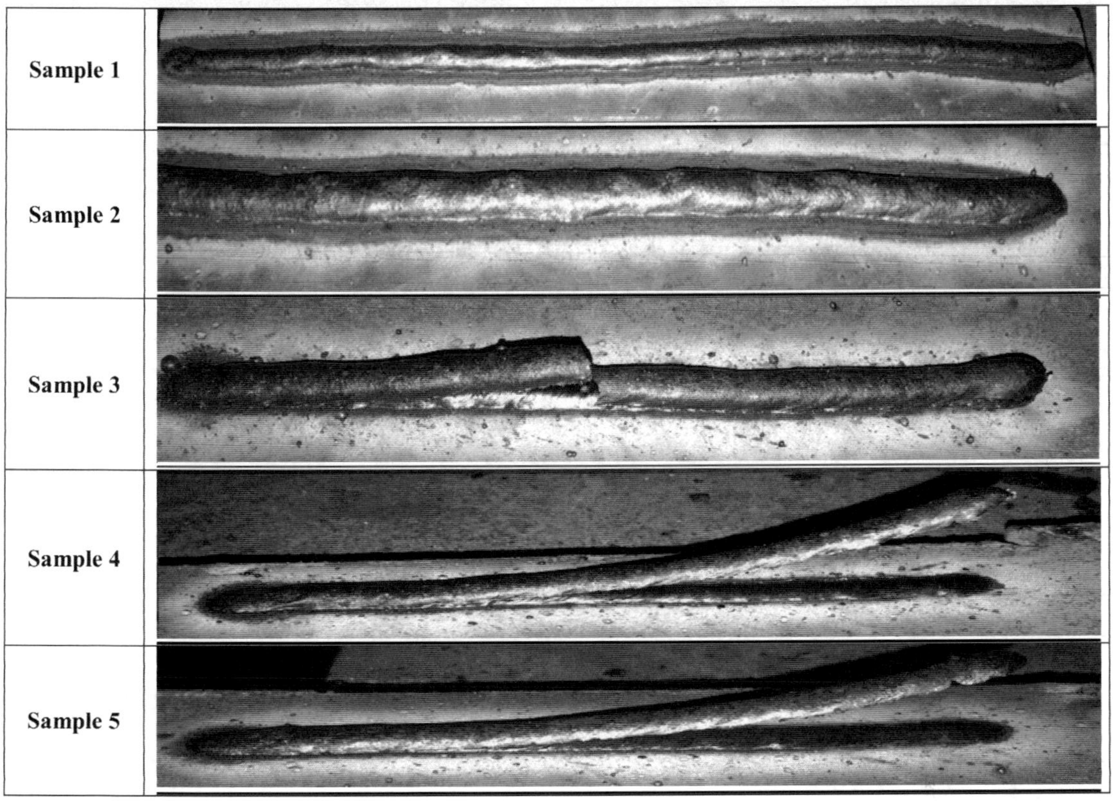

Figure 2. Slag and welding profile

Table 8. Chemical analysis for weld metal

Sample No.	$ZrSiO_4$ %	Chemical composition %				
		C	Mn	Si	P	S
1	0	0.56	1.24	0.514	0.018	0.013
2	1	0.55	1.26	0.567	0.019	0.011
3	3	0.61	1.16	0.48	0.016	0.013
4	5	0.57	1.19	0.436	0.018	0.014
5	7	0.64	1.22	0.472	0.020	0.016

The weld metal analysis of the sample number 1 with comparison of samples 2,3,4,5 showed no considerable changes in chemical analysis where this proofed that there was not any estimated mobility of zirconium silicate or it dissociated fraction towards the weld metal and almost was oriented to the slag phase.

IOP Conf. Series: Materials Science and Engineering **975** (2020) 012016 doi:10.1088/1757-899X/975/1/012016

Table 9. Chemical analysis for slag

Sample No.	$ZrSiO_4$ %	Chemical composition %								
		CaO	SiO_2	Al_2O_3	TiO_2	Fe_2O_3	MnO	ZrO_2	K_2O	Na_2O
1	0	53.52	21.11	1.55	10.76	4.29	5.12	0.025	2.11	1.48
2	1	52.12	22.08	1.58	10.56	4.14	5.88	0.67	1.99	0.96
3	3	49.24	22.99	1.6	10.44	4.11	5.78	2.08	1.94	1.01
4	5	47.13	23.91	1.77	10.58	4.08	5.68	3.45	1.98	1.4
5	7	45.66	25.8	1.53	10.28	4.17	5.56	4.88	1.28	0.78

4.3.2. Mechanical analysis of weld metal.

The mechanical test of the weld metal for each test sample shown in table 10 was conducted to assure and investigate the impact of addition the zirconium silicate, where the chemical analysis which confirmed that there was no any considerable change in the chemical analysis between sample No 1 and other samples with variable percentage of zirconium silicate the addition of zirconium didn't affected on the deposition and fluidity of the weld metal which didn't impacted negatively to the mechanical analysis of the weld metal ,the ultimate tensile strength of sample No1 ,2,3,4,5 was 540,543,537,544,543 MPa respectively which means that there was no any considerable effect of addition zirconium silicate on the mechanical test of weld metal .

Table 10. Tensile test of weld metal

Sample No.	$ZrSiO_4$ %	Yield strength N/mm^2	Ultimate tensile strength N/mm^2	Elongation %
1	0	429	540	34
2	1	430	543	34
3	3	422	537	32
4	5	433	544	33
5	7	428	543	34

5. Conclusion.

This research targeted the study of the effect of adding zirconium ore from Egyptian run of mine, without any preparation or processing, except grinding and sizing, the study has been accomplished by two routes; one is basic and other is secondary for assurance and confirmation the first one ,the first rout is conducting thermodynamic analysis for showing and tracking the phases of dissociation the zirconium silicate in which forms have been dissociated and how these forms behaves with other additives. Thermodynamic analysis showed that zirconium silicate has been dissociated to zirconia (ZrO_2) at temperature 1200°C with Delta G of -2 KJ and continued spontaneously for dissociation to temperature up to 2000°C with Delta G of -18.261 KJ. Also, thermodynamic analysis confirmed that the optimum quantity of zirconium silicate which shall be added is not more than 4 -5 % where any increase of quantity will lead to form calcium zirconate ($CaZrO_3$) on the credit balance of zirconia.

Thermodynamic analysis revealed any suspicious of the probability that zirconium dioxide (ZrO_2) may be reduced by manganese, Silicon or iron , which already existed at the recipe at the form of Ferro manganese and Ferro silicon where Gibbs's free energy calculations showed that there is no any negativity of the reduction reaction can be occurred with an elevated temperature up to 2000°C.

The chemical analyses of the slag phases of test samples showed that zircon phases have been oriented to the slag and contributed effectively as high temperature refractory material prevented any chemical or physical bonding between slag and metal phases, which led to prompt, easy slag removal after each pass deposited.
The chemical analysis also confirmed that there is no any evidence of zircon or its phases entered to weld metal.
The mechanical test of the weld metal showed that there is no any positive or negative effect on the mechanical value of ultimate tensile test of test samples and all their value was average as 540MPa.

Acknowledgment
The research team would like to present deep thanks to the company Fusion Tech® for their support during the manufacturing welding electrodes, where their team of quality and production was promptly response to use the facility of production and extrusion machine during the time of our internship, their contribution to increase the capacity of research and development for their product is highly appreciated.

6. References

[1] J. F. Lancaster, 1999, Metallurgy of welding (6th ed.). Cambridge, Abington Publishing, pp. 244, 249.
[2] Welding, 1993, Brazing and Soldering, ASM hand book, volume 6, pp. 248-250.
[3] ASM Handbook, 2000 (Mechanical Testing and Evaluation) volume 8, pp.366.
[4] J. DU PLESSIS, M. DU TOIT, AND P. C. PISTORIUS, 2007, The influence of fluoride and calcite fluxadditions on the hydrogen content of shielded metal arc welds was examined, 273-280s, VOL. 86
[5] K. Shinozaki, L. KE AND T.H. North, 1992, Cracking sensitivity appears related to an excess of 50% delta ferrite in the weld, Welding Research Supplement, November, 387-396s
[6] Ahmed Abdelkareem, 2019, New Extraction Technique of Zirconium and Hafnium from Zircon Mineral ,Arab J. Nucl. Sci. Appl., Vol. 52, 2, 201-208)
[7] Levine, S.R., et al., 2002, Evaluation of ultra-high temperature ceramics for aeropropulsion use. Journal of the European Ceramic Society. 22(14-15): p. 2757-2767.
[8] Sandhi, A., R.F. Reidy, and T.W. Scharf, 2013, Understanding the mechanism of zirconia to zirconium carbide conversion for ultra-high-temperature ceramic applications. American Ceramic Society Bulletin. 92(6): p. 56-56.
[9] Abdelkader, A.M. and D.J. Fray, 2013, Synthesis of self-passivated, and carbide-stabilized zirconium nanopowder. Journal of Nano particle Research. 15(12).
[10] Zhang, S.C., G.E. Hilmas, and W.G. Fahrenholtz, 2006, Pressureless densification of zirconium diboride with boron carbide additions. Journal of the American Ceramic Society. 89(5): p. 1544-1550.
[11] Ahrenholtz, W.G., et al., 2007, Refractory diborides of zirconium and hafnium. Journal of the American Ceramic Society. 90(5): p. 1347-1364.
[12] Heck, K., J.R. Blackford, and R.F. Singer, 1999, Castability of directionally solidified nickel base superalloys. Materials Science and Technology. 15(2): p. 213-220.
[13] Abdelkader, A.M. and E. El-Kashif, 2007, Calciothermic Reduction of Zirconium Oxide in Molten CaCl₂. ISIJ International. 47(1): p. 25-31.
[14] M.A.Abostat, Y.Saleh, H.I.Mira, M.Amin, M.M kazindar and B.M.Ahmed , 2018, characterization kinitics and thermodynamics of biosynthesized uranium nanoparticles , artif.cells . nanomedibio.tech. , (46) , 147- 154

Radiological and environmental studies on the metamorphosed sandstones at Wadi Sikait area, southeastern desert, Egypt

A. H. El-Farrash[**], M. G. El-Feky[*], S. F. Hassan[*], Soliman A. Abu Elatta[*], and M. Refaat[*]

[*]Nuclear materials Authority, Cairo, Egypt
[**]Corresponding author, nuclear materials Authority, Cairo, Egypt

Abstract. Wadi Sikait area lies in the southern part of the Eastern Desert of Egypt along the upper stream of Wadi.Sikait. The exposed rock units in Wadi Sikait area are ophiolitic mélange roccks, metamorphosed sandstones, and porphyritic granites, lamprophyre dykes, in addition to fluorite and quartz veins. The activity concentrations of various radionuclides are measured by NaI-Tl activated detector and the radioactive mineralization were determined by Alpha track technique using CR-39 films. Radiometric investigations indicate that ^{238}U activity concentrations range between 37.2 and 520.4 with average 148.8 $BqKg^{-1}$, ^{232}Th activity concentrations vary between 8.08 and 366.99 $BqKg^{-1}$ with 76.75 $BqKg^{-1}$ as an average. The studied metamorphosed sandstones have higher values relative the world concentration levels. High uranium content of the metamorphosed sandstone is attributed to presence radioactive minerals like uranophane and autonite, in addition to accessory minerals like allanite, zircon and monazite. Other U-bearing minerals are also recorded as biotite, muscovite, iron oxides and clays. Absorbed Dose Rate (D), annual effective dose equivalent (AEDE), radium equivalent activity (Raeq), external (H_{ex}) and internal (H_{in}) hazard index, in addition to activity gamma index (I_γ) caused by gamma emitting natural radionuclide are determined from the obtained values of ^{226}Ra, ^{232}Th and ^{40}K. Fairly, many of the studied metamorphosed sandstone do not satisfy the universal standards.
Keywords. Radiological, Environment, Uranium, Thorium, Egypt

1. Introduction

The natural environmental radiation depends mainly on geological and geographical conditions (Florou and Kritidis, 1992). Higher radiation levels are associated with igneous rocks, such as granite, and lower levels with sedimentary rocks. There are exceptions, however, as some shales and phosphate rocks have relatively high content of radionuclides (UNSCEAR, 1993). The study area lies in the southern part of the Eastern Desert of Egypt along the upper stream of Wadi.Sikait area is covered by moderate to high mountains with rugged topography. The high peak is Gabal Sikait (769 m a.s.l), it is formed of ultramafic rocks, which are often associated with beryl mineralization. The Sikait area was geologically investigated by Hassan and Hashad (1990); Hashed and El Redy (1979); Hegazy (1984); Mohamed and Hassanen (1997); Saleh (1998); Ibrahim et al., (1999) and Omar (1999) among others. Assaf et al. (2000) studied the polyphase folding in Nugrus-Sikait area and reported that the lithological constitution of the area comprise a sequence of dismembered ophiolites, ophiolitic mélange association and arc assemblage. The older rocks are intensively deformed and intruded by intracratonic association, which, within the mapped area, is ophiolitic mélange, metamorphosed sandstones with porphyritic granites. Detailed spectrometric and radiometric studies (Ibrahim et al. 2007) for metamorphosed sandstones site at Wadi.Sikait indicate that eU range from 15 to 100 ppm, but chemically range from (60 – 480 ppm); whereas eTh up to 85 ppm. Also they reported that the emplacement of both lamprophyre dykes and porphyritic granites may play an important role as a heat source, which lead to U– mobilization from hot granitic magma, transported (along deep fault and banding) and redeposit in metamorphosed sandstones under suitable conditions.

The metamorphosed sandstones occur in two locations in Wadi. Sikait. The first location (Sikait-1) lies at the upper stream of Wadi. Sikait and the second location (Sikait-2) occur west the bend of Wadi. Sikait figure 1 , figure 2a and 2b. The metamorphosed sandstone rocks are fine- to medium- grained, white color, highly sheared, sometimes bedded, cross- cut by lamprophyre dykes (NW-SE and NNE-SSW), quartz veins and left strike slip faults (NW-SE , NNE- SSW& N-S) figure 2 a. Sikait-1 is the largest outcrop of metamorphosed sandstones at Wadi. Sikait with low to moderate peaks, elongated in NW-SE direction (1.4 km in length and range in width from 120 m to 300 m). These rocks are varying in color from pale white to milky white figure and show relics of primary bedding; banding and obvious foliations in NW-SE with angle of dip 35° /SW. Metamorphosed sandstones in this location are dissected by three types of left strike slip faults, N-S and NNE-SSW and NW-SE figure, so they are highly tectonized. The NW-SE is the largest and oldest one. These faults especially NNE- SSW characterized by mylonitization and many types of alterations as silicification and Fe-Mn oxy-hydroxides figure. Mineralization occurs along zones of these faults visible by naked eyes.

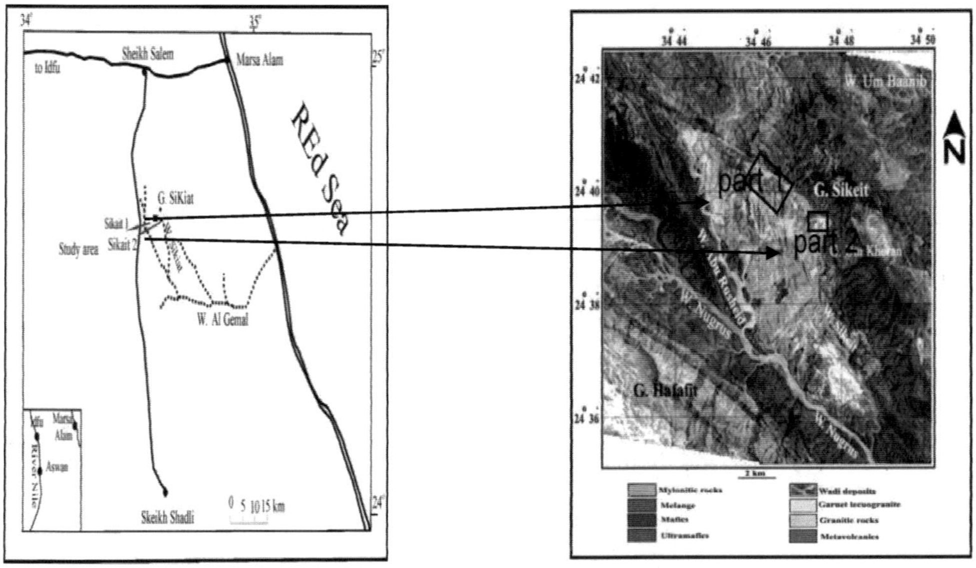

Figure 1. Location map of the study area and false color composite map TM-image that showing the location of metamorphosed sandstone.

Sikait-2 covers a small area where its length is about 450 m and maximum width about 230 m, forming low terrain, highly sheared, dissected by branches of strike slip fault running in N-S with left movements and frequently curved to N direction. Branch zones of the strike slip fault characterized silicification alteration. Many mineralizations are associated with quartz found along zones of fault branches as fluorite at west this location and wolframite and cassiterite which visible by naked eyes. Metamorphosed sandstones are generally uniform in texture and composed from fused quartz grains. Semi-angular and elongated rock fragments of older rocks are enclosed in the metamorphosed sandstones. The present work aims to study the radioactivity and environmental impacts of Wadi. Sikait metamorphosed sandstone that used as building material or as ornamental stones.

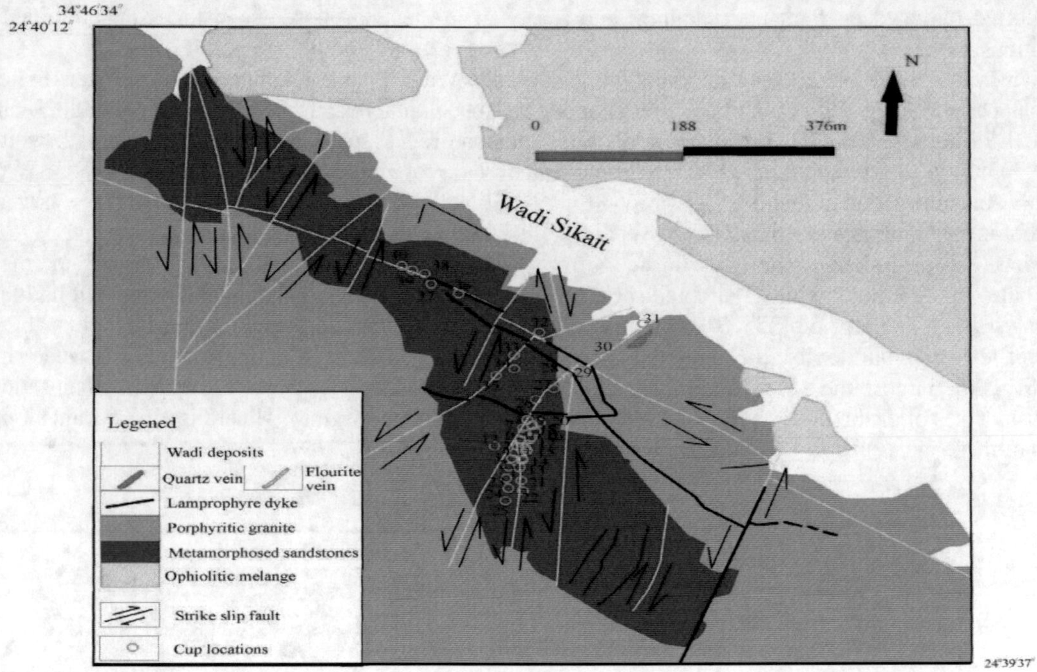

a

Figure 2 a. Detailed geologic map of Sikait-1modified after Ibrahim et al. (2010).

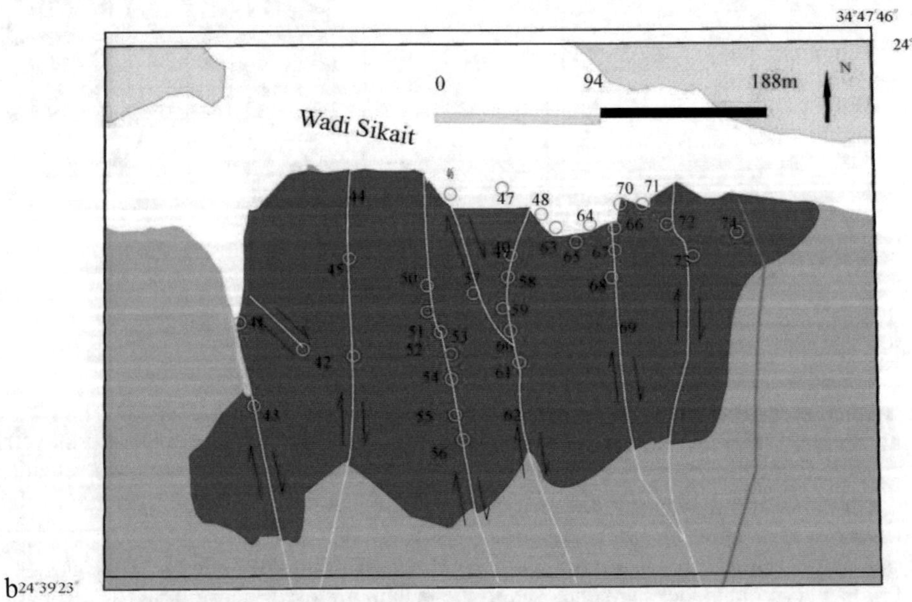

b

Figure 2 b. Detailed geologic map of Sikait-2 modified after after Ibrahim et al. (2010).

2. Sampling and analytical techniques

Representative grab samples were collected fromthe two-mica granite (seven samples) and from the metamorphosed sandstone to represent the highest values of anomalous field radioactivity. These samples were prepared for gamma-ray spectrometric analysis in order to determine their uranium, thorium, radium and potassium contents by using a multichanel analyzer equipped with a γ-ray detector (Gamma-Spectrometer technique). The instrument used in determination of the four radioactive elements consists of a bicron scintillation detector NaI (Tl) 76×76 mm, hermetically sealed with the photomultiplier tube in aluminum housing. The tube is protected by a copper cylinder measured at 0.6 cm in thickness against induced X-ray and a chamber of lead bricks against environmental radiation. Uranium, thorium, radium and potassium are measured by using four energy regions representing Th-234, Pb-212, Pb-214 and K-40 at 93, 239, 352 and 1460 kV for uranium, thorium, radium and potassium, respectively. The measurements were carried out in plastic sample containers, cylindrical in shape, 212.6 cm3 in volume, 9.5 cm in average diameter and 3 cm in height. The rock sample was crushed as fine as about 1 mm in grain size, and then the container was filled with about 300–400 gm of the crushed sample, sealed well and left for at least 21 days to accumulate free radon to attain radioactive equilibrium. The relation between the percentage of Rn-222 accumulation and time increased till reaching the steady state in about 38 days (Matolin, 1991).

3. Results and discussion

Firstly, the metamorphosed sandstone were investigated by spectrometric method and then were confirmed by α-track as a suitable technique for tracing radioactive and radioelement-bearing minerals.

3.1 Particle track analyses

Some samples from the different mineralized parts of metamorphosed sedimentary rocks were submitted to α-track analyses as a mean of micromapping the radioelement distribution. It is denoted that mineralization is mainly located along grain boundaries, in addition to being contained in crystal lattice of allanite, zircon and monazite figure 3a and 3b. It is clear that uranium mineralization increase in the highly fractured portions of the rock with remarkable epidotization and sericitization of plagioclase. The biotite and muscovitc grains are enriched in uranium mineralization along their outer boundaries and their cleavage planes figure 3a, 3b, 3c and 3d). It appears that biotite acts as reductant precipitating the uranium. Chlorites, sericeitized plagioclase and highly fractured quartz grains also contain uranium mineralization figure 3e and3 f. Most of the rock forming minerals is stained with iron oxides which occasionally adsorb uranium causing intense uranium mineralization figure 3g and 3h.

The highly uranium content of the metamorphosed sandstone is attributed to the presence of radioactive minerals like uranophane and autonite figure 3a and 3b) In addition accessory minerals like allanite, zircon and monazite figure 3c, 3d, 3e and 3f. Other U-bearing minerals are also recorded as biotite, muscovite, iron oxides and clays figure 3i and 3j. In the figure 3a and 3b, radioactive- mineralization along grain boundaries and in crystal lattice of allanite, zircon and monazite and its corresponding dense alpha track image showing radioactive minerals. Figure 3c, 3d, 3e and 3f) are biotite and muscovitc grains which enriched in uranium mineralization along their outer boundaries and their cleavage planes and its corresponding dense alpha track image showing radioactive minerals. Figure 3g and 3h are Chlorites, sericeitized plagioclase and highly fractured quartz grains also contain uranium mineralization. Figure 3i and 3j) are Iron oxides stained with uranium causing intense uranium mineralization and its corresponding dense alpha track image showing dense and disseminated radioactive minerals.

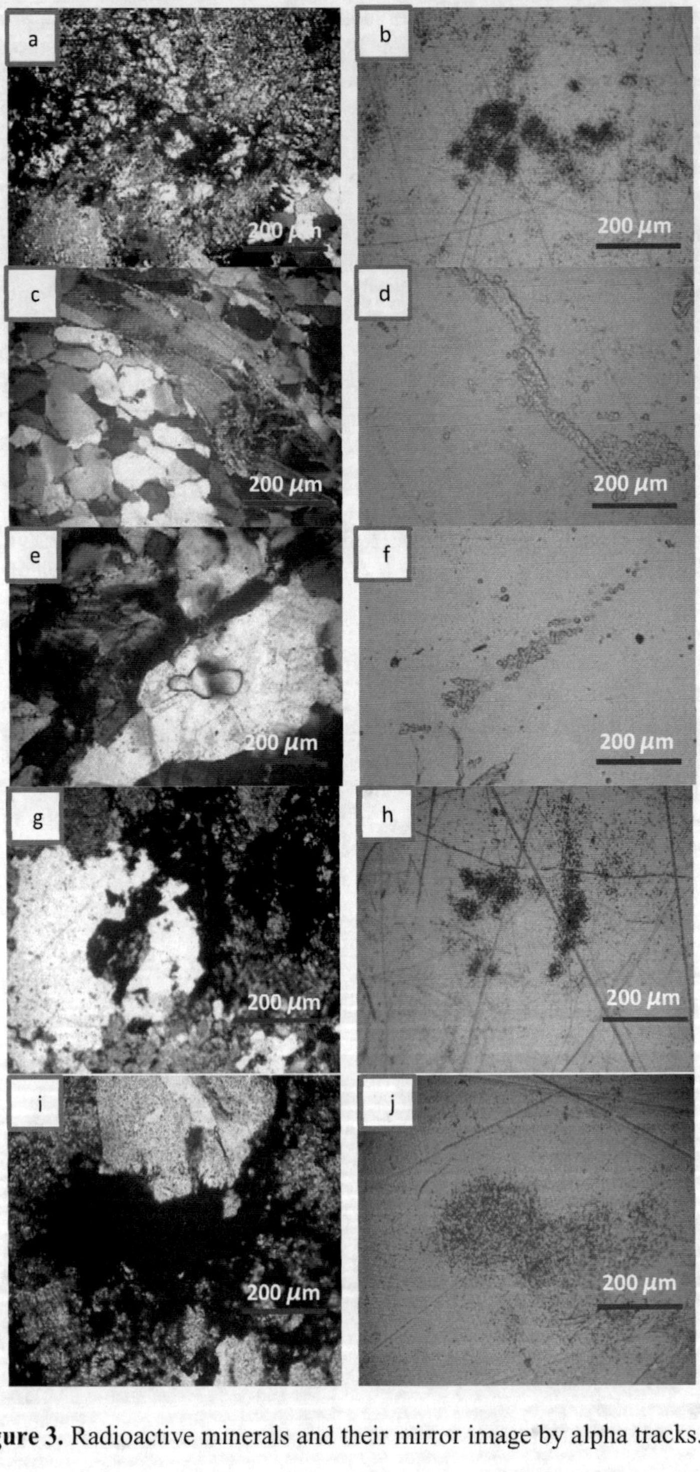

Figure 3. Radioactive minerals and their mirror image by alpha tracks.

3.2. *Activity concentrations the metamorphosed sandstone*

Activity concentrations of ^{238} U, 232Th, ^{226}Ra, ^{40}K were measured in the metamorphosed sandstone samples as in table (). The ^{238}U activity concentrations range between 37.2 and 520.4 with average 148.8 BqKg^{-1}, ^{232}Th activity concentrations vary between 8.08 and 366.99 BqKg^{-1} with 76.75 BqKg^{-1} as an average. ^{226}Ra ranges between 11.1 and 624.04 BqKg^{-1} with average 118.05 BqKg^{-1}. ^{40}K ranges between 11.1 and 1774.71 BqKg^{-1} with average 995.42 BqKg^{-1}. The world concentration limits of ^{238}U, ^{232}Th and ^{40}K are equal to 35, 30 and 400 BqKg^{-1}, respectively [94)]. The studied metamorphosed sandstones have higher values relative that recorded in the UNSCEAR (2000).

3.3. *Assessment of Hazard Indices*

3.3.1 *Absorbed Dose Rate in Air (D)*

The absorbed gamma dose rates in air at 1m above the ground surface for the uniform distribution of radionuclides (^{226}Ra, ^{232}Th and ^{40}K) were calculated by using equation 1 on the basis of guide lines provided by UNSCEAR (2000) and Ȍrgün et al. (2007).

$$D\ (\text{nGy h}^{-1}) = 0.462A_{Ra} + 0.604A_{Th} + 0.0417A_{K} \quad (1)$$

Where: A_{Ra}, A_{Th}, and A_K are the average specific activities of ^{226}Ra, ^{232}Th and ^{40}K in Bq/kg, respectively. The estimated indoor gamma dose rate values for metamorphosed sandstones samples are shown in table 1. The DR values for metamorphosed sandstones samples range from 22.07 to 463.98 nGy h^{-1} with a mean of 156.60 nGy h^{-1}. The mean DR values for all metamorphosed sandstone samples exceed the worldwide average value of soils (55 nGy h^{-1}), (Yang et al. 2005).

3.3.2 *Annual Effective Dose Equivalent (AEDE)*

The annual effective dose equivalent (AEDE) was calculated from the absorbed dose by applying the dose conversion factor of 0.7 Sv/Gy and the outdoor occupancy factor of 0.2, (UNSCEAR, 2000; Ȍrgün et al. 2007).

Furthermore, the average values of annual effective dose for all metamorphosed sandstone samples were also listed. The values obtained varied between 0.0.3 and 0.57 mSvy^{-1}. The mean value 0.19 found to be less that than 0.48 mSvy^{-1} recommended by UNSCEAR (2000) as the worldwide average of the annual effective dose.

3.3.3. *Radium Equivalent Activity (Ra_{eq}).*

The radium equivalent activity for the samples was calculated. The exposure to radiation (Tufail et al. 1992) can be defined in terms of the radium equivalent activity (Ra_{eq}), which can be expressed by the following equation:

$$Ra_{eq} = A_{Ra} + 10/7A_{Th} + 10/130A_{K} \quad (2)$$

Where: A_{Ra}, A_{Th} and A_K are the specific activities of Ra, Th and K, respectively, in Bq/kg. The maximum value of Ra_{eq} in metamorphosed sandstone samples must be less than 370 Bqkg^{-1} for safe use, i.e., to keep the external dose below 1.5 mSv y^{-1} (UNSCEAR, 1993; UNSCEAR, 2000). The range and the mean value of Ra_{eq} of metamorphosed sandstone samples are presented in table 1. The mean Raeq values for metamorphosed sandstone samples (304.26 Bqkg-1) are below the criterion limit of 370 Bqkg^{-1} except for some samples in the N-S shear zone figure 3. From the results it is evident that there are considerable variations in the Ra_{eq} of the same rock type originating from different shear zones. This fact is important from the point of view of selecting suitable materials for use in building and construction especially concerning those which have large variations in their activities. Large variation in radium equivalent activities may suggest that it is advisable to monitor the radioactivity levels of materials from a new source before adopting it for use as a building material (Kumar et al. 1999; Ngachin et al. 2007). The recommended maximum level of Radium equivalents for building materials used for homes is 370 $Bqkg^{-1}$ and for industries is 370- 740 $Bqkg^{-1}$, (Oresegun and Babalola, 1998).

3.3.4. External and Internal Hazard Index (Hex and Hin).

To limit the annual external gamma-ray dose (Saito, and Jacob, 1995; Saito et al. 1998; UNSCEAR, 2000) to 1.5 Gy for the samples under investigation, the external hazard index (Hex) is given by the following equation:

$$H_{ex} = A_{Ra}/370 + A_{Th}/259 + A_K/4810 \quad (3)$$

The internal exposure to ^{222}Rn and its radioactive progeny is controlled by the internal hazard index (H_{in}), which is given by

$$H_{in} = A_{Ra}/185 + A_{Th}/259 + A_K/4810 \quad (4)$$

These indices must be less than unity in order to keep the radiation hazard insignificant. Values of eU, eTh and Ra(eU) in ppm as well as K in %, were converted to activity concentration, Bq kg^{-1}, using the conversion factors given by Polish Central Laboratory for Radiological Protection (Malczewski et al. 2004).

The value of this index must be less than unity for the radiation hazard to be negligible, i.e. the radiation exposure due to radioactivity in construction materials must be limited to 1.5 mSv y^{-1}. The values of external and internal hazard indices H_{ex} and H_{in} for the studied metamorphosed sandstone samples range between 0.13 and 2.77 and 0.23 to 3.76 respectively. External and internal hazard indices are higher than unity for some of the studied samples indicating that many of these samples cannot be used as building or decorative material of dwelling.

3.3.5. Activity Concentration Index (Iγ).

Another radiation hazard index called the representative level index Iγ is defined as follows (NEA-OECD and Nuclear Energy Agency, 1979).

$$I\gamma = ARa/150 + ATh/100 + AK/1500 \quad (5)$$

Where: A_{Ra}, A_{Th} and A_K are the activity concentrations of 226Ra, 232Th and 40K, respectively in Bq/kg (Abbady et al. 2005). The safety value for this index is ≤ 1 (El-Galy et al. 2008; El-Aassy et al. 2012).

The values of the range and mean gamma-index (Iγ) are presented in table 1, for raw materials. It is observed that the mean Iγ values (2.42) of the metamorphosed sandstone samples which are higher dose criterion (0.3 mSv/y) and correspond to an activity concentration index of $2 \leq I\gamma \leq 6$ proposed by EC (1999), for materials used in bulk construction.

Contribution of ^{238}U, ^{232}Th and ^{40}K radionuclides for the absorbed dose rate within the studied rocks are plotted in figure 4. It is clear that ^{40}K plays the main and most important role in dose rate contribution.

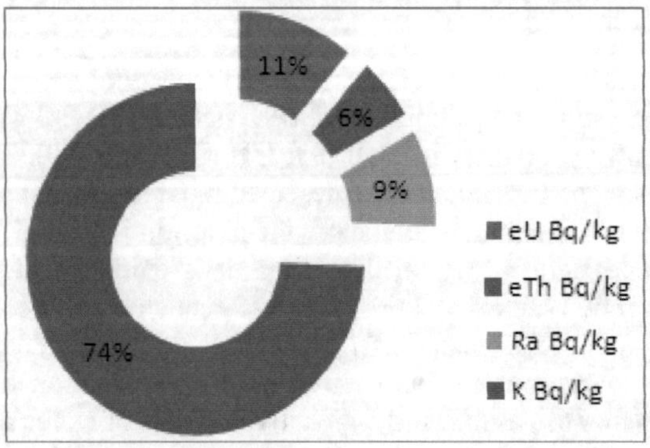

Figure 4. Contribution of ^{238}U, ^{232}Th and ^{40}K radionuclides for the absorbed dose rate within the studied rock samples.

Table 1. Results of radionuclide concentrations, the dose rate (D), the annual effective dose equivalent (AEDE), radium equivalent activity (Ra_{eq}), external (H_{ex}), internal (H_{in}) hazard indices and gamma index (I_γ) for metamorphosed sandstone samples.

S. No.	eU Bq/kg	eTh Bq/kg	Ra Bq/kg	K Bq/kg	Abs. Dose nGy/h	Eff. Dose mSv	Ra_{eq}	H_{ex}	H_{in}	I_γ
1	86.80	44.44	44.40	1208.18	117.32	0.14	200.82	0.66	0.89	1.83
2	49.60	36.36	55.50	1048.55	88.60	0.11	188.10	0.49	0.63	1.39
3	62.00	28.28	44.40	873.27	82.14	0.10	151.97	0.46	0.63	1.28
4	106.39	67.43	88.91	845.10	125.12	0.15	250.24	0.72	1.01	1.95
5	62.00	8.08	55.50	478.89	53.49	0.07	103.88	0.30	0.47	0.81
6	99.20	40.40	44.40	1079.85	115.26	0.14	185.18	0.65	0.92	1.79
7	37.20	52.52	55.50	1452.32	109.47	0.13	242.25	0.61	0.71	1.74
8	364.31	366.99	624.04	1774.71	463.98	0.57	1284.83	2.77	3.76	7.28
9	86.80	20.20	44.40	1026.64	95.11	0.12	152.23	0.53	0.76	1.47
10	134.54	80.96	94.13	1136.19	158.44	0.19	297.19	0.91	1.28	2.46
11	103.17	40.32	108.56	738.68	102.82	0.13	222.98	0.59	0.87	1.58
12	130.32	81.33	110.67	1161.23	157.75	0.19	316.17	0.91	1.26	2.46
13	243.41	157.12	191.70	1264.52	260.08	0.32	513.42	1.53	2.19	4.04
14	37.20	8.08	11.1	11.1	22.07	0.03	11.54	0.13	0.23	0.33
15	89.40	68.11	79.48	920.22	120.82	0.15	247.57	0.70	0.94	1.89
16	248.00	181.80	133.20	112.68	229.08	0.28	401.58	1.40	2.07	3.55
17	114.08	87.26	97.13	754.33	136.87	0.17	279.81	0.80	1.11	2.14
18	234.36	208.22	130.43	237.88	243.96	0.30	446.18	1.49	2.12	3.80
19	248.00	163.82	158.18	410.03	230.62	0.28	423.75	1.39	2.06	3.56
20	74.40	60.60	77.70	870.14	107.26	0.13	231.21	0.62	0.82	1.68
21	520.43	21.98	536.13	694.86	282.69	0.35	620.98	1.64	3.04	4.15
22	198.40	68.68	99.90	973.43	173.74	0.21	272.89	1.00	1.54	2.66
23	124.00	60.60	111.00	867.01	130.04	0.16	264.26	0.75	1.08	2.01
24	212.54	120.84	169.05	1690.20	241.66	0.30	471.69	1.39	1.97	3.75
25	308.14	11.47	179.93	729.29	179.70	0.22	252.42	1.03	1.86	2.66
26	62.00	84.84	88.80	1302.08	134.18	0.16	310.16	0.77	0.93	2.13
27	86.80	8.08	55.50	1599.43	111.68	0.14	190.08	0.60	0.83	1.73
28	86.80	8.08	55.50	1414.76	103.98	0.13	175.87	0.56	0.79	1.60
29	210.80	64.64	177.60	109.55	141.00	0.17	278.37	0.84	1.41	2.12
30	104.28	50.90	100.12	970.30	119.39	0.15	247.48	0.68	0.96	1.85
31	186.00	88.88	77.70	635.39	166.11	0.20	253.55	0.98	1.48	2.55
32	86.80	52.52	66.60	1230.09	123.12	0.15	236.25	0.69	0.93	1.92
33	186.00	76.76	88.80	1333.38	187.90	0.23	301.02	1.08	1.58	2.90
34	111.60	76.76	99.90	1395.98	156.13	0.19	316.94	0.89	1.19	2.44

IOP Conf. Series: Materials Science and Engineering **975** (2020) 012017 doi:10.1088/1757-899X/975/1/012017

Table 1. Continuous.

S. No.	eU Bq/kg	eTh Bq/kg	Ra Bq/kg	K Bq/kg	Abs. Dose nGy/h	Eff. Dose mSv	Ra_{eq}	H_{ex}	H_{in}	I_{γ}
35	148.80	84.84	88.80	1295.82	174.02	0.21	309.68	1.00	1.40	2.70
36	124.00	72.72	99.90	1295.82	155.25	0.19	303.46	0.89	1.22	2.42
37	124.00	68.68	77.70	1280.17	152.15	0.19	274.29	0.87	1.20	2.37
38	148.80	64.64	77.70	1543.09	172.14	0.21	288.74	0.97	1.37	2.67
39	111.60	88.88	122.10	1001.60	147.01	0.18	326.12	0.85	1.15	2.30
40	198.40	92.92	111.00	1061.07	192.03	0.24	325.36	1.12	1.65	2.96
av	148.8	76.75	118.05	995.42	156.60	0.19	304.26	0.91	1.31	2.42
Min	37.2	8.08	0	0	22.067	0.03	11.543	0.13	0.23	0.33
Max	520.4	366.99	624.04	1774.71	463.98	0.57	1284.83	2.77	3.76	7.28

4. Conclusion

The field mapping indicates that metamorphosed sandstones occur in two sites at Wadi. Sikait (Sikait-1 and Sikait-2) and the exposed rocks at this area can be arranged based on field observations and structural relations from older to younger as follows:

1. Veins (fluorite and quartz) youngest
2. Lamprophyre dykes
3. Porphyritic granites
4. Metamorphosed sandstones
5. Ophiolitic mélange oldest

High uranium content of the metamorphosed sandstone is attributed to presence radioactive minerals like uranophane and autonite, in addition to accessory minerals like allanite, zircon and monazite. Other U-bearing minerals are also recorded as biotite, muscovite, iron oxides and clays. Absorbed Dose Rate (D), annual effective dose equivalent (AEDE), radium equivalent activity (Ra_{eq}), external (H_{ex}) and internal (H_{in}) hazard index, in addition to activity gamma index (I_{γ}) caused by gamma emitting natural radionuclide are determined from the obtained values of ^{226}Ra, ^{232}Th and ^{40}K. Fairly, many of the studied metamorphosed sandstone samples are higher than the international limits.

5. References

[1] Assaf, H. S., Ibrahim, M. E., Zalata, A.A., El-Metwally, A. and saleh, G. M. (2000): Polyphase folding in Nugrus- Sikeit area, south Eastern Desert, Egypt. JKAW: Earth Sci., 12, 1-16 p.

[2] EC (European Commission) (1999): Radiation protection 112. Radiological protection principles concerning the natural radioactivity of building materials. Directorate-General Environment, Nuclear Safety and Civil Protection.

[3] El-Aassy I. E., Afaf A. Nada, El-Galy, M.M., El-Feky M.G., Abdel Maksoud, T.M., Talaat, Sh.M., Ibrahim, E.M. (2012): Behavior and environmental impacts of radionuclides during the hydrometallurgy of calcareous and argillaceous rocks, Southwestern Sinai, Egypt. Applied Radiation and Isotopes, 70, 1024–1033.

[4] El-Galy, M.M., El-Mezayen, A.M., Said, A.F., El Mowafy, A.A., Mohamed, M.S. (2008): Distribution and environmental impacts of some radionuclides in sedimentary rocks at Wadi Naseib area, Southwest Sinai, Egypt. Jour. Environ. Radioact. 99, 1075–1082.

[5] Florou, H., kritidis, p. (1992): Gamma radiation measurements and dose rate in the coastal areas of a volcanic island, Aegean Sea, Greece.radiat. Port. Dosim. 45(1/4), 277-279.

[6] Hashad A.H. and El Reedy M.W.M. (1979) Geochronology of the anorogenic alkalic rocks, South Eastern Desert, Egypt [J]. *Ann. Geol. Surv. Egypt*. 9, 81–101.

[7] Hassan, M.A. and Hashad, A.H., (1990): Precambrian of Egypt. In: Said R. (ed.). The geology of Egypt, Balkema, Rotterdam, p. 201-245

[8] Hegazy, H.M. (1984): Geology of Wadi El Gemal area, Eastern Desert, Egypt. Ph. D. Thesis, Assiut Univ., Egypt. 271p.

[9] Ibrahim, M. E., Saleh, G. M. and Ibrahim, W. S. (2010): Low grade metamorphosed sandstone-type uranium deposit, Wadi Sikait, South Eastern Desert, Egypt, Journal of Geology and Mining Research, 2(6), 129-141.

[10] Ibrahim, M. E., Abdel-Wahed, A. A., Oraby, F., Abu El-Hassan, El Galy, M.M. and Watanabe, K. (2007): factors controlling mineralization in lamprophyre dyke, Abu Rushied area, Eastern Desert, Egypt. 5 th International Conf. on the geology of Africa. Vol. (1), 79- 92. Assiut – Egypt.

[11] Ibrahim, M. E.,Amer, T. E.and Saleh, G. M. (1999): New occurrence of some nuclear materials and gold mineralization at Wadi Sikait area , south Eastern Dessert, Egypt .First seminar on Nuclear Raw Materials and their technology, Cairo, Egypt, 271-284p.

[12] Kumar, V., Ramachandran, T. V., and Prasad, R. (1999): Natural radioactivity of Indian building materials and by-products. Applied Radiation and Isotopes, 51, 93- 96.

[13] Malczewski D.,Teper L. and Dorda J. (2004): "Assessment of natural and anthropogenic radioactivity levels in rocks and soils in the environs of Swieradow Zdroj in studentes, Poland, by insitu gamma-ray spectrometry". Journal of Environmental Radioactivity, V.73, P. 233-245.

[14] Matolin M.M. (1991): "Construction and Use of Spectrometric Calibration Pads", Laboratory of Gamma-ray Spectrometry, N.M.A., Egypt [R]. A report to the Government of the Arab Republic of Egypt. Project Egy. 4, 030–03, IAEA.

[15] Mohamed F.H., and Hassanen M.A. (1997). Geochemistry and petrogenesis of Sikait leucogranite, Egypt: an example of S-type granite in a metapelitic sequence [J].Geol Rundsch. 86, 81-92.

[16] NEA-OECD, Nuclear Energy Agency, (1979): Exposure to Radiation from Natural Radioactivity in Building Materials. Report by NEA Group of Experts. OECD, Paris.

[17] Ngachin, M., Garavaglia, M., Giovani, C., Kwato Njock, M.G and Nourreddined, A. (2007): Assessment of natural radioactivity and associated radiation hazards in some Cameroonian building materials, Radiat. Meas., 42, 61 – 67.

[18] Omar, S. A. (1995): "Geological and geochemical features of the radioactive occurrences South G. Um Anab granitic masses, Eastern Desert, Egypt". M. Sc. Thesis, Cairo Univ., 164p.

[19] Oresegun, M. O., & Babalola, A. I. (1998): Annual indoor dose burden estimates in dwellings built in Nigeria with radioactive U-Th rich tailings. In Proceedings of an international conference on radiation protection in nuclear energy (Vol. 2). ; (pp. 159e166).Vienna, Austria: IAEA.

[20] Örgün ,Y., Altınsoy N., S¸ Ahin, S.Y., Gu¨ngo¨r, Y., Gu¨ltekin, A.H., Karahan, G., Karacık, Z., (2007): Natural and anthropogenic radionuclides in rocks and beach sands from Ezine region (Canakkale), Western Anatolia, Turkey. Appl. Radiat. Isot. 65:739–747

[21] Saito, K., Petoussi, H., Zanki, M. (1998): Calculation of the effective dose and its variation from environmental gamma ray sources. Health Phys. 74, 698–706.

[22] Saito, K., Jacob, P. (1995): Gamma ray fields in the air due to sources in the ground. Radiat. Prot. Dosim. 58, 29-45.

[23] Saleh, G. M., (1998): The potentiality of uranium occurrences in Wadi-Nugrus area, South Eastern Desert: Unpubl. Ph.D. thesis, Faculty of Science, Mansoura University, 171 p.

[24] Tufail, M., Ahmad, N., Mirza, S.M., Mirza, N.M., Khan, H.A. (1992): Regular paper, natural radioactivity from the building materials used in Islamabad and Rawalpindi, Pakistan. Sci. Total Environ. 121, 283–291.

[25] UNSCEAR (2000): United Nations Scientific Committee on the effects of atomic radiation, sources and effects of ionizing radiation. Report to General Assembly, with Scientific Annexes United Nations. United Nations, New York.

[26] UNSCEAR (1993): Sources and effects of ionising radiation. Report to general assembly, with scientific annexes, United Nations, New York.
[27] Yang, Y., Wu, X., Jiang, Z., Wang, W., Lu, J., Lin, J., Wang, L., Hsia, Y. (2005): Radioactivity concentration in soils of the Xiazhuang granite area. China. Appl. Radiat. Isot. 63, 255–259.

IOP Conf. Series: Materials Science and Engineering **975** (2020) 012018 doi:10.1088/1757-899X/975/1/012018

Mineralogical and radiometric studies of granitic rocks in Wadi Sabbagh area, south Sinai, Egypt

Mira, H. I.[2]., El-Gharbawy, R.I.[1], Elmowafy, A.A.[2], Osman, A.F.[1]and El maadawy, W.M.[2]

[1] Geology Department, Faculty of Science, Ain Shams University, Egypt.
[2] Research sector, Nuclear Materials Authority, Cairo, Egypt.

Abstract. Wadi Sabbagh area is located at southern part of Sinai Peninsula. It is constrained by Latitudes 28° 05′ and 28° 15′ N and Longitudes 34° 00′ and 34° 15′ E. Gneisses, migmatites, metasediments, syn-and post-orogenic granites, in addition to pegmatites and post-granitic dykes are the main rock units encountered in this area. Geological, petrological and geochemical studies indicate that there are two distinct suites of granitoids: older tonalite to granodiorite assemblage and younger two phases (monzo- to syenogranites and alkaline granites). The field radiometric survey and lab investigations indicate that, the anomalous uranium contents are mainly restricted to the pegmatites and altered monzogranites. The eU contents of pegmatites reach up to 168 ppm with an average 117 ppm, while the eTh reach up to 257 ppm with an average 175 ppm. The eTh/eU ratios range from 1.1 to 2.4 with an average 1.6. In the altered monzogranites, eU reaches up to 305 ppm with an average of 126 ppm. The eTh reaches up to 382 ppm with an average of 166 ppm. The average values of Ra and K are 60 ppm and 2.93 wt percentage, respectively. The average values eTh/eU ratios in the examined anomalies monzgranites (1.56) indicating addition of uranium. The high eU and eTh contents may be attributed to the secondary ascending hydrothermal solutions that lead to the accumulation of radioactive mineralizations mostly along fractures, faults and shear districts. The recognized radioactive minerals in the studied pegmatites are thorite, uranothorite, zircon, fluorite, columbite, samarskite, monazite, xenotime and allanite, whereas thorite, uranothorite, zircon and fluorite are the foremost radioactive minerals in the altered younger granites. Additional accessory minerals such as ilmenite, magnetite, hematite, pyrite and rutile were found in the studied pegmatites. Hematite, magnetite and pyrite are the foremost accessories accompanying the radioactive minerals in the altered younger granites.
Keywords. Sabbagh, alkaline granite, monzogranite, pegmatite, dykes, radioactive,hydrothermal solutions.

1. Introduction

Wadi Sabbagh area occupies a part of south Sinai, Egypt figure 1, that featured by the presence of subparallel ridges and low to moderately elevated hills, mostly formed of older and younger granites. Gneisses, migmatites, metasediments, syn- and post-orogenic granites, in addition to pegmatites and post-granitic dykes are the main rock units encountered in this area.

IOP Conf. Series: Materials Science and Engineering **975** (2020) 012018 doi:10.1088/1757-899X/975/1/012018

The younger granites, which crop out as elongated belt trending MW-SE and NW-SE directions, occupy about 93% of the total exposed area. Their contacts with all the surrounding rock units are generally sharp and well defined.

The present study deals with field, mineralogy and radioactivity of the younger granites and associated pegmatites in order to identify the radioactive minerals responsible for mineralization.

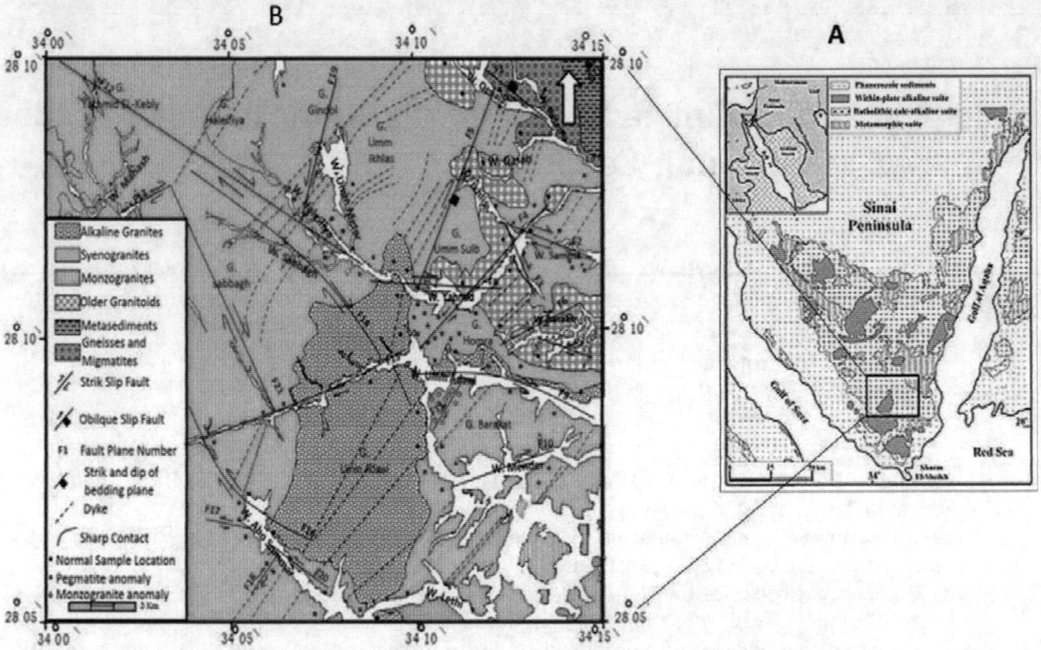

Figure1. A) Simplified geologic map of the late Proterozoic rocks in south Sinai [1] inset a location map of the study area, B) Geologic map of Wadi Sabbagh area, southeastern Sinai, Egypt**.**

2. Methodology

2.1 Field measurements

In situ gamma ray spectrometry measurements have been carried out using a GS-256 spectrometer (designed by Geofyzika Brno-Czech Republic) with a 3" ×3" sodium iodide (Thalium) [NaI (Tl)] crystal detector. The device, having an automatic dead-time correction and an internal ^{137}Cs source allows the spectrometer to automatically maintain system gain stability, is measured over a large body of water. Before field measurements, the spectrometer is calibrated on concrete pads containing known concentrations of U, Th and K. This calibration provides the stripping ratios and sensitivities required for correcting the measured eU, eTh and K.

The measurements are based on the detection of γ-radiation emitted in the decay of ^{214}Bi (^{238}U series) at 1.76 Mev, ^{208}Tl (^{232}Th series) at 2.41 MeV, and the primary decay of potassium ^{40}K (1.46) MeV is measured directly.

The determinations of uranium and thorium are based on the assumption that the daughter nuclides are in equilibrium with the parent nuclides that is none of the intermediate steps in the decay series has been disrupted. Consequently, the deduced amounts of uranium and thorium are equivalent to what would be in equilibrium with the measured radioactivity of the bismuth or thallium isotopes. Therefore, the results of our γ-ray analyses are expressed as equivalent uranium (eU) and equivalent thorium (eTh) [2].

2.2 Lab measurement

Representative grab samples were collected from the studied rocks of Wadi Sabbagh representing the highest values of anomalous field radioactivity. These samples were prepared for gamma-ray spectrometric analysis in order to determine their uranium, thorium, radium and potassium contents by using multichannel analyzer of gamma-ray detector (Gamma-Spectrometer technique). The instrument used in determination of the four radioactive elements consists of a Bicron scintillation detector NaI (Tl) 76x76 mm, hermetically sealed with the photomultiplier tube in aluminum housing. The tube is protected by a copper cylinder protection of thickness 0.6 cm against induced X ray and a chamber of lead bricks against environmental radiation. Uranium, thorium, radium and potassium are measured by using four energy regions representing ^{234}Th, ^{212}Pb, ^{214}Pb and ^{40}K at 93 kV, 239 kV, 352 kV, and 1460 kV for uranium, thorium, radium and potassium, respectively. The measurements were carried out in sample plastic containers, cylindrical in shape, 212.6 cm^3 volumes with 9.5 cm average diameter and 3 cm height. The rock sample is crushed to about 1 mm grain size, and then the container is filled with about 300-400 gm of the crushed sample sealed well and left for at least 21 days to accumulate free radon to attain radioactive equilibrium. The relation between the percentage of ^{222}Rn accumulation and time increase till reaching the steady stage after about 38 days [3]. Mineral separation and identification were carried out at the Laboratories of the Nuclear Materials Authority (NMA). The heavy liquids separation technique using bromoform of specific gravity 2.85 gm cm^{-3} was used to concentrate the heavy minerals. Then, the separation of magnetite was achieved by hand magnet. The heavy mineral fractions were passed through a Frantz isodynamic magnetic to separate the remaining magnetite and produce several magnetic fractions at 0.2, 0.5, 0.7, 1, and 1.5 amperes. Each of these fractions contained its own characteristic minerals. Mineral identification was performed through environmental scanning electron microscope (ESEM) and X-ray diffraction (XRD).

2.3 Field radiometric survey

The field radiometric survey stated that the exhibited rock-units of the studied area are commonly different in their radioactivity concentrations, amongst the granitic rocks, Wadi Umm Adawi monzogranites have the highest radioactive levels, varied between 1000 and 2000 cps. This reveals the function of post magmatic alterations, in addition to fracturing, joints and shearing enriched the radioactivity. The pegmatites exhibit high values of radioactivity (600 to 1200 cps.)

2.4. Lab radioactive survey

2.4.1 Older granites.

The older granites are relatively depleted in eU, eTh, Ra and K contents as compared with the younger granites, the average contents of eU in the tonalite and granodiorite are 3.4 and 5.2 ppm, respectively, whereas the average contents of eTh in these rocks are 8.2 ppm and 11.4 ppm, respectively table 1. All these averages are lower than twice the Clark value (< 8 ppm U and/or 20 ppm Th). The eU content of the two varieties of older granites rest inside the range of acidic intrusives (1-6 ppm) given by Adams et al., 1956 and the low content of K (1.6 and 1.2%) may be related to the low content of K-feldspar minerals in these rocks [4-5].

2.4.2 Younger granites.

Equivalent uranium content ranges from 4-10 ppm in the monzogranites, 5-14 ppm in syenogranites and from 7-15 ppm in alkaline granites (with an average 6.0 ppm, 9.8 ppm and 11.8 ppm, respectively), which are slightly higher than the average of granitic rocks (U; 4 ppm) [6] table 2. On the other hand, the equivalent thorium content of the younger granites varies from 8 ppm to 19 ppm, 11 ppm to 32 ppm and 17 ppm to 28 ppm with an average 13.5, 19.3 and 21.1 ppm for monzogranites, syenogranites and alkaline granites, respectively. The average of thorium content for the studied monzogranites is lesser than the average of granitic rocks (Th; 19 ppm), whereas syenogranites and alkaline granites average contents are slightly higher than the average of granitic rocks.

Table (1): eU, eTh, Ra and K contents and eU/eTh, eTh/eU and eU/Ra ratios of the studied older granites compared with different local and World averages (10 Samples).

Rock type	Sample no.	eU (ppm)	eTh (ppm)	Ra (ppm)	K%	eU/eTh	eTh/eU	eU/Ra
a)Tonalite	O1	4	11	3	1.95	0.4	2.8	1.3
	O2	3	9	3	1.18	0.3	3.0	1.0
	O3	5	8	4	1.55	0.6	1.6	1.3
	O4	2	5	1	1.47	0.4	2.5	2.0
	O5	3	8	2	1.82	0.4	2.7	1.5
	Average	3.4	8.2	2.6	1.6	0.4	2.5	1.4
b)Granodiorite	O6	4	8	3	1.08	0.5	2.0	1.3
	O7	5	10	4	1.12	0.5	2.0	1.3
	O8	6	13	6	1.07	0.5	2.2	1.0
	O9	3	7	2	1.42	0.4	2.3	1.5
	O10	8	19	7	1.2	0.4	2.4	1.1
	Average	5.2	11.4	4.4	1.2	0.5	2.2	1.2
Older granites of(El Sayed, 1993)		8	17	-	-	-	2.12	-
Older granites of (Sallam, 2006)		2.8	9.3	-	-	-	3.8	-
Older granites of (Abbas, 2008)		3.3	5.8	-	-	-	2.3	-
Acidic intrusive of (Adams et al., 1956)		1-6	1-25	-	-	-	2.6	-

2.4.3 Dykes

The acidic dykes are relatively more enriched in the radionuclides, if compared with the basic dykes. They have eU average content reach up to 9 ppm with an average value 6 ppm, the eTh reaches up to 19 ppm with an average value 12 ppm and the average contents of Ra and K are 5 ppm and 3.7 %, respectively table 3. On the other hand, the average values of eU in the studied acidic dykes show slightly decrease in U contents, but increase in Th contents in comparison with West Dahab area **[7]** and it is in the range of acidic intrusive and effusive given by Adams et al., 1956 [8] table 3.

The field radiometric survey, using γ- gun spectrometry, showed two anomalous zones in the examined pegmatite pockets and veins hosted in syenogranites at the upstream of Wadi Umm Adawi and Wadi Umm Meter, in addition anomalous monzogranites at the downstream of Wadi Umm Adawi figure 1. Thirty representative samples (15 from the pegmatites and 15 from the monzogranites) collected from these regions for the radiometric measurements, table 4 and 5.

Table 2. eU, eTh, Ra and k concentration and eU/eTh, eTh/eU and eU/Ra ratios in the studied younger granites compared with different local and World rocks. (30 samples; 10 monzogranites, 10 syenogranites and 10 alkaline granites)

Rock type	S.N	eU(ppm)	eTh(ppm)	Ra(ppm)	K%	eU/eTh	eTh/eU	eU/Ra
a)Monzogranites	Y1	4	8	2	3.06	0.5	2.0	2.0
	Y2	4	12	5	3.69	0.3	3.0	0.8
	Y3	7	19	5	4.85	0.4	2.7	1.4
	Y4	4	8	3	4.47	0.5	2.0	1.3
	Y5	4	13	6	4.57	0.3	3.3	0.7
	Y6	8	18	5	3.53	0.4	2.3	1.6
	Y7	10	17	4	3.21	0.6	1.7	2.5
	Y8	8	17	4	4.16	0.5	2.1	2.0
	Y9	5	12	5	3.48	0.4	2.4	1.0
	Y10	6	11	5	3.89	0.5	1.8	1.2
	Average	6.0	13.5	4.4	3.9	0.4	2.3	1.5
b)Syenogranites	Y11	7	24	7	3.74	0.29	3.4	1.0
	Y12	5	15	4	4.12	0.33	3.0	1.3
	Y13	14	21	8	4.32	0.67	1.5	1.8
	Y14	10	17	8	3.91	0.59	1.7	1.3
	Y15	12	18	9	4.4	0.67	1.5	1.3
	Y16	8	11	7	3.75	0.73	1.4	1.1
	Y17	8	15	6	3.65	0.53	1.9	1.3
	Y18	9	16	8	4.15	0.56	1.8	1.1
	Y19	11	24	7	3.88	0.46	2.2	1.6
	Y20	14	32	8	3.67	0.44	2.3	1.8
	Average	9.8	19.3	8	3.78	0.53	2.1	1.4
C)Alkaline granites	Y21	15	21	12	3.22	0.35	2.9	1.8
	Y22	12	21	8	4.08	0.57	1.8	1.5
	Y23	10	18	7	4.11	0.56	1.8	1.4
	Y24	16	24	12	3.65	0.67	1.5	1.3
	Y25	8	17	6	4.15	0.47	2.1	1.3
	Y26	7	20	4	4.66	0.35	2.9	1.8
	Y27	9	19	8	3.47	0.47	2.1	1.1
	Y28	17	25	11	4.31	0.68	1.5	1.5
	Y29	11	18	9	3.54	0.61	1.6	1.2
	Y30	13	28	10	3.16	0.46	2.2	1.3
	Average	11.8	21.1	8.7	3.84	0.52	2.0	1.4
Younger granites of (Abbas, 2008)		10.6	27.1	-	-	-	2.3	-
Younger granites of (Sallam, 2006)		7.5	17.5	-	-	-	2.8	-
Granitic rocks of (Rogers and Adams, 1967)		2.2-15	-	-	-	-	-	-
Acidic intrusive of (Adams et al., 1956)		1-6	1-25	-	-	-	2.6	-

IOP Conf. Series: Materials Science and Engineering **975** (2020) 012018 doi:10.1088/1757-899X/975/1/012018

Table (3): eU, eTh, Ra and K concentrations and eU/eTh, eTh/eU and eU/Ra ratios in the studied acidic and basic dykes compared with different localities. (11 samples; 8 acidic dykes and 3 basic dykes)

Rock type	Sample no.	eU (ppm)	eTh (ppm)	Ra (ppm)	K%	eU/eTh	eTh/eU	eU/Ra
Acidic dykes	GD1	7	14	5	3.97	0.50	2.0	1.4
	GD2	4	7	3	3.75	0.57	1.75	1.3
	GD3	5	8	4	4.4	0.63	1.6	1.3
	GD4	8	10	5	3.56	0.8	1.3	1.6
	GD5	9	16	7	2.79	0.56	1.8	1.3
	GD6	3	11	2	3.68	0.27	3.7	1.5
	GD7	8	19	6	3.45	0.42	2.4	1.3
	GD8	7	14	5	3.97	0.5	2.0	1.4
	Average	6	12	5	3.70	0.53	2.05	1.4
Basic dykes	GD9	3	2	2	0.68	1.50	0.67	1.5
	GD10	2	5	1	0.95	0.40	2.50	2.0
	GD11	3	7	2	1.79	0.43	2.33	1.5
	Average	3	5	2	1.14	0.78	1.8	1.7
Dykes of El (Sayed, 1993)	Acidic	7	11	-	-	-	1.5	-
	Basic	4	6	-	-	-	1.5	-
Acidic effusiveof (Adams et al., 1956)		2-7	9-25	-	-	-	4.7	-
Acidic intrusive of (Adams et al., 1956)		1-6	1-25	-	-	-	2.6	-
Basic effusiveof (Adams et al., 1956)		0.2-4	0.5-10	-	-	-	3.7	-

2.4.4 Anomalous pegmatites

The anomalous pegmatites are highly enriched in eU, eTh and Ra contents compared to their hosting syenogranites. Their eU content reaches up to 168 with an average of 117 ppm. The eTh content reaches up to 257 with an average of 175 ppm, whereas the Ra content reaches up to 148 with an average of 89 ppm (Table 4). These averages of uranium and thorium contents are higher than those of the world uraniferous pegmatites (av= 28 and 21 ppm for U and Th, respectively) [9]. These averages are also higher than those reported for the Egyptian uraniferous pegmatites (av= 33 and 28 ppm for U and Th, respectively), [10]. The low values of eTh/eU ratios (av = 1.6) of the measured samples suggests that the examined pegmatites are uraniferous [11]. When the average values of both eU and Th in these pegmatites are compared with the average of the Gabal Sheikh El-Arab area, South Sinai, Egypt given by sayed, 1993; Sallam, 2006, the former have lower averages of both eU and eTh table 4.

IOP Conf. Series: Materials Science and Engineering **975** (2020) 012018 doi:10.1088/1757-899X/975/1/012018

Table 4. eU, eTh, Ra and K concentrations and eU/eTh, eTh/eU and eU/Ra ratios in the studied anomalous pegmatites compared with different local and World rocks. (15 samples)

Sample no.	eU (ppm)	eTh (ppm)	Ra (ppm)	K%	eU/eTh	eTh/eU	eU/Ra
P1	155	182	76	0.72	0.85	1.2	2.0
P2	90	150	75	3.32	0.60	1.7	1.2
P3	117	153	74	1.6	0.76	1.3	1.6
P4	126	144	94	0.37	0.88	1.1	1.3
P5	168	251	148	1.39	0.67	1.5	1.1
P6	130	156	117	4.09	0.83	1.2	1.1
P7	136	208	121	1.25	0.65	1.5	1.1
P8	123	163	112	3.71	0.75	1.3	1.1
P9	132	178	102	1.75	0.74	1.4	1.3
P10	95	146	69	0.95	0.65	1.5	1.4
P11	111	148	92	0.59	0.75	1.3	1.2
P12	115	169	80	3.86	0.68	1.5	1.4
P13	80	178	67	0.75	0.45	2.2	1.2
P14	114	257	91	0.62	0.44	2.3	1.3
P15	60	141	13	0.65	0.43	2.4	4.6
Average	117	175	89	1.7	0.68	1.6	1.5
Pegmatites of (Sallam, 2006)	157	242.3	-	-	-	1.6	-
Pegmatites of (El Sayed,1993)	162	80	-	-	-	0.49	-

2.4.5 Anomalous monzogranites

The anomalous monzogranite samples at Wadi Umm Adawi area display the greatest values of eU and eTh contents, the eU reachs up to 305 ppm with an average value 126 ppm, the eTh reaches up to 382 ppm with an average value 166 ppm (Table 5). The average values of Ra and K are 60 ppm and 2.93 %, respectively. High eU and eTh contents might be related to the presence of titanite, metamict zircon, opaques, allanite, thorite and uranothorite. The average values eTh/eU ratios in the studied anomalies monzogranites (1.56) exceed those of Sallam, 2006 (0.5) [12] and decrease than those of Abbas, 2008 [13] (4.3), table 5.

IOP Conf. Series: Materials Science and Engineering **975** (2020) 012018 doi:10.1088/1757-899X/975/1/012018

Table 5. eU, eTh, Ra and K concentrations and eU/eTh, eTh/eU and eU/Ra ratios in the studied anomalous monzogranites compared with different localities. (15 samples)

Sample no.	eU (ppm)	eTh (ppm)	Ra (ppm)	K%	eU/eTh	eTh/eU	eU/Ra
GR1	305	285	127	0.86	1.07	0.93	2.4
GR2	66	126	50	1.22	0.52	1.9	1.3
GR3	94	109	76	2.16	0.86	1.16	1.2
GR4	228	175	63	5.41	1.3	0.77	3.6
GR5	261	195	58	2.46	1.34	0.75	4.5
GR6	251	382	100	1.87	0.66	1.52	2.5
GR7	55	94	29	2.15	0.59	1.71	1.9
GR8	77	235	65	5.09	0.33	3.05	1.2
GR9	130	100	74	1.67	1.3	0.77	1.8
GR10	75	163	68	2.82	0.46	2.17	1.1
GR11	28	31	14	5.88	0.9	1.11	2.0
GR12	41	60	20	2.59	0.68	1.46	2.1
GR13	21	37	8	5.59	0.57	1.76	2.6
GR14	165	188	93	1.67	0.88	1.14	1.8
GR15	97	315	54	2.52	0.31	3.25	1.8
Average	126	166	60	2.93	0.78	1.56	2.1
Anomalous granites of (Abbas, 2008)	376.3	1624.7	-	-	-	4.3	-
Anomalous granites of (Sallam, 2006)	65.3	31.3	-	-	-	0.5	-

3. Results and Discussions

3.1. eU vs eTh

Figure 2, shows moderately positive correlation for older, younger granites and acidic dykes, while anomalous monzogranites and anomalous pegmatites show moderately to weak positive correlation between eU and eTh figure 2, indicating that the magmatic processes played an important role in the concentration of the radioelements in these rocks.

3.2. eU vs eU/eTh

The eU-eU/eTh variation diagram figure 3 clarifies that the eU/eTh ratios increase with increasing eU content, suggesting post magmatic redistribution of uranium [14], the older, younger granites and acidic dykes display weak positive correlation, while both anomalous pegmatites and monzogranites illustrate moderately positive correlation.

3.3. eTh vs eU/eTh

The eTh and eU/eTh binary relation figure 4 exhibit negative correlation, proposes that the distribution of uranium and thorium remained is partly controlled by magmatic processes, [14 and 15].

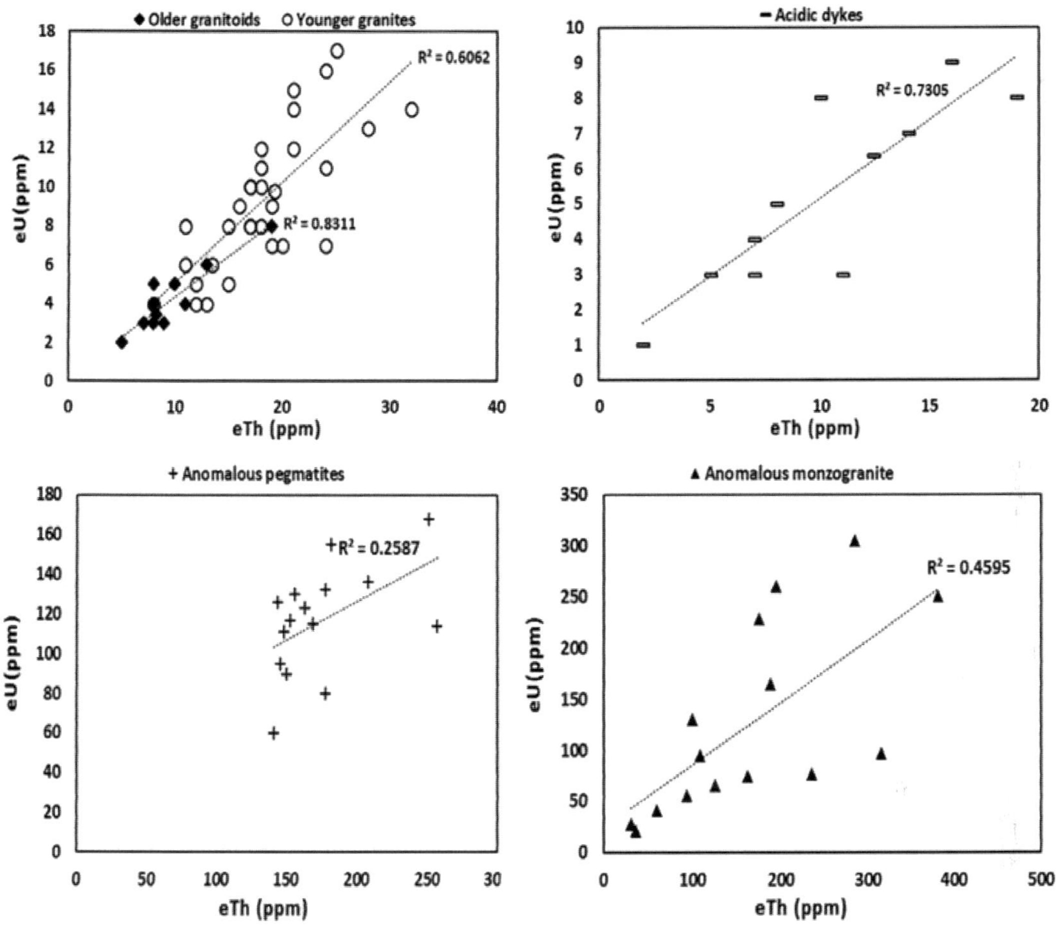

Figure 2. Variation diagrams of eU vs eThin studied rock samples.

3.4. eU vs K

The eU-K variation diagram shows scatter relationship between the eU and K designates the redistribution and mobilization of uranium [14]. This variety of the inter-elements relationship recommends that the radioactivity could be linked to syngenitic and epigenetic processes.

3.5. Mineralogical studies

The mineralogical studies are carried out on some representative samples of the mineralized uraniferous pegmatite and monzogranite. These samples have high uranium and/or thorium.

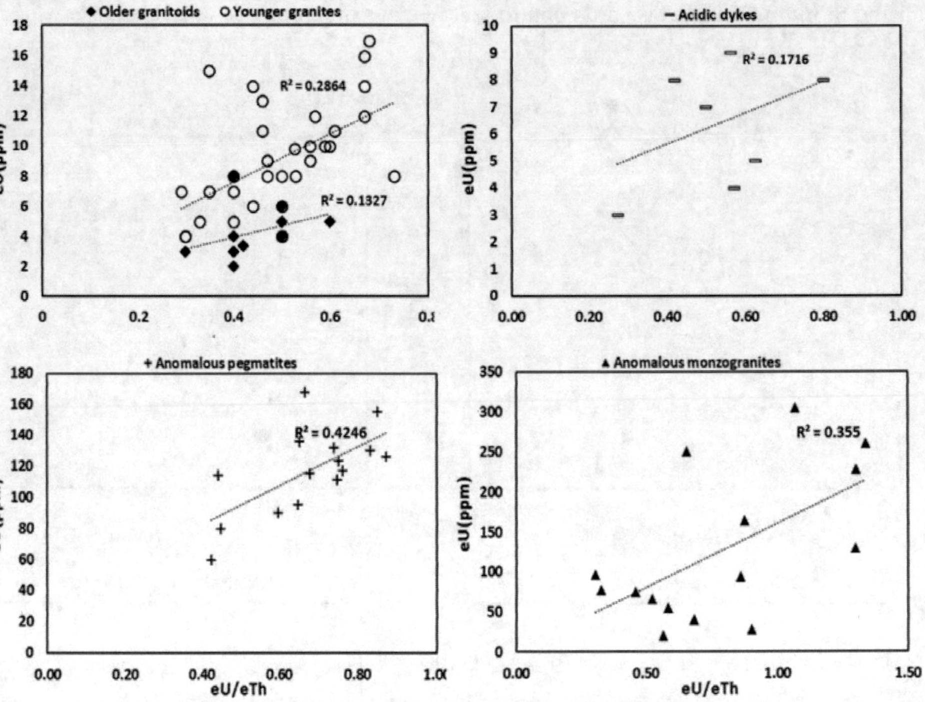

Figure 3. Variation diagrams of eU vs eU/eTh in studied rock samples.

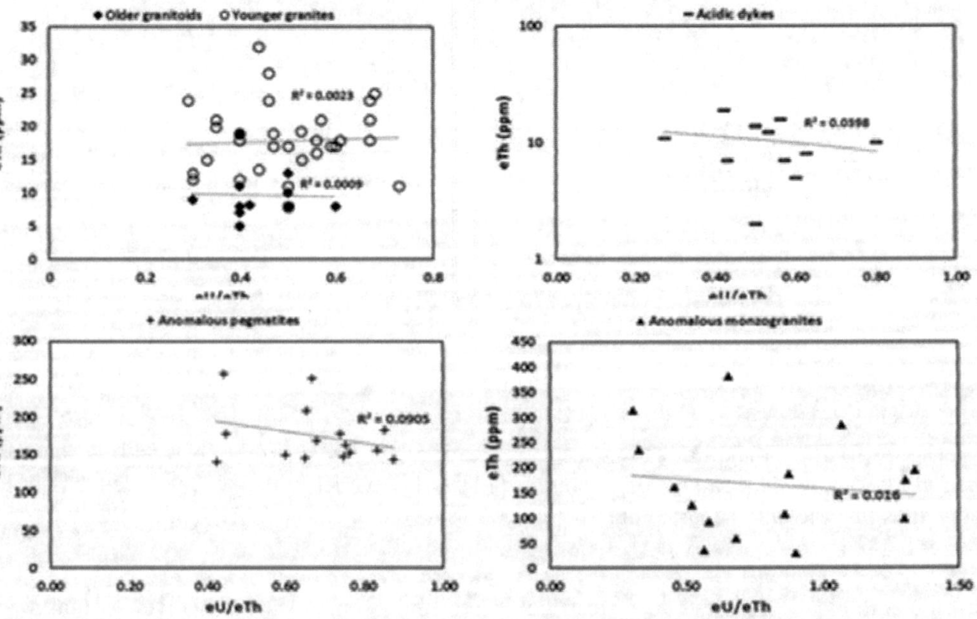

Figure 4. Variation diagrams of eTh vs eU/eThin studied rock samples.

IOP Conf. Series: Materials Science and Engineering **975** (2020) 012018 doi:10.1088/1757-899X/975/1/012018

contents. Twenty-one radioactive samples selected for mineralogical studies. The identification of the heavy minerals in these fractions carried out using X-ray diffraction (XRD) and scaning electronic microscope (SEM) technique

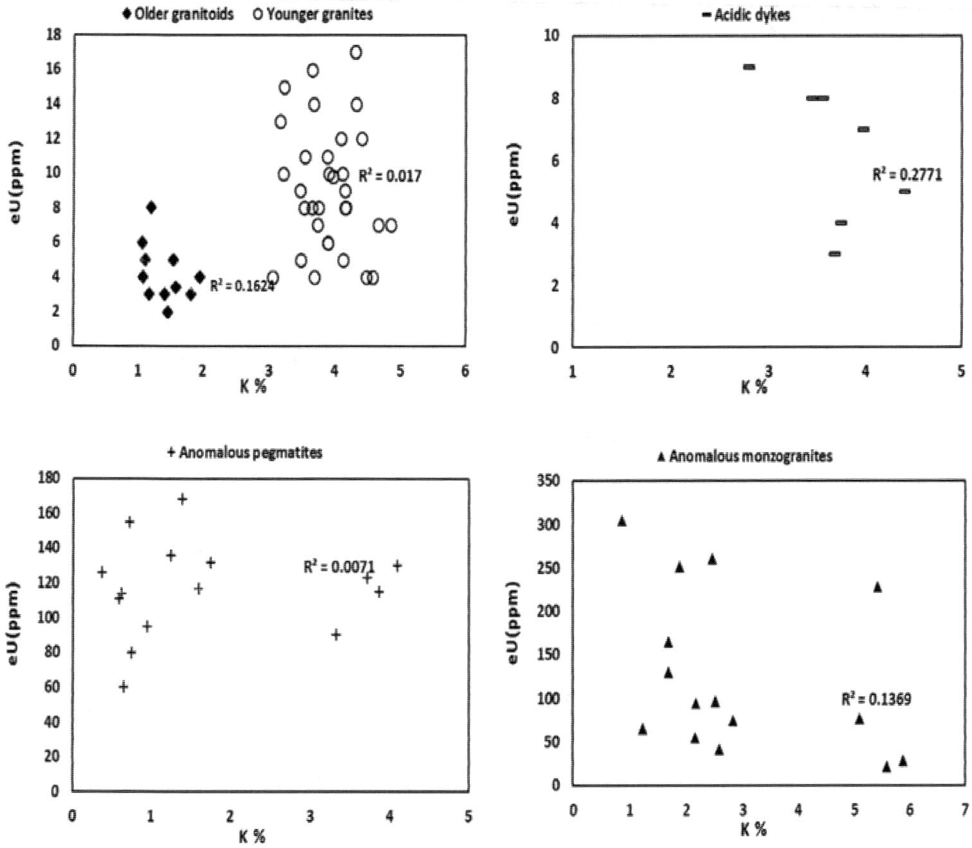

Figure 5. Variation diagrams of eU vs K in studied rock samples.

The data of both XRD and EDX analyses manifested the presence of significant radioactive minerals in assosciätion with non-radioactive ones. The identified radioactive minerals in the anomalous pegmatites are thorite, uranothorite, zircon, fluorite, columbite, samarskite, monazite, xenotime and allanite. However, thorite, uranothorite, zircon and fluorite are the main radioactive minerals in the anomalous monzogranite samples. Other accessory minerals such as ilmenite, magnetite, hematite, pyrite and rutile were identified in the studied pegmatite. On the other hand, hematite, magnetite and pyrite are the foremost accessories accompanying the radioactive minerals in the anomalous monzogranites.

3.6. *Radioactive Minerals*

Thorite [$ThSiO_4$] mineral grains noted in the pegmatites and in the monzogranites, appears as subhedral to anhedral form varying in colour from brownish black to black with greasy luster figure 6A. figure 6B showed the XRD patttern of picked separated grains of thorite and were also confirmed by EDX analyses which reflecting the chemical composition of thorite figures. 7 and 8.

IOP Conf. Series: Materials Science and Engineering **975** (2020) 012018 doi:10.1088/1757-899X/975/1/012018

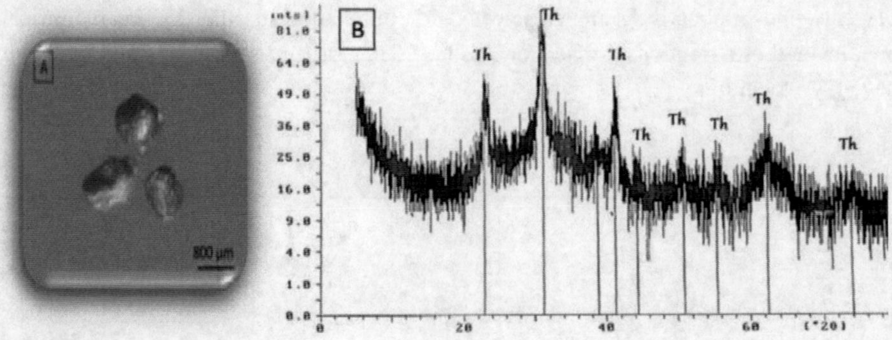

Figure 6. Photomicrograph of thorite grains(A) and X-ray diffraction pattern of thorite (B).

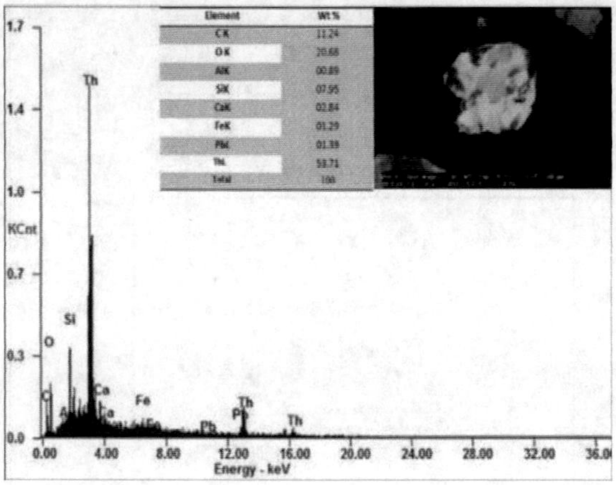

Figure 7. EDX analysis and BSE image of thorite in pegmatite.

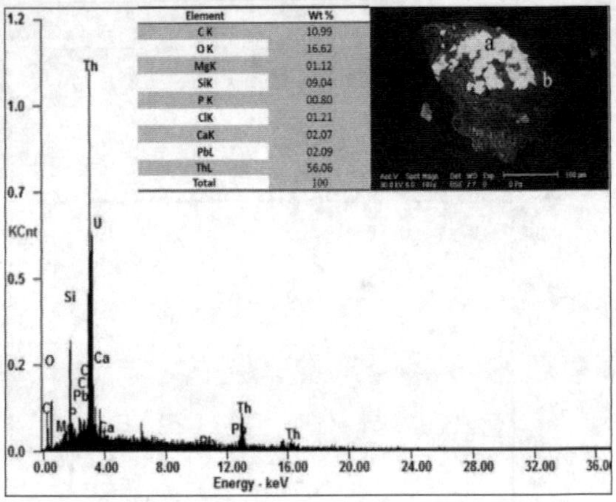

Figure 8. EDAX and BSE image shows inclusion of thorite (a) in flourite grain (b), anomalous monzogranite.

Uranothorite[$(U,Th)SiO_4$] is noted as subhedral to anhedral grains in the pegmatites, yellowish brown to brown with vitreous and resinous luster figure 9, and found in the monzogranites as inclusions in zircon and flourite. The uranothorite grains are highly metamicted, due to crystal lattice annihilation resulting from the influence of its own radioelement contents. Occasionally, the influence of this metamictization destroy the crystal lattice completely but leave the outward appearance unchanged. The existence of Y and P may assign to the association of xenotime as solid solution with uranothorite, the obtained EDAX spectrum figures 10 and 11 reveals the chemical composition of uranothorite.

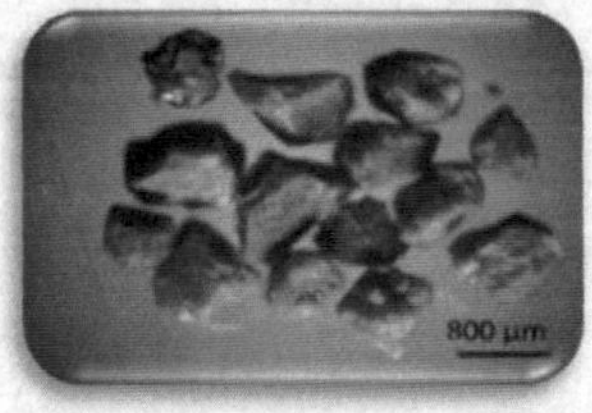

Figure 9. Photomicrograph of uranothorite grains.

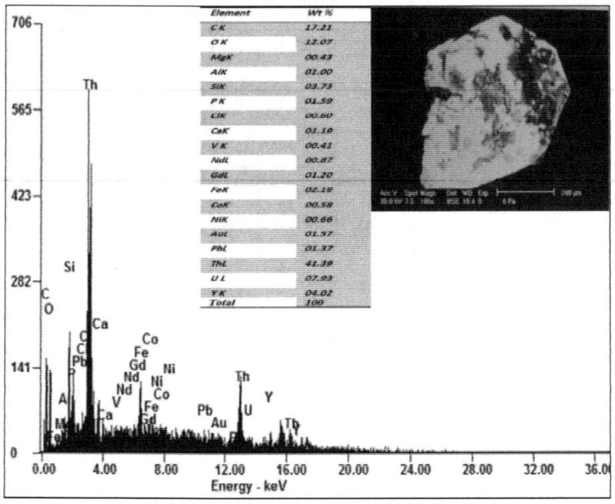

Figure 10. EDAX and BSE image shows uranothorite, pegmatite.

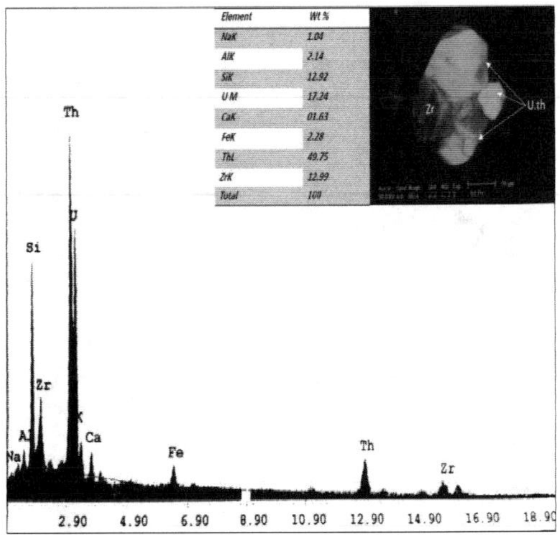

Figure 11. EDAX and BSE image shows uranothorite in zircon grain, anomalous monzogranite.

3.7. Radioelements-bearing minerals

Zircon [$ZrSiO_4$] is the most widespred mineral in pegmatite and anomalous monzogranite, it forms euhedral to anhedral prismatic grains, characterized by bipyramidal ends, varying in colour from yellow, yellowish red to reddish brown and sometimes has colorless to pale yellow colors figure 12. The X-ray diffraction pattern confirms that the picked zircon grains match with the ASTM card No. (6-0266) figure 12E. Also, Scanning Electron Microscope analyses carried out on particular zircon mineral grains picked from pegmatites and alterd younger granites, the back scattered electron microscope images of euhedral zircon grains and the EDX analyses confirm zircon composition [16] figure 13.

Fluorite[CaF_2] is found in all rock units especially in fluorite-bearing quartz vein cutting Gabal Homra pegmatite and Gabal Barakat monzogranite, the seperated grains appear as euhdral to subhedral cubic crystals, varying in colour from colorless to violet and occasionally black in pegmatites (Figures 14A and B) and colorless to deep violet with vitreous luster in altered younger granites figures 14C and D. The X-ray diffraction pattern displays that the picked fluorite grains match with the ASTM card No. (4-864) figure 15. On the other hand, Environmental Scanning Electron Microscope (ESEM) analyses match the chemical composition of fluorite figures 16 and 17.

Columbite [(Fe, Mn) Nb_2O_6] verified in the studied pegmatite by using ESEM analyses. It is characterized as blackish to dark brown prismatic crystals figures 18. Semiquantitative analyses display that columbite has prominent Nb/(Nb+Ta) and could be classed as ferrocolumbite in expression of quadrilateral end members; $FeNb_2O_6$, $MnNb_2O_6$ and $MnTa_2O_6$. The U-bearing columbite is Ta poor figure 19.

IOP Conf. Series: Materials Science and Engineering **975** (2020) 012018 doi:10.1088/1757-899X/975/1/012018

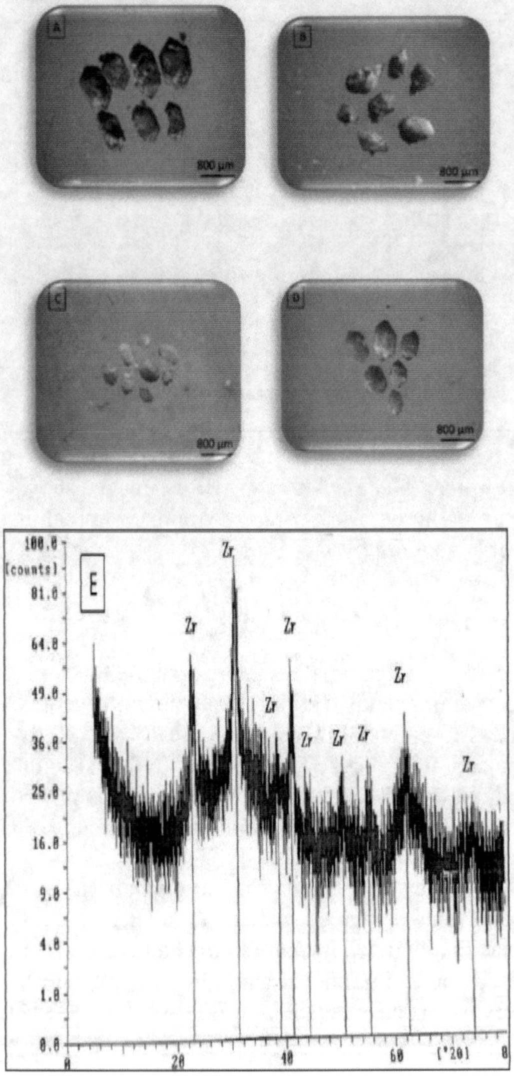

Figure 12. Photomicrographs of zircon grains from pegmatite (A&B), anomalous monzogranite (C&D)and X-ray diffraction pattern (E).

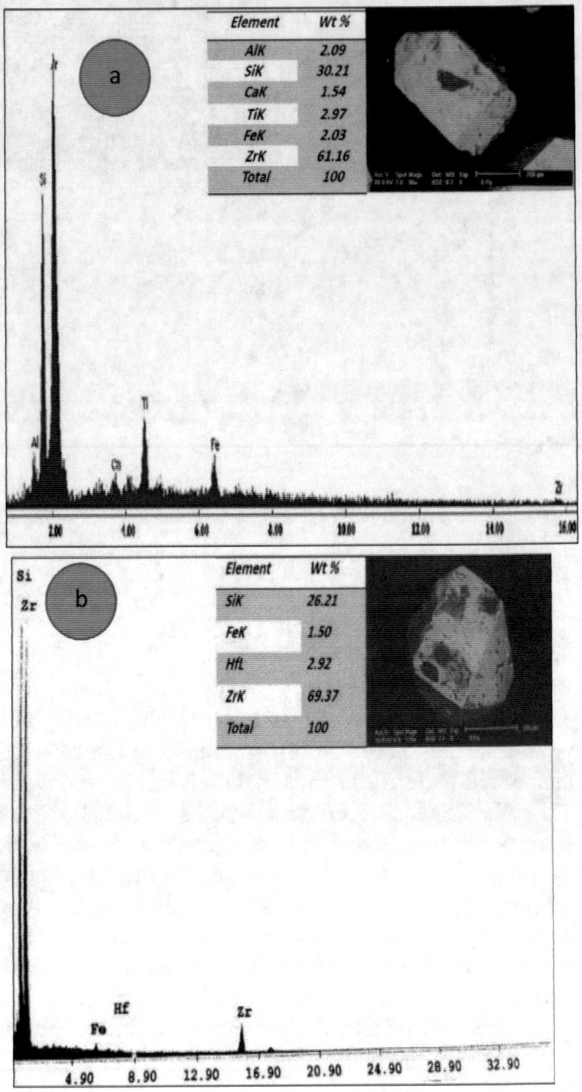

Figure 13. ESEM analyses of zircon grains, pegmatites (a) and anomalous monzogranite (b).

Figure 14. Photomicrographs of flourite grains from pegmatite (A&B),anomalous monzogranite (C&D).

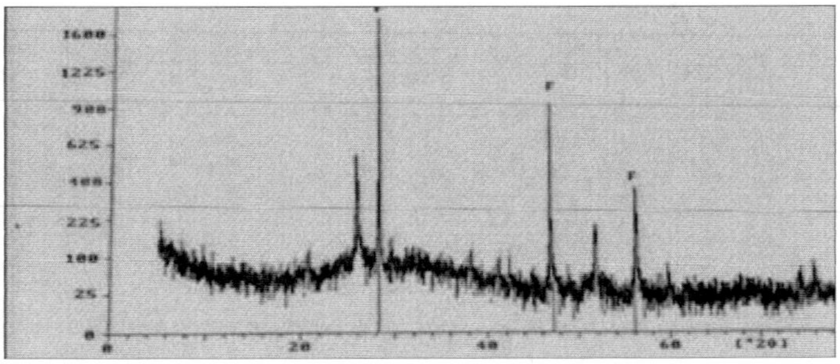

Figure 15. X-ray diffraction pattern of flourite grains, pegmatites (E).

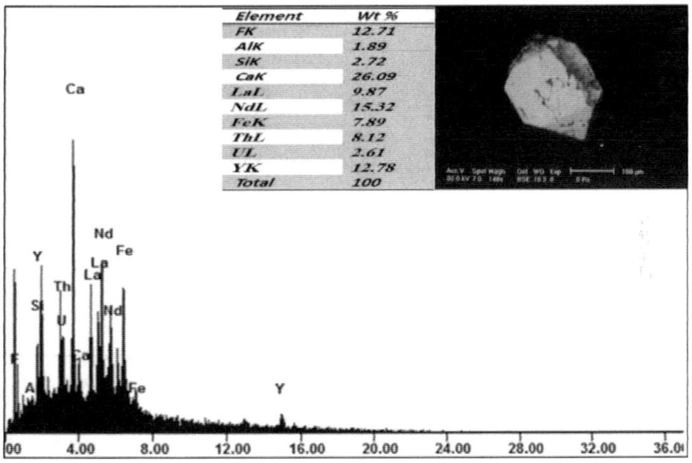

Figure 16. ESEM analyses of flourite grain, pegmatites

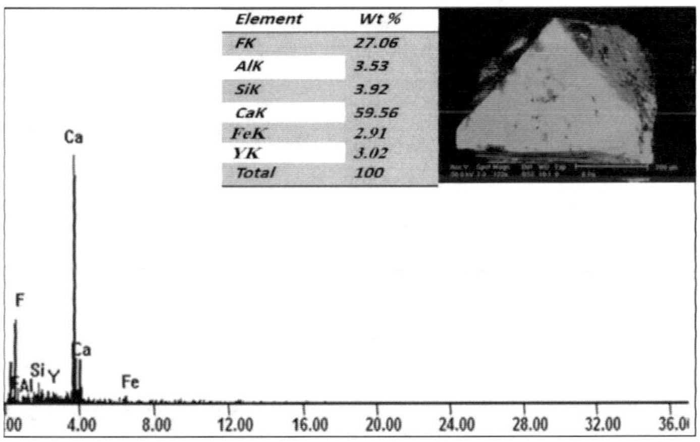

Figure 17. ESEM analyses of flourite grain, altered younger granites .

IOP Conf. Series: Materials Science and Engineering **975** (2020) 012018 doi:10.1088/1757-899X/975/1/012018

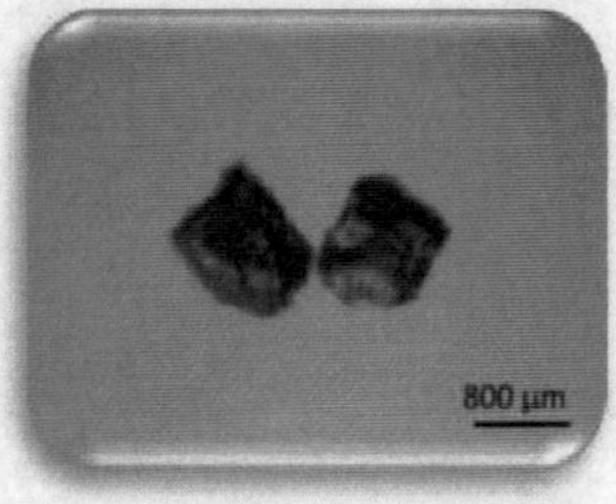

Figure 18. Photomicrograph of columbite grains, pegmatite.

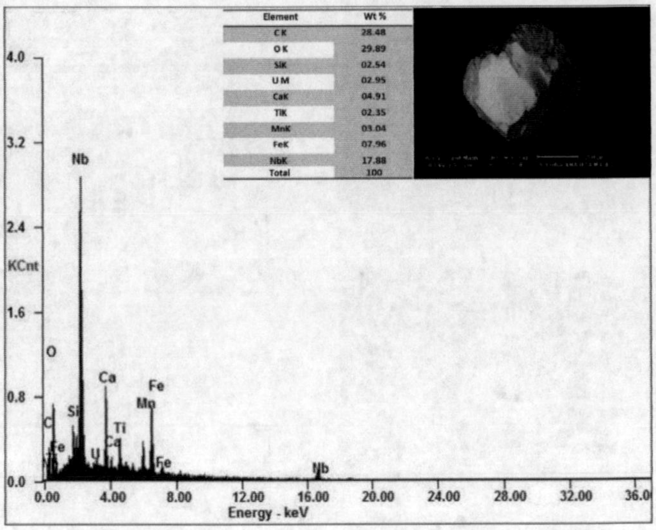

Figure 19. ESEM analyses of columbite grain, pegmatite.

Samarskite{$(Y,Er,HRE)_4[(Nb,Ta)_2O_7]_3$} grains occur as granular massive crystals of reddish orange color (Fig. 20). The recognized mineral is confirmed by ESEM analysis, also the acquired semi quantitative analysis demonstrate that this mineral is chiefly composed of Nb, Th, U, Y, Ca Ti, Ta in addition to REEs figure 21.

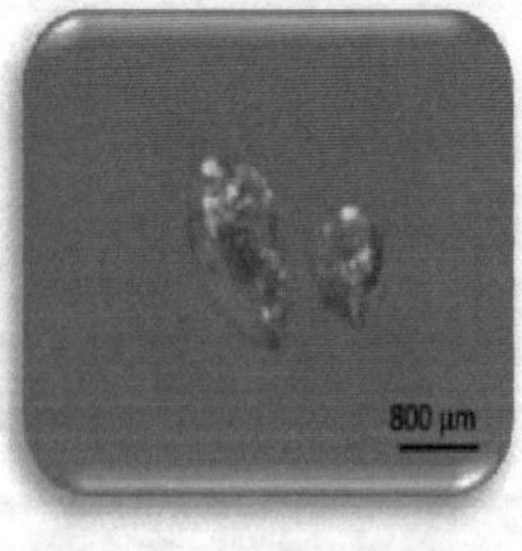

Figure 20. Photomicrograph of samarskite grains, pegmatite.

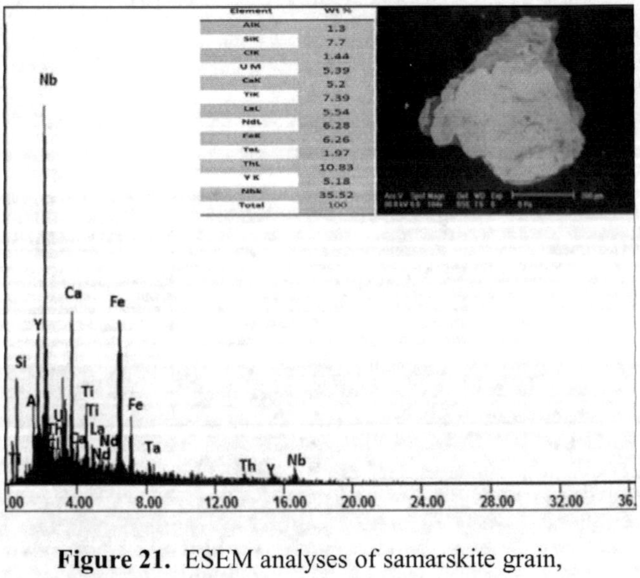

Figure 21. ESEM analyses of samarskite grain, pegmatite.

IOP Conf. Series: Materials Science and Engineering **975** (2020) 012018 doi:10.1088/1757-899X/975/1/012018

Monazite[(Ce,La,Y,Th)(PO_4)] is picked up from pegmatite of Gabal Homra. It appears as oval yellow grains with vitreous luster figure 22. The EDX analysis shows the presence of Ce, La, Nd, Pr, Sm, Gd, U, Th and P figure 23.

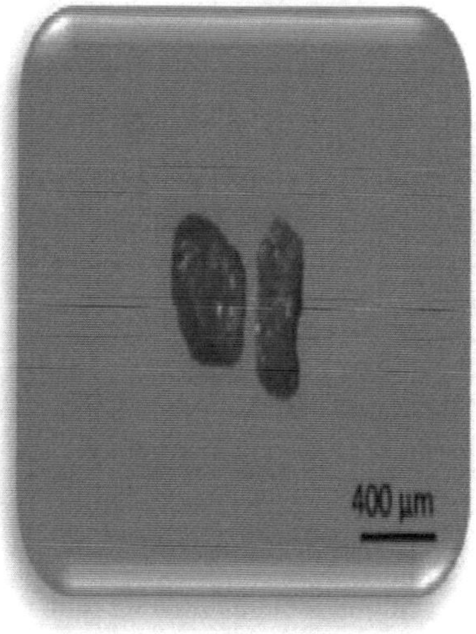

Figure 22. Photomicrograph of monazite grains, pegmatite.

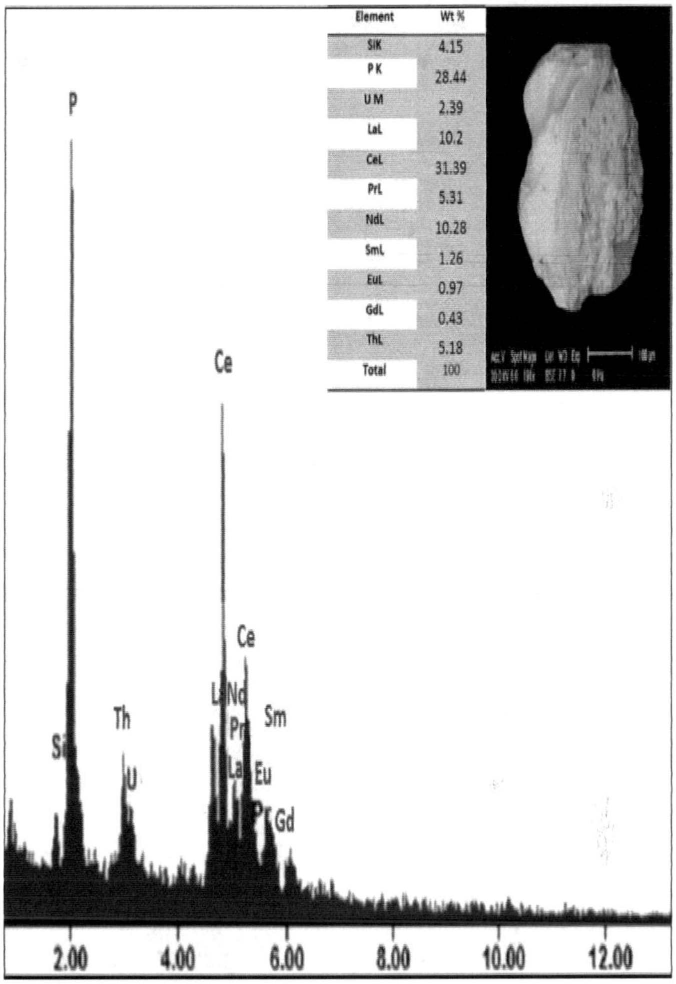

Figure 23. ESEM analyses of monazite grain, pegmatite.

Xenotime[Y (PO_4)] is found as grains in pegmatites. It is uraniferous, brownish in colour, infrequently pleochroic from pale pink to pale yellow, exemplified by short prismatic crystals shape figure 24. The EDX data signify the presence of Y, P, U, Th and REE figure 25.
Xenotime[Y (PO_4)] is found as grains in pegmatites. It is uraniferous, brownish in colour, infrequently pleochroic from pale pink to pale yellow, exemplified by short prismatic crystals shape figure 24. The EDX data signify the presence of Y, P, U, Th and REE figure 25.
Allanite[(Ca, Ce, La)$_2$ (Al, Fe^{3+}, Fe^{2+})$_3$ O. SiO_4. Si_2O_7. OH] occurs in anomalous monzogranite. It forms subhedral to anhedral grains with submetallic luster, varying in colour from brownish black to black figure 26. It is composed of Ce, La, Nd, Y, Sm, Gd, U, Th, Fe, Ca, Al and Si figure 27.

IOP Conf. Series: Materials Science and Engineering **975** (2020) 012018 doi:10.1088/1757-899X/975/1/012018

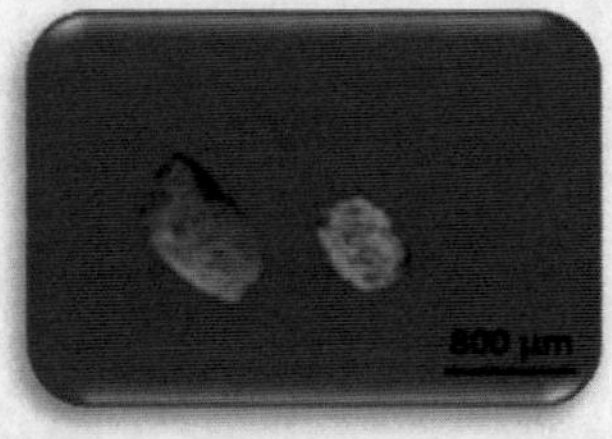

Figure 24. Photomicrograph of xenotime grains, pegmatite.

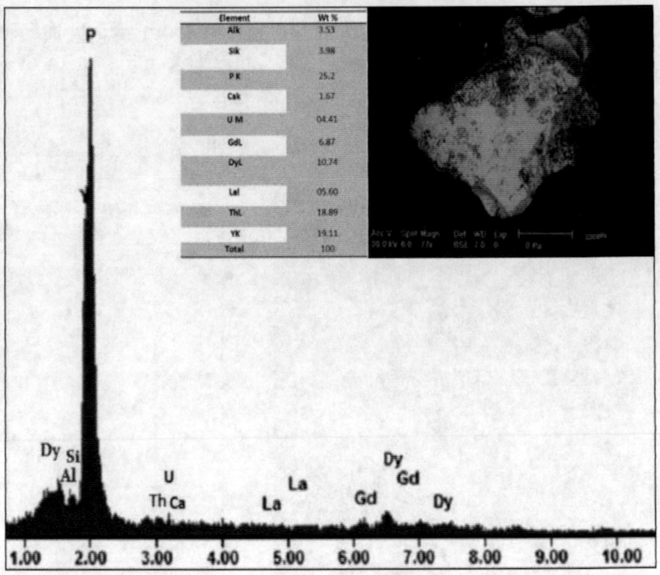

Figure 25. ESEM analyses of xenotime grain, pegmatite.

Figure 26. Photomicrograph of allanite grains, anomalous monzogranite.

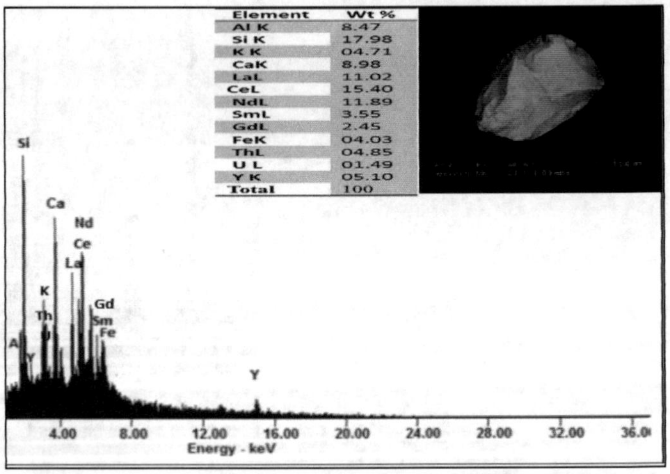

Figure 27. ESEM analyses of allanite grain, anomalous monzogranite.

3.8. Non-radioactive minerals

Pyrite[FeS_2] and hematized pyrite are pick up from both pegmatite and monzogranite. The separated pyrite grains appear as cubic crystals varying in colour from pale brass yellow to brass yellow with metallic luster figure 28A and D. Hematized pyrite appears combining pyrite crystals and broadcasted in the wall rocks mainly along faults and fractures zones figure 28B and C. The X-ray diffraction pattern confirms that the separated pyrite grains are matched with the ASTM card No. (0026-0801) and (0083-0983), figure 28E.

IOP Conf. Series: Materials Science and Engineering **975** (2020) 012018 doi:10.1088/1757-899X/975/1/012018

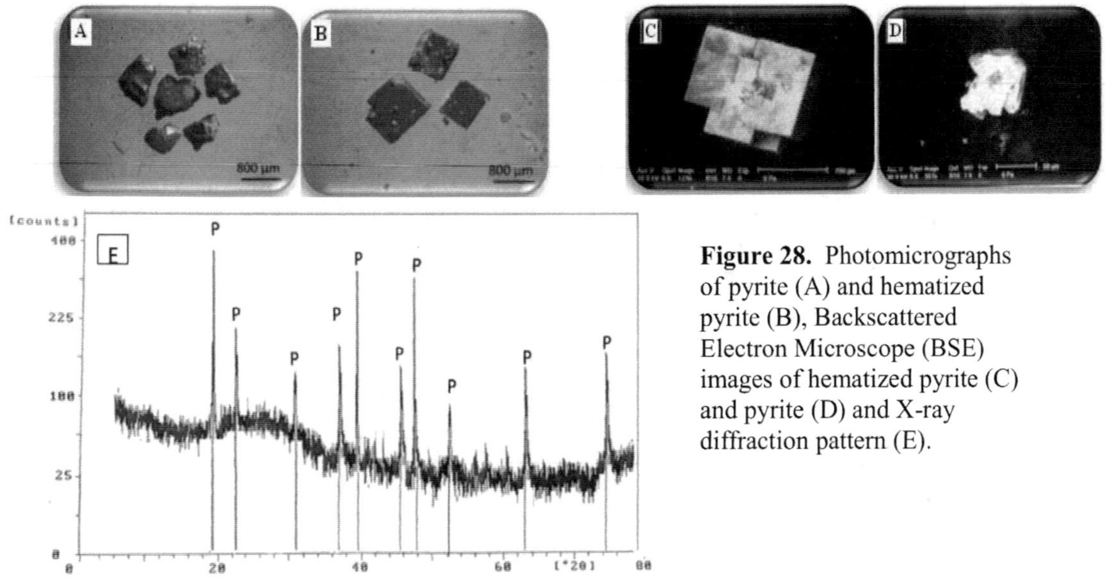

Figure 28. Photomicrographs of pyrite (A) and hematized pyrite (B), Backscattered Electron Microscope (BSE) images of hematized pyrite (C) and pyrite (D) and X-ray diffraction pattern (E).

Goethite [FeO (OH)] is detected in pegmatite and monzogranite as separated massive grains varying in color from dark brown to black and display cloudy luster figure 29A. The X-ray diffraction pattern displays that the goethite grains agree with with the ASTM card No. (0017-0536) and hematite grains agree with the ASTM card No. (0079-0007) figure 29E.

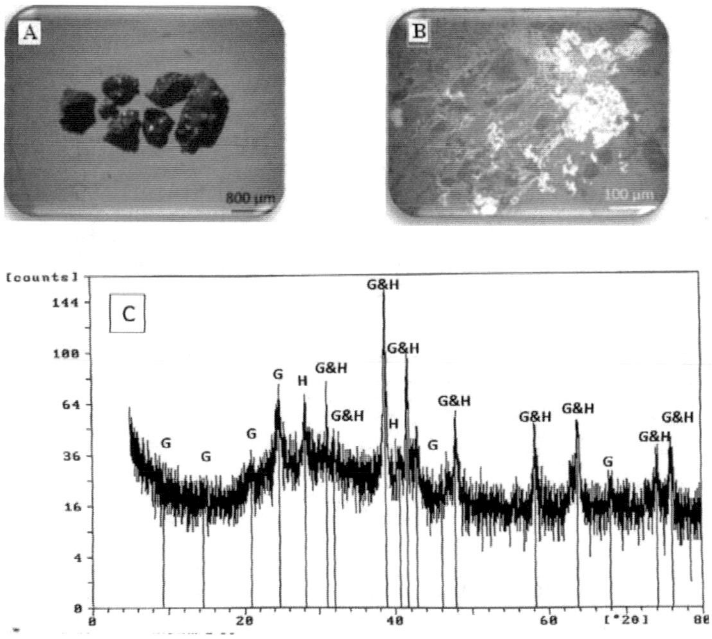

Figure 29. Photomicrograph of geothite (A), microphotograph of hematite and X-ray diffraction pattern (E) for geothite (G) and hematite (H).

Hematite (Fe_2O_3) is found as thin or thick tabular reddish brown crystals of cloudy luster in compact or fibrous form figure 29B. It was recognized by XRD analyses associated with goethite figure 29E.
Magnetite (Fe_3O_4) occurs in both monzogranite and pegmatite. It is partially altered to hematite figure 30A). The X-ray diffraction pattern displays that the magnetite grains agree with the ASTM card No. (0079-0417) and ilmentite grains agree with the ASTM card No. (0003-0741), figure 30C.

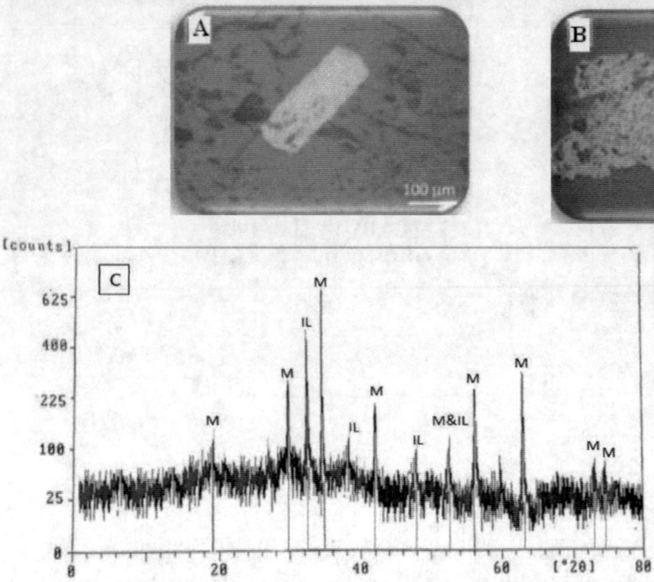

Figure 30. (A) micro-photographs of magnetite (M), (B) ilmenite (IL) and X-ray diffraction pattern (c).

Ilmenite ($FeTiO_3$) is mainly reported as an accessory mineral in the pegmatite, black in colour with metallic luster, unstable Ti-bearing mineral appearing in tabular crystals or platy shape (Figure 30B). This mineral is recognized by XRD pattern associated with magnetite figure 30C. It is predominantly constituted of Fe_2O_3 and TiO_2 with little amounts of Si, Al, U and Mg oxides.
Rutile (TiO_2) differs in colour from yellowish red, red, brown to black colours figure 31A. The EDX records indicates that rutile is essentially composed of Ti, Fe, Nb, Ca, Al and Si figure 31C.

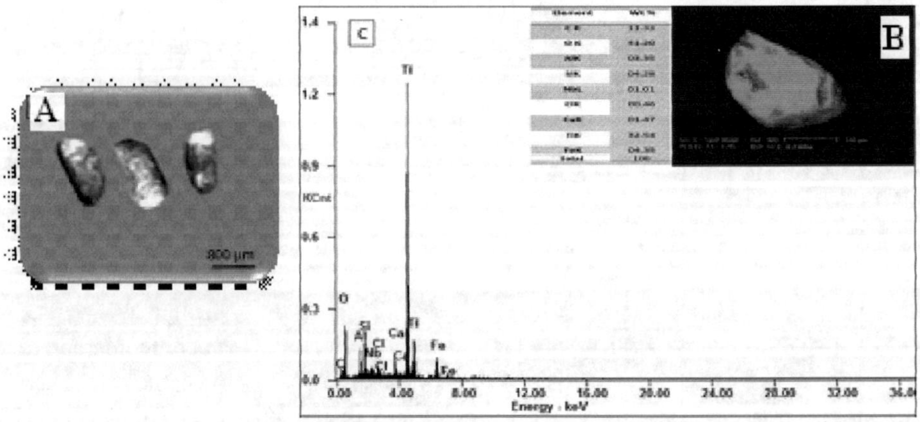

Figure 31. Photomicrograph of rutile (R) grains from pegmatites (A), X-ray diffraction pattern (B) and ESEM analyses (C).

4. Conclusions

Wadi Sabbagh area is situated at southern part of Sinai Peninsula, occupies about 469 Km^2, embodied by moderate and high relief mountains of rugged topography mainly of older and younger granites. The pegmatites broadly distributed in the examined area.

The field radiometric survey indicates that, monzgranites verify the highest values of radioactivity reflecting the role of post-magmatic processes. Two anomalous zones are recognized. The first anomaly confind to the pegmatite bodies hosted in syenogranites. The laboratory radiometric measurements of pegmatites revealed that, the eU reachs up to 168 ppm, while the eTh reachs up to 257 ppm. The average values of Ra and K are 89 ppm and 1.7 wt percentage, respectively. This means that the studied pegmatites are uraniferous. The second anomaly restricted to the anomalous monzogranite at Wadi Adawi area, its eU reachs up to 305 ppm and eTh to 382 ppm. The post magmatic processes could explain the high value of uranium and thorium which lead to the later hydrothermal alteration, which occur mostly along weak planes (fractures, faults and shear districts). The inter-elements variation diagrams recommends that the radioactivity of the concerned rocks most possibly linked to syngenetic and epigenetic foundations. The recognized radioactive minerals in the considered pegmatites are thorite, uranothorite, zircon, fluorite, columbite, samarskite, monazite, xenotime and allanite, whereas thorite, uranothorite, zircon and fluorite are the main radioactive minerals in the altered monzogranite. Additional accessory minerals such as ilmenite, magnetite, hematite, pyrite and rutile were found in the studied pegmatites. Hematite, magnetite and pyrite are the foremost accessories assosciating the radioactive minerals in the monzogranite.

5. References

[1] *Be'eri-Shlevin Y., Katzir Y., Whitehouse M. (2009a):*Post-collisional tectonomagmatic evolution in the northern Arabian-Nubian Shield (ANS): time constraints from ion-probe U–Pb dating of zircon. Journal of Geological Society 166, 71–85.

[2] Abd El Hadi A. Abbas, Mahmoud R. Khattab, Marwa M. Abdel Azeem (2018): Natural radionuclides distribution and environmental impacts of ferruginous sand-siltstone (raw material) and their manufactured Ahmer oxide used as wall paints. Environmental Forensics, 19: 217-224.

[3] Matolin M (1991) Construction and use of spectrometric calibration pads. A report to the Arab republic of Egypt. EGY 4/030-03, Laboratory of Gamma ray spectrometry.

[4] El Nahas, H.A., El Feky, M.G., Mira, H (2011): Mineralogy, M-type tetrad effect and radioactivity of altered granites at the G. Abu Garadi shear zone, central Eastern Desert, Egypt. Chinese journal of Geochemistry, 30, 153- 158.

[5] Khamis, H. A., Ebyan, O. A. and Abed, N. S. (2019): Mineralogy and geochemistry of a new uranium occurrence at the decant of Wadi El Reddah, Northeastern Desert, Egypt. Nuclear Sciences Scientific Journal, in press.

[6] Rogers, J. J. W. and Adams, J. A. S. (1967): "Uranium". In: Wedepohl, K. H. (ed.), Handbook of geochemistry, 2 (1), Springer-Verlag, Berlin.

[7] EL Sayed, A. A. (1993): "Geology and radioactivity of west Dahab area, Southeastern Sinai, Egypt". M. Sc. Thesis, El Mansoura Univ, 174p.

[8] Adams, J. A. S., Osmand, J. K. and Rogers, J. J. W. (1956): "The geochemistry of thorium and uranium. In: Physics and Chemistry of the Earth", 3, pp. 298-348, Pergamon Press, New York.

[9] Ford, K. L. (1982):Uraniferous pegmatites of the sharbot lake area, Ontario; in Uranium in Granites, ed. Y.T.Maurice, Geol. Surv. Can., paper 81-23, p. 125-138.

[10] Moharem, A. F. (2004): Uranium distribution in the Um Samra-Um Bakra granitic plutons and associated pegmatites, Central Eastern Desert, Egypt. 7. Arab Conference on the Peaceful Uses of Atomic Energy, Sana'a (YE). Vol. 1.

IOP Conf. Series: Materials Science and Engineering **975** (2020) 012018 doi:10.1088/1757-899X/975/1/012018

[11] Darnely, A. G. (1982): “"Hot granite": Some general remarks”. In Maurice, Y. T. (ed.), Uranium in granites. Geol. Sur. of Canada. Paper 81-23, pp. 1-10.

[12] Sallam, O.R. (2006): Geology and Uranium Potentiality of GabalShiekhEl-Arab Area, South Sinai, Egypt. Ain Shams Univ., Egypt.

[13] Abbas, A. A. (2008): Geology and radioelements distribution in the basement rocks of Wadi Um Adawi area, south Sinai, Egypt. Ph.D. Thesis Ain ShamsUniv., Egypt.

[14] *Charbonneau, B. W. (1982):* “Radiometric study of three radioactive granites in the Canadian Shield: Elliot Lake, Ontario; Fort Smith, Fury and Hecla, NWT. In: Uranium in granites (ed), Y.T.Maurice”. Geol. Surv. Canada, pp. 88-23, pp. 91-99.

[15] El Gharbawy R.I. and El Maadawy W.M. (2012): Geochemistry of the uranium-thurium-bearing granitic rocks and pegmatites of Wadi Haleifiya area, Southeastern Sinai, Egypt. Chin.J.Geochem., 31, 242–259.

[16] El Mezayen, A. M., M. A. Heikal1, El-Feky, M. G., Shahin, H. A., Abu Zeid, I. K., Lasheen1, S. R. (2019): Petrology, geochemistry, radioactivity, and M–W type rare earth element tetrads of El Sela altered granites, south eastern desert, Egypt. Acta Geochim., 38(1), 95–119.

Sub-channel thermal-hydraulic analysis for VVER- 1000 generation III PWR

Alsherief M Almessallmy[1], Mohamed K Shaat[2] and Saeed A Hassanien[1]

[1]Egyptian Armed Forces, Cairo, Egypt
[2]Egyptian Atomic Energy Authority, Cairo, Egypt

Abstract. The sub-channel thermal-hydraulic analysis of a nuclear reactor is essential for assessing of its safety aspects. In this paper, the VVER-1000 has been selected as an example of the third generation reactors since it meets most of the international safety standards and because it has been taken as a base for designing the VVER-1200 which is belonging to the III+ generation. A steady state mathematical model has been proposed and solved to validate and assure that the hottest channel temperature limits are satisfied. The various temperature distributions, the critical heat flux and the departure from nucleate boiling ratio (DNBR) for the hottest channel were evaluated. Also, a transient state model has also been presented and solved using the finite difference method with the aid of MATLAB algorithm. An exponential loss of flow rate of the reactor core coolant was triggered from the steady state conditions. We assumed that the neutron flux and the generated power were unchanged during the postulated event. The average core coolant flow time constant was treated as a single parameter expressing the rapidity of the event. A value of 250 seconds time constant was assumed for slow transient, whereas 10 seconds was assumed for fast one. The reactor core was assumed to be protected through the reactor control system and mitigated according to the regular emergency operating procedures. The time dependent temperature distributions were calculated for the cladding of the hottest coolant channel. For each value of the temperature, the response time required for reaching unsafe conditions was evaluated, discussed and presented.
Keywords: Thermal-hydraulics, steady-state, transient state, hottest channel, MATLAB, DNBR, loss of flow, time constant, postulated event.

1. Introduction

The objective of thermal hydraulic analysis is to ensure the operational temperature in the core does not exceed the design limit of temperature. If it can be ensured the hottest channel exhibits the temperature below the core design limit, the remaining channels then will presumably fall within this limit. Hence, particular attention in this research has been paid on the hottest channel to ensure its temperature remains below the design limit. During the continuous operation of the reactor at high power levels, the nuclear heat generated can be permitted to flow through the following auxiliary systems:

a. During the reactor operation, the forced circulation coolant system pumps pool water down through the fuel elements to remove the fission heat from the reactor.
b. A water-to-water heat exchanger transfers the generated heat to a secondary water coolant system, and the primary water is returned to the reactor core.
c. The secondary water carries heated water to the cooling tower which dissipates the heat to the atmosphere. Water from the cooling tower is recirculating through the secondary system.

To ensure reactor safety, the reactor must have sufficient margins during normal operation as well as in all possible accidental events, such as reactivity-induced accidents (RIAs), loss-of-coolant accidents

(LOCAs), loss-of-flow accidents (LOFAs), etc. This paper focuses on LOFA analysis of selected example of VVER-1000 PWR reactor with the aim of investigating the impact of an unprotected LOFA on the reactor safety. A LOFA occurs in a reactor due to many causes, such as loss of off-site power, pump failure, heat exchanger blockage, pipe blockage, etc. The danger of a LOFA is that it could lessen fuel integrity due to overheating that arises from a low heat transfer coefficient in the reactor core. As a LOFA is classified by regulatory bodies as a design-basis accident, necessary precautions need to be taken during the design stage to ensure that the primary coolant system has adequate safety margins against the onset of flow instability and also against occurrence of the nucleate boiling. This paper provides the result of a sub-channel thermal-hydraulic analysis of the typical VVER-1000 reactor core at steady-state and transient conditions using an analysis approach that required predicting the heat produced from the burn-up of the fuel as it transferred through the fuel and cladding materials reaching to the coolant that passes beside the outer surface of the cladding.
The heat transfer coefficient of the light water which is used as a coolant is calculated. Also the equivalent diameter of the coolant channel, Reynolds number, Prandlt number, Nusselt number and coolant temperature are calculated at the average coolant temperature and normal operating coolant pressure. A code which was designed to match with the MATLAB environment has been used to solve both the steady-state and the transient models.

2. Reactor Description and Analytical Approach

2.1. Reactor Description

VVER-1000 reactor is a Russian-type pressurized water reactor. The major difference between the VVER and a Western PWR, in the present study, is the fuel assembly design and the core geometry as the fuel rod of the VVER type includes a central hole as shown in figure1. The specifications of reactor which is selected in this paper are presented in table.1[1]. Specifically, the reactor under study is made up of 163 hexagonal fuel assemblies of three different enrichment, (1.6%, 2.4% & 3.6). In the core analysis, the individual sub-channels are lumped together to give an equivalent flow area and the fuel rods are modeled using a single rod to represent the average behavior of all rods in each channel. In this analysis, a thermal-hydraulic grid of 280 control volume, each assembly-size of 23.6 cm hexagonal horizontal cross-section and axial height of 353 cm was considered.

Thermal-hydraulic calculations for; temperature, enthalpy and critical heat flux (CHF) and also, the Minimum Departure from Nucleate Boiling Ratio (MDNBR) of the hottest channel in the reactor core were calculated. In this case, the Dittus-Boelter correlation is used for heat transfer coefficient calculation in both sub-cooled and saturated nucleate boiling region. In sub-cooled condition, a partition of the total heat flux, from the heated fuel rods, induces the liquid and therefore produces vapor. The latter is formed by adding single-phase convection from the wall to the liquid film, to the nucleate boiling term while the former is remaining convective heat flux from the liquid film to the coolant bulk. In saturated conditions, however, the total heat flux (including direct heating of the coolant) is converted into vapor, so holding the liquid enthalpy at the saturation value [2].

2.2 Steady-State Model

A mathematical model that describes the heat transfer process from the core to the coolant taking into account the geometric specifications of the hexagonal fuel assemblies and its effect on the fluid dynamic calculations of the coolant is discussed through the following analytical equations. The nodalization of the hottest channel that is chosen to execute the calculations is shown in figure 2. In order to investigate the axial temperature distribution of fuel, clad and coolant along the hottest channel, the core power density firstly calculated using the following formula:

$$\left(q'''_{max}\right)_{actual} = \frac{2*1.16\,P\,E_d}{3\,H\,a^2\,n\,E_R} \tag{1}$$

IOP Conf. Series: Materials Science and Engineering **975** (2020) 012019 doi:10.1088/1757-899X/975/1/012019

Where: q'''_{max} .. is the actual maximum heat generated per unit volume of the hottest channel, E_d... is the energy per fission deposited in fuel elements, E_R... is the energy recovered per fission, n... is the no. of fuel rods, a.... Is the fuel radius, H... is the extrapolated height of the fuel rod; P... is the reactor power.

Table 1. Reactor specifications

Reactor core operating conditions	Value
Reactor Pressure (MPa)	15.7
Reactor thermal power (MWt)	3000
Inlet coolant flow rate (kg/s)	18600
Coolant inlet temperature (°c)	289.7
Coolant outlet temperature (°c)	320
Fuel assembly	
Fuel assembly form	Hexagonal
Number of fuel assemblies in the core	163
No. of fuel rods in the fuel assembly	312
No. of other tubes	19
Fuel rod	
Central hole of fuel pellet diameter (mm)	2.3
Fuel pellet outside diameter (mm)	7.53
Cladding outside diameter (mm)	9.1
Fuel pellet material	UO2
Cladding material	Alloy-110
Fuel rods pitch (mm)	12.75
Maximal temperature of fuel rod cladding external surface (°c)	350

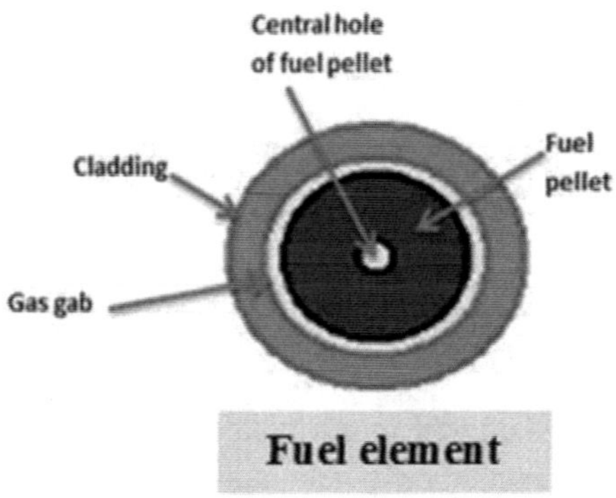

Figure 1. plan view of the fuel rod.

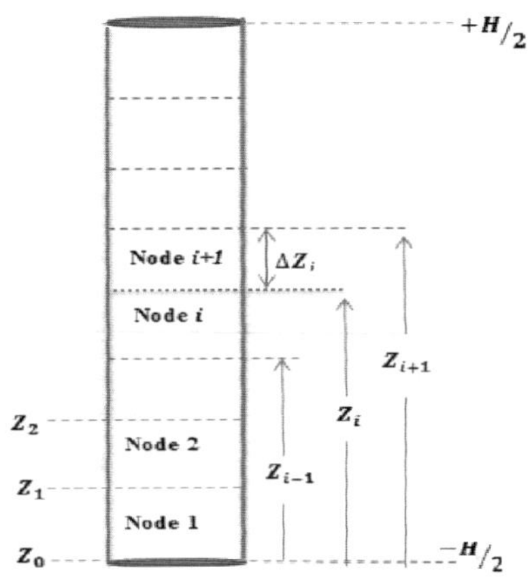

Figure 2. Nodalization of the hottest channel

Calculation of coolant temperature along the fuel rod of the hottest channel is determined as follows:

$$T_w(z) = T_{w_0} + \frac{q'''_{max} V_f}{\pi\omega C_p}\left[1 + \sin\frac{\pi z}{H}\right] \quad (2)$$

Where;

T_{w_0} .. Is the inlet coolant temperature, ºC,
V_f .. Is the volume of the fueled portion of the rod, m^3,
z .. Is the height of the fuel rod, m,
ω .. Is the rate of flow of the coolant through the coolant channel, $kg.s^{-1}$,
C_p .. Is the specific heat of the coolant, $J.kg^{-1}.\,^{o}C^{-1}$.

And In order to calculate the cladding temperature, the heat transfer coefficient is required to be determined through the Dittus - Boelter correlation as follows [3] :

$$h = C\left(\frac{K}{D_e}\right) R_e^{\,0.8}\, P_r^{\,0.333} \quad (3)$$

$$R_e = \frac{\rho\, V\, D_e}{\mu} \quad (4)$$

$$P_r = \frac{C_p\, \mu}{K} \quad (5)$$

where;
C .. is a constant that can be given by C= 0.026 $(\frac{P}{D})$ - 0.024 for triangular lattices with $1.1 \le \frac{P}{D} \le 1.5$. The quantities P and D are, respectively, the lattice pitch and rod diameter,
μ .. is the water viscosity $kg.s^{-1}\, m^{-1}$,
K .. is the water thermal conductivity, $W.m^{-1}\,^{o}C^{-1}$.
C_p .. is the specific heat of the coolant, $J.kg^{-1}.\,^{o}C^{-1}$.
D_e .. is the equivalent diameter of the coolant channel, m.

Hence, the cladding outside temperature distribution can be calculated from the following equations[4]:

$$T_{co}(Z) - T_w(z) = \frac{q'_{max}}{2\pi R_{co} h} \quad (6)$$

where; R_{co} .. is the clad outside radius, m.

Conduction heat transfer through the cladding material and gap is calculated using Equations (7) and (8), respectively:

$$T_{ci}(z) - T_{co}(z) = \frac{q'_{max}}{2\pi K_c}\ln\left(\frac{R_{co}}{R_{ci}}\right) \quad (7)$$

$$T_{fo}(z) - T_{ci}(z) = \frac{q'_{max}}{2\pi R_g h_{gap}} \quad (8)$$

Where;
$T_{ci}(z)$.. is the clad inner surface temperature at any height (z).
$T_{co}(z)$.. is the clad outer surface temperature at any height (z).
$T_{fo}(z)$.. is the fuel outer surface temperature at any height (z).
R_{ci} .. is the clad inside radius, m.
K_c.. is the clad thermal conductivity, $W.m^{-1}\,^{o}C^{-1}$.

R_g .. is the mean radius in the gap, m.
h_{gap} .. is the effective gap conductance, W.m^{-2} °C^{-1}.

Moreover, conduction heat transfer in the annular fuel is found by Equation 9:

$$\int_{T_{fo}}^{T_{fi}} K_f(t)dt = \frac{q'_{max}}{4\pi}\left[1-\frac{\ln\left(\frac{R_{fo}}{R_v}\right)^2}{\left(\frac{R_{fo}}{R_v}\right)^2-1}\right] \tag{9}$$

Where; $T_{fi}(z)$.. is the fuel inner surface temperature at any height z.

As the determination of both the maximum fuel and cladding temperatures along the fuel rod of the hottest channel is of great importance in the thermal analysis of the reactor core and for the thermal design specifications, the following equations are utilized to investigate these temperatures [5]:

$$T_{w,max} = T_{w_0} + \frac{2q'''_{max}V_f}{\pi\omega C_p} \tag{10}$$

$$T_{c,max} = T_{w_0} + q'''_{max}V_f\frac{1}{h*A}\left[\frac{1+\sqrt{1+\alpha^2}}{\alpha}\right], \text{ Where } \alpha = \pi\ \omega\ C_p\frac{1}{h*A} \tag{11}$$

$$T_{f,max} = T_{w_0} + q'''_{max}V_f \mathrm{R}_{\mathrm{th}_f}\left[\frac{1+\sqrt{1+\beta^2}}{\beta}\right] \tag{12}$$

$$\text{As } \beta = \pi\ \omega\ C_p \mathrm{R}_{\mathrm{th}_f} \text{ and } \mathrm{R}_{\mathrm{th}_f} = \frac{1}{4\pi H_e K_f} + \frac{\ln(1+{}^{b}/_{a})}{2\pi H_e K_c} + \frac{1}{h*2\pi(a+b)*H_e} \tag{13}$$

Where;
V_f .. is the volume of the fuel portion, m^3.
A .. is the is the area of the fuel rod, m^2.
a .. is the fuel thickness, m.
b .. is the clad thickness, m.
K_f .. is the fuel thermal conductivity,W.m^{-1} °C^{-1}.

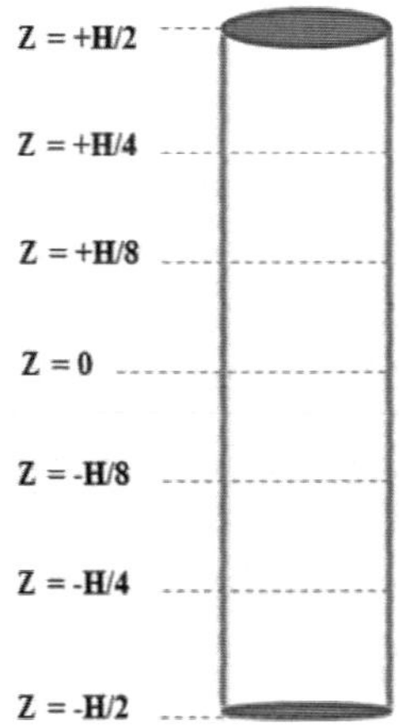

Figure 3. Fuel rod segments

The position of these temperatures along the fuel rod can be determined through the segments of the fuel rod as shown in fig.3 and can be calculated as follows:

$$z_{c,max} = \frac{H_e}{\pi} \tan^{-1}\left(\frac{H_e * h * C}{\pi \omega C_p}\right), \text{ Where } C = 2\pi(a+b) \tag{14}$$

$$z_{f,max} = \frac{H_e}{\pi} cot^{-1}\left(\pi\, \omega\, C_p\, \mathrm{R}_{\mathrm{th}_f}\right) \tag{15}$$

Calculation of DNBR is can be investigated by the determination of the critical and actual heat flux using the most widely used correlation of Jens and Lottes [5]as follows:

$$\text{DNBR} = \frac{q''_c}{q''_{actual}} \tag{16}$$

$$q''_c = C \times 10^6 \left(\frac{G}{10^6}\right)^m \Delta\, T_{sub}{}^{0.22} \tag{17}$$

And C = 0.445 and m = 0.5 for this case. And the T_{sub} can be defined as the difference between the saturation temperature of the coolant and the average coolant temperature along the fuel rod of the hottest channel,

$$q''_{actual} = \frac{q'''_{max} V_f}{C_c} \tag{18}$$

Where; Cc is the circumference of a heated rod.

2.3 Transient-State Model

The transient-state model that describes the thermal-hydraulic behavior of the hottest channel through the loss of flow anticipated transient without scram (ATWS) scenario assumed is described below. The transient-state model is divided into three heat balance equations as given in reference [6]. It simulates the effect of variations of the coolant flow rate on the transfer of the heat generated in the fuel pellets to the coolant through the cladding material. While the power density distribution in hottest channel is assumed to be constant as found under the steady-state operating conditions.

2.3.1 Time dependent coolant temperature distribution.

The time dependent coolant temperature $T_w(z,t)$in the hottest channel can be written as follows:

$$A_w \rho_w C_w \frac{\partial T_w}{\partial t} = 2\pi r_c h_{cw}(T_c - T_b) - \dot{m} C_w \frac{\partial T_w}{\partial z} \tag{19}$$

Where:

T_b .. coolant average temperature within the segment dz.°C,
A .. cross-sectional area, m^2,
ρ .. density, $kg.m^{-3}$,
C .. specific heat, $J.kg^{-1}.°C^{-1}$,
r_c .. the outer clade radius, m,
h_{cw} .. cladding-coolant heat transfer coefficient, $W.m^{-2}.\ °C^{-1}$.

2.3.2 Time dependent cladding temperature distribution

The time dependent cladding temperature $T_c(z,t)$in the hottest channel can be written as follows:

$$A_c \rho_c C_c \frac{\partial T_c}{\partial t} = 2\pi r_f h_{fc}(T_f - T_c) - 2\pi r_c h_{cw}(T_c - T_b) \tag{20}$$

Where; h_{fc} is Fuel-cladding heat transfer coefficient = $4\,{}^{K_f}/_{r_f}$, $W.m^{-2}.\ °C^{-1}$ [6].

2.3.3 Time dependent fuel temperature distribution.

The time dependent cladding temperature $T_f(z,t)$in the hottest channel can be written as follows:

$$A_f \rho_f C_f \frac{\partial T_f}{\partial t} = P(z) - 2\pi r_f h_{fc}(T_f - T_c) \qquad (21)$$

Where; $P(z)$ is the linear thermal power rate (W.m^{-1}) and equals $Q(z) = A_f\, q'''_{max} \int_{z_{i-1}}^{z_i} \cos\frac{\pi z}{H} dz$.

3. Results and discussion

3.1 Steady - state

By solving the previous equations analytically, coolant, cladding and fuel temperatures at different axial heights of the fuel rod (fuel rod is divided to seven segments) can be obtained as shown in Table 2.

Table.2: coolant, cladding and fuel temperatures at different axial heights of the fuel rod.

z	T_{fi}, °C	T_{fo}, °C	T_{ci} , °C	T_{co} , °C	T_w, °C
+H/2	363.1	337.1	331.7	330.5	330
+H/4	1274.1	530.3	375.8	340.3	323.8
+H/8	1547	584.6	384.7	338.7	317.4
Zero	1637.2	598.2	382.4	332.8	309.8
-H/8	1531.8	569.4	369.5	323.5	302.2
-H/4	1246.1	502.3	347.8	312.3	295.8
-H/2	322.9	296.9	291.5	290.3	289.7

And from equation (10) the maximum coolant temperature along the fuel rod can be determined to be equal to (330 °c) and it could be at the upper end of the fuel rod exactly at (3.53m) from the bottom of the fuel rod. And so from both equations (11) and (14) the maximum cladding outer surface temperature obtained along the fuel rod is equal to (340.3 °c) and can positioned at about (83 cm) from the center of the fuel rod (about 260 cm from the bottom of the rod).And from equations (12), (13) and (15) the maximum fuel inner surface temperature along the rod could be equal (1637.2 °c) and could be at 1.6 cm from the center of the rod (about 178 cm from the bottom of the rod) [7] as shown in table 3.

The model of the previous equations is simulated using MATLAB software in order to verify the results found in the literature that shows a good agreement. And this also done for the results of the maximum coolant, maximum cladding and maximum fuel temperatures and their positions along the fuel rod for the hottest channel that is chosen for the reason of that if the thermal design specifications is determined upon which the reactor integrity during steady state operation conditions could be guaranteed [8]. The following analysis of the results obtained from MATLAB simulation would show more explanation for the importance of the thermal analysis of the reactor core with respect to the safety point of view and would also show how near the results are from that found in other references [2].

Table 3. Maximum coolant, cladding and fuel temperatures and their heights from the bottom of the fuel rod.

T	$T,^{\circ}C$	H, cm
$T_{b,max}$	330	353
$T_{c,max}$	340.3	258
$T_{m,max}$	1637.2	178.3

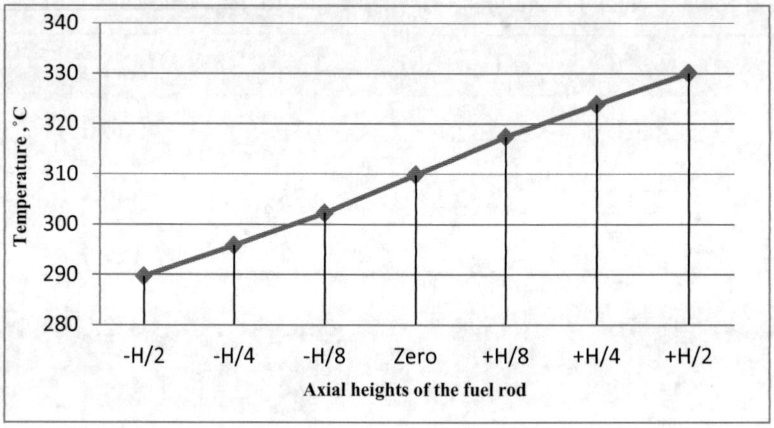

Figure 4. Coolant axial temperature axial distribution

As shown in figure 4, the coolant temperature goes increasing while moving from the bottom side to the upper side of the fuel rod transferring all the heat released from the nuclear fission inside the fuel rod to the coolant `s secondary circuit [9].

It can be noticed from the figures (5) and (6) that both cladding and fuel temperatures rises along the channel and reach maximum values in the upper part of the channel beyond the midpoint of the fuel. There are two reasons that the maximum temperature of the fuel occurs there, rather than at mid channel, where q''' is the greatest.

First, the temperature of the coolant continues to increase past the midpoint. Second, the heat flux q'' is determined only by the value of q''', and this, being a cosine function, decreases very slowly towards z = o. But q'', in turn, specifies the temperature difference (T_f –T_b). Therefore, with $T_b$increasing, T_f must also increase to provide the actual value of q'' Further along the channel, q'' begins to drop more rapidly, and T_f eventually decreases. It is this combined effect of a rising T_b and a decreasing q'' that gives rise to and determines the position of the maximum fuel (and cladding) temperature [5].

By Solving equations (16), (17) and (18) the value of the DNBR that is found to be equal about 3.7 can be obtained to be found more greater than the limiting ratio of 1.73 for the case of VVER-1000 reactors and that can be acceptable indication for the thermal design of the core and the thermal-hydraulic analysis`s results of the reactor in steady-state conditions to be so far away from problems that could occur due to the formation of the nucleate boiling on the outer surface of the fuel cladding as a result of the non-homogenous distribution of the temperatures along the fuel rods of the core.

IOP Conf. Series: Materials Science and Engineering **975** (2020) 012019 doi:10.1088/1757-899X/975/1/012019

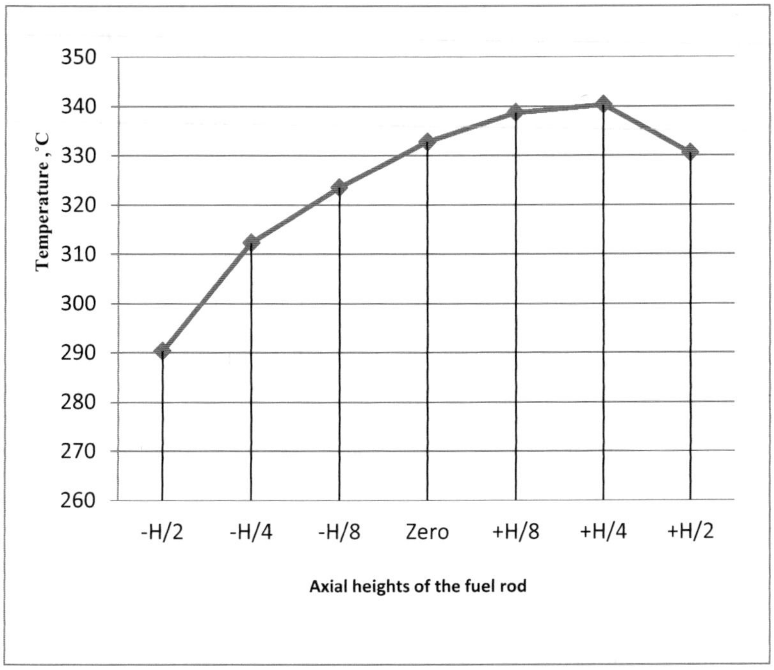

Figure 5. Cladding outer surface temperature axial distribution

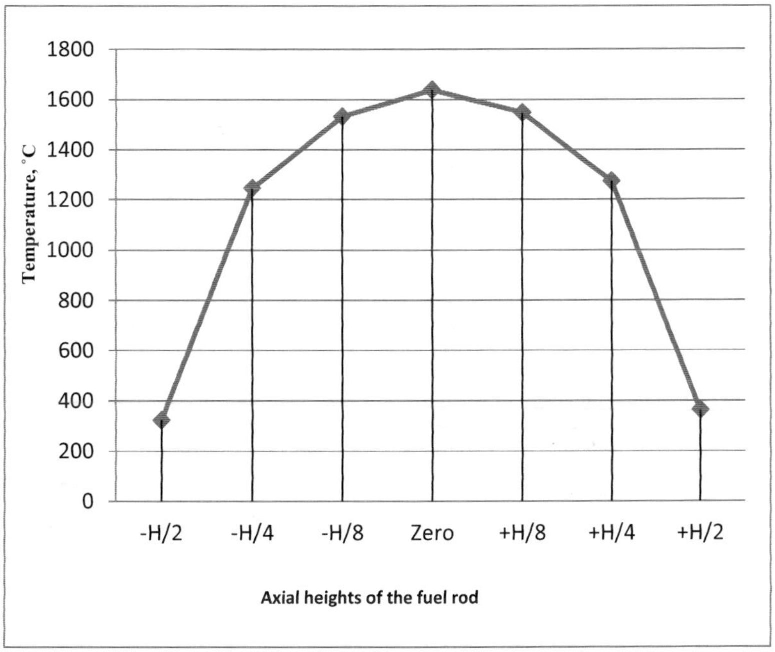

Figure 6. Fuel inner surface temperature axial distribution

IOP Conf. Series: Materials Science and Engineering **975** (2020) 012019 doi:10.1088/1757-899X/975/1/012019

3.2 Transient - state

A scenario of occasional malfunction in the primary loop causing an exponential loss of coolant in the hottest channel has been postulated and formulated as follows:

$$\dot{m}\,(t) = \dot{m}_o\, e^{-t/\tau}\,,$$

Where;

τ .. is the time constant,

t .. is the transient time .

According to the IAEA (IAEA, 1992a), the reactor controller shall activate certain protection criteria in case of LOFA. These criteria requires that the reactor must be shut down when the total flow of the coolant passing through it becomes less than 85% of its nominal value. And according to the publications concerning the adopted design safety margins of the reactor core integrity, the following limitations are to be considered throughout the postulated scenario:

a. $T_{c,max}$ < 352.15 °C (at which nucleate boiling occurs).

b. $T_{f,max}$ < 1883.5 °C.

3.2.1 Slow Loss of Flow transient (SLOFT), τ = 250 s.

The safety systems of the reactor are assumed to be not working during the SLOT scenario and so, the peak cladding and fuel temperatures are shown in the following figures:

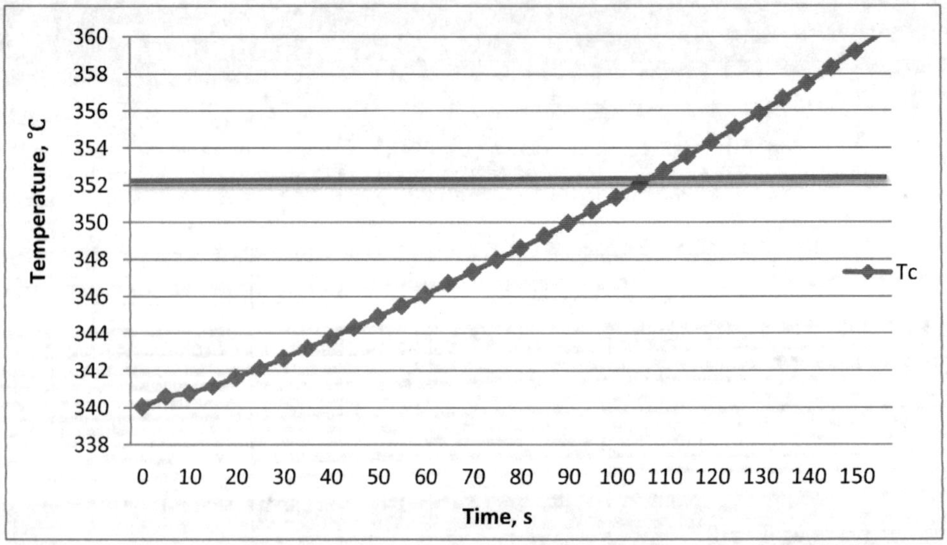

Figure 7. maximum clad outer surface temperature at τ = 250 s (SLOFT)

Through the SLOFT scenario the maximum clad outer surface temperature that located at a distance about 83 cm from the center of the fuel rod to be not exceeding 352° C (625 K) is shown to be occurred after about 110 s from the initiation of the transient as shown in figure 7.

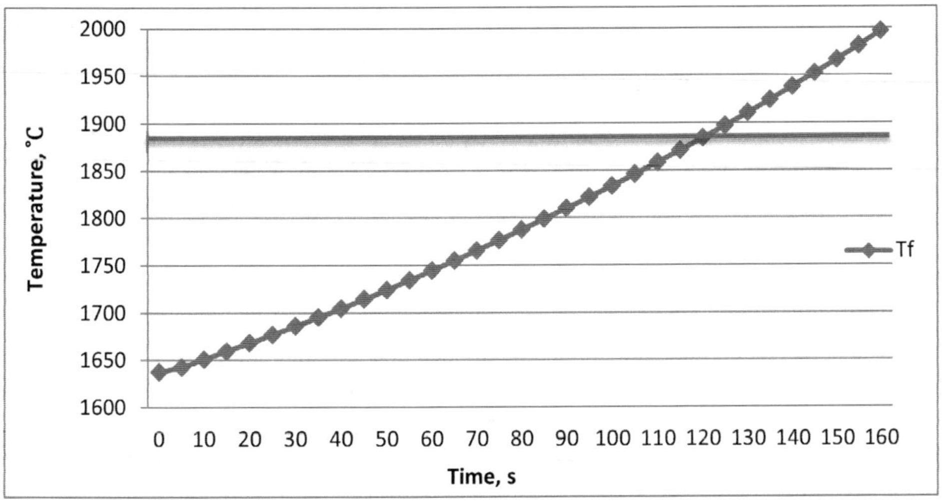

Figure 8. Maximum fuel inner surface temperature at τ = 250 s (SLOFT)

As shown in figure 8 and according to the references that determine the maximum fuel inner surface temperature that located at a distance about 1.5 cm from the center of the fuel rod to be not exceeding 1883.5° C (2156.5 K) is shown to be occurred after about 120 s from the initiation of the transient.

There is no doubt that maximum fuel temperature is a very important parameter for code validation because of that Fuel temperature responds to many effects that takes place during irradiation, such as fuel conductivity degradation, fuel densification, swelling, relocation, and other phenomena affecting the gap conductance value.

3.2.2 Fast loss of flow transient (FLOFT), τ = 10 s.

The safety systems of the reactor are assumed to be not working during the SLOT scenario and so, the peak cladding and fuel temperatures are shown in the following figures:

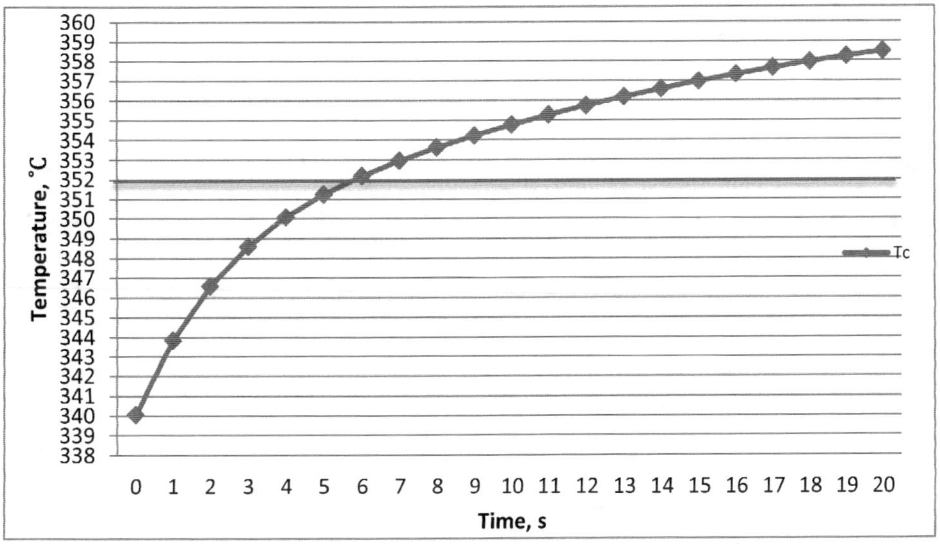

Figure 9. Maximum clad outer surface temperature at τ = 10 s (FLOFT)

According to the references that determine the maximum clad outer surface temperature that located at a distance about 83 cm from the center of the fuel rod (the orange color) to be not exceeding 352° C (625 K) is shown to be occurred after about 6 s from the initiation of the transient as shown in figure 9.

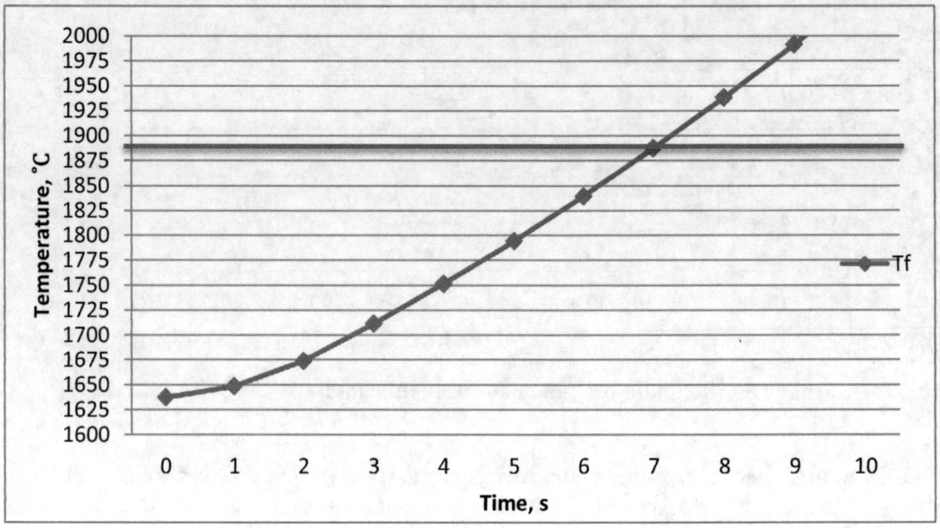

Figure 10. Maximum fuel inner surface temperature at τ =10 s (FLOFT)

On the other side the maximum fuel inner surface temperature is shown to be occurred at about 7 s from the initiation time of the transient that looks like to be reasonable due to the extreme scenario assumed of the rapid loss of flow that of course should resulting in the rapid increase in the fuel inner surface temperature as shown in figure10.

4. Conclusion

Steady-state and the transient state models were solved using finite difference method and MATLAB code. The coolant, cladding and fuel axial temperature distributions for the hottest channel in the reactor core of VVER-1000 NPP were calculated in case of loss of coolant flow event. The results were in good agreement with those found in the literature .The operator and /or the safety systems have always a short period of time for intervention and actuation to preserve the fuel integrity. Calculation prevailed that the available period for the adopted exponential loss of flow scenarios is directly related to the average time constant of the core coolant flow. For the postulated fast loss of flow it was about 6 s which means that the maximum time available for the intervention of both safety systems or the operator is about 7 seconds in order to prevent the melting of the fuel, while in case of the slow loss of flow it was about 120 s.

5. References

[1] Tikhonov, N., *WWER-1000 reactor simulator.* General Energy Technologies, Moscow Engineering and Physics institute, Moscow, Russia, 2011.

[2] *<Modified-COBRA-EN-code-to-investigate-thermal-hydraulic-analysis-of-the-Iranian-VVER-1000-core_2010_Progress-in-Nuclear-Energy.pdf>.*

[3] Othman, R., *Steady state and transient analysis of heat conduction in nuclear fuel elements.* Masters degree project, KTH-Stockholm Sweden, 2004.

[4] Arshi, S.S., S. Mirvakili, and F. Faghihi, *Modified COBRA-EN code to investigate thermal-hydraulic analysis of the Iranian VVER-1000 core.* Progress in Nuclear Energy, 2010. **52**(6): p. 589-595.
[5] Lamarsh, J.R. and A.J. Baratta, *Introduction to nuclear engineering.* Vol. 3. 2001: Prentice Hall Upper Saddle River, NJ.
[6] Tylee, J.L., *Simple Reactor Model Simulation of a LOFT ATWS Event.* Nuclear Technology, 1983. **61**(1): p. 25-32.
[7] Hashemi-Tilehnoee, M., et al., *Sub-channel analysis in hot fuel assembly's of VVER-1000 reactor using drift-flux model.* Indian Journal of Science and Technology, 2015. **8**(33).
[8] Aybar, H.S. and P. Ortego, *A review of nuclear fuel performance codes.* Progress in Nuclear Energy, 2005. **46**(2): p. 127-141.
[9] El-Wakil, M.M., *Nuclear heat transport.* 1971: International Textbook Company New York.

Effect of the formation parameters on the thermal properties of $La_2Zr_2O_7$

Ahmed Abdelkareem
Nuclear Materials Authority, Cairo, Egypt
Email address: kareemnma@yahoo.com

Abstract Lanthanum zirconates has been suggested as a thermal barrier coating for many high temperature applications. This study is focusing on the effect of different preparation parameters on the thermal conductivity of $La_2Zr_2O_7$. We have used solid state reaction to form the zirconates powder was very good homogeneity. We have studied the effect of the intimal ZrO_2/La_2O_3 ratio, and found the 50% to be the optimum. The effect of the sintering temperature has also been studied and 1100°c was chosen as the optimum sintering temperature. The sintering time was also studied, where 3 hours was proved to be enough to obtain a homogeneous single phase compound with affordable method.

Keywords: Ceramic composites, Sintering, Thermal properties, Thermodynamics and kinetics of processes in materials

1. Introduction

Rare earth zirconates have received considerable attention due to its wide industrial applications such as solid oxide fuel cell electrode materials, fluorescence screens, thermal – barrier coatings, high temperature heating elements, and oxidation catalysts.[1-3] Of particular interest, lanthanum zirconate, $La_2Zr_2O_7$, LZ has been suggested as an alternative to the traditional NiCoCrAlY coating on super alloys as a thermal barrier in modern gas-turbine engines.[2, 4] LZ exhibits high thermal stability nearly up to its melting point at 2300 ^oC. The thermal conductivity of LZ is as low as 1.56 W m^{-1} K^{-1} that is much lower than that of yettria stabilised zirconia (2.12 W m^{-1} K^{-1}).[5-7] LZ has low sintering rate, which helps on the formation of films without damaging the substrate. However, LZ suffers from low thermal expansion coefficient (9×10^{-6} K^{-1}), which makes the coating more susceptible to thermal stress due to the large thermal expansion mismatch with the metallic substrates.[6, 8, 9]

There have been several attempts to minimise the effect of the thermal expansion mismatch between the ceramics and the substrate. Using nano particles to form a more robust structures of the coating has shown great success.[10] Doping the LZ with other rare earth elements also proven to be successful strategy to increase the thermal expansion coefficient and subsequently improve the coating stability.[11, 12] Other techniques such as graded coatings and double-ceramic-layer coatings in which LZ is coated on YSZ inner layer, has also been used industrially.[13-16] Amongst the other thermal properties modifiers, doping with other rare earth is gaining more attention since it usually associated with less problem during the fabrication.

The present paper has a more economic strategy of using the other rare earth oxides exists in the natural mineral to increase the thermal expansion coefficient. The paper also is investigating different parameters that might affect the thermal conductivity of the LZ. The present work adopted simple solid-state reaction process to prepare the RE-doped LZ.

IOP Conf. Series: Materials Science and Engineering **975** (2020) 012020 doi:10.1088/1757-899X/975/1/012020

2. Experimental work

2.1. Preformation calculations

Before proceeding to the formation of the LZ from natural ore, a comprehensive array of calculation was carried out in order to have a pre-estimation of the materials preparation conditions HSC chemistry-5 was used to calculate the thermodynamic parameters at different reaction temperatures. The calculations showed that it is possible to form the single phase compound even at room temperature. Although increasing the sintering temperature seems to have a small effect on the delta G (which reflect the spontaneity of the reaction), the reaction kinetic seems to increase significantly with the temperature. From the resulted change in enthalpy values we deduce that the reaction is exothermic, the entropy values behaved as positive values at temperature range from zero to 1100 °C, and started to be negative at 1200 °C ,which means that the system is not controllable at temperature more than 1200 °C, so it is recommended to apply the sintering of the reactants at temperature less than 1200 °C

$$La_2O_3 + 2ZrO_2 \longrightarrow La_2Zr_2O_7$$

Table 1: Showing the values of free energy, enthalpy, entropy and equilibrium constant for the lanthanum zirconate at different temperatures

T, °C	ΔH, kJ	ΔS, J/K	ΔG, kJ	K	Log(K)
0	-86.800	9.960	-89.521	1.320 E+17	17.121
100	-86.800	9.960	-90.517	4.697 E+12	12.672
200	-86.800	9.960	-91.513	1.270 E+10	10.104
300	-86.800	9.960	-92.509	2.701 E+8	8.432
400	-86.801	9.959	-93.505	1.804 E+7	7.256
500	-86.801	9.959	-94.501	2.427 E+6	6.385
600	-86.801	9.959	-95.497	5.169 E+5	5.713
700	-86.801	9.959	-96.492	1.513 E+5	5.180
800	-86.801	9.959	-97.488	5.566 E+4	4.746
900	-86.801	9.958	-98.484	2.429 E+4	4.385
1000	-86.802	9.958	-99.480	1.207 E+4	4.082
1100	-86.802	9.958	-100.476	6.644 E+3	3.822
1200	-103.442	-1.558	-101.146	3.861 E+3	3.587
1300	-102.746	-1.102	-101.012	2.261 E+3	3.354
1400	-101.855	-0.553	-100.929	1.416 E+3	3.151
1500	-100.770	0.076	-100.904	9.392 E+2	2.973

2.2. Experimental and methodology procedure

The experimental work and methodology reperents on the materials processing, dehydration, hydpressing, sintering and instrumental investigation by XRD,SEM and thermal conductivity measurement device. The starting materails to be processesd; are lanathunm oxide and zirconium oxide with a commercial grdae brand name ALDRICH ™ with and purity 99%with using screen anaylysis for contrroling mesh size up to -63 micron, both two materials were mixed and grinded inside discmill for two hours in dry atmosphere ,after finishing mixing by mechanical grinding the loosed sample was released off and wieghed to be 10 grams ,then the smaple was dried inside oven to a temperatute up to 100 °c ,the dried sample was mixed with bindary materials paraffin wax with a weight perecntage up to 1.5%; estimated as function of the total weight of the sample ,then the sample was mounted inside a die with a dimension 23mm diameter and 12 mm height, the pressing applied load estmiated to be 15 ton , this procedure was repeated with different materials balance, where we used lanthanum different weighing perecentges , where the mixing process used 20%,

IOP Conf. Series: Materials Science and Engineering **975** (2020) 012020 doi:10.1088/1757-899X/975/1/012020

30%,40%,50%, and 60% lanathanum oxide and the balance was zirconium oxide, ie the mixixng process was repeated five tiems with different starting materials weight balance.eache weight of mixed materaisl was repeated also five times to prduce five samples with the same materials balance to study subsequently the effect of sintering temeperature.the pressed samples was sintered at different temperatures to test the effect of sintering temperatures ,the sintering was done at muffle furnace at temperature 900°C,950 °C,1000 °C,1050 °C, and 1100 °C, eventually each five samples was sintered at these five sintering temperature. The sequence of the process is illustarting in flowsheet at figure (1).

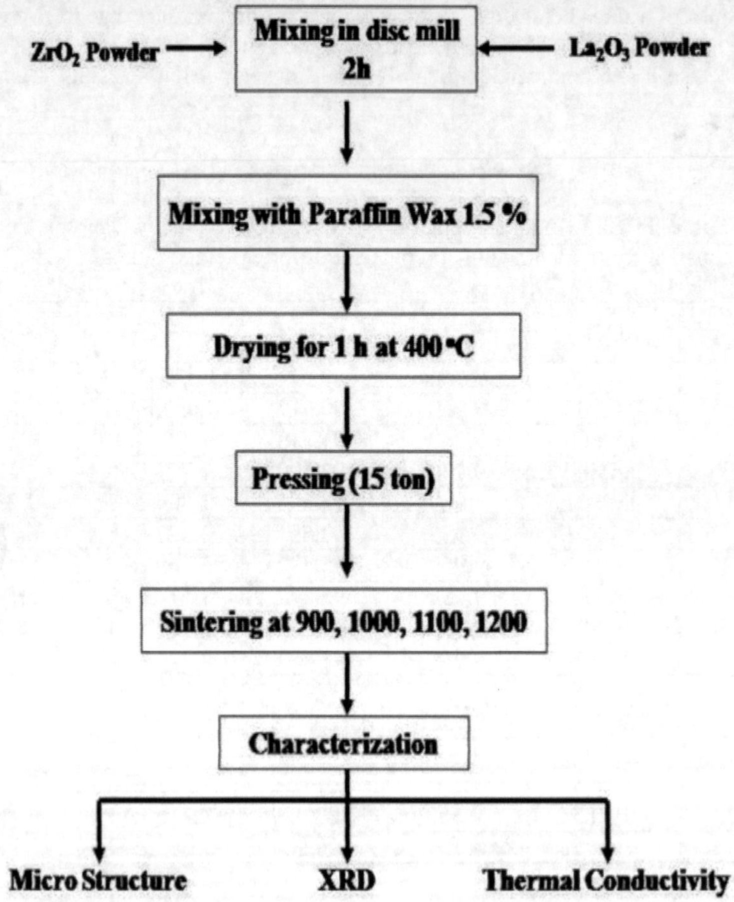

Figure1. Simple flow sheet for pressing and sintering process

Each produced sample was investigated by X-ray powder diffraction (XRD) measurements were carried out with a Philips x-ray generator model PW 3710=31 a diffractometer with automatic sample changer model PW1775 (scintillation counter, Cu-target tube and Ni-filter at 40KV and 30 mA). Energy dispersive x-ray measurements were measured using environmental scanning electron microscope (ESEM) EXL 130 and attached by energy dispersive spectrometry (EDX) unit system.

Thermal conductivity of each test sample was measure and detected by diffuse meter measure flash used to measure the heat transferred through the sample thickness for this purpose the apparatus theheat flux diffusion froDXF 200 employs a High Speed Xenon-pulse Delivery source (HSXD) each sample was mounted and measure the heat flux diffusion and the calculation of heat diffusion showed the that the thermal conductivity varied from sample to another.

3. Ruselts and Discution

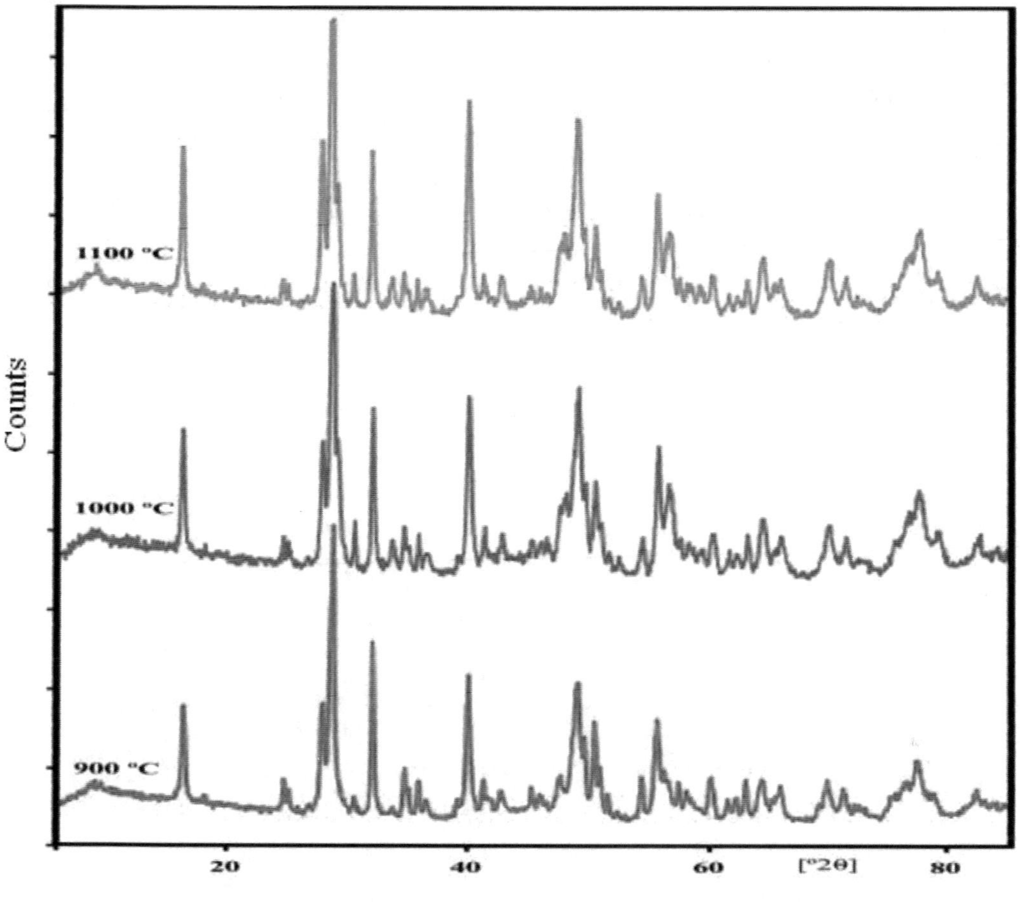

Figure 2. XRD patterns of $La_2Zr_2O_7$ at diffeent sintiring tempereture

The XRD patterns of $La_2Zr_2O_7$ ceramics under deffernt sintring conditions are shown in figure 2. All the defraction peaks can be fited with the typical pyrochlor LZ pattern (JCPDS 73 -0444),[17, 18] sugesting scuccesful formation of the single phase oxide. The intensity of the peaks and the width at half length decreased with increasing the sintering temprture, which is a sign of increasing the crystalinity. The lattice parameter evaluated from the X ray deffrection pattern are a= 1.0823 nm , b= 1.0823 nm , c = 1.0823nm, which is very close to the reported values of the LZ lattace.[17] This bascially means that the presence of impurties did not change the cyrstal stracture or reduced the ordering of the atoms. We can then conclude that the other RE impurties are occuping the La^{+3} ions posation, i.e. they are packed in the vacancy of the octahedron constituted by $6ZrO_6$ in the pyrochlor crystal structure. Indeed this substitution of the La^{+3} ions with lower valence ions creates oxygen vecancies, which is curcial for controling the conductivity proprties and the thermal expansion coffecient.[19]

Figure 3: Shows microstructures using SEM of $La_2Zr_2O_7$ ceramics at content of La_2O_3 50% in matrix under different sintering conditions : a)- at 900 °C, b)- at 1000 °C and c)- at 1100 °C all for 3 h

The changes in the microstucture of lanthanum zirconate with the sintering temperetures are evaluated using scanning electron microscope. Figure (3) showes that the particles size are growing with increasing the sintering temprture. For the low temprture sinrtering, the grain boundry is very clear, indicating very low sinterability of the powder. Increasing the sintering temprture leads to aggrigation of the particles into biger culsters. The grain boundry almost vanished for the sample heated at 1200 indicating the small particles are in growing state to a larger particle. The bulk densty of the particles also increased with increasing the temprture, which propably due to the compacting of the particles and the reduction of the between – the-particles porous. [20]

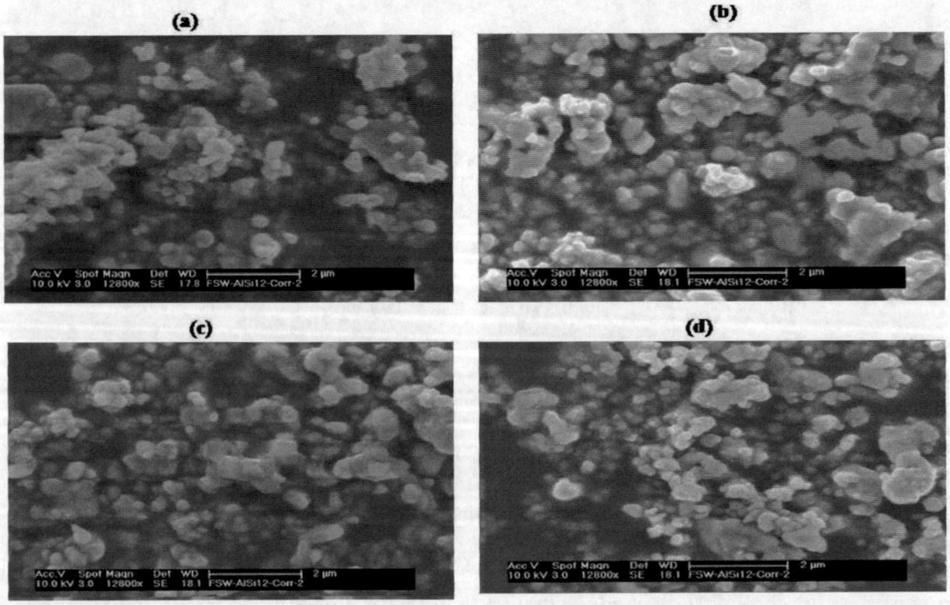

Figure 4: Shows the microstructure using SEM of $La_2Zr_2O_7$ at sintring temperture 1100 c at 3h and different content of La_2O_3 in matrix: a)- at 30% , b)- at 40 % , c) – at 50% and d)- at 60%.

Figure 4 showes the SEM images of the LZ powderd with various La_2O_3 conetent all sintered at 1100 oC. It seems like the precentage of the La_2O_3 is less pronounce than the sintering temprture on the microstructure of the prepared LZ powder. This is mainly because both componant of the single oxides phase have very similar melting point and particle size. There are slightly tendency to form biger aggrigation with increasing the rare earth content. La2O3 may dispropontiate to LaO_2 and LaO, which may accelarte the reaction with the zirconium oxide to form non-stoichiometeric oxide phase, which usually have lower melting point than the more ordered oxide phase ($La_2Zr_2O_7$).

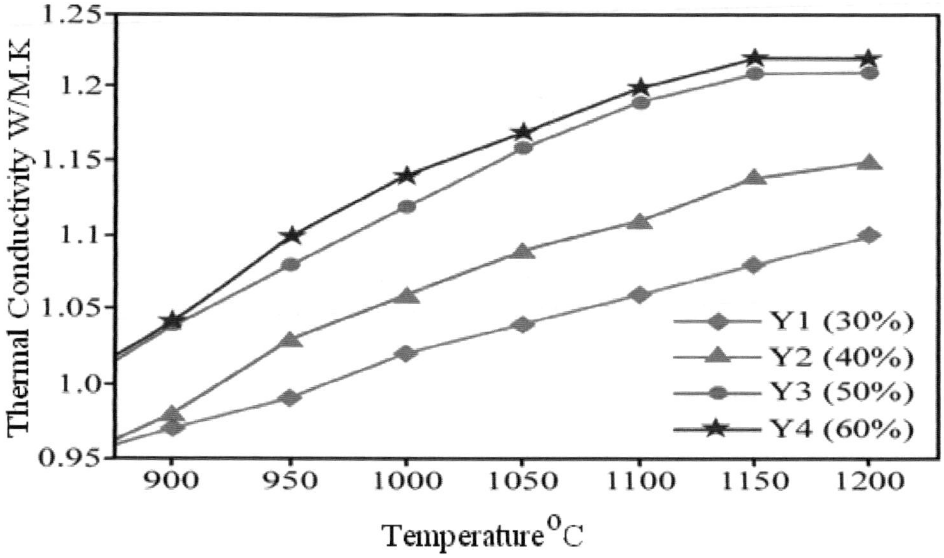

Figure 5. Thermal conductivity at various sintering temperature and different of La_2O_3 content

The thermal conductivity of the sample with various La_2O_3 content sintered at different temperature are illustrated in figure 5. The thermal conductivity measurements were conducted using the diffuse-meter technique. Clearly the thermal conductivity increased with increasing the temperature due to the decrease of the boundary effect with the growth of the particles size. The conductivity increased also with increasing the La_2O_3 content. By almost of 30 % at any temperature, we believe that the increase of the thermal conductivity is related to introducing lattice defects, including oxygen vacancies, which can changes the energy state of the binary oxide. Interestingly the thermal conductivity value measured in the current work is much higher than that reported in the literature.[21] We believe the enhanced thermal conductivity is due to the increase of the oxygen vacancies and other disorders in the LZ lattice with increasing the doping by the other RE oxides.[22]

4. Conclusion

We have successfully produced pure phase of $La_2Zr_2O_7$using facial and simple solid stat reaction.The results of the current work showed that the most homogeneous phase can be produced by mixing the stoichiometric amount of the La_2O_3 and ZrO_2. The morphology of the particles was found to be dependent on the sintering temperature, with tendency of the particles to form large clusters and grow in size. Interestingly, we found that the impurities associated with La_2O_3 have a significant role on increasing the density of the oxygen vacancies and other defects in the $La_2Zr_2O_7$cubic lttice. The slighly lacttice dystortion has a significant role on increasing thermal conductivity to reach a maximum value of 1.22 W/m.k.

5. References

[1] Cao, X.Q., R. Vassen, and D. Stoever, 2004, *Ceramic materials for thermal barrier coatings.* Journal of the European Ceramic Society,. **24**(1): p. 1-10.

[2] Wu, J., et al., 2002, *Low-Thermal-Conductivity Rare-Earth Zirconates for Potential Thermal-Barrier-Coating Applications.* Journal of the American Ceramic Society,. **85**(12): p. 3031-3035.

[3] Shimada, T., et al., 2004, *The high temperature proton conductor BaZr0.4Ce0.4In0.2O3−α.* Journal of Power Sources,. **131**(1–2): p. 289-292.

[4] Xu, Z., et al., 2010, *Hot corrosion behavior of rare earth zirconates and yttria partially stabilized zirconia thermal barrier coatings.* Surface and Coatings Technology,. **204**(21): p. 3652-3661.

[5] Shimamura, K., et al., 2007, *Thermophysical Properties of Rare-Earth-Stabilized Zirconia and Zirconate Pyrochlores as Surrogates for Actinide-Doped Zirconia.* International Journal of Thermophysics,. **28**(3): p. 1074-1084.

[6] Joulia, A., M. Vardelle, and S. Rossignol, 2013, *Synthesis and thermal stability of Re2Zr2O7, (Re = La, Gd) and La2(Zr1−xCex)2O7−δ compounds under reducing and oxidant atmospheres for thermal barrier coatings.* Journal of the European Ceramic Society,. **33**(13–14): p. 2633-2644.

[7] Nair, J., et al., 1998, *Preparation and characterization of lanthanum zirconate.* Journal of Materials Science,. **33**(18): p. 4517-4523.

[8] Kutty, K.V.G., et al., 1994, *Thermal expansion behaviour of some rare earth oxide pyrochlores.* Materials Research Bulletin,. **29**(7): p. 759-766.

[9] He, L., et al., 2009, *Adhesive strength of new thermal barrier coatings of rare earth zirconates.* Vacuum,. **83**(11): p. 1388-1392.

[10] Dhas, N.A. and K.C. Patil, 1993, *Combustion synthesis and properties of fine-particle rare-earth-metal zirconates, Ln2Zr2O7.* Journal of Materials Chemistry,. **3**(12): p. 1289-1294.

[11] Davies, R.A., M.S. Islam, and J.D. Gale, 1999, *Dopant and proton incorporation in perovskite-type zirconates.* Solid State Ionics,. **126**(3–4): p. 323-335.

[12] M.F.Chera, A.K. Sakr, H.I. Mira and S.A.Mohamed **, 2019,** *Removal of REEs from aqueoussoiution using aminophenol anchored treated chinese waste clay.* Materials Science and Engineering **,. 5(1).**

[13] Wang, C., et al., 2015, *Optimized functionally graded La2Zr2O7/8YSZ thermal barrier coatings fabricated by suspension plasma spraying.* Journal of Alloys and Compounds,. **649**: p. 1182-1190.

[14] Wang, L., et al., 2012, *Thermal shock behavior of 8YSZ and double-ceramic-layer La 2Zr 2O 7/8YSZ thermal barrier coatings fabricated by atmospheric plasma spraying.* Ceramics International,. **38**(5): p. 3595-3606.

[15] Wang, L., Y. Wang, and X. Sun, 2013, *Preparation of 8YSZ and double-ceramic-layer La2Zr2O7/8YSZ thermal barrier coatings on GH4169 superalloy substrates and their static high temperature oxidation behaviour*, in *Materials Science Forum*.. p. 617-632.

[16] Xu, Z., et al., 2009, *Double-ceramic-layer thermal barrier coatings of La2Zr2O7/YSZ deposited by electron beam-physical vapor deposition.* Journal of Alloys and Compounds,. **473**(1-2): p. 509-515.

[17] de Camargo, A.S.S., et al., 2004, *Structural and spectroscopic properties of rare-earth (Nd3+, Er3+, and Yb3+) doped transparent lead lanthanum zirconate titanate ceramics.* Journal of Applied Physics,. **95**(4): p. 2135-2140.

[18] Tong, Y., et al., 2008, *Preparation and characterization of Ln2Zr2O7 (Ln = La and Nd) nanocrystals and their photocatalytic properties.* Journal of Alloys and Compounds,. **465**(1–2): p. 280-284.

[19] Scheetz, B.E. and W.B. White, 1979, *Characterization of Anion Disorder in Zirconate A2B2O7 Compounds by Raman Spectroscopy.* Journal of the American Ceramic Society,. **62**(9-10): p. 468-470.

[20] Abdelkader, A.M. and D.J. Fray, 2013, *Synthesis of self-passivated, and carbide-stabilized zirconium nanopowder.* Journal of nanoparticle research,. **15**(12): p. 2112.

[21] Suresh, G., et al., 1997, *Investigation of the thermal conductivity of selected compounds of gadolinium and lanthanum.* Journal of Nuclear Materials,. **249**(2): p. 259-261.

[22] Wan, C., et al., 2011, *Order–Disorder Transition and Unconventional Thermal Conductivities of the (Sm1−xYbx)2Zr2O7 Series.* Journal of the American Ceramic Society,. **94**(2): p. 592-596.

Bioleaching of uranium from el-sella ore material using epicoccum nigrum

Azhar M Elsayad[1], *Shimaa S Hussien[2], Mohamed G Mahfouz[2], Abdou A El Mougith[1] and Wesam A Hassanien[1]

[1]Department of Botany, Faculty of Science, Zagazig University, Egypt
[2]Ore Processing Department, Nuclear Materials Authority, Cairo, Egypt
Corresponding author e-mail: mash_1997@yahoo.com (Shimaa S. Hussien)

Abstract. Uranium (U) is one of the strategic elements and essential for many applications as a fuel in nuclear power plants and nuclear weapons. Bioleaching of uranium was carried out with Epicoccum nigrum isolated from El-Sella ore material. El-Sella area is located in the southern eastern desert of Egypt. The maximum bioleaching of U(VI) was found to be 76.6 % at these optimum conditions: 9 days incubation period, 0.5% pulp density, 30 °C incubation temperature at shaking speed 175 rpm. Glucose and NH4Cl were considered the best carbon and nitrogen sources for fungal growth. E. nigrum exhibits good potential in generating varieties of organic acids which effective for bioleaching of uranium. It produced 112.3 and 23.5 µg/ml from gallic and ellagic acids in the culture filtrate under the optimum conditions. This work addresses the area of beneficiation of the used mineral to solubilize U(VI) through the biotechnological route in Egypt, where the bioleaching method is more effective than the chemical one using organic acids.
Keywords: Bioleaching; uranium; organic acids; Epicoccum nigrum.

1. Introduction

Bioleaching has been successfully used in commercial metal mining for decades. It uses microbes to biosolubilize metal-containing inorganic compounds such as metal oxides and sulfides [1]. Benefiting from lower operational costs and energy requirements than do hydrometallurgical and pyro-metallurgical processes in metal recovery [2]. The use of microorganism to dissolve metals from ores can be environmentally friendly and a low cost alternative to traditional methods that lead to high cost because of either energy consumption or bioremediation of polluted environment [3]. Several mechanisms may be involved in bioleaching these include, acidolysis, complexolysis, redoxolysis and bioaccumulation. Acidolysis is the main principal mechanism in bioleaching of metals where, the fungus and bacterium produce varieties of organic acids as citric acid, oxalic and gluconic acids during the bioleaching [4]. Fungus has four advantages in the leaching of uranium ore [5]. Firstly, chemo-organo-heterotrophic fungi can use utilize organic carbon source as substrate and can dissolve uranium at high pH. And the secreted organic acids can react with calcium, aluminum, iron, and other elements in gangue to form complexes with higher solubility. Secondly, in addition to acidic ore, fungi can transform uranium oxides, carbonates, and phosphates to form uranium complexes with carboxylic acids. Thirdly, fungi are the heterotrophic microorganisms characterized by rapid growth rate, large biomass and short extraction cycle **[6]**. Fourthly, fungal leaching has low anti-corrosion requirement for equipment, since the organic acids produced by fungi are mainly weak acids, which show less environmental hazards than H_2SO_4 and other strong acids and can be degraded by environmental microorganisms. As an environment-friendly method, fungal leaching is of huge development potential and wide application prospect and has been extensively studied [7]. Hussien *et al.*[4] observed that the highest percentages of bioleached uranium from the tested samples directly by *Penicillium purpurogenium* were found to be 72.49, 55.60 % at a pulp density 300 g/L after 9 days of incubation at 30 °C. The main objective of this study is the use of *Epicoccum nigrum* for uranium solubilization from El-Sella ore material.

2. Materials and Methods

2.1. *Characterization of the tested sample*

El-Sella area is located in the southern eastern desert of Egypt, between Latitudes 22°14′ 30′′ and 22° 18′ 36′′N and Longitudes 36° 11′ 45′′ and 36° 16′ 30′′ E, at a distance of about 60 km south west of Abu Ramad according to [8].

2.2. *Microorganisms and growth conditions*

Seven fungal species were isolated from El-Sella ore material using Czapek′s-Dox agar medium by serial dilution method. Aspergillus terrus, A. niger sp1, A. niger sp2, A. sulfurous, Penicillium chrysogenum, P. notatum and Epicocuum nigrum. The fungal strain Epicoccum nigrum was found to be one of the most acid producer fungal isolate through lowering the pH of the growth medium. E. nigrum was sub-culturing using Modified Czapek's- Dox agar (MCDA) for 7-10 days at 30 °C. It was identified as described by **[9]**. E. nigrum was grown in the absence and the presence of different weights of El-Sella ore material. E. nigrum cultures were incubated at 28 °C for 7 days then the dry weights were determined as mg/mL.

2.3 Microbial leaching techniques

A. Direct bioleaching process: In direct bioleaching, 100 ml of sterilized Modified Czapek's–Dox broth (MCDB) containing 1g of El-Sella ore material inoculated with 1mL (approximately 1×10^7 spores/mL) fungal spore suspension. Culture was incubated in a rotary shaking incubator at 175 rpm at 28°Cfor 7days [10].

B. Indirect bioleaching process: The production of organic acids can be performed by an indirect bioleaching process. *E. nigrum* was cultivated in the corresponding medium under the previous growth conditions. All cultures were incubated in a rotary shaking incubator at 175 rpm at 28°C for 7 days .After separation of the fungal biomass, 1g of sterilized El-Sella ore material was added to the filtrates (1 g/100 ml) at shaking 150 rpm for 24 h. For each bioleaching the final pH of filtrates and percentages of bioleached uranium in supernatant was determined [11].

2.4 Estimation of uranium (VI) by arsenazo III

The bioleached liquors were filtered through a filter paper (Whatman No. 41) and the concentration of U(VI) in the solution was measured before and after equilibrium by Metertech Ino model Sp-5001, UV- Visible spectrophotometer) using arsenazo III **[12]**

2.5 Production and estimation of organic acids by high performance liquid chromatography (HPLC)

The efficiency of the tested strain to produce organic acids was examined by adding 0.5 g/L of $CaCO_3$ to the growth agar media [13].Diameters of the clear zone were determined. The organic acids produced in the presence of El- Sella ore material with the organism were estimated using high-performance liquid chromatography (HPLC).Samples of 10 µL were injected onto C18 column (4.6 mm*250mm in length) at a flow rate of 3.0 mL /min (at room temperature).The mobile phase consisted of acetonitrile and water (80:20) was detected with UV detector at 254 nm for gallic and ellagic acids estimation [14].

2.6 Factors affecting the uranium bioleaching from El-Sella ore material

The optimum conditions for maximum of uranium bioleaching from El-Sella ore material were evaluated. Sample concentration from (0.1-1) g/50 mL, pulp density from (0.1-1.5) %, incubation periods from (5-12) days, incubation temperatures frpm (20-40) °C, shaking speed from (120-175) rpm and different carbon and nitrogen sources were examined for direct bioleaching of uranium. Erlenmeyer flasks (250 ml), each containing 50 ml of the used MCDB medium in presence of 0.3 g of ore sample .These conical flasks were autoclaved then inoculated with 1 mL of fungal spore suspension.

3 Results

3.1. Ore sample characterization

The chemical composition of El-Sella ore material was shown in table 1. The results indicated that U(VI) concentrate was reached 1173.4 ppm.

Table 1. Chemical analysis of El-Sella ore material and the residue of leaching.

Wt. Percent, %	Oxides											
	SiO_2	Al_2O_3	Fe_2O_3	CaO	MgO	MnO	K_2O	Na_2O	TiO_2	P_2O_5	LOI	Total
Sample	68.76	15.07	5.21	1.03	0.53	0.04	1.25	0.37	1.07	0.57	5.78	99.86
Residue	62.64	12.41	3.07	3.28	2.48	0.03	1.79	0.96	1.89	0.47	1.43	99.45

3.2. Effect of El-Sella ore material on the growth of E. nigrum

This experiment was carried out to study the effect of sample weights on the fungal biomass. At the end of the incubation periods 9 days results were obtained and are shown in figure 1. The dry weights of the tested organism were clearly decreased as the weight of El-Sella ore material was increased as compared to control.

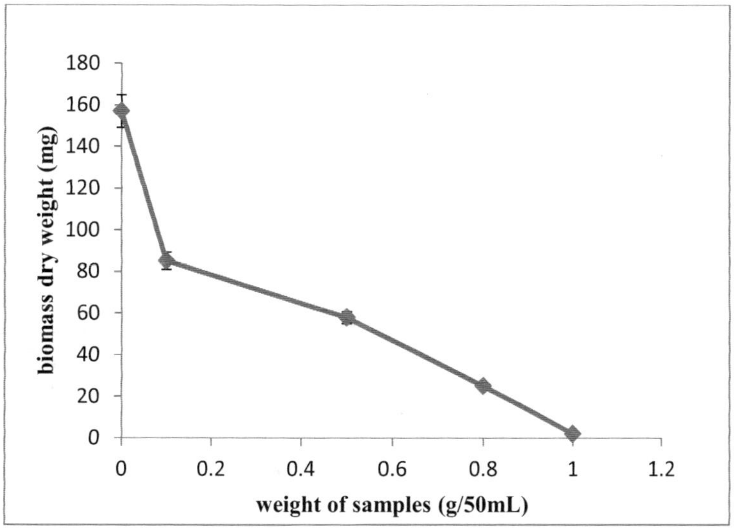

Figure 1. Growth of *E.nigrum* in the presences of different weight of El-Sella ore material.

3.3 Microbial leaching techniques

This experiment was carried out to study the bioleaching technique by direct and indirect method. The result obtained in Table 2, indicated that the highest bioleaching percentage was obtained by direct method. The bioleaching efficiency of uranium was about 41.3 % by direct method.

Table2. Uranium bioleaching percentages by *E. nigrum* using direct and indirect methods.

Fungus isolate	% of bioleached uranium			
	Final pH	Direct method	Final pH	Indirect method
E. nigrum	5.1	41.3±0.14	5.31	30.9±0.91

3.4 Production and estimation of organic acids by high performance liquid chromatography (HPLC)

Table 3, illustrates the total acid produced by microorganisms using 2 testing methods: the first one indicated the diameter of the clear zones formed in the growth medium MCDA containing calcium carbonate as an indicator for acid production after 9 days incubation period. In the second method, organic acids produced in the fungal and bacterial filtrates were analyzed using HPLC in presence of El-Sella ore material after 9 days. It was found that the main acids produced by *E. nigrum* were gallic acid 112.9 (μg/ml) followed by ellagic acid 23.5 (μg/ml).

Table 3. Acid productions of the microorganisms after 7 days for clear zone, while after 9 days for HPLC.

fungus	Diameter of clear zone (cm)	Organic acid produced (μg/ml)	Final pH filtrate
E. nigrum	7.6	Gallic acid 112.9 Ellagic acid 23.5	4.3

3.5. Factors affecting the uranium bioleaching from El-Sella ore material by E. nigrum.

The effect of different factors as weight of samples (g/50ml), pulp density, incubation periods (days), incubation temperatures (°C), shaking speed (rpm), on bioleaching of uranium from the tested sample were evaluated.

The maximum bioleaching percentages of U(VI) from El-Sella ore material was found to be about 76.6 % at weight of sample (0.5g/50 mL), pulp density (0.3 %), incubation period (9 days), incubation temperature (30°C), shaking speed (175 rpm) as shown in figures (2-6).

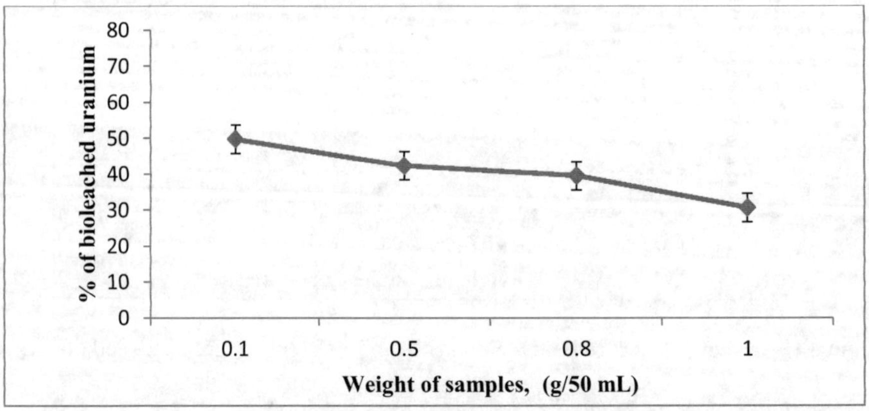

Figure 2. Effect of sample weights (g/50mL) on bioleaching of uranium from El-Sella ore material by *E. nigrum.*

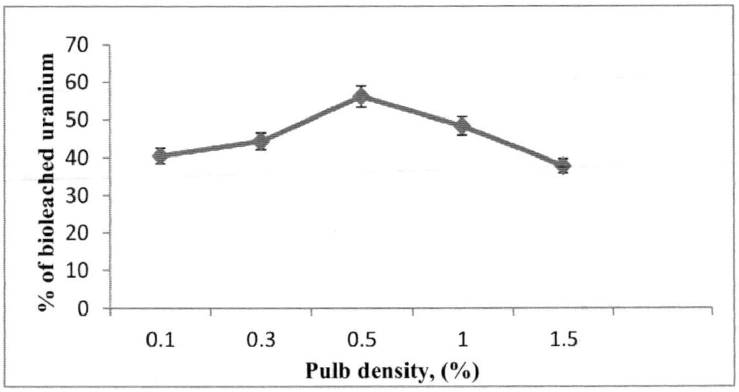

Figure 3. Effect of pulp density on bioleaching of uranium from El-Sella ore material by *E. nigrum.*

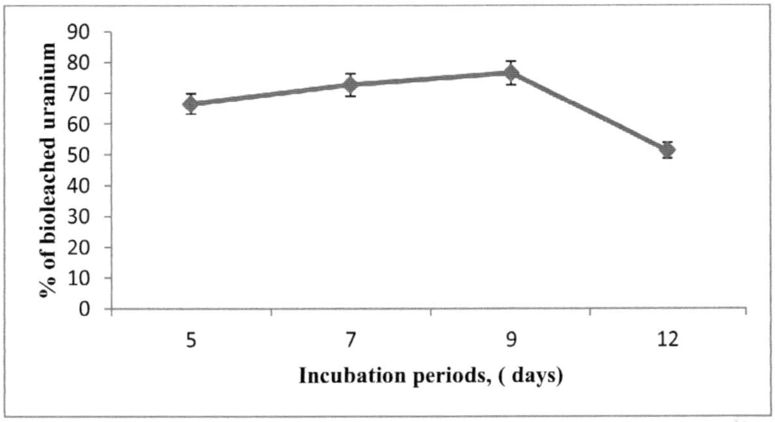

Figure 4. Effect of incubation periods (day) on bioleaching of uranium from El-Sella ore material by *E. nigrum.*

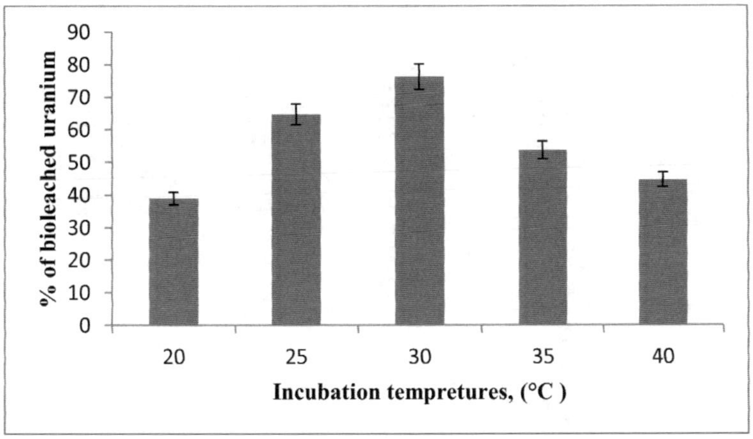

Figure 5. Effect of incubation temperatures (°C) on bioleaching of uranium from El-Sella ore material by *E. nigrum.*

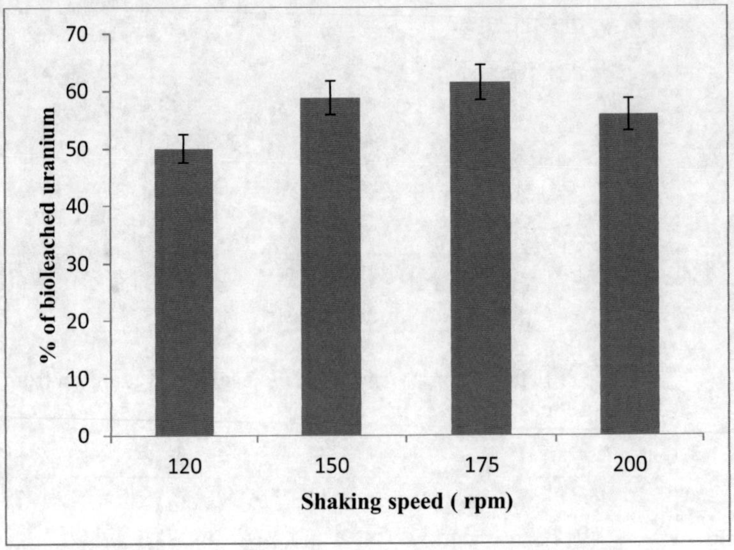

Figure 6. Effect of shaking speed (rpm) on bioleaching of uranium from El-Sella ore material by *E. nigrum.*

Also, glucose and ammonium chloride were found to be the best carbon and nitrogen source, respectively, as appeared in figures (7-8).

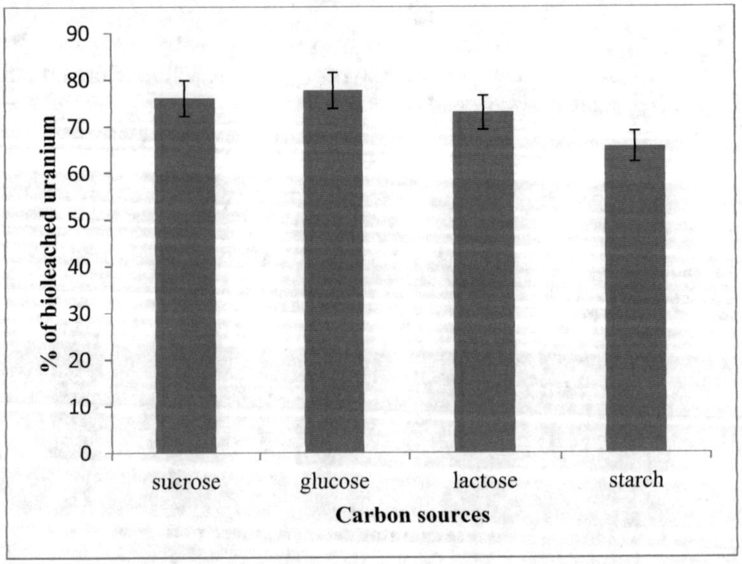

Figure 7. Effect of different carbon sources on bioleaching of uranium from El-Sella ore material by *E. nigrum.*

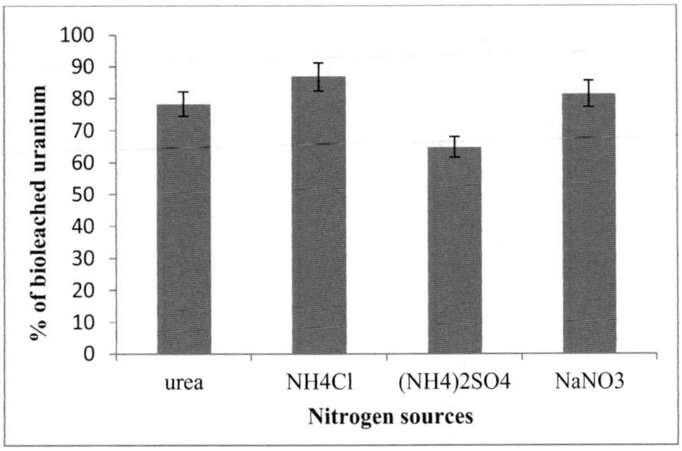

Figure 8. Effect of different nitrogen sources on bioleaching of uranium from El-Sella ore material by *E. nigrum.*

4 Discussion

The sample was taken from El-Sella area in the eastern desert in Egypt for determination of uranium. The fungal isolate was identified according to their features and their frequency of occurrence according to [9,15]. *E. nigrum* is capable of dissolving heavy metals from the tested ore material and surface of the rocks as previously mentioned by [16]. It was observed that there was a gradual decrease in the biomass of *E. nigrum* as the concentration of U(VI) concentrate increased in the growth medium. These obtained results may be due to the toxic effect of heavy metals present in the tested ore material. During the growth studies of bioleaching process by *E. nigrum*, glucose was considered the substrate undergo microbial oxidations which resulted in the production of organic acids as gallic acid and ellagic acid. Organic acids play a fundamental role in the environmental mobility of metal ions. Concerning the growth medium of *E. nigrum,* it was found that glucose as carbon and energy while ammonium chloride as nitrogen were considered the best sources for maximum growth conditions. Hussien [17] evaluates that the maximum bioleaching of uranium from El-Sella mineralisation by *Aspergillus clavatus* was achieved under two-step bioleaching process when a total concentration of 3% (w/v) El-Sella mineralisation was applied. The decrease in pH was observed due to the organic acids production via in complete oxidation by tannase enzyme produces gallic and ellagic acids as in Equations 1 and 4.

$$\underset{\text{(Tannic acid)}}{C_{27}H_{24}O_{18}} \xrightarrow[C_6H_{12}O_6]{\text{Tinnase enzyme}} \underset{\text{Gallic acid}}{3(C_7H_6O_5)} \quad (1)$$

$$\underset{\text{Gallic acid}}{C_7H_6O_5} \xrightarrow[-H]{\text{Dissociation}} \underset{\text{llate}}{C_7H_5O_5^{-1}} \quad {}_{P}Ka = 4.4 \quad (2)$$

$$2C_7H_5O_5^{-1} + UO_2^{+2} \longrightarrow (C_7\,H_5O_5)_2UO_2 \quad (3)$$

Gallate Uranyl gallate

$$C_{27}H_{24}O_{18} \xrightarrow{\text{Tinnase enzyme}} C_{14}H_6O_8 \quad (4)$$

(Tannic acid) ellagic acid

$$C_{14}H_6O_8 \xrightarrow{\text{lactonization}} C_{14}H_4O_8^{-2} \quad {}_pKa = 5.54 \quad (5)$$

Ellagic acid Ellagate

$$C_{14}H_4O_8^{-2} + UO_2^{+2} \longrightarrow C_{14}H_4O_8UO_2 \quad (6)$$

Ellagate Uranyle ellagate

Organic acids promote mineral dissolution by (1) donating H^+ to proton-promoted dissolution processes, (2) forming inner-sphere surface complexes that dislodge structural metals from the mineral surface , and (3) the formation of aqueous metal-ligand complexes that reduce the relative solution saturation with respect to minerals undergoing dissolution [18]. Generally, the bioleaching (direct or indirect) process of U(VI) from the tested sample using *E. nigrum* will be more effective on presence of some organic acids as gallic and illagic acids. These obtained results are in agreement with that reported by [4]. They concluded that bioleaching using organic acids produced by organisms would become more effective using organic acid as a leaching agent, due to higher leaching efficiency and lower capital cost.

5 Conclusions

The present investigation has demonstrated the effect of different weight sample on growth, acids production and leaching behaviour of *E. nigrum*. Tested organism was produced organic acids result in high bioleaching efficiency of U(VI) from tested samples. In direct process, the highest U(VI) dissolution was76.6 % from El-Sella ore material by *E. nigrum*. The bioleaching process showed better U(VI)bioleaching and reduced cost. These results indicate the challenges but also the promising aspects of biological leaching as alternative process for recovering U(VI) from El-Sella ore material.

6. Reference

[1] Gu T, Rastegar S O, Mousavi SM, Li M and Zhou M 2018 Advances in bioleaching for recovery of metals and bioremediation of fuel ash and sewage sludge. Bioresource Technology, Vol. 261, PP. 428-440.

[2] Wu W, Liu X, Zhang X, Li X, Qiu Y, Zhu M and Tan W 2019 Mechanism underlying the bioleaching process of $LiCoO_2$ by sulfur-oxidizing and iron-oxidizing bacteria. Journal of Bioscience and Bioengineering.Vol,128(3),PP. 344-354.

[3] Haghshenas DF, Keshavarz Alamdari E, Amouei Torkmahalleh M, Bonakdarpour B and Nasernejad B 2009 Adaptation of Acidithiobacillus ferrooxidans to high grade sphalerite concentrate. Mineral Engineering, Vol. 22, PP. 1299–1306.

[4] Hussien S, Desouky OA and Mohamadey SE 2016 Microbial Leaching of Uranium from Low Grade Ore and Waste Sample of Northern Part of GabalGatter, Egypt Using Penicillium Purpurogenium and Pseudomonas fluorescens. Journal of Progressive Research in Biology (JPRB), Vol.(3)1, PP.127-141.

[5] Saqib M, Claire SJ, David LP, Keith AM, Ping L, Karen SG, 2013 Fungi out compete bacteria under increased uranium concentration in culture media. Journal of Environmental Radioactivity,Vol.120, PP. 39–44.

[6] Desouky OA, El-Mougith AA, Hassanien WA, Awadalla GS, Hussien SS, 2011 Extraction of some strategicelements from thorium-uranium concentrates using bioproducts of Aspergillus ficuum and Pseudomonas aeruginosa. Arabian Journal of Chemistry, doi: 10.1016/ j.arabjc.2011.08.010.

[7] Ilyas S, Sarwar S, Bhatti HN, 2012 Effect of organic acids produced by Penicillium notatum on the extraction of metals ions from Brown Shale. Journal of Chemical Society of Pakistan, Vol.34, PP.1040–1047.

[8] Ibrahim TM, KG Ali, IM Gaafar, SM Masoud, SA Omer, HA Shahin, MM Haridy, EM Aly, M Biomy, A Abdel Gawad, A Abo Donia, M Abdel-Monsef, M Awad. Extensive studies on G. El-Sella Shear zone South Eastern Desert, Egypt. Int. report, N.M.A., Cairo, Egypt (2005) 91.

[9] Mostafa, H., 2006, Zytgomycetes. Fungi of Egypt AUMC descriptions NO.1. (Assiut: Assiut University Mycological Centre (AUMC)).

[10] Wang J, Li A, Xu L and Zhou Y, 2009 Adsorption of tannic and gallic acids on a new polymeric adsorbent and the effect of Cu(II) on their removal. Journal of Hazardous Materials,Vol.169, PP.794–800.

[11] Ren WX, Li PJ, Yong G and Li XJ 2009 Biological leaching of heavy metals from a contaminated soil by Aspergillus niger. Journal of Hazardous Materials, Vol.167,PP. 164-169.

[12] Marczenko Z, 1976 Spectrophotometric determination of elements,(New York: John Wiley and Sons).

[13] Tomas MJE, Simon PMD, Congergado F, and Suarez FG 1980 Methods to assess antagonism of soil microorganisms towards fungal spore germination. Soil Biology and Biochemistry, Vol. 12, PP. 197-198.

[14] Chen GH, Lin Y, Hsu W, Hsieh S and Tzen JTC 2015 Significant elevation of antiviral activity of strictinin from Pu'er tea after thermal degradation to ellagic acid and gallic acid . Journal of Food and Drug Analysis, Vol. 23,PP. 116 -123.

[15] Moubasher A 1993 Soil fungi in Qatar and other arab countries .Published by the center for scientific and applied research, (Qatar: University of Qatar).

[16] Castro MI, Fietto JLR, Viera RX, Tropia MJM, Campos LMM, Paniago EB and Brandão RL 2000 Bioleaching of zinc and nickel from silicates using Aspergillus niger cultures. Hydrometallurgy,Vol.57, PP. 39-49.

[17] Goyne WK, LS Brantley and J Chorover 2010 Rare earth element release from phosphate minerals in thepresence of organic acids. Chemical Geology, Vol. 278, PP.1-14.

[18] Hussien SS 2020 Microbial leaching of El-Sella mineralisation by Aspergillus clavatus- a fact of fungal–uranium interface. International Journal of Environmental Studies, Vol. 77(2), PP.275–296

Acknowledgement
The authors gratefully acknowledgement the assistance or encouragement from colleagues, special work by technical staff or financial support from organizations of the Faculty of Science, Zagazig University and the Nuclear Materials Authority, Cairo, Egypt.

Commercial approach for highly pure thorium from Egyptian monazite mineral acid process

A I L Abd El Fatah

Nuclear Materials Authority, Cairo, Egypt
Email address:Nuclic_science@yahoo.com

Abstract. Thorium has a special interest in the future nuclear energy as a fuel especially the breeder reactors. The present research has been fulfilled throughout successive research schedule manner on separation and purification of thorium from Egyptian beach monazite mineral after its acid digestion. Monazite mineral was digested with sulfuric acid 96% at 220°C for 2.5h and dissolute with ice water (1:15), the clear solution precipitated with ammonia solution 16.5% at pH 1to produce thorium pyrophosphate with 45% thorium content in case of ideal optimized whole process. The purification was carried out after the caustafication and leaching with hydrochloric acid followed by the addition of oxalic acid at pH 0.3-0.4 to produce thorium oxalate with 91.68 % purity. Thorium oxalate was transformed to oxide by caustafication then washed and separated and dissolute by hydrochloric acid. Highly pure thorium salt was attained with the addition of thiosulphate salt on the produced faint chloride solution under 0.1 to maximum 0.5 phosphate content to produce 99% purity.
Keywords: monazite mineral, thorium, separation, purification.

1. Introduction

Among contentious commercial nuclear power issues are the questions of what to do with long-lived radioactive waste and how to minimize weapon prolife radiation dangers. The substitution of thorium for uranium as fuel in nuclear reactors has significant potential for minimizing both problems. Thorium is three times more abundant in nature than uranium. Whereas uranium has to be imported in some countries (e.g. Egypt), there is enough thorium in such countries (e.g. USA, Egypt…) alone to provide adequate grid power for many centuries. A well-designed thorium reactor could produce electricity less expensively than an existed generation coal- fired plant or a current-generation uranium-fueled nuclear reactor. Importantly, thorium reactors produce substantially less long-lived radioactive waste than uranium reactors [1].

Thorium is the most abundant natural radioactive element and was discovered in 1828 by J.J. Berzelius and it exists in earth' crust with concentration of 8 ppm and has half-life of about 14.05 billion years with alpha decay. With respect to Thorium geological origin, in typical igneous rocks, thorium is mainly present in the accessory mineral, particularly thorite, monazite, allanite, zircon, titanite, anatase, rutile and perovskite. The same is true for sedimentary and metamorphic rocks but in some sedimentary rocks the resistates such as monazite, zircon and titanite may be quit abundant. Thorium is the largest of the tetravalent cations with an ionic radius comparable to that of U^{4+} and Ce^{4+}. The similarities in ionic size, outer electron configuration and bond character are the main reasons for the close relationship between the crystal chemistry of Th, Ce,

U and Zr. The large highly charged Th^{4+} ion has a marked tendency to form complex ions with others in solutions [2].

After mining, concentration and mineral physical beneficiation (separation), the monazite then undergoes attacking either by acid (without further mineral pretreatment) or alkali (necessitate mineral content >97% & grinding to -325 mesh size) to prepare aqueous solution containing all the mineral elemental contents. Starting from solutions produced from mineral solubilization, many schemes may be applied to attain a proper regime to produce ThO_2 powder as final target with highest possible purity and recovery to be ready for fuel fabrication.

During a previous work in the same context to separate and purify the uranium from the Egyptian monazite by applying oxalic- nitrate- TBP extraction method. One of the most important results of this separation was the acquisition of pure thorium. This work required the purification by an organic solvent (TBP), which represent economic burdens on the separation process, which would be avoided in this paper [3].

In another attempt in the same context to reach the highest concentration of thorium by conventional chemical methods (economical, available and easy to scale up), thorium was obtained with 93.88% recovery and 90.2% purity without further treatment with resin or solvent [4].

The present work is to provide an improved commercial method for the recovery of thorium values as an extremely pure product from an acidic aqueous solution containing said values together with rare earth values without further complicated, tedious and economically burdens solvent extraction and ion exchange processes.

2. Experimental methods

2.1. Chemicals and reagents

All chemicals used in the preoperative work were included the following: Sodium hydroxide, oxalic acid, sodium thiosulfate, hydrochloric acid, sulfuric acid, ammonia solution. The reagents were prepared by distilled water only for analysis and the final purification, while the separation and concentration processes tap water was used for commercial concepts during up-scaling proposal. The pH of the solutions was readjusted using 0.1M HCl and 0.1M NaOH.

2.2. Analytical techniques (instrumentation and methods)

The most common analytical techniques used to determine thorium in its various media (mineral, processes solution and solid products) were:-

X-ray fluorescence (EDXRF) using Rigaku spectrometer NEX CG with polarized optics [5], energy dispersive X-ray fluorescence [6, 7] and gravimetric [8].

Nuclear Materials Authority labs, Cairo, Egypt, were used to monitor the analysis through the processing of monazite.

2.3. Hydrometallurgical procedures

A process for separating Th(IV) from the sulfuric acid solution of monazite mineral was carried out as the following process.

Monazite mineral digestion and dissolution:-

The mineral was digested in concentrated sulphuric acid after addition of 0.05% of BaCl2at $220^{°C}$ for about 2.5 hours and the reaction products were leached by ice water by the ratio (1:15) and reduction of acidity to pH 1 to precipitate thorium pyrophosphate by ammonia solution (16.5%).

There are three main reactions occurring in the mineral digestion reaction beaker as follows:

$$2REPO_4 + 3H_2SO_4 \longrightarrow (RE)_2(SO_4)_3 + 2H_3PO_4 \quad (1)$$

$$Th_3(PO_4)_4 + 6H_2SO4 \longrightarrow 3Th(SO_4)_2 + 4H_3PO_4 \quad (2)$$

$$U_3(PO_4)_4 + 6H_2SO_4 \longrightarrow 3U(SO_4)_2 + 4H_3PO_4 \quad (3)$$

2.3.1. Caustafication of thorium pyrophosphate cake.

About 1 Kg of wet thorium precipitate was reacted with sodium hydroxide in stiochiometric ratio applying external heat (90ºC – 1h). The obtained thorium hydroxide was filtered of and washed thoroaly then dissolved in concentrated hydrochloric acid to produce thorium chloride solution. The produced solution will be the starting liquor for the different thorium purification regimes.

2.3.2. Oxalate precipitation

Thorium oxalate precipitate was produced with the addition of different amounts of oxalic acid (12.5, 6.25, 3.125, 2.5, 1.67, 1.25, 0.83 and 0.625 g) to thorium chloride solution (200 ml for each experiment) at different pH (0.3-0.4, 0.8, 1.2).

2.3.3. Thiosulphate precipitation

Thorium oxalate precipitate was converted to hydroxide by boiling with caustic soda solution, filters, washes thoroaly then dissolved by hydrochloric acid and boil(to reduce Ce(IV) to Ce (III)) to faintly acid solution, sodium thiosulfate was added in different amounts.

3. Results And Discussion

3.1. Monazite sand composition

A sample of monazite mineral after a proper sampling technique (which carried out as the following: making a pile of conical shape of all amount of the sample then spread out into a circular for through mixing. The process may be repeated if necessary. The sample is made again a pile and flattened, then divided into quarters; two quarters diagonally opposite are taken while the other two are rejected. The two quarters are collected together and subjected to coning and quartering until a lot of material of the desired size is obtained). The obtained sample was sent to NMA Analytical Lab for X-ray fluorescence analysis. This to investigate the chemical composition of the main targeted constituents in the high-grade monazite sand sample (about 85% monazite) produced during physical beneficiation of Rashid black sand, Egypt for more abundant Heavy Economic Minerals (HEMs) e.g. zircon, illmenite, rutile [9].The thorium, rare earth elements and uranium concentrations were indicated in Table 1.

Table 1. REEs, Th and U concentration percents in monazite mineral (85% purity) analyzed by XRF.

Metal Oxide	**%**
$(REE)_2O_3$	58.7
ThO_2	4.6
U_3O_8	0.5

3.2. Solubilization of monazite by sulfuric acid

Monazite mineral was digested by concentrated sulfuric acid. About 85 % of the sample was dissolved forming acidic leach liquor via dilution using ice water with about 15 % residue part which was filtered off.

This process consists essentially of heating the ore with concentrated acid to convert thorium and lanthanon to a hydrous sulfates and extracting the reaction product with water. This product consists of hydrous sulfates in a mixture of phosphoric and sulfuric acids, excess sulfuric acid being need to give complete reaction. When it is extracted with water, the results obtained depend on the acidity of the final solution, which is controlled by the amounts of excess sulfuric acid and of water. If the acidity is low, thorium phosphate is precipitated but lanthanon remains in solution, lanthanon phosphate being soluble at much lower acidities than thorium phosphate. At higher acidities, both thorium and lanthanon remains in solution (monazite sulfate solution). At very high acidities, lanthanon sulphate is precipitated by the high sulfate concentration, although thorium sulphate can also be "salted out" of solution by increase in sulfuric acid concentration, its precipitation is inhibited by phosphoric acid. After the initial reaction product has been extracted with water, there for, a solution of both thorium and lanthanon was targeted and prepared. This process is a temperature, acid/ore ratio and dilution ratio dependent to control the percent dissolution of each main element of mineral component.

3.3. Separation of crude thorium concentrate

Separation of thorium, rare earths and uranium from the clear monazite sulfate solution obtained after mineral solubilization, via precipitation is a highly sensitive pH control process affected by the neutralizing agent used to control the pH [10]. The clear monazite sulfate solution is treated with ammonium hydroxide under acidity control. The acidity of the solution is gradually decreased with precipitant spray over all solution surface area while stirring until substantially all of the thorium is precipitated and most of the rare earths remain in the solution. There is no sharp pH value to precipitate each constituent individually but rather a range of acidities through which each content can be precipitated. Therefore, instead of sharp separation of each constituent, the bulk of thorium separates from the bulk of the rare earths.

The clear monazite sulfate solution was precipitated at three stages with ammonium hydroxide solution; first at pH 1.0 precipitation of thorium then at 2.5-3.0 to produce rare earth elements precipitate then finally at pH 5.5-6 to precipitate uranium.

Thorium and uranium are most probably precipitated as thorium-pyrophosphate and uranium-phosphate respectively while rare earths are precipitated as rare earths ammonium double sulphate as shown in the following equations

$$Th^{+4} + 2H_2PO^{4-} \longrightarrow ThP_2O_7 + H_2O + 2H^+ \quad (4)$$

$$3U^{+4} + 4H2PO4- \longrightarrow U_3(PO_4)_4 + 8H^+ \quad (5)$$

$$(RE)_2(SO_4)_3 + (NH_4)_2SO_4 \longrightarrow (RE)_2(SO_4)_3.(NH_4)_2SO_4.2H_2O \quad (6)$$

The most usual starting point for thorium purification is a crude hydroxide or phosphate material in which phosphate will range from 5 - 40% P_2O_5/ThO_2 and lanthanon from 25-100 % Ln_2O_3/ThO_2. The sequence of precipitation intended mainly to remove lanthanon and phosphate, and the principal impurities (iron, titanium...).

3.4. Purification of thorium

Many processes of purification were carried out on the first crude cake of thorium phosphate which precipitated at pH 1-1.5 from the mother acid leach solution derived after mineral digestion and dissolution as stated below.

3.4.1. Caustafication-hydrochloric acid leaching process.

Thorium and total rare earth content in the crude thorium pyrophosphate concentrate is given in table 2. From the table it's clear that, the crude cake contains 49.96% thorium and 43.13% total rare earth elements. The precipitate was reacted with sodium hydroxide in stiochiometric ratio to

IOP Conf. Series: Materials Science and Engineering **975** (2020) 012022 doi:10.1088/1757-899X/975/1/012022

produce the hydrated oxides of the elements of the cake (Th and REEs) as in the following equation:

$$ThP_2O_7 + 6\ NaOH \longrightarrow ThO_2.3H_2O + 2\ Na_3(PO_4) \quad (7)$$

Table 2. Thorium and total rare earth percents in the crude thorium pyrophosphate cake precipitated at 1.5 after mineral digestion and solubilization.

Elements	%
Th	49.96
ΣREEs	43.13

Thorium hydrated oxide precipitate was washed thoroughly with water till neutralization. The wet precipitate was leached with hydrochloric acid to produce thorium chloride as the following equation:

$$ThO_2 + 4HCl \longrightarrow ThCl_4 + 2H_2O \quad (8)$$

The produced residue after filtration was analyzed and its composition with respect to Th and REEs is given in table 3., which contains 18.74% thorium and 78.72 % rare earth elements. EDXRF analysis provides that the most of 78.72% REEs consists of light group.

Table 3. Thorium and total rare earth percents in the residue after hydrochloric acid leaching of thorium hydrated oxide.

Elements	%
Th	18.74
REEs	78.72

At low pH values hydrous thorium (IV) oxides are quite soluble, but at pH values higher than 3-4 solubilities decrease significantly. Ryan &Rai (1987) [11] found the solubility values of amorphous ThO_2 to decrease steeply from 10^{-3}-10^{-2} M to 10^{-9}-10^{-10}M between pH 3.6 and pH 4.7.The solubility of crystalline thorium oxide, ThO_2, is extremely low, Baes &Mesmer (1986) [12] report the calculated minimum value above pH 6 to be 10-9.6 M. ThO_2 is also known to equilibrate very slowly with the aqueous phase. Thus, in this work it was regarded reasonable to use amorphous hydrous Th(IV) oxide, which is the primary solid phase precipitated under oversaturation conditions, and which equilibrates more rapidly with the aqueous phase.

Complexing agents, such as Cl- ions, affect solubility properties according to Felmy et al.[13], who found solubility values of hydrous Th(IV) oxide in chloride media to be higher when compared to those of Ryan & Rai (1987), who had a pure Th(IV)- H_2O system. The solubility values obtained by Felmy et al. (1991) increased with increasing chloride concentration ($[Cl^-]$ = from 0.6 M up to 3 M).

As a result of the previous work, most of thorium content was dissolved after addition of hot hydrochloric acid in stiochiometric ratio 1(precipitate):4 (acid) with 15 minutes of stirring. The un-dissolved part contains mainly light rare earth elements may be attributed to the oxidized part of cerium which may resists reduction and dissolution by hydrochloric acid. The weight percent

IOP Conf. Series: Materials Science and Engineering **975** (2020) 012022 doi:10.1088/1757-899X/975/1/012022

of un-dissolved part was calculated in relation to the mother cake amounted to 0.1g which represents 0.23% of the mother cake.

3.4.2. Oxalic acid purification

The produced filtrate after hydrochloric acid leaching was precipitated with oxalic acid as solid particles to produce thorium oxalate according to the following equation:

$$ThCl_4 + 2\ C_2O_4H_2 \longleftrightarrow Th\,(C_2O_4)_2 + 4HCl \quad (9)$$

During oxalic acid addition, a thorium solid phase precipitated under acidic conditions at below pH 3 was attributed to the formation of $Th(C_2O_4)_2$ as revealed from thermal analysis of the solid phase [14-18].

Oxalic acid concentration (g/experiment) the effect was inspected and tabulated in table 4. The results in table 4 represent the effect of oxalic acid variation, added in grams, per 200 ml of thorium chloride solution (eight experiments labeled from A to H). The produced oxalate precipitate was ignited and analyzed for thorium and total REEs contents in each one.

Table 4. Thorium and rare earth concentration after oxalic acid addition to thorium chloride solution which analyzed by EDXRF.

oxalic acid addition in grams	% Th	% REEs
12.5	(A) 45.37	54.63
6.25	(B) 47.04	52.96
3.125	(C) 51.08	48.92
2.5	(D) 70.28	22.35
1.67	(E) 62.86	30.88
1.25	(F) 65.25	26.01
0.83	(G) 63.62	29.26
0.625	(H) 64.61	28.71

From table 4., it was found that, in the first three oxalic acid addition which equals 12.5, 6.25 and 3.125 there is there is a slight increase in thorium concentration(A, B and C) which equal 45.37, 47.04 and 51.08 % respectively, on other hand rare earth's concentration decrease from 54.63 to 48.92 %. At 2.5g oxalic acid addition (D), the high concentration of thorium reaches to 70.28% with 22.35% of REEs. The last four additions (E, F, G and H), thorium concentration decreases again in the presence of different amounts of REEs. So, the best addition of oxalic acid for pure thorium precipitation was 2.5g which equal 70.28% with small amounts of REEs.

As a good results of thorium oxalate precipitation, there was another experiments were done applying controlled different pH values. Taking the filtrate produced after caustafication and opening with hydrochloric acid for the first ore and precipitate by oxalic acid higher obtained values at pH 0.3-0.4 and 0.8. Table 5 was shown the results of thorium oxide precipitate after ignition at 750°C.

Table 5. thorium and rare earth content after oxalic acid addition to thorium chloride solution at different pH.

pH	% Th	% REEs
0.3-0.4	90.24	9.76
0.8	41.72	51.45

From this table it was found that thorium oxide produced with good purity equal 90.24% with 9.76% of REEs. At pH 0.8 the purity of thorium decreases to the half which equals 41.72 % due to the precipitation of rare earth elements most properly the heaviest members.

After the aforementioned thorium purification trials, one can conclude that, thorium oxalate is precipitated from acid solutions by addition of either oxalic acid or sodium oxalate. The precipitation gives a good degree of separation from phosphate, this being better the more strongly acid the initial solution. Light REEs oxalates are more soluble in acid solution than thorium oxalate and, if just enough reagents are added to precipitate all the thorium, some HREEs will be left in solution. An initial REEs level of 40% RE_2O_3/ ThO_2 can be reduced in this way to 10% but the technique is of no value in reducing trace contaminations.

3.4.3. Thiosulfate purification

In this part thorium oxalate precipitation which produced after oxalic acid addition to thorium chloride solution was leached with hydrochloric acid according to the following equation:

$$Th\,(C_2O_4)_2 + 4HCl \longrightarrow Th\,Cl_4 + 2\,C_2O_4H_2 \qquad (10)$$

The un-reacted part which weight 0.1g (0.28% of mother oxalate precipitate) was analyzed as shown in table 6.

Table 6. The analysis of the residue of thorium oxalate precipitate after leached with HCl.

Elements	%
Th	91.68
LREEs	8.32

Thorium was precipitated again with the addition of sodium thiosulfate to the produced filtrate with 99% purity without rare earth elements.

Addition of sodium thiosulfate to a slightly acid thorium solution precipitates thorium thiosulphate. The REEs compounds were decomposed by acid (boil chloride solution before addition of precipitating agent to reducible Ce (IV) to Ce (III)). The presence of phosphate not only reduces the efficiency of REEs separation but also adversely affects the nature of the precipitate.

A maximum phosphate level of 0.5% P_2O_5/ThO_2can be tolerated; it should preferably be less than 0.1%. The thiosulphate is soluble in HCl, giving a solution of $ThCl_4$with a precipitate of sulphur.

So, if these conditions were obtained before oxalic treatment, thiosulfate precipitation process can be carried out directly after washing caustafication and dissolution of the precipitate with reasonable purification results with respect to trivalent REEs under lower acidity.

After this satisfies results, the experiments were done again at 2 Kg of thorium pyrophosphate to calculate the recovery percents of thorium. 1 Kg was done with the stiochiometric ratios of each reaction, the other one was done with the double stiochiometric ratios of each reaction. It was found that, the precipitation of thorium with stiochiometric ratio for all the reactions, starting from sodium hydroxide reaction with the first crude cake (thorium pyrophosphate), hydrochloric acid leaching, oxalic acid addition, hydrochloric acid leaching finished with sodium thiosulphate precipitation. The produced thorium precipitate has the following results as shown in table 7.

Table 7. The final thorium precipitate with stiochiometric precipitation results after sodium hydroxide reaction with the first crude thorium cake (thorium pyrophosphate), hydrochloric acid leaching, oxalic acid addition, hydrochloric acid leaching finished with sodium thiosulphate precipitation analyzed by EDXRF.

Elements	%
Th	78.69
LREEs	11.62

From table 7., thorium has 78.69% purity in the presence of rare earth elements. Thorium percent decreases from 99 to 78.69 % due to increase the heat of reaction with the precipitation of rare earth elements.

On other hand, the precipitation of thorium with the double stiochiometric ratio of all precipitants follows the same steps (starting from sodium hydroxide reaction, hydrochloric acid leaching, oxalic acid addition, hydrochloric acid leaching finished with sodium thiosulphate addition).The results of the final product were shown in table 8.

From table 8. it was found that thorium purity decrease to 71.82 from 78.69 after final precipitation with the double stiochiometric ratio of sodium thiosulphate. This decrease was attributed to the increase of rare earths which equal 17.57%. From all the previous results the precipitation with the stiochiometric results was better than double stiochiometric results but the purity of final product still not satisfied. The good results obtained after addition of small amounts of sodium thiosulphate which give 99% purity of the final product.

Table 8. Analysis of thorium precipitate with double stoichiometric precipitation results after sodium hydroxide reaction, hydrochloric acid leaching, oxalic acid addition, hydrochloric acid leaching finished with sodium thiosulfate addition by EDXRF.

Elements	%
Th	71.82
LREEs	17.57

Figure 1 and 2 show the shape of the precipitate after oxalic acid and thiosulfate purification processes respectively.

Figure 1. Thorium oxalate precipitation after oxalic acid addition.

Figure 2. The high pure Thorium after sodium thiosulphate precipitation.

3.5 Up-scaling experiment

After all these experimental work, the procedures were applied at 1Kg monazite mineral with 85% purity. First, monazite was digested with concentrated sulfuric acid at 250°C with stirring for 2.5 h. Dissolution for the paste with 15L of ice water with stirring for 2 hours, decantation then filtration. The filtrate was precipitated with ammonia solution at pH 1 to produce thorium pyrophosphate with 45% purity. The precipitate was reacted with sodium hydroxide to give thorium oxide precipitate which leached with concentrated hydrochloric acid and filtered. Thorium chloride filtrate was precipitated with oxalic acid at pH 0.3-0.4 producing thorium oxalate with 87.3% purity and 82.35% recovery. But the application of thiosulphate precipitation cannot carried out at the high amounts due to the precipitation of light rare earth elements with thorium precipitate which lead to the decrease the purity and recovery of the final product of thorium and the phosphorous over the tolerate values. The flow chart that summarizes the whole process for obtaining pure thorium oxide after monazite dissolution and thorium precipitation followed by purification can be represented in figure 3. The calculated percent recovery is related to the started Th content in the leach liqueur after mineral dissolution.

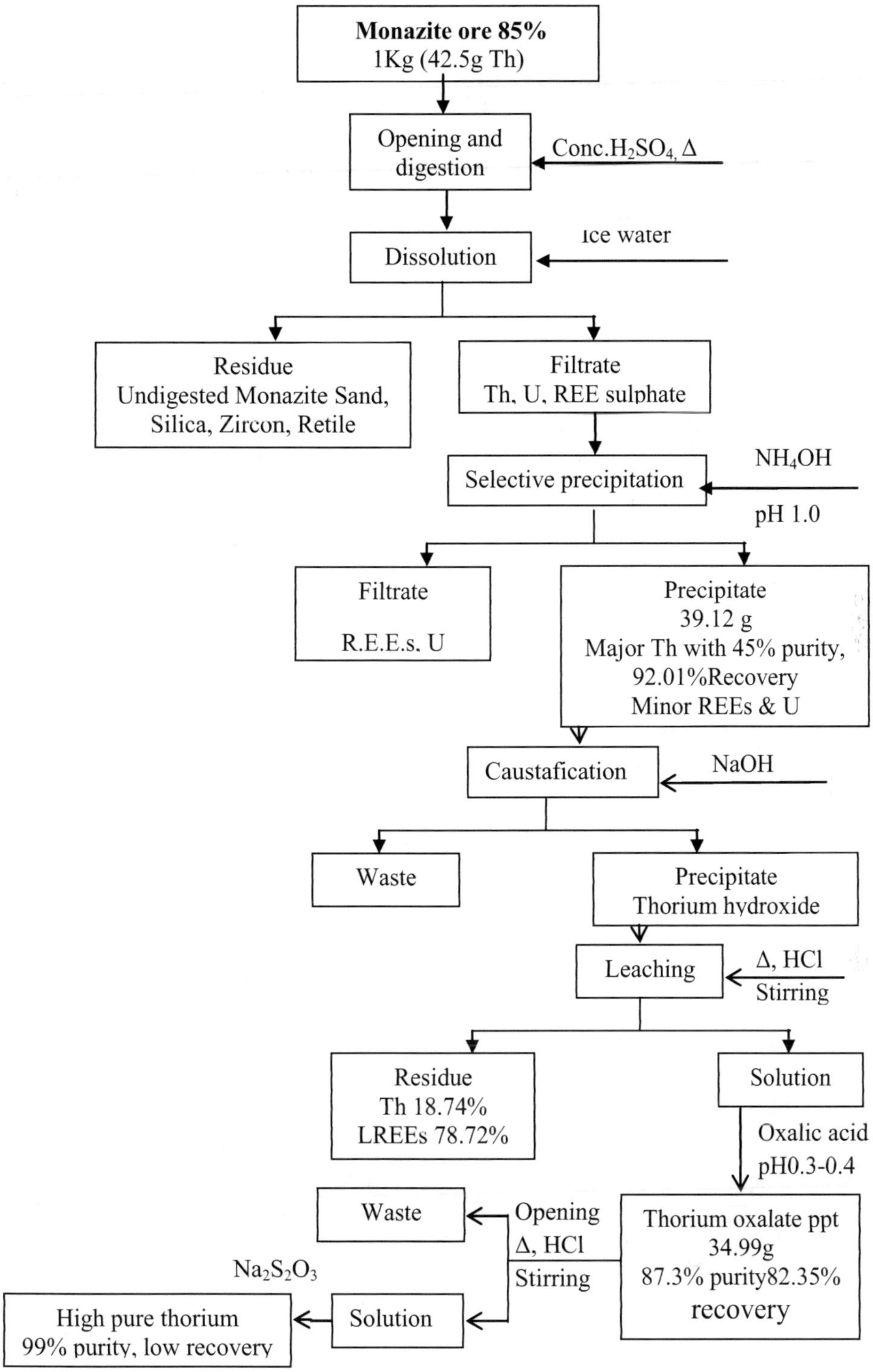

Figure 3. Flow chart of Egyptian monazite processes for thorium separation and high purification.

IOP Conf. Series: Materials Science and Engineering **975** (2020) 012022 doi:10.1088/1757-899X/975/1/012022

4. Conclusion

The technical viability of economic process for the purification of thorium from acidified monazite leach liquor was attained and verified to some extent. This process was dependent on the pH, amount of oxalic acid, amount of thiosulfate salt. Thorium was undergoes to three stages, first thorium pyrophosphate with 45% purity after mineral digestion, dissolution and filtration then precipitation by ammonia solution. The second stage, 91.68% purity of thorium oxalate was attained after caustafication, leaching and oxalic acid addition to thorium pyrophosphate precipitate at pH 03-0.4. Finally, the high purification of thorium with 99% purity after leaching by hydrochloric acid and thiosulfate salt addition to thorium oxalate product with low recovery.

5. References

[1] Marvin Baker Schaffer 2013 "Abundant thorium as an alternative nuclear fuel Important waste disposal and weapon proliferation advantages" Energy Policy 60, 4–12

[2] Terrence P. Mernagh and YanisMiezitis 2008 " A Review of the Geochemical Processes Controlling the Distribution of the Thorium in the Earth's Crust and Australia's Thorium Resources"Australian Government, Geoscience Australia. Applying Geoscience to Australia's most important Challenges.

[3] Rabie. K. A.; S. M. Abdel Wahaab; K. F. Mahmoud; A.E.M.Hussein and Abd El-Fatah. A. I. L. 2013 "Monazite-Uranium Separation and Purification Applying Oxalic-Nitrate- TBP Extraction" Arab Journal of Nuclear Science and Applications, 46(1), (30-42).

[4] Mostafa M.H.Khalil, K.A.Rabi, M.S.Atrees, Abd El Fatah A.I.L, M. Demerdash, R. Roshdi, 2017 "Development of Thorium separation, up grading and Purification Processes for Its pre-conceptual High Grade Production from Crude Monazite Mineral" ISOTOPE & RAD. RES., 49(1), 123-137.

[5] Abostate, M., Salah, Y., Mira, H., Amin, M., Alkazindar, M. and Ahmed, B. M., 2018 "Characterization, Kinetics and thermodynamics of biosynthesized uranium nanoparticles (UNPs). Artificial cells, Nanomedicine, and Biotechnology, 46(1): 147-159.

[6] Mirashi, N.N., Sangita Dhara, S. Sanjay Kumar, S. Sanjay Kumar, 2010, Energy dispersive X-Ray fluorescence determination of thorium in phosphoric acid solutions, Spectrochimica Acta Part B Atomic Spectroscopy 65(7):579-582 .

[7] Mostafa M.H.Khalil, K.A.Rabi, M.S.Atrees, Abd El Fatah A.I.L, M. Demerdash, R. Roshdi, 2017 "Development of Thorium separation, up grading and Purification Processes for Its pre-conceptual High Grade Production from Crude Monazite Mineral" ISOTOPE & RAD. RES., 49(1), 123-137.

[8] Arora, H.C., and Rao, G.N. , 1981, Mn Manganese: Coordination Compounds, p 252.

[9] Abdelhay m. S. El shafey ; 2015, ".upgrading and mineralogical studies for some economic and nuclear element (especially monazite) bearing minerals of rosetta black deposits, egypt. Zagazig University, faculty of science, geology department.

[10] Mostafa M.H.Khalil, K.A.Rabi, M.S.Atrees, Abd El Fatah A.I.L, M. Demerdash, R. Roshdi, 2017 "Development of Thorium separation, up grading and Purification Processes for Its pre-conceptual High Grade Production from Crude Monazite Mineral" ISOTOPE & RAD. RES., 49(1), 123-137.

[11] Ryan, J. L. & Rai, D. 1987, Thorium (IV) hydrous oxide solubility. Inorganic Chemistry 26, pp. 4140-4142.

[12] Baes, C. F Jr. & Mesmer, R. E. 1986, The hydrolysis of cations. Malabar, Florida, USA: Krieger Publishing Company, 489 p.

[13] Felmy, R. A., Rai, D. & Mason, M J 1991, The solubility of hydrous thorium(IV) oxide in chloride media: development of aqueous ion-interaction model. RadiochimicaActa 55, pp. 177-185.

[14] Felmy, R. A., Rai, D. & Mason, M J 1993, Solid phase precipitates and anionic aqueous thorium fluoride complexes in the Na- NH4 - Th- F- H20 system to high concentration. RadiochimicaActa

62, pp. 133-139.
[15] T. Kobayashi, T. Sasaki, I. Takagi, H. Moriyama, ''Solid phase precipitates in (Zr, Th)-OH-(oxalate, malonate) ternary aqueous system,'' Radiochim. Acta (to be published).
[16] R. W. M. D'Eye, P. G. Sellman, 1955 ''The thermal decomposition of thorium oxalate,'' J. Inorg. Nucl. Chem., 1, 143.
[17] V. M. Padmanabhan, S. C. Saraiya, A. K. Sundaram, 1960 ''Thermal decomposition of some oxalates,'' J. Inorg. Nucl. Chem., 12, 356.
[18] M. Aybers, 1998 ''Kinetic study of the thermal decomposition of thorium oxalate dihydrate,'' J. Nucl. Mater., 252, 28.

Acknowledgment
The author would like to express her appreciation to the efforts of Prof. **Rabie. K. A.,** for Suggesting the idea of the present work, faithful discussion and critical reading of the manuscript.

IOP Conf. Series: Materials Science and Engineering **975** (2020) 012023 doi:10.1088/1757-899X/975/1/012023

Bio-sorption of uranium by saragassum detifolium from the trachytic sills of Wadi El Atshan area, central eastern desert, Egypt

N SH Gad[1] and N M Farag[2]
[1, 2] Nuclear Materials Authority, Cairo, Egypt.
norashenouda@yahoo.com

Abstract In wadi El Atshan area, Central Eastern Desert of Egypt, the main rock types exposed in that area are the older granites, Hammamat sedimentary rocks, younger granites, trachytic sills and dikes. The Field studies and radiometric survey indicate that the trachytic sills are the main rock type in the investigated area which contain secondary uranium minerals. Sargassum detifolium is used to capture uranium from the trachytic sills. The analysis indicates that this alga absorbs the uranium content by 100% in two samples where happened in the other samples as well as, it changes the chemical composition of some minerals content, while some elements changed while other elements are decreased, Sargassum detifolium has the ability to biosorbe uranium content about 75% as in uranotile mineral {CaU2 $(SiO_4)_2(OH)_6.3H_2O$}.
Keyword: *Sargassum* detifolium, trachytic rocks, uranium.

1. Introduction:

From all microorganisms algae are cheap and efficient there for many researches use it in Bio-sorption of metal which occurs by ion exchange method on the cell surface. Absorption of some metals like La, Cd, and Pb by algae is listed in table 1.

Table 1. Bio-sorption uptake by different types of Algal species.

References	Bio-sorption capacity (mg/g)	Algal species	Metal removed
(5)	140	Spirogyra sp.	Pb
(4)	34.10	Sargassum Detofiulum	Zn
(1)	181.82	Ulva lactuca sp.	Pb
(6)	84.7	Sargassum sp.	Cd
(2)	67.93	Spirulina platensis	Cu
(11)	80.47	Sargassum Detofiulum	La
(11)	84.6	Sargassum Detofiulum	Ce
(11)	86.6	Sargassum Detofiulum	Pr
(11)	86	Sargassum Detofiulum	Nd
(11)	88.7	Sargassum Detofiulum	Sm
(11)	70	Sargassum Detofiulum	Eu
(11)	77.4	Sargassum Detofiulum	Gd
(11)	77.1	Sargassum Detofiulum	Tb
(11)	72.9	Ulva lactuca	La

Bio-sorption by algae requires high metal uptake. The most effective and promising for bio sorption is the brown algae. This behavior is due to its basic biochemical constitution which responsible for the high performance. It is proven that the properties of their cell wall constituents are responsible for metal uptake. Bio-sorption of the metallic cations occurs essentially on the surface of the algal cell wall. There are many chemical groups involved in metallic cation bio-sorption like sulfhydryl, amino, Carboxyl and sulfonate. These groups are part of the algal cell wall structural polymers namely, polysaccharides (alginic acid, sulfated polysaccharides), proteins, and peptide-glycans. Ion exchange is one of the main bio-sorption mechanisms for metal uptake by algae. However other binding mechanisms like micro-precipitation and complexation are also involved in the process of metal uptake [3].

2. General Geology:

El- Atshan area occurs within the central part of the Eastern Desert of Egypt, covering about 120 km^2 of crystalline basement rocks (Fig. 1). It is covered by moderate to low elevated hills of rugged topography, the area mainly includes Hammamat sedimentary rocks that are intruded by syn- and late tectonic Older and Younger granites. All rocks are dissected and traversed by numerous post-orogenic trachyte dykes and sills figure1.

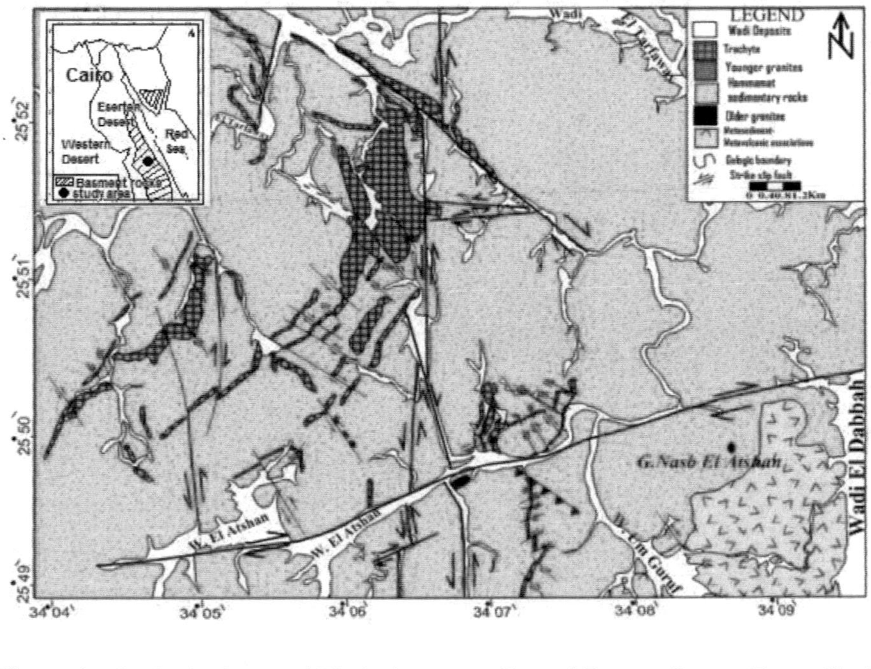

Figure 1. Geological map of El- Atshan area, Central Eastern Desert, Egypt.(Scale 1:40,000). Showing location of study area (after 10)

3. Materials and methods

3.1. Preparation of algal biomass:

Samples of Sargassum *detifolium w*as collected from the beaches of Hurghada. The biomass was washed extensively with de-ionized (DI) water and then dried in oven at 50 °C for 24 hr. The dried biomass was then grounded using blender and sieved subsequently to obtain particles size in the range of 0.85-1mm.

The Bio-sorption experiments were conducted in 250 ml Erlenmeyer flask where 0.5g dry weight of Sargassum detifolium was added with trachyte of El-Atshan samples as outlined in the work of [7]. The flasks were placed on a shaker with constant shaking for 100 rpm, then they were incubated at 30

°C for 5 days. With variable uranium concentrations, the algal biomass were washed several time to remove any sample particles and then examined using (ESEM) and chemically treated by 0.1N Hcl remove Uranium from the biomass, The samples were examined under Infrared Spectroscopy (IR) using the model Naxux 670 FTIR, in the Central National Research (CNR), Environmental scanning electron microscope (ESEM), field imission gun (FIG) in the Central National Research (CNR) and U measured by Flame Atomic Absorption Spectrometry (FAAS), UNICAM in NMA to determine the Bio-sorption efficiency.

Table 2. Ranges of gamma ray's radioactivity of the different rock types exposed at El-Atshan area.

Field measurements (ppm)			Rock Type
Min.	Max.	Av.	
130	3200	1640	Anomalous trachyte
41	85	63	Trachyte
100	135	117.5	Younger Granite
55	980	517.5	Normal Hammamat sedimentary rocks
15	30	22.5	Hammamat sedimentary rocks
30	62	46	Older Granite

Min. = minimum Max. = maximum Av. = Average

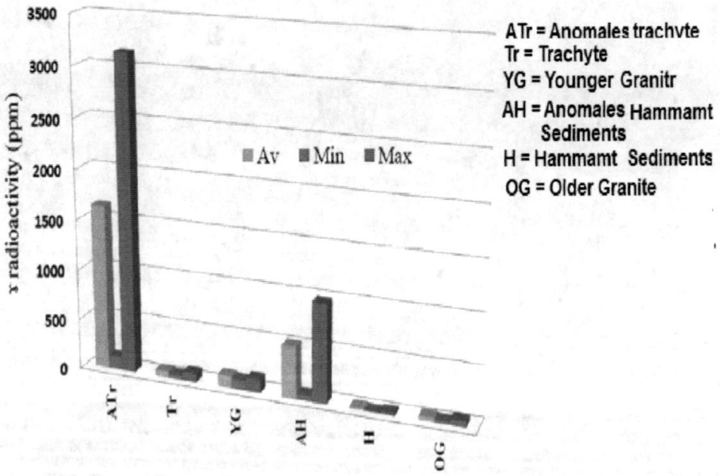

Figure 2. Histogram showing the distribution of normal gamma rays radioactivity in different rock types exposed at El-Atshan area.

4. Bio-sorption of uranium from trachyte of El-Atshan area

Sargassum detifolium has high capacity for Bio-sorption [12; 14; 15; 16; 17]. Further studies of this biosorbent revealed that the ion-exchange nature of the high metal uptakes by Sargassum, for example, of cadmium [14; 15; 16], chromium [8], copper/lead [9] and uranium [14]. Sargassum contains high amounts of alginate, well locked within its cellular structures, with abundant carboxylic groups capable of capturing cations present in solution. It is especially the Glucuronic acid in alginate that offers these functional groups [13] for ion exchange. Because the alginate matrix is present as a gel phase, this material is well penetrable for tiny metallic cations making it thus an acceptable biosorbent with a high sorption potential [11].

IOP Conf. Series: Materials Science and Engineering **975** (2020) 012023 doi:10.1088/1757-899X/975/1/012023

5. Discussion and Result:

5.1. Chemical Analysis

Selected samples from EL Atshan area were chosen to carry out treatment of up taking uranium from them. *Sargassum detifolium* were used for this work and lead to decreasing of uranium content in samples shown in table 3, materials and method previously mentioned in the chapter of introduction. The results of chemical analysis for uranium after the action of Sargassum detifolium are given on table 3 the results will be discussed on basis of category of the sample numbers as will be mentioned here.

5.1.1 Sargassum Detifolium:

Sargassum has high capacity for Bio-sorption [12; 14; 15; 17]. Further studies of this biosorbent revealed the ion-exchange nature and the high metal uptakes by *Sargassum*, for example, uranium [14] Chemical analysis for uranium after the action of *Sargassum* detifolium on the fourteen samples gave highest retention effect for uranium in samples A11, A14 and A40. the original uranium content in the three samples was (2002,2002& 2402 ppm) respectively *Sargassum* show higher adsorption in samples A11 than A40 and finally sample A14 (96%, 95% & 87%) while samples A12 (250ppm), A97 (200ppm), A26 (150ppm) and A13 (125ppm) they show moderate concentration of uranium action of *Sargassum* on them verified that it take 78% of the uranium content in samples (A12 , A13, A26 and A99) as shown in figure 3 and table 3. Samples A38, A99and A60 show the same content of uranium (100 ppm). The ability of adsorbing of uranium in the three previous samples are different from A38 which show the maximum adsorption by Sargassum detifolium about 93% while in sample A60 take 75% of uranium and the lowest one is sample A99 about 69%. The content of uranium in samples A15 and A61 is about 75 ppm and 74 ppm respectively. Sargassum take 93% of uranium in the two samples. Finally, the uranium content both in samples A29 and A100 Sargassum was decreased to zero.

Table 3. Chemical analysis of U (ppm) data for El-Atshan samples after treatment by Sargassum detifolium.

Sample number	U (ppm) in mother sample	U (ppm) after treatment by Sargassum
A11	2002	75
A12	250	55
A13	125	27
A14	2002	255
A15	75	5
A26	150	34
A29	25	0
A38	100	7
A40	2402	105
A60	100	25
A61	74	5
A97	200	43
A99	100	31
A100	50	0

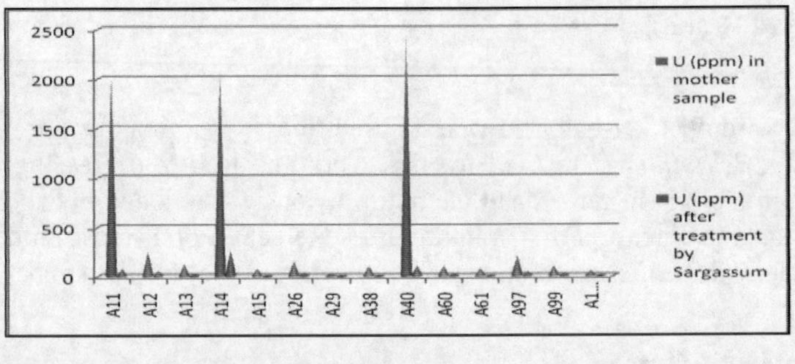

Figure 3. Chemical analysis of U (ppm) data for El-Atshan samples after treatment by *Sargassum detifolium.*

5.2 IR Analysis

IR analysis for, *Sargassum detifolium* before and after treatment in figures 4- to14 gave the characteristic wavelength for sample A11 show different functional group NH , CH aliphatic , , coupling amide and C-O-C after treatment NH, , coupling amide disappeared with the appearance of C=C, OH and CO amide as shown in figure 4 sample A12 treated with Sargassum show OH ,coupling amide , SO_2 and C-O-C after treatment coupling amide and SO_2 disappeared with the appearance of NH,CH aliphatic, , C=C and CO amide as shown in figure 5 sample A14 treated with *Sargassum detifolium* was mention with high amount of uranium show the presence of NH, coupling amide and C-O-C after treatment NH and coupling amide disappeared with the appearance of OH, C=C, CO amide and SO_2 as shown in figure 6. Sample A15 and A29 functional group before treatment with Sargassum detifolium are OH, NH, C=C and C-O-C after treatment NH, C=C disappeared with the appearance of CO amide as seen in figures 7and 8. Sample A38 show the presence of OH , CH aliphatic , , coupling amide and C-O-C after treatment coupling amide disappeared with the appearance of C=C, SO_2, CO amide as seen in figure 9, sample A40 show NH, OH, ,C=c, coupling amide and C-O-C after treatment coupling amide disappeared with the appearance of CH aliphatic , SO_2 and CO amide , CH aliphatic , SO_2 and CO amide as show in figure (10) , sample A60 show the appearance of NH, CH aliphatic , ,CO amide , SO_2 and C-O-C after treatment NH, CO amide , SO_2 disappeared with the appearance of OH and coupling amide as seen in figure 11, sample A61 treated with *Sargassum detifolium* show the presence of OH, ,C=C, CO amide and C-O-C after treatment OH, C=C and CO amide disappeared with the appearance of CH aliphatic , NH and coupling amide as seen in figure 12. Finally, samples A99 and A100 show the appearance of OH, CH aliphatic, coupling amide and C-O-C after treatment NH appeared as shown in figures 13 and 14 respectively.

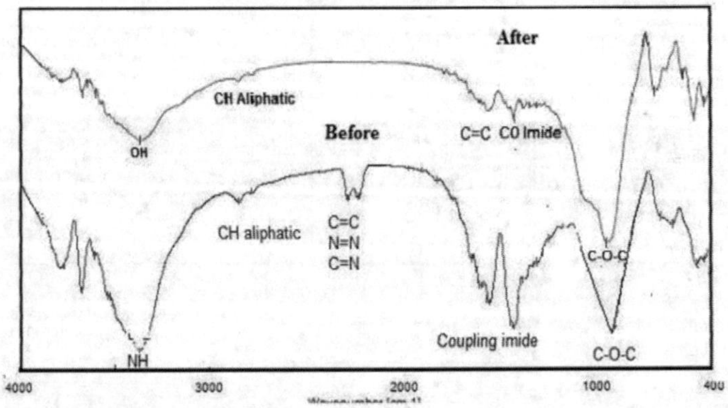

Figure 4. IR analysis data for sample A11 before and after treatment with Sargassum detifolium

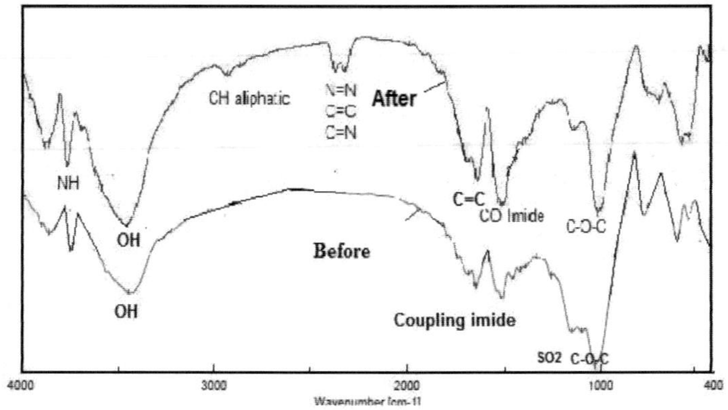

Figure 5. IR analysis data for sample A12 before and after treatment with Sargassum detifolium

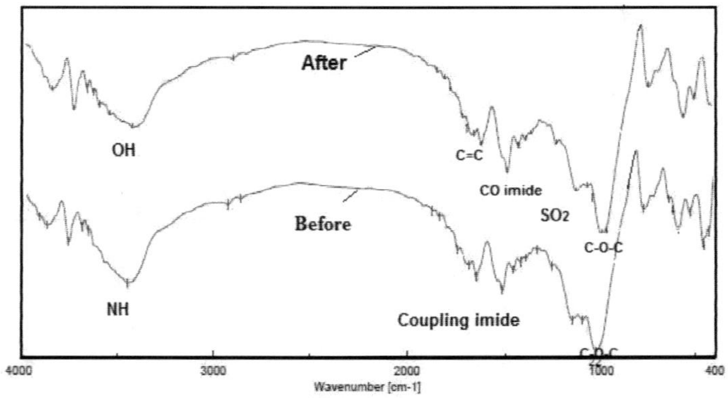

Figure 6. IR analysis data for sample A14 before and after treatment with Sargassum detifolium

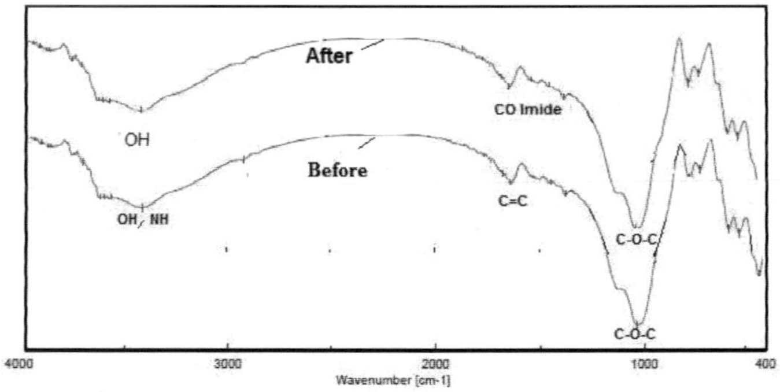

Figure 7. IR analysis data for sample A15 before and after treatment with *Sargassum detifolium*

IOP Conf. Series: Materials Science and Engineering **975** (2020) 012023 doi:10.1088/1757-899X/975/1/012023

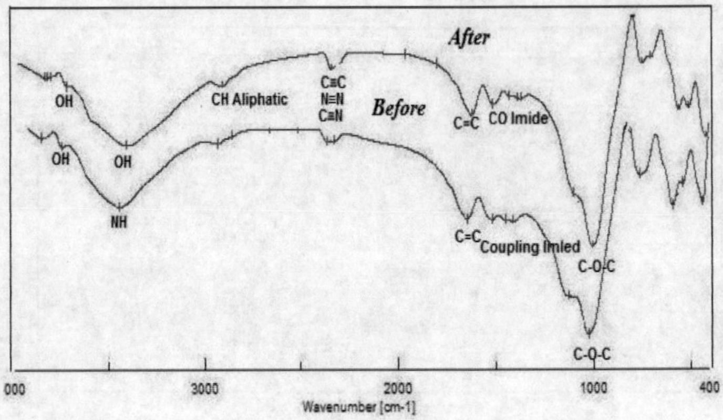

Figure 8. IR analysis data for sample A38 before and after treatment with Sargassum detifolium

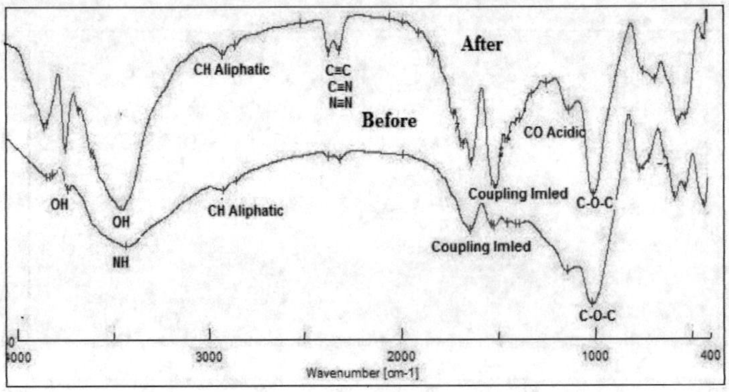

Figure 9. IR analysis data for sample A29 before and after treatment with Sargassum detifolium

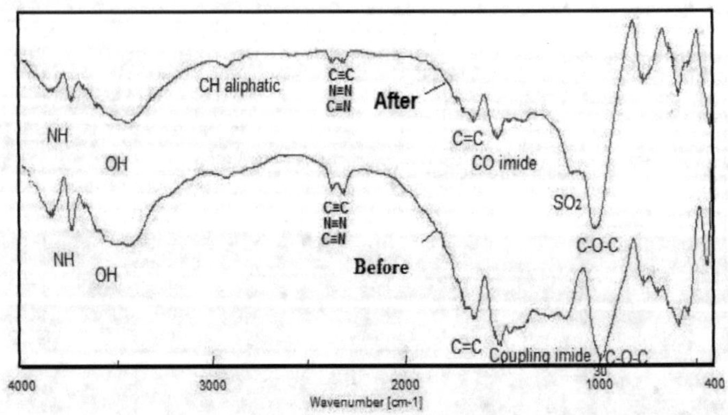

Figure 10. IR analysis data for sample A40 before and after treatment with *Sargassum detifolium*

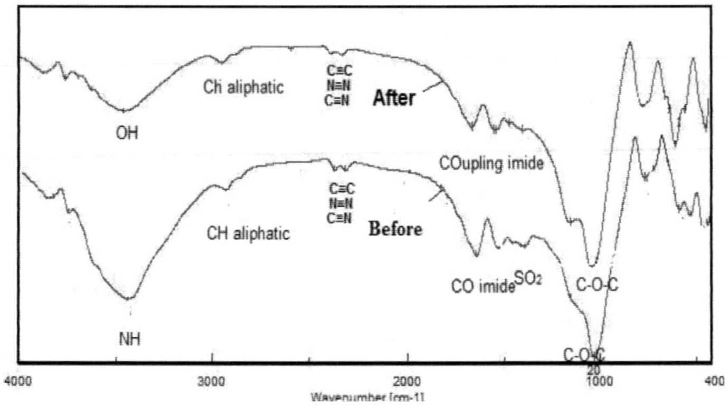

Figure11. IR analysis data for sample A60 before and after treatment with Sargassum detifolium

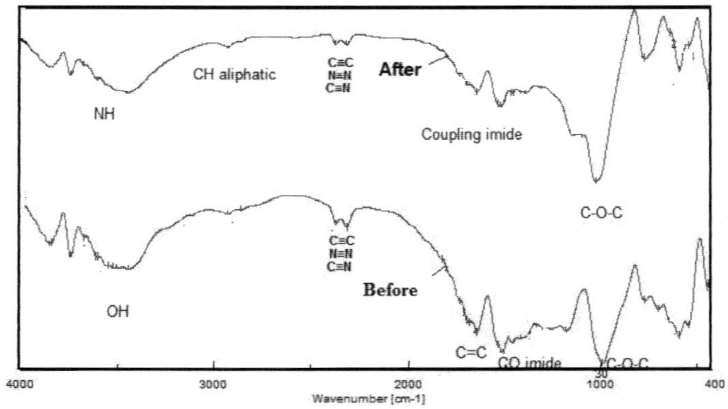

Figure 12. IR analysis data for sample A61 before and after treatment with Sargassum detifolium

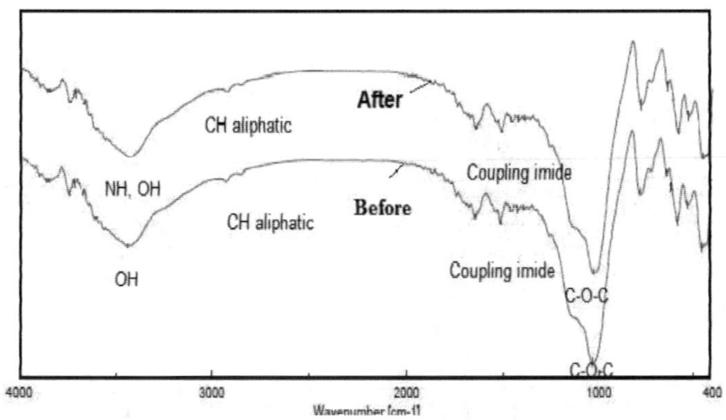

Figure 13. IR analysis data for sample A99 before and after treatment with Sargassum detifolium

IOP Conf. Series: Materials Science and Engineering **975** (2020) 012023 doi:10.1088/1757-899X/975/1/012023

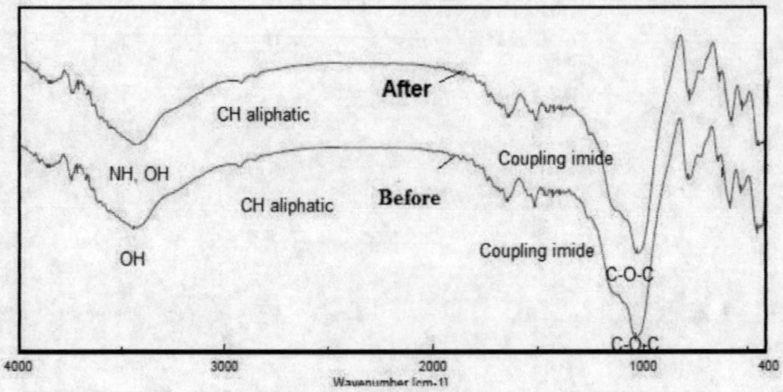

Figure 14. IR analysis data for sample A100 before and after treatment with Sargassum detifolium

5.3. Field Emission Gun (FIG) data analysis

The Environmental Scanning Electron Microscopy (ESEM) model (FIG) gave the *Sargassum* change in the chemical composition of the mineral content; some elements disappeared or decreased while other elements increased.

Uranotile {$CaU_2(SiO_4)_2(OH)_6.3H_2O$} with bright color figure 15 and EDX figure 16 show U concentration of (61.8%) before treatment with Sargassum detifolium and after treatment the bright color disappear figure17, and the U content decreased to 15.55% figure18.

Figure 15. FIG image showing uranotile mineral before treatment by Sargassum

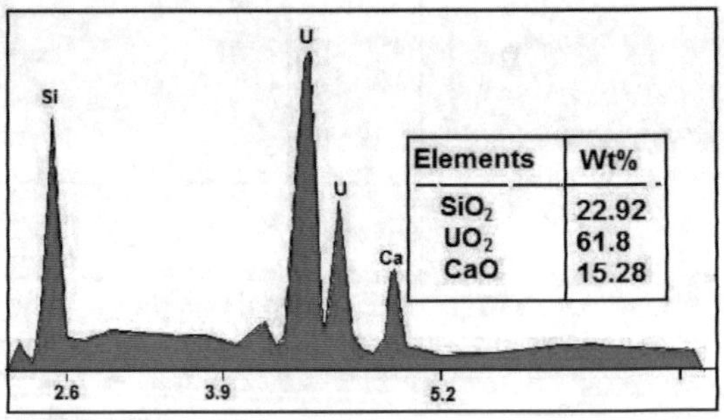

Figure 16. EDX chart for uranotile before treatment by Sargassum detifolium

Figure 17. FIG image showing disappear bright color (uranotile) after treatment

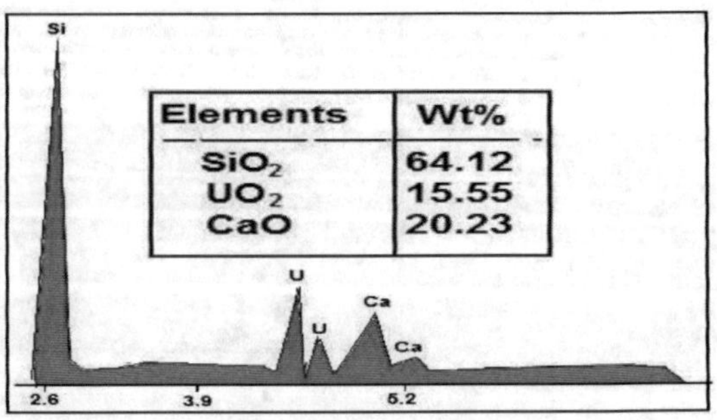

Figure 18. EDX chart for uranotile showing decreasing of UO_2after

IOP Conf. Series: Materials Science and Engineering **975** (2020) 012023 doi:10.1088/1757-899X/975/1/012023

6. Conclusion

Sargassum detifolium has been known for its high cation sorption as shown in the uranium mineralized samples in trachytic rocks from wadi El Atshan area, this area is characterized by its high U of secondary uranium mineralization such as uranophane and uranotile.

Chemical analysis of the fourteen samples gave highest retention effect for uranium in samples (A11,A14,A15,A38 and A40) uranium content Bio-sorption reach to (96%, 95%,93%,93% & 87%) respectively, while samples (A12 , A97 , A26 and A13) they show moderate Bio-sorption by Sargassum detifolium about 78% of the uranium content in the four samples , A60 sample show low Bio-sorption capacity about 75% . on the other hand, samples A29 and A100 show vanishing of the uranium content by Sargassum detifolium.

7. Reference

[1] Bulgariu Laura, Catalin Balan, Dumitru Bulgariu and Matei Macoveanu 2014 Valorisation of romanian peat for the removal of some heavy metals from aqueous media Desalination and Water Treatment Vol 52 p 31-33

[2] Çelekli M, Yavuzatmaca and H Bozkurt 2010 An eco-friendly process: predictive modeling of copper adsorption from aqueous solution on *Spirulina platensis J Hazard Mater* vol 173 p123-129

[3] Davis J A and Kent D B 2003 Surface Complexation Modeling in Aqueous Geochemistry In Michael Mineral-Water Interface Geochemistry *Mineralogical Society of America, Washington* DC

[4] Freitas OM M , Martins RJE , Delerue-Matos CM and Boaventura RAR 2008 Removal of Cd(II) Zn(II) and Pb(II) from aqueous solutions by brown marine macro algae *Kinetic modelling J Hazard Mater* vol 153 493-501

[5] Gupta VK and Rastogi A 2008 Bio-sorption of lead from aqueous solutions by green algae Spirogyra species *kinetics and equilibrium studies J . Hazard Mater* vol 152 p 407-414

[6] Hajar 2009 Bio-sorption of Cadmium from aqueous solution using dead biomass of Brown Algea *Sargassum Sp chemical engineering transactions* vol17 p1173-1178

[7] Kato K , Maruta F, Kon N and Tonouchi S 2003 Distributions and behaviors of C137s and stable Cs in seawater and marine algae on the coast of Niigata Prefecture, *Annual Report* Niigata Prefectural Institute of Environmental Radiation Monitoring Niigata Japan

[8] Kratochvil D Pimentel P and Volesky B 1998 Removal of trivalent and hexavalent chromium by seaweed biosorbent *Environ Sci Technol* vol 32 p 2693–2698

[9] Matheickal J T and Yu Q M 1999 Bio-sorption of lead (II) and copper (II) from queous solutions by pre-treated biomass of Australian marine algae Biores *Technol* vol 69 p 223–229

[10] Mohamed M Hamdy and Adel F Sadek 2017 Phanerzoic extensional faulting and alteration control on uranium minerlization in trachytes of the central eastern desert of Egypt vol 136 p 282-304

[11] Rabie K A Nora Sh Gad and Abd-Elfatah A I 2017 Ulva lactuca and Sargassum Detofiulum Bio-sorption susbitibility towards monazite mineral rare earth elements group *Isotope and Radiation Research* vol 49 p 139-153

[12] Schiewer S 1996 Multi-Metal Ion Exchange in Bio-sorption McGill University *Chemical Engineering Montreal Canada*

[13] Siegel B Z and Siegel S M 1973 The chemical composition of algal cell walls *CRC Crit Rev Microbiol* vol 3 p 1–26

[14] Yang J and Volesky B 1999a Bio-sorption of uranium on Sargassum biomass *Water Res* vol 33 3357–3363

[15] Yang J and Volesky B 1999b Bio-sorption of uranium on Sargassum biomass *Water Research* vol 33 3357–3363

IOP Conf. Series: Materials Science and Engineering **975** (2020) 012023 doi:10.1088/1757-899X/975/1/012023

[16] Yang J and Volesky B 1999c Cadmium Bio-sorption rate in protonated Sargassum biomass *Environ Sci Technol* vol 33 p 751–757
[17] Yang J 2000 Bio-sorption of Uranium and Cadmium on Sargassum Seaweed Biomass *McGill University Chemical Engineering Montreal Canada* p116

Acknowledgment
The authors are grateful to NMA , Egypt to the facilities in laboratory services for this research work.

AUTHOR INDEX